P9-DCP-133

ANALYTIC GEOMETRY

ANALYTIC GEOMETRY

Third Edition

Ross R. Middlemiss
Professor of Mathematics
Washington University

John L. Marks
Professor of Mathematics and Education
San Jose State College

James R. Smart
Professor of Mathematics
San Jose State College

McGraw-Hill Book Company
New York St. Louis San Francisco Toronto London Sydney

ANALYTIC GEOMETRY

Library of Congress Catalog Card Number 68-15472
41896

1234567890 HDMM 7543210698

PREFACE

The teaching of mathematics in colleges and secondary schools has undergone significant changes since the second edition of this text was published. Within the last two decades, analytic geometry was still considered as merely a tool for the calculus, but lately it has emerged as an independent subject of intrinsic value. While still thought of as necessary preparation for calculus, analytic geometry is now recognized as a prerequisite for higher geometry and for modern abstract algebra. Consequently it was deemed desirable that the content, vocabulary, and philosophy of the third edition include modern trends to the extent that they provide the foundations for an effective course. It is equally desirable that this new edition retain the body of content known as analytic geometry. This point of view—selecting the best from a comprehensive text, providing more attention to concepts, including a wide variety of applications, and giving greater attention to proofs—has resulted in a text which presents the material of traditional analytic geometry in a modern up-to-date form.

Many features of earlier editions have been retained. These include an extensive and clear discussion of polynominals and rational fractional functions and of the graphs and properties of the exponential, logarithmic, and trigonometric functions; exercises through which students practice critical thinking and careful analysis; a brief section on hyperbolic functions; and a study of those properties of the trigonometric functions that make them important in scientific work.

The concepts of vectors should be an integral part of any course in analytic geometry today; vectors provide an alternative and often simple procedure for deriving certain equations, and they have wide application in problems from the physical world. In the third edition, vectors and vector methods are introduced and used; they do not dominate the text, but serve as a supplement to basic procedures.

A second major change in content is the increased emphasis on mathematical sentences in the form of inequalities. A careful study of the graphs of inequalities and sets of inequalities leads to new and interesting applications such as linear programming. A third change is the development of the theory of direction cosines of a line in a plane. This leads to the analysis of motion along a line. Also in keeping with the current trend in mathematics and science texts, the three-dimensional coordinate axes have been drawn as a right-handed system.

In general, the order of topics in this edition is based on the type of equations under consideration. For example, the straight line is studied before conics. Changes in the order of topics include the earlier use of parametric equations and polar coordinates. Traditionally, these two topics appear in separate chapters late in an analytic geometry text, and hence can be neglected. In the new edition, they are integral parts of various chapters. They provide alternative approaches and serve as tools for use in applied problems throughout the text. Another change in the arrangement of topics is the combining of the work on the circle and the ellipse in one chapter to make more apparent the role of the circle as a special ellipse.

Precise mathematical vocabulary and symbolism is used consistently, in keeping with the level of maturity of the students. Cumbersome notation has not been introduced just because it is modern. This edition makes far more use of the absolute value notation and of inequalities and introduces set notation when helpful. It shows that geometric transformations can be expressed as sets of algebraic equations. It considers a function in the modern sense, as an ordered pair with a unique element in the range for each element in the domain.

New end-of-chapter features for the student include a concise review of the important concepts in the chapter, a set of review problems, and finally a set of stimulating problems for extended study, taking the student beyond the usual scope of a course in analytical geometry. Answers to selected odd-numbered problems are given at the back of the text.

Included in a teacher's manual which supplements this text are suggestions for teaching various topics, suggested course outlines and time schedules, sample tests for each chapter, and answers to selected even-numbered problems in the text.

The text is appropriate for a three- or four-semester-hour course in analytic geometry at the freshman level of college or for a year's course in analytic geometry in the senior high school. The first eleven chapters may be used for a shorter course without curve fitting or the analytic geometry of three dimensions. The text may be used along with a calculus text for a combined course in analytic geometry and calculus.

Thanks are still due Professor C. A. Hutchinson of the University of Colorado, G. C. Helme of Pratt Institute, S. E. Warschawski of the University of California at San Diego, and J. R. Britton of the University of West Florida for their valuable contributions to the first two editions.

The authors would like to thank Professor Howard Myers of San Jose State College for his many contributions to the third edition. The third edition was first used in preliminary form with freshman students at San Jose State College; their suggestions, which helped produce a more teachable text, are greatly appreciated. Finally, it should be noted that, although this revision of "Analytic Geometry" relies heavily on the two previous editions written by Professor Middlemiss, the third edition is primarily the work of authors Marks and Smart, who assume full responsibility for the content.

Ross R. Middlemiss
John L. Marks
James R. Smart

CONTENTS

CHAPTER 4 *The Line*

CHAPTER 5 *Transformation of Coordinates*

CHAPTER 6 *The Circle and the Ellipse*

CHAPTER 7 *The Parabola and the Hyperbola*

CHAPTER 8 *Polynomials and Rational Fractional Functions*

CHAPTER 9 *Algebraic Curves of Higher Degree*

CHAPTER 10 *The Trigonometric Curves*

CHAPTER 11 *The Exponential and Logarithmic Curves*

CHAPTER 12 *Curve Fitting*

CHAPTER 13 *Analytic Geometry of Three Dimensions: Definitions and Formulas*

1 RECTANGULAR COORDINATES, FUNDAMENTAL DEFINITIONS, AND THEOREMS

1-1 THE SET OF REAL NUMBERS

The numbers that are called *natural numbers* or *positive integers* are the numbers 1, 2, 3, 4, ... These, together with the corresponding negative numbers −1, −2, −3, . . . and the number *zero*, constitute the *integers*. The set of integers may thus be indicated by

$$I = \{. . . , -4, -3, -2, -1, 0, 1, 2, 3, 4, . . .\}$$

If p and q are two integers and $q \neq 0$, then the quotient p/q is called a *rational number*. The word "rational" comes from the idea of ratio—the *ratio of two integers*.

Now take a straight line (extending indefinitely far in both directions) and associate a particular point on it with each rational number, as follows (see Fig. 1-1). Choose any point on the line and associate with it the number zero; call this point the *origin*. Now choose a unit of measurement and mark off this unit an indefinite number of times in both directions from the origin; associate with these points the numbers 1, 2, 3, . . . and −1, −2, −3, . . . (it is purely a matter of convention that

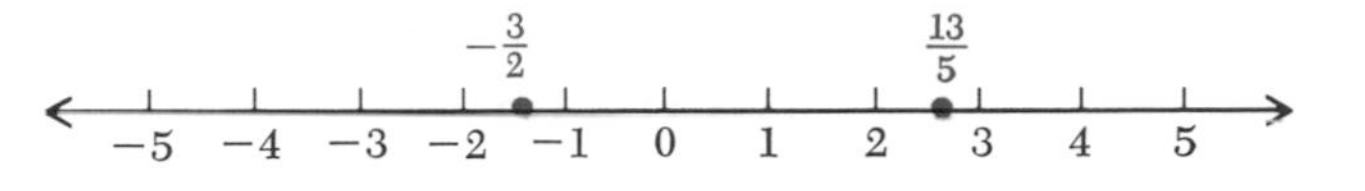

Fig. 1-1 Number line with some rational numbers indicated.

the positive numbers are located on the right and the negative ones are located on the left). Now it is possible to associate a point of the line with every rational number p/q; for example, with $-1\frac{1}{2}$, or $-\frac{3}{2}$, associate the point that is halfway between -1 and -2; with $2\frac{3}{5}$, or $\frac{13}{5}$, associate the point that is three-fifths of the way from 2 to 3, and so on.

When this is done, the points of the line that have been associated with rational numbers are very close together, yet in spite of this fact if any two are chosen (as close together as we please, but not coincident), then there are infinitely many other points associated with rational numbers between them. This can be seen as follows. Think of the points associated with $\frac{43}{65}$ and $\frac{44}{65}$. They are quite close together, but the point associated with $\frac{1}{2}(\frac{43}{65} + \frac{44}{65})$, or $\frac{87}{130}$, lies midway between them. The point associated with $\frac{1}{2}(\frac{43}{65} + \frac{87}{130})$ also lies between them, and so on.

One would at first be inclined to think that all the points on the line have been named when a point has been associated with each rational number in this manner. This, however, is not the case. There are many points on the line that have not yet been associated with any number—for example, the point $\sqrt{2}$. It lies somewhere between the points associated with $\frac{141}{100}$ and $\frac{142}{100}$. (Recall from algebra that there is no rational number p/q which is equal to $\sqrt{2}$, so if there is a point on the line for $\sqrt{2}$, it must be distinct from all those points associated with rational numbers.) The situation is similar for points named by $4\sqrt[3]{17}$ or $\frac{1}{2}\sqrt[3]{5} - \sqrt{23}$. The numbers that may be associated with points on the line but are not rational numbers are called *irrational numbers*.

The union of the sets of rational and irrational numbers is the set of *real numbers*. The real numbers fill the line, and there is a one-to-one correspondence between the real numbers and the points of the line in Fig. 1-1; that is, corresponding to each point there is a particular real number (rational or irrational), and corresponding to each real number there is a particular point.

The real number *a is greater than* the real number b (written $a > b$) if $a - b$ is a *positive* number. The same relationship is expressed by saying that *b is less than a* (written $b < a$). When the real numbers are associated with the points of a line in the manner of Fig. 1-1, the statement $a > b$ always implies that the point associated with a lies to the right of that associated with b.

1-2 ABSOLUTE VALUE

The *absolute value* of a real number a is defined as the number a itself if a is positive, or the positive number $-a$ if a is negative, or zero if a is zero. The symbol $|a|$ denotes the absolute value of a. Thus $|-5| = 5$, $|4| = 4$, $|-\frac{2}{3}| = \frac{2}{3}$, $|3 - 12| = 9$. In general,

$$|x| = \begin{cases} x & \text{if } x > 0 \\ -x & \text{if } x < 0 \\ 0 & \text{if } x = 0 \end{cases}$$

and

$$|x - 4| = \begin{cases} x - 4 & \text{if } x - 4 > 0 \\ -(x - 4) & \text{if } x - 4 < 0 \\ 0 & \text{if } x - 4 = 0 \end{cases}$$

It should be specifically noted that for all real values of a, $|a|$ represents a nonnegative number.

Recall that the symbol $\sqrt{a}$, where a is a positive number, is defined as the positive number whose square is a. Thus $\sqrt{16} = 4$, not -4, and not ± 4. Similarly, $\sqrt{9} = 3$, but $-\sqrt{9} = -3$. *Observe now that* $\sqrt{a^2} = a$ *if* a *is positive, but* $\sqrt{a^2} = -a$ *if* a *is negative.* Thus if $a = -5$, $\sqrt{a^2} = \sqrt{(-5)^2} = -(-5) = -a$. Similarly,

$$\sqrt{(x - 4)^2} = \begin{cases} x - 4 & \text{if } x - 4 \geqslant 0 \\ -(x - 4) & \text{if } x - 4 < 0 \end{cases}$$

All cases are covered by writing

$$\sqrt{(x - 4)^2} = |x - 4|$$

Example 1

Find the solution set for P if $P = \{x \mid |x + 2| = 5\}$.

Solution The problem is read "find all real numbers x such that the absolute value of $x + 2$ is equal to 5." Intelligent application of trial and error shows that $x = 3$ and $x = -7$ make $|x + 2| = 5$ true. That these are the only members of the solution set is indicated by the solution which begins by replacing $|x + 2|$ with an equivalent expression. $|x + 2| = 5$ has the same solution set as $x + 2 = 5$ if $x + 2 \geqslant 0$ or as $-(x + 2) = 5$ if $x + 2 < 0$. This is equivalent to

$$x = 3 \qquad \text{if } x \geqslant -2$$

or

$$x = -7 \qquad \text{if } x < -2$$

which is simplified to $x = 3$ or $x = -7$; thus $P = \{3, -7\}$.

Example 2

Find the members of set Q if $Q = \{x \mid |x-3| > 4\}$.

Solution After trying various replacements for x, it seems that any number greater than 7 or less than -1 makes the sentence true. The solution may be shown in the following form:

$$\begin{aligned}\{x \mid |x-3| > 4\} &= \{x \mid x-3 > 4 \text{ if } x-3 \geq 0 \quad \text{or} \\ &\qquad -(x-3) > 4 \text{ if } x-3 < 0\} \\ &= \{x \mid x > 7 \text{ if } x \geq 3 \quad \text{or} \quad x < -1 \text{ if } x < 3\} \\ &= \{x \mid x > 7 \quad \text{or} \quad x < -1\}\end{aligned}$$

PROBLEMS

1. Let A and B be the points of the line in Fig. 1-1 that are associated with $\frac{5}{7}$ and $\frac{6}{7}$, respectively. Determine at least three rational numbers associated with points that are located between A and B. Would it be possible to find 5,000 such numbers?
2. Determine three rational numbers that are greater than $\frac{13}{5}$ and less than $\frac{14}{5}$.
3. Let x be defined as $\frac{173}{100} < x < \frac{174}{100}$. Name two rational and two irrational numbers which x may represent.
4. Between the two rational numbers p/q and $(p+1)/q$ find (*a*) one rational number; (*b*) two rational numbers; (*c*) three rational numbers. (*d*) State a general rule for determining n (where n is a positive integer) rational numbers between p/q and $(p+1)/q$.
5. (*a*) Show by means of an example that the sum of two irrational numbers a and b may be a rational number. (*b*) The sum of a rational and an irrational number must be what kind of number?
6. In the quadratic equation $ax^2 + bx + c = 0$ let a, b, and c be rational numbers. Is it possible that one root may be rational and the other irrational? Why?
7. Between what two numbers does x lie if (*a*) $|x| < 5$; (*b*) $|x-3| < 5$?
8. Write a statement involving absolute value that is equivalent to $-3 < x < 5$.
9. Between what two real numbers does x lie if $5 < 2x - 3 < 10$?
10. For what values of x is $|x-5|$ equal to $x-5$? For what values of x is it equal to $5-x$?
11. For what values of x is $|2x-7|$ equal to $2x-7$? For what values of x is it equal to $7-2x$?
12. Find the solution set for each of the following:

    (*a*) $\{x \mid |x-2| = 5\}$ (*b*) $\{n \mid \left|\frac{n}{n-2}\right| < 2\}$ (*c*) $\{x \mid \left|\frac{1}{x}\right| = 4\}$

    (*d*) $\{y \mid |y| + y \geq 4\}$ (*e*) $\{x \mid |3-x| > 4\}$ (*f*) $\{z \mid |2z| + |z| < -1\}$

13. Solve for x:

    (*a*) $|x+4| = 7$ (*b*) $|\frac{2}{7}x| - 3 > 8$ (*c*) $|x| - x = 1$

(*d*) $\left|3+\frac{4}{x}\right| > 2$ (*e*) $\frac{1}{|x|} < 3$ (*f*) $|x| = x - 2$

14. Simplify the expression $3x + |x+y| + |x-y|$ if x and y are positive numbers and $x < y$.
15. Simplify the expression $3x + |x-1| + |2-x|$ if $1 < x < 2$.
16. (*a*) Simplify the expression $3x + |x| + |3-2x|$ if x is a negative number. (*b*) Simplify it if x is a number between 0 and 1.

1-3 DIRECTED DISTANCES

Let A and B be two points on line AB, written $\overleftrightarrow{AB}$, in Fig. 1-2. Let the distance between them be 2 units; that is, the length of line segment AB, written $\overline{AB}$, is 2 units. Let it be agreed that the distance from A to B is $+2$ and that the distance from B to A is -2 (or vice versa). Under this agreement to regard segments measured in one direction along a line as positive and those measured in the opposite direction as negative, the line is called a *directed line* (that is, it has been given a positive and negative direction), and the distances are called *directed distances*. The distance from A to B is denoted by AB, and that from B to A by BA. The absolute value of AB or BA is called the *undirected distance* between two points. It may be denoted by the symbol $|AB|$ or $|BA|$.

Fig. 1-2 Segment $\overline{AB}$ on line $\overleftrightarrow{AB}$.

A fundamental property of a directed line is that if A, B, and C are any three points on a directed line, then

$$AC = AB + BC$$

regardless of the relative positions of these points (Fig. 1-3).

When the real numbers are associated with points of a line, as in Fig. 1-4, the line is regarded as a directed line on which distances measured from left to right are positive. This line is often called an *axis of real numbers*, and the number associated with each point is called the *coordinate* of that point. *This coordinate is the directed distance from the origin to the point.* Thus point D has the coordinate -3 (since $OD = -3$),

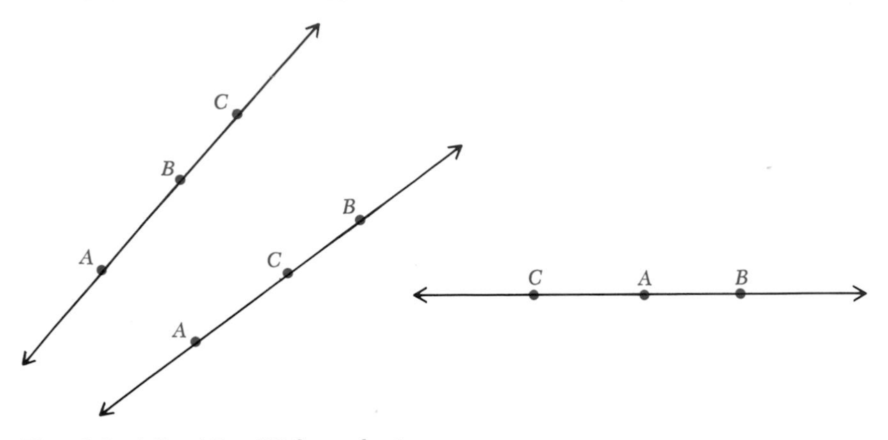

Fig. 1-3 $AC = AB + BC$ for each picture.

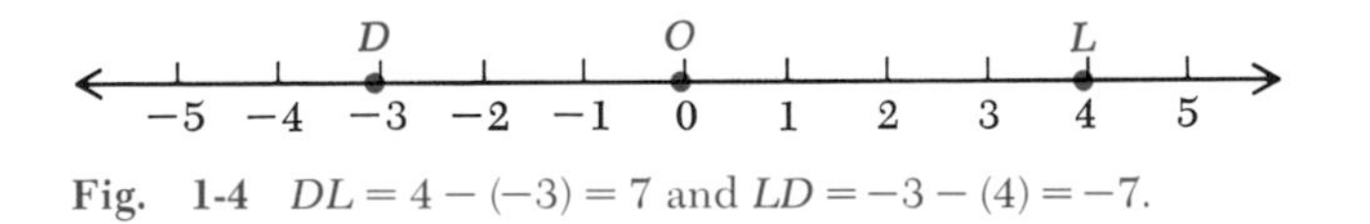

Fig. 1-4 $DL = 4 - (-3) = 7$ and $LD = -3 - (4) = -7$.

and point L has the coordinate 4 (since $OL = 4$). Observe that

$$DL = 4 - (-3) = 7$$

and

$$LD = -3 - (4) = -7$$

If P_1 and P_2 are any two points on the axis with coordinates x_1 and x_2, respectively, then

$$P_1P_2 = x_2 - x_1$$

Furthermore, the directed distance from P_2 to P_1 is

$$P_2P_1 = x_1 - x_2$$

and the undirected distance between the two points is

$$|x_2 - x_1| = |x_1 - x_2|$$

1-4 RECTANGULAR COORDINATES

Let $\overleftrightarrow{OA}$ and $\overleftrightarrow{OB}$ in Fig. 1-5 be two axes which intersect at right angles at a common origin. These lines are called the *rectangular-coordinate axes;* $\overleftrightarrow{OA}$ is called the *x axis* and $\overleftrightarrow{OB}$ the *y axis.* (By convention the positive direction is taken to the right on the x axis and upward on the y axis.)

Let P be any point in the plane of the axes. Its directed distance from the y axis and along the x axis is called its *x coordinate*, or *abscissa*. This is equal to either of the directed segments OA or BP. Similarly, its directed distance from the x axis and along the y axis is called its *y coordinate*, or *ordinate*. This is equal to either of the directed segments OB or AP. The abscissa and ordinate together are called the *rectangular*, or *cartesian*, *coordinates* of P. The coordinates of a point are written as an *ordered pair*, enclosed in parentheses, with the x coordinate written first. Thus the coordinates of P in Fig. 1-5 are (2.5, 3.5) and those of P and Q in Fig. 1-6 are, respectively, (3, 4) and (−2, −5).

The process of locating and marking a point whose coordinates are given is called *graphing* or *plotting* the point. Plotting is facilitated by the use of paper that is ruled in small squares, as in Fig. 1-6. Such paper is called *rectangular-coordinate paper*.

It was mentioned in Sec. 1-1 that there is a one-to-one correspondence between the real numbers and the points of a line. In the rectangular-coordinate system there is a similar relationship between pairs of real numbers and the points of a plane; that is, to each pair of real numbers (x, y) there corresponds a definite point of the plane, and to each point of the plane there corresponds a unique pair of coordinates (x, y).

The coordinate axes partition the plane into four disjoint regions called *quadrants*, which are numbered as in Fig. 1-7. This figure shows the signs of the coordinates for each quadrant. The quadrants may be

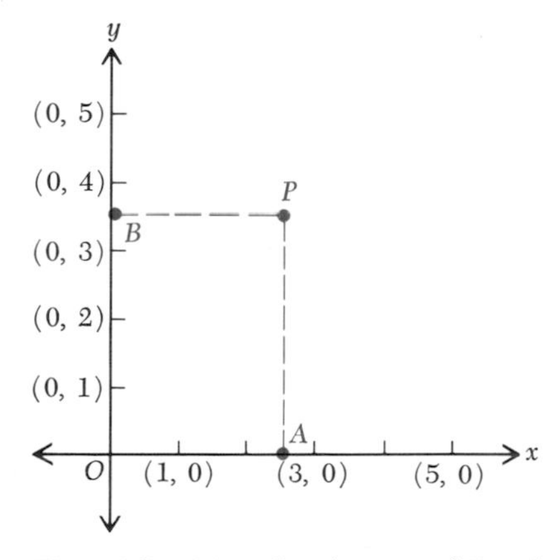

Fig. 1-5 OA is the abscissa of P and OB is the ordinate of P.

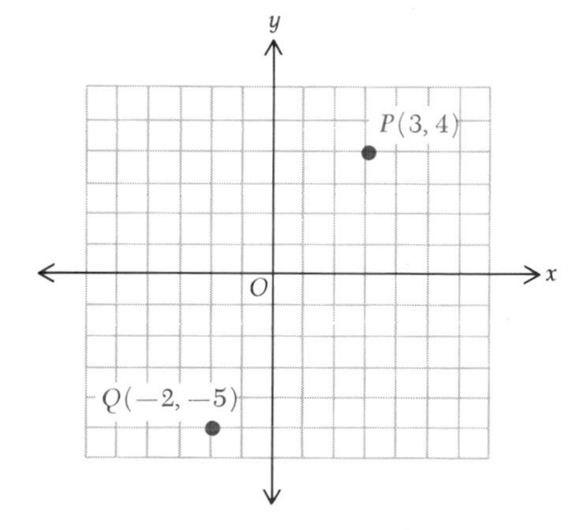

Fig. 1-6 Points $P(3, 4)$ and $Q(-2, -5)$ graphed on rectangular-coordinate paper.

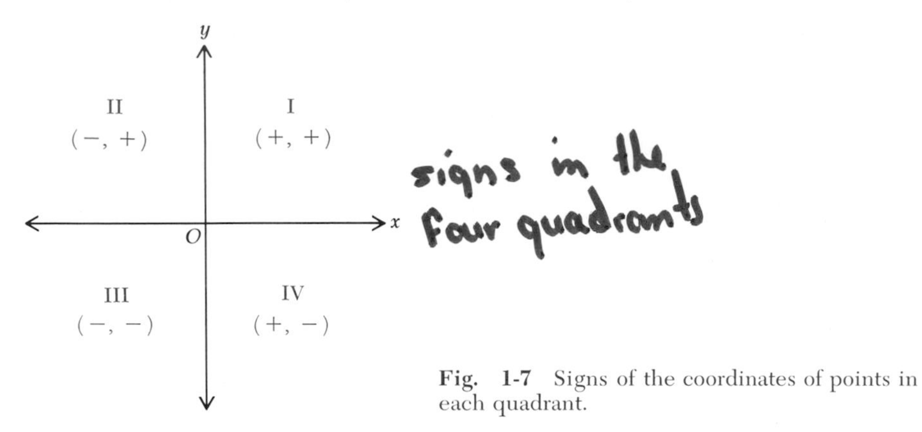

Fig. 1-7 Signs of the coordinates of points in each quadrant.

defined in terms of half-planes. For example, quadrant I is the intersection of the half-plane above the x axis and the half-plane to the right of the y axis. According to these definitions the coordinate axes are in none of the quadrants.

1-5 THE DISTANCE BETWEEN TWO POINTS

The distance between two points in a plane may be considered in three cases:

1. $P_1(x_1, y_1)$ and $P_2(x_2, y_1)$ are two distinct points on the same horizontal line (Fig. 1-8*a*). Then the *directed distance* P_1 to P_2 is

$$P_1P_2 = x_2 - x_1$$

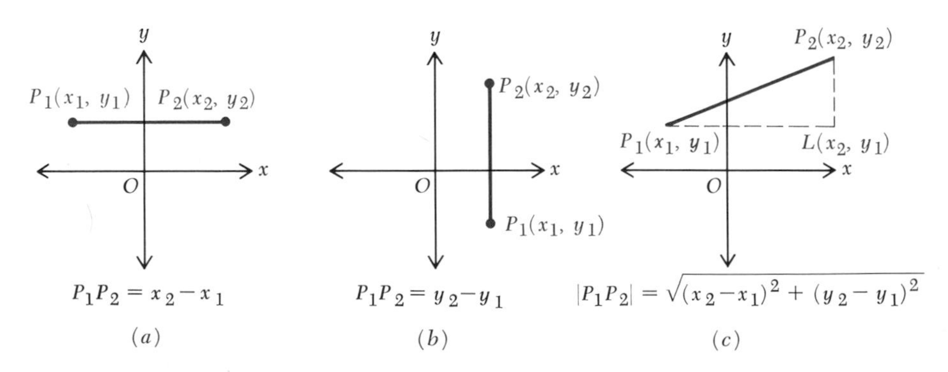

Fig. 1-8 The distance between two points in a plane.

The *undirected distance* between P_1 and P_2 is

$$|P_1P_2| - |x_2 - x_1| = |x_1 - x_2|$$

2. $P_1(x_1, y_1)$ and $P_2(x_1, y_2)$ are two distinct points on the same vertical line (Fig. 1-8*b*). Then the *directed distance* P_1 to P_2 is

$$P_1P_2 = y_2 - y_1$$

The *undirected distance* between P_1 and P_2 is

$$|P_1P_2| = |y_2 - y_1| = |y_1 - y_2|$$

3. $P_1(x_1, y_1)$ and $P(x_2, y_2)$ are two distinct points not on the same vertical or horizontal line (Fig. 1-8*c*). If a line is drawn through P_1 parallel to the x axis and another line is drawn through P_2 parallel to the y axis, these lines will meet at point $L(x_2, y_1)$, forming the right triangle P_1LP_2. Then $|P_1P_2|$ is the measure of the hypotenuse of a right triangle whose legs $\overline{P_1L}$ and $\overline{LP_2}$ measure $|x_2 - x_1|$ and $|y_2 - y_1|$, respectively. Therefore

$$(P_1P_2)^2 = (x_2 - x_1)^2 + (y_2 - y_1)^2$$

If d represents the undirected distance between the two points P_1 and P_2, then

$$d = \sqrt{(x_2 - x_1)^2 + (y_2 - y_1)^2} \tag{1-1}$$

Example 1

The distance between $A(-3, 6)$ and $B(3, -2)$ in Fig. 1-9 is calculated by Eq. (1-1) as

$$\begin{aligned} d &= \sqrt{[3 - (-3)]^2 + [-2 - 6]^2} \\ &= \sqrt{6^2 + (-8)^2} = 10 \end{aligned}$$

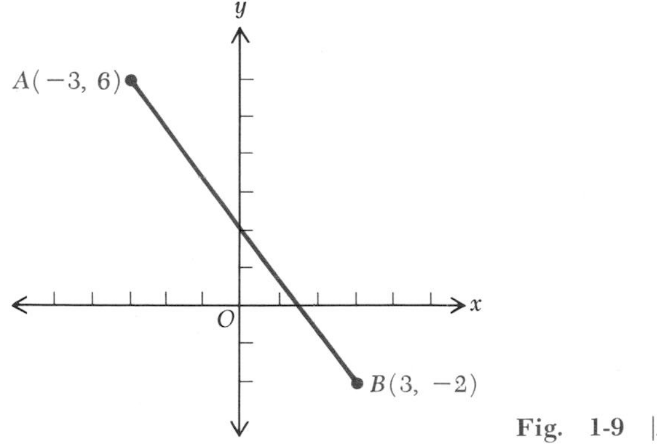

Fig. 1-9 $|AB| = \sqrt{[3 - (-3)]^2 + (-2 - 6)^2}$.

HISTORICAL NOTE

Analytic geometry is actually a method rather than a branch of geometry. In essence, the method reduces a problem in geometry to an algebraic problem by establishing a correspondence between a curve and a definite equation.

René Descartes (1596–1650) is usually considered the inventor of analytic geometry, but at least one other mathematician, Pierre Fermat, simultaneously and independently studied curves by means of their algebraic equations. There are several legends that describe how the idea of analytic geometry may have occurred to Descartes. According to one the thought came to him in a dream. Another story, possibly as apocryphal as that of Isaac Newton and the falling apple, is that the concept was suggested as he watched a fly crawling on the ceiling of a room near a corner. He thought he could represent the fly's path by describing algebraically the distance of the fly from the two intersecting walls and the ceiling.

Descartes' writings were not a systematic development of the methods of analytic geometry, but historians have constructed the methods as we know them from certain of his statements. Actually, coordinate axes are nowhere described explicitly in his writings, but his ideas formed the basis for the familiar analytic geometry of today.

Descartes is respected for many other mathematical ideas and inventions. Those relating to analytic geometry are the use of exponential notation, such as x^3 to replace $x \cdot x \cdot x$, which led to the use of expressions such as x^n, and the systematic use of the first letters of the alphabet, a, b, and c, to represent constants and the final letters, x, y, and z, to represent variables.

This powerful tool, the method of analytic geometry, was one of the inventions that led to extensive mathematical discoveries, including the calculus, in the seventeenth and eighteenth centuries.

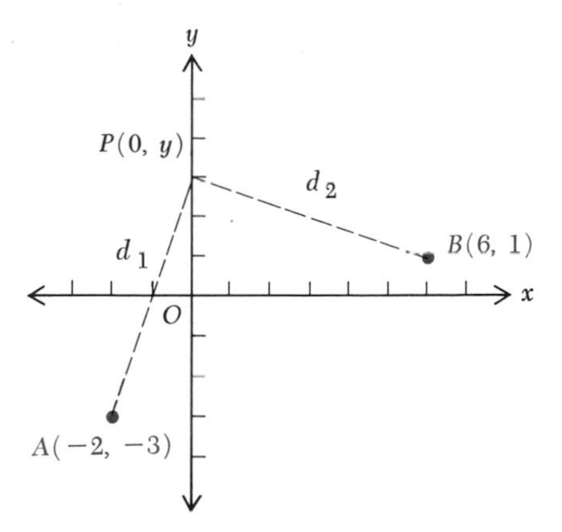

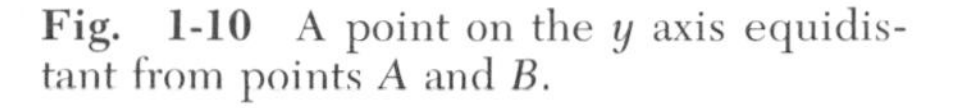
Fig. 1-10 A point on the y axis equidistant from points A and B.

Example 2

Find the point on the y axis that is equidistant from $A(-2, -3)$ and $B(6, 1)$.

Solution Since the required point is on the y axis (see Fig. 1-10), its x coordinate is zero, and it may be designated as $P(0, y)$. Let the undirected distances of P to A and P to B be d_1 and d_2, respectively. Then, from Eq. (1-1),

$$d_1 = \sqrt{[0 - (-2)]^2 + [y - (-3)]^2}$$
$$d_2 = \sqrt{(0 - 6)^2 + (y - 1)^2}$$

The conditions of the problem require that y be determined such that $d_1 = d_2$. Hence

$$\sqrt{(0 + 2)^2 + (y + 3)^2} = \sqrt{(0 - 6)^2 + (y - 1)^2}$$

and
$$4 + y^2 + 6y + 9 = 36 + y^2 - 2y + 1$$

$$8y = 24$$

or
$$y = 3$$

The required point is thus $P(0, 3)$.

PROBLEMS

1. Show on a line the relative positions of points P_1, P_2, and P_3 which meet each of the following conditions:

   (a) $P_1P_3 = P_1P_2 + P_2P_3$ (b) $P_1P_3 = P_1P_2 - P_2P_3$

2. Each of the following segments $\overline{AB}$ is parallel to the x axis. Draw each segment and find the directed distance from the first point to the second:

   (*a*) $A(2, 3), B(5, 3)$ (*b*) $A(0, -1), B(-3, -1)$
   (*c*) $A(5, 1), B(-3, 1)$ (*d*) $A(4, 4), B(-2, 4)$
   (*e*) $A(-2, 0), B(3, 0)$ (*f*) $A(-7, -4), B(-1, -4)$

3. Each of the following segments $\overline{AB}$ is parallel to the y axis. Draw each segment and find the directed distance from the first point to the second:

   (*a*) $A(5, -2), B(5, 5)$ (*b*) $A(0, 1), B(0, -5)$
   (*c*) $A(-3, -7), B(-3, -1)$ (*d*) $A(4, 7), B(4, -2)$
   (*e*) $A(1, -8), B(1, -6)$ (*f*) $A(-3, 2), B(-3, 0)$

4. Show that the directed distance from $A(2, 3)$ to $B(x, 3)$ is $x - 2$ for all values of x. What is the directed distance from B to A?
5. Plot the points $A(2, 5)$ and $B(2, y)$, where $y < 0$. Is the directed distance BA equal to $5 + y$, $5 - y$, or $y - 5$?
6. Describe each of quadrants II, III, and IV in terms of the intersection of half-planes.
7. For each of the following find the undirected distance between A and B:

   (*a*) $A(3, 4), B(-2, 7)$ (*b*) $A(-4, -3), B(5, 1)$
   (*c*) $A(-4, 1), B(6, 2)$ (*d*) $A(-8, -1), B(-2, 4)$
   (*e*) $A(-1, 2), B(-2, 3)$ (*f*) $A(-7, -3), B(-1, -2)$

8. Which of the following expressions represent the undirected distance between the two points $A(2, 3)$ and $B(x, 3)$?

   (*a*) $x - 2$ (*b*) $|x - 2|$ (*c*) $\sqrt{(x - 2)^2}$ (*d*) $2 - x$ (*e*) $|2 - x|$

9. Show that the triangle with vertices at $A(1, 3)$, $B(3, -1)$, and $C(-5, -5)$ is a right triangle.
10. A quadrilateral $ABCD$ has vertices $A(-1, 3)$, $B(5, 7)$, $C(9, 1)$, and $D(3, -3)$. Prove either that it is or that it is not a rectangle.
11. One vertex A of a triangle is the origin and the other two are the points $B(1, 12)$ and $C(11, -1)$. Prove that the triangle is or is not a right triangle.
12. Show that the triangle whose vertices are $P(-1, 0)$, $Q(7, -8)$, and $R(-2, -9)$ is isosceles.
13. Show that $C(4, 1)$ is the center of the circle that passes through $D(0, -2)$, $E(7, -3)$, and $F(8, -2)$.
14. Find the center of the circle that is circumscribed about the triangle with vertices $A(2, 2)$, $B(2, -8)$, $C(-1, 6)$.
15. Find the point on the x axis that is equidistant from $P(-4, 6)$ and $Q(14, -2)$.
16. A point P lies on a line that is parallel to the x axis and 2 units below it. P is twice as far from $A(6, 2)$ as from the origin. Find its coordinates.
17. The ordinate of a certain point P is twice the abscissa. This point P is equidistant from $A(-3, 1)$ and $B(8, -2)$. Find its coordinates.

18. The sum of the undirected distances from a point P to $A(8, 0)$ and $B(-8, 0)$ is 20. The abscissa of P is 6. Find its ordinate.

1-6 POLAR COORDINATES

The rectangular coordinate system in the plane is one in which a point is located by specifying its directed distances from two fixed lines, the x and y axes. Now consider a coordinate system in which a point is located by specifying its *distance* and *direction* from a fixed point. This system is essentially involved in saying that a town B is 30 miles northeast of another town A, where the distance (30 miles) and direction (northeast) of B from A are given instead of the distance east and the distance north.

Let O be a fixed point in a plane and let ray OA, written $\overrightarrow{OA}$, be a fixed ray in the plane through O, as shown in Fig. 1-11. The fixed point (here, the origin) is called the *pole*, and the ray $\overrightarrow{OA}$ is called the *polar axis*. It may be regarded as the positive half of the x axis of the rectangular-coordinate system, when changing from one system to the other.

Consider now any point P (other than O) in the plane. Its position can be specified by giving its directed distance OP from O and the measure of the angle that $\overrightarrow{OP}$ makes with $\overrightarrow{OA}$. Then OP is called the *radius vector* of P and is denoted by the Greek letter ρ. The angle AOP is called the *vectorial angle* of P and is denoted by the Greek letter θ. The two numbers ρ and θ are called the *polar coordinates* of P and are written as the ordered pair (ρ, θ).

In plotting a point whose coordinates (ρ, θ) are given, ρ is the directed distance from the origin to the point. It is further agreed that if θ is a positive number, the angle of θ rad is measured from $\overrightarrow{OA}$ as the initial side in the counterclockwise direction. If θ is negative, the angle is to be measured in the clockwise direction. (It may, of course, be specified that the number θ is to be interpreted as the number of degrees instead of the number of radians in angle AOP.) It is also agreed to associate the origin with coordinates $(0, \theta)$, where θ is any real number.

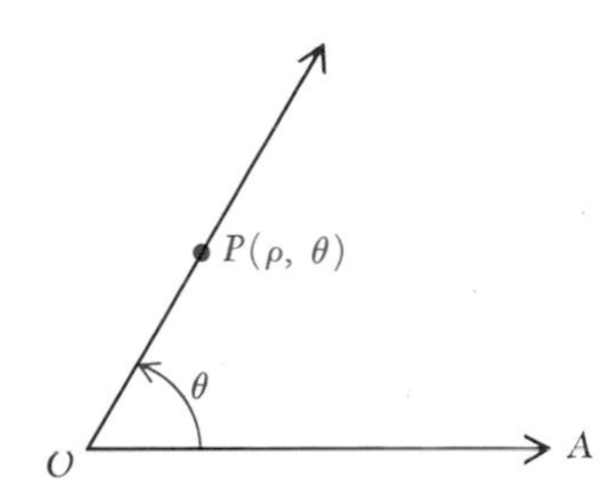

Fig. 1-11 Polar coordinates: $OP = \rho$, angle $AOP = \theta$; P has coordinates (ρ, θ).

HISTORICAL NOTE

The invention of this graphing system, the polar coordinate system, is generally credited to Jacques Bernoulli (about 1690). As is the case with other mathematical ideas, applications were suggested prior to this time, and other mathematicians developed the concept as part of analytic geometry. Because this system has angle measure as one of its component parts, its extensive application depends on trigonometry. Descartes, for example, did not make use of polar coordinates because accepted definitions of the trigonometric functions at that time were in terms of right triangles only; there was no provision in such definitions for angles greater than 90° or less than zero.

Some mathematical ideas are more easily investigated in terms of polar coordinates. For example, polar coordinates often provide the simplest way of expressing a relationship which specifies the outline of a closed curve.

Thus there is defined a coordinate system which associates a definite point of the plane with every ordered pair of real numbers (ρ, θ). In Fig. 1-12 the points whose coordinates are $(4, \frac{1}{6}\pi)$, $(3, \pi)$, and $(2, -\frac{2}{3}\pi)$ are plotted.

Recall that in the rectangular coordinate system there is a one-to-one correspondence between the points of the plane and the ordered pairs of numbers (x, y); to each pair of coordinates there corresponds a definite point, and to each point there corresponds one and only one pair of coordinates. In the case of polar coordinates the situation is somewhat different. To a given pair of polar coordinates there corresponds a definite point of the plane, but to a given point there correspond indefinitely many pairs of coordinates. Thus the point with coordinates $(3, \pi)$ in Fig. 1-12 also has the coordinates $(3, -\pi)$, $(3, 3\pi)$, $(3, -3\pi)$, and so on. In general, if a point P has coordinates (ρ, θ), then it also has the coordinates $(\rho, \theta + 2n\pi)$, where ρ is positive and n is any integer.

When the coordinates (ρ, θ) are given, and ρ is a negative number, the corresponding point is plotted as follows. After constructing θ in the usual way, extend its terminal ray through the origin and locate the point on this extension of the ray at a distance $|\rho|$ from the origin. The point $(-4, \frac{1}{4}\pi)$ is plotted in Fig. 1-13. This interpretation amounts to taking the point associated with the number pair (k, θ), where $k > 0$, and agreeing also to associate with it the number pair $(-k, \theta + \pi)$. In general, point $P(\rho, \theta)$ for $\rho > 0$ also has coordinates $(-\rho, \theta + (2n + 1)\pi)$ for n an integer.

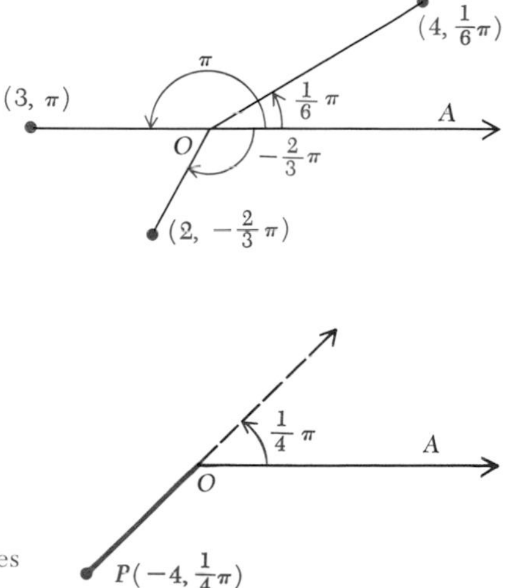

Fig. 1-12 Graph of points $(4, \frac{1}{6}\pi)$, $(3, \pi)$, and $(2, -\frac{2}{3}\pi)$.

Fig. 1-13 Graph of $(-4, \frac{1}{4}\pi)$ in polar coordinates (with ρ negative).

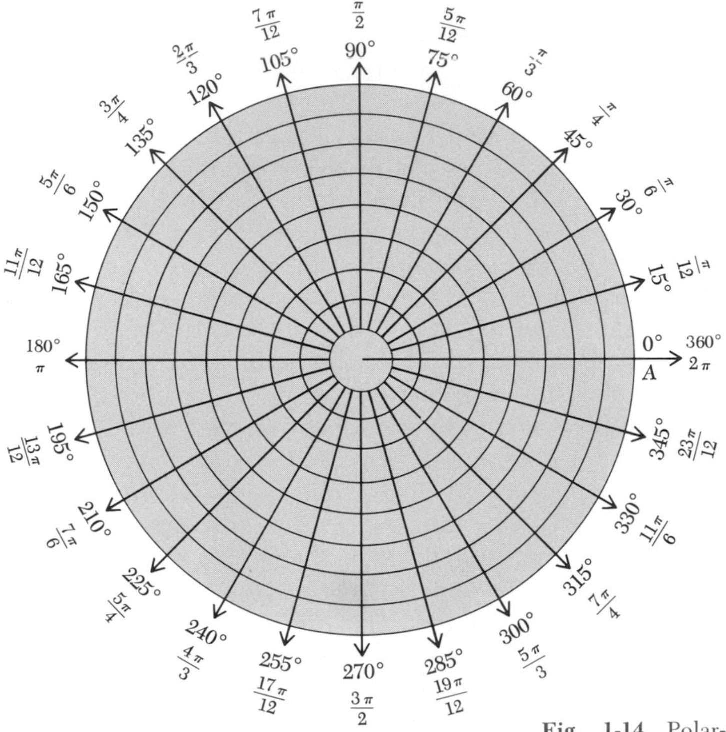

Fig. 1-14 Polar-coordinate paper.

Plotting in polar coordinates is facilitated by the use of polar-coordinate paper. This paper is ruled off in rays and concentric circles, as indicated in Fig. 1-14.

PROBLEMS

1. Plot the following points, which are given as polar coordinates, and list four other pairs of coordinates which specify each point:

   (*a*) $(3, 90°)$ (*b*) $(2\frac{1}{2}, \frac{5}{6}\pi)$ (*c*) $(-3, 450°)$ (*d*) $(2, -\frac{5}{3}\pi)$

2. Plot the points whose polar coordinates are as follows:

   (*a*) $(-5, \pi)$ (*b*) $(3, 1)$ (*c*) $(3, 4\pi)$
   (*d*) $(-5, 3)$ (*e*) $(-5, -\frac{19}{6}\pi)$

3. Each two of the ordered pairs given below represent two vertices of an equilateral triangle. What are the coordinates of the third vertex? Use $\rho > 0$ and $0 \leq \theta < 2\pi$. (Is there more than one answer for each problem?)

   (*a*) $(3, 0°)$, $(0, 0°)$ (*b*) $(5, 45°)$, $(0, 0°)$ (*c*) $(5, \pi/2)$, $(5, \pi/6)$

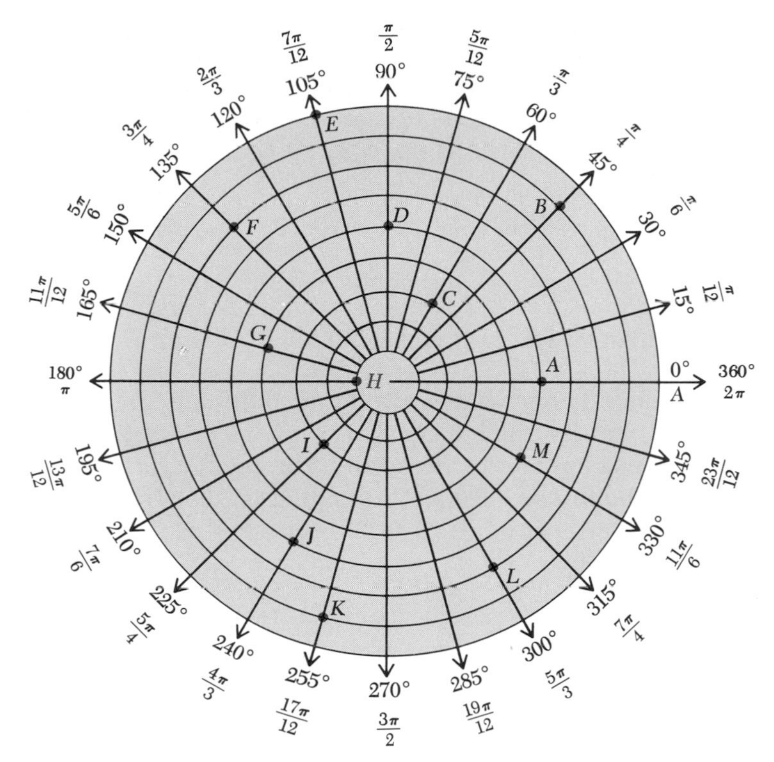

Fig. 1-15

4. For each of the points plotted in Fig. 1-15 list one ordered pair of polar coordinates under each of the following restrictions:

   (*a*) $\rho > 0$; $0 \leqslant \theta < 2\pi$ (*b*) $\rho > 0$; $-2\pi < \theta \leqslant 0$
   (*c*) $\rho < 0$; $0° \leqslant \theta < 360°$ (*d*) $0 \leqslant \theta < \pi$

5. The coordinates of the end points of the base of an isosceles triangle are given below, along with the ρ coordinate of the third vertex. What is the θ coordinate of the third vertex?

   (*a*) $(3, 0)$, $(3, \pi)$, $(a, ?)$ (*b*) $(6, 0)$, $(6, \pi/2)$, $(a, ?)$

6. Three vertices of a square are $(5, \pi/4)$, $(5, 3\pi/4)$, and $(5, 7\pi/4)$. (*a*) What are the coordinates of the fourth vertex? (*b*) What are the coordinates of intersection of the diagonals?
7. Give the rectangular coordinates of each of the following points whose polar coordinates are given (pole and origin are same point, and the polar axis coincides with the positive "half" of the x axis):

   (*a*) $(3, \pi/2)$ (*b*) $(-7, -\pi)$ (*c*) $(3, 3\pi)$
   (*d*) $(-3, -5\pi/4)$ (*e*) $(5, 7\pi/4)$ (*f*) $(2, \pi/4)$

8. The polar coordinates of three vertices of a rectangle are (6, 0), (3, $\pi/2$), and (0, 0). What are the polar coordinates of the fourth vertex?
9. Suppose ρ and θ are restricted by the requirements listed below. For each case, would there be a one-to-one correspondence between the points of the plane and the number pair (ρ, θ)?

   (*a*) $\rho \geqslant 0$; no restriction on θ (*b*) $\rho \geqslant 0$; $0 \leqslant \theta < 2\pi$
   (*c*) no restriction on ρ; $0 \leqslant \theta < 2\pi$ (*d*) $\rho \leqslant 0$; $0 \leqslant \theta < 2\pi$

1-7 DIVISION OF A LINE SEGMENT

It is frequently necessary to find the coordinates of a point P_0 which lies on a segment $\overline{P_1P_2}$ such that P_1P_0 is some given fraction of P_1P_2. As a preliminary example let us determine the point P_0 in Fig. 1-16 that is two-thirds of the distance from P_1 to P_2, that is, such that $P_1P_0 = \frac{2}{3}P_1P_2$. If the required coordinates are (x_0, y_0), then

$$\begin{aligned} x_0 = QM_0 &= QP_1 + \tfrac{2}{3}P_1M_2 \\ &= QP_1 + \tfrac{2}{3}(QM_2 - QP_1) \\ &= x_1 + \tfrac{2}{3}(x_2 - x_1) \end{aligned}$$

With a similar figure it can be shown that

$$y_0 = y_1 + \tfrac{2}{3}(y_2 - y_1)$$

The more general situation, in which $\frac{2}{3}$ is replaced by any real number k, is covered by the following theorem.

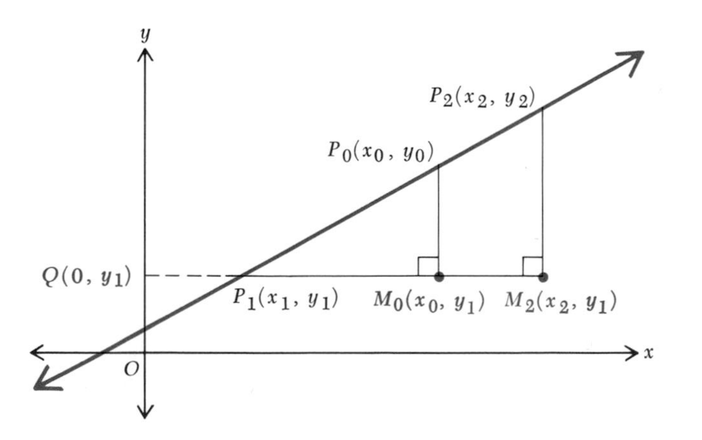

Fig. 1-16 Division of a line segment: $P_1P_0/P_1P_2 = \frac{2}{3}$ and $P_1M_0/P_1M_2 = \frac{2}{3}$.

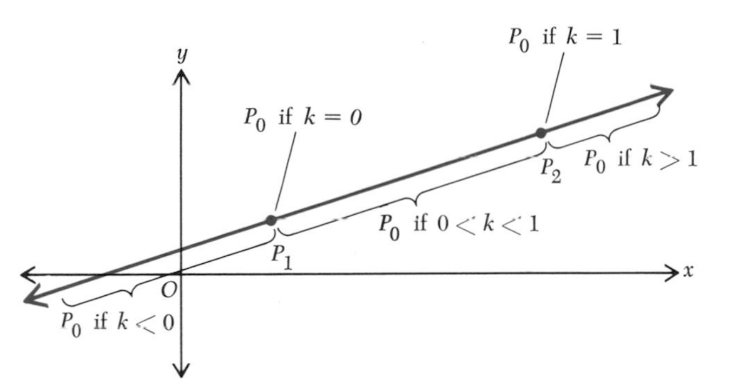

Fig. 1-17 Ratio of division of a line segment: $P_1P_0/P_1P_2 = k$.

Theorem 1-1
If $P_0(x_0, y_0)$ is the point on $\overleftrightarrow{P_1P_2}$ with $P_1(x_1, y_1)$ and $P_2(x_2, y_2)$ such that $P_1P_0/P_1P_2 = k$, for k a real number, then

$$x_0 = x_1 + k(x_2 - x_1) \qquad (1\text{-}2)$$
$$y_0 = y_1 + k(y_2 - y_1)$$

This theorem is used most often where k is a fraction between 0 and 1; for a value of k in this range any point on the segment between P_1 and P_2 can be found. The theorem is valid, however, in a broader sense. In fact (see Fig. 1-17),

1. If $0 < k < 1$, P_0 lies between P_1 and P_2.
2. If $k = 0$ or 1, P_0 coincides with P_1 or P_2, *respectively*.
3. If $k > 1$, P_0 lies on $\overrightarrow{P_1P_2}$ but not on $\overline{P_1P_2}$.
4. If $k < 0$, P_0 lies on $\overrightarrow{P_2P_1}$ but not on $\overline{P_2P_1}$; in this case the directed segments P_1P_0 and P_1P_2 have opposite signs.

It can be proved that by a suitable choice of k the coordinates of any point on $\overleftrightarrow{P_1P_2}$ may be found.

As a special case of the above result, the midpoint $P_0(x_0, y_0)$ of a given segment $\overline{P_1P_2}$ with $P_1(x_1, y_1)$ and $P_2(x_2, y_2)$ may be determined by taking $k = \frac{1}{2}$:

$$x_0 = x_1 + \tfrac{1}{2}(x_2 - x_1) = \tfrac{1}{2}(x_1 + x_2)$$
$$y_0 = y_1 + \tfrac{1}{2}(y_2 - y_1) = \tfrac{1}{2}(y_1 + y_2)$$

Example
Find the coordinates of the point P_0 that is five-sixths of the way from $A(-2, 5)$ to $B(10, 1)$ (see Fig. 1-18).

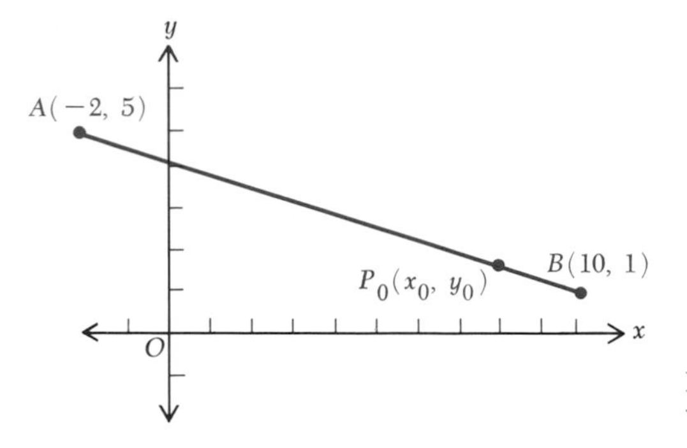

Fig. 1-18 Point P_0 five-sixths of the way from A to B.

Solution Using $k = \frac{5}{6}$ in Eqs. (1-2), the coordinates are

$$x_0 = -2 + \tfrac{5}{6}[10 - (-2)] = 8$$

and

$$y_0 = 5 + \tfrac{5}{6}(1 - 5) = \tfrac{5}{3}$$

The required point is then $P_0(8, \frac{5}{3})$.

1-8 AREA OF A TRIANGULAR REGION

When the coordinates of three points which are vertices of a triangle are known, the area of the enclosed region may be determined. For the case shown in Fig. 1-19 the area of the triangular region $P_1P_2P_3$ can be obtained by adding the measures of the trapezoidal regions $M_1M_3P_3P_1$ and $M_3M_2P_2P_3$ and then subtracting the measure of the trapezoidal region $M_1M_2P_2P_1$. Now, since the area of the interior of a trapezoid is equal to one-half the sum of the measures of the parallel sides times the measure of the altitude,

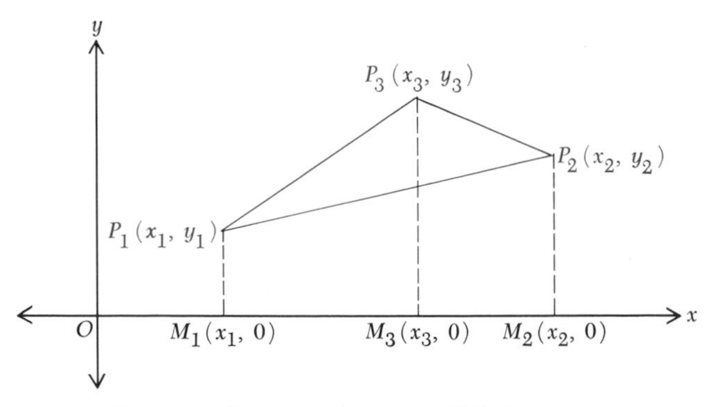

Fig. 1-19 Area of a triangular region $P_1P_2P_3$.

$$\begin{aligned}
\text{Area } M_1M_3P_3P_1 &= \tfrac{1}{2}(y_1 + y_3)(x_3 - x_1) \\
\text{Area } M_3M_2P_2P_3 &= \tfrac{1}{2}(y_3 + y_2)(x_2 - x_3) \\
\text{Area } M_1M_2P_2P_1 &= \tfrac{1}{2}(y_1 + y_2)(x_2 - x_1)
\end{aligned}$$

If the area of the triangular region is denoted by $P_1P_2P_3$, then

$$\begin{aligned}
P_1P_2P_3 &= \tfrac{1}{2}(y_1 + y_3)(x_3 - x_1) + \tfrac{1}{2}(y_3 + y_2)(x_2 - x_3) - \tfrac{1}{2}(y_1 + y_2)(x_2 - x_1) \\
&= \tfrac{1}{2}(y_1x_3 - y_3x_1 + y_3x_2 - y_2x_3 + y_2x_1 - y_1x_2)
\end{aligned}$$

The final expression in parentheses is precisely the expansion of the determinant

$$\begin{vmatrix} x_1 & y_1 & 1 \\ x_2 & y_2 & 1 \\ x_3 & y_3 & 1 \end{vmatrix}$$

Therefore, at least for the case shown in the figure, the area of the interior of the triangle is equal to one-half the value of this determinant. For other positions of the vertices relative to the coordinate axes the proof may differ somewhat in the details, but the result is the same.

Theorem 1-2

Let $P_1(x_1, y_1)$, $P_2(x_2, y_2)$, and $P_3(x_3, y_3)$ be vertices of a triangle, lettered so that in traversing the boundary from P_1 to P_2 to P_3 to P_1 the interior is always on the left. Then the area of the triangle, A, is given by

$$A = \frac{1}{2}\begin{vmatrix} x_1 & y_1 & 1 \\ x_2 & y_2 & 1 \\ x_3 & y_3 & 1 \end{vmatrix} \tag{1-3}$$

If the order of the points is such that the above traverse of the boundary has the opposite sense, then the formula yields a negative number whose absolute value is equal to the area.

In practice, the determinant

$$\begin{vmatrix} a & b & c \\ d & e & f \\ g & h & i \end{vmatrix}$$

is often evaluated according to the following rule:

$$\begin{aligned}
\begin{vmatrix} a & b & c \\ d & e & f \\ g & h & i \end{vmatrix} &= a\begin{vmatrix} e & f \\ h & i \end{vmatrix} - b\begin{vmatrix} d & f \\ g & i \end{vmatrix} + c\begin{vmatrix} d & e \\ g & h \end{vmatrix} \\
&= a(ei - fh) - b(di - fg) + c(dh - eg)
\end{aligned}$$

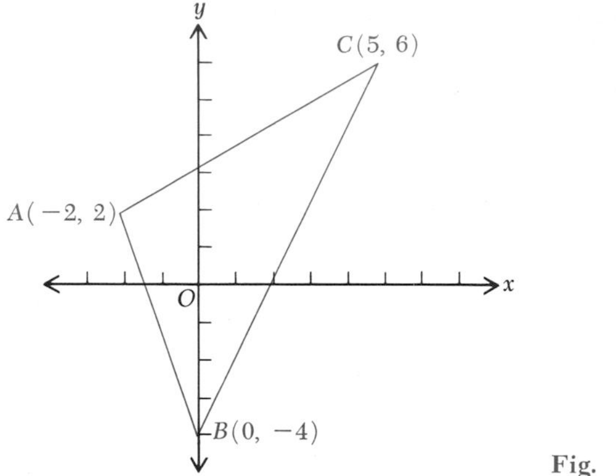

Fig. 1-20

Example

The area Q of the triangular region ABC in Fig. 1-20 is, from Eq. (1-3),

$$Q = \frac{1}{2}\begin{vmatrix} -2 & 2 & 1 \\ 0 & -4 & 1 \\ 5 & 6 & 1 \end{vmatrix} = \frac{1}{2}\left[-2\begin{vmatrix} -4 & 1 \\ 6 & 1 \end{vmatrix} - 2\begin{vmatrix} 0 & 1 \\ 5 & 1 \end{vmatrix} + 1\begin{vmatrix} 0 & -4 \\ 5 & 6 \end{vmatrix}\right]$$

$$= \tfrac{1}{2}[-2(-4-6) - 2(0-5) + 1(0+20)]$$
$$= \tfrac{1}{2}[20 + 10 + 20] = 25$$

PROBLEMS

1. Draw $\overline{AB}$; find the coordinates of the midpoint of $\overline{AB}$ and plot it in each of the following cases:

   (a) $A(4, -8)$, $B(-2, 0)$ (b) $A(-5, 2)$, $B(-1, 5)$
   (c) $A(3\sqrt{2}, -5\sqrt{3})$, $B(-2\sqrt{2}, -\sqrt{3})$

2. Find the coordinates of the points that divide the segment $\overline{AB}$ with end points $A(-6, -3)$ and $B(3, 1)$ into three congruent parts.
3. Find the coordinates of the points that divide the segment $\overline{PQ}$ with end points $P(8, -3)$ and $Q(-2, 6)$ into four congruent parts.
4. In each of the following cases find the coordinates of point P such that $AP/AB = k$; draw $\overleftrightarrow{AB}$ and graph P:

   (a) $A(2, -4)$, $B(8, 12)$; $k = \frac{3}{4}$ (b) $A(-4, 6)$, $B(11, 1)$; $k = \frac{2}{5}$
   (c) $A(-1, -1)$, $B(2, 3)$; $k = 2$ (d) $A(6, -2)$, $B(-2, 4)$; $k = \frac{3}{2}$
   (e) $A(1, -2)$, $B(6, 3)$; $k = -\frac{3}{2}$ (f) $A(3, -6)$, $B(0, 0)$; $k = -\frac{2}{3}$

5. Draw a figure similar to Fig. 1-16 and show that $y_0 = y_1 + \frac{2}{3}(y_2 - y_1)$.
6. Prove Theorem 1-1.
7. Draw the triangle with vertices at $A(-3, -4)$, $B(6, -2)$, and $C(3, 8)$. Find

the coordinates of the point on each median that is two-thirds of the way from the vertex to the midpoint of the opposite side.

8. Draw the triangle with vertices at $A(x_1, y_1)$, $B(x_2, y_2)$, and $C(x_3, y_3)$. Show that the point on each median which is at a distance from the vertex equal to two-thirds the length of the median has the coordinates

$$\left(\frac{x_1 + x_2 + x_3}{3}, \frac{y_1 + y_2 + y_3}{3}\right)$$

9. In each of the following cases draw the triangle whose vertices are the three given points and compute its area:

   (*a*) $(-4, 1)$, $(0, -5)$, $(3, 2)$ (*b*) $(1, -6)$, $(6, 1)$, $(-2, 5)$
   (*c*) $(0, 3)$, $(8, -1)$, $(0, -7)$ (*d*) $(1, -1)$, $(6, -2)$, $(3, -5)$

10. In each of the following cases draw the parallelogram *ABCD* with the given vertices and find its area:

    (*a*) $A(-2, -3)$, $B(6, -1)$, $C(7, 4)$, $D(-1, 2)$
    (*b*) $A(-3, 1)$, $B(1, -2)$, $C(7, 1)$, $D(3, 4)$

11. Find the area of each of the following polygons (*Hint:* Apply the formula for the area of a triangular region):

    (*a*) Polygon *ABCD*, with vertices at $A(-2, -3)$, $B(1, 7)$, $C(8, 5)$, $D(0, -1)$
    (*b*) Polygon *ABCDE*, with vertices at $A(0, 0)$, $B(5, 1)$, $C(7, 6)$, $D(2, 8)$, $E(0, 3)$

12. The triangle *OAB* in Fig. 1-21 has one vertex at the origin. Show that its area is given by the absolute value of

$$\frac{1}{2}\begin{vmatrix} x_1 & y_1 \\ x_2 & y_2 \end{vmatrix}$$

13. Show that the points $P(-3, -4)$, $Q(0, -2)$, and $R(9, 4)$ all lie on the same straight line.

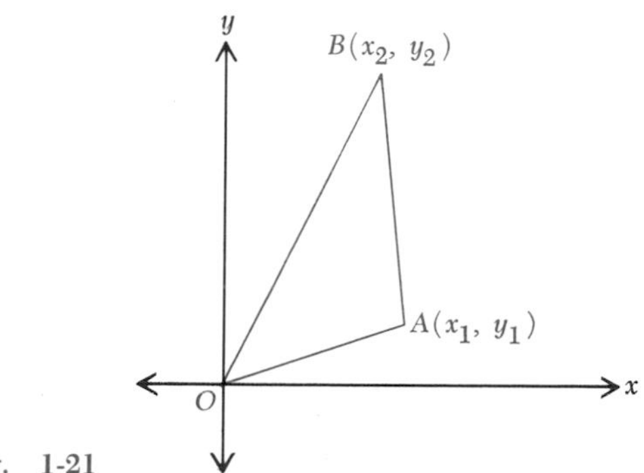

Fig. 1-21

14. In the triangle with vertices at $A(-2, -6)$, $B(10, -1)$, and $C(2, 8)$ find the length of the altitude drawn from C. *Hint:* Find the area and use the fact that the measure of area equals one-half the product of the measures of base and altitude.
15. Find the shortest distance from the origin to the line that passes through $P(3, 5)$ and $Q(6, 9)$.
16. Consider the line determined by the points $A(2, 3)$ and $B(6, 5)$. For what value of k in Theorem 1-1 does one obtain the point on the line that has 7.5 as its abscissa?
17. Let the coordinates of P_1 and P_2 in Theorem 1-1 be rational numbers. What points on the line determined by P_1 and P_2 will be obtained by using for k all rational numbers satisfying the condition $0 \leqslant k \leqslant 1$?

SUMMARY

Assuming a one-to-one correspondence between the real numbers and points on a number line, a rectangular, or cartesian, coordinate system can be established by drawing two number lines perpendicular to each other at the point 0 on each. These two lines are called the x and y axes. The distance x of any point from the y axis is the abscissa of the point, and the distance y from the x axis is called the ordinate of the point. The numbers x and y of the ordered pair (x, y) are the coordinates of the point.

The directed distance P_1P_2 from $P_1(x_1, y_1)$ to $P_2(x_2, y_1)$ (two points on a line parallel to the x axis) is

$$P_1P_2 = x_2 - x_1$$

Similarly,

$$P_2P_1 = x_1 - x_2$$

The undirected distance $|P_1P_2|$ between P_1 and P_2 is

$$|P_1P_2| = |x_2 - x_1| = |x_1 - x_2|$$

The directed distance P_1P_2 from $P_1(x_1, y_1)$ to $P_2(x_1, y_2)$ (two points on a line parallel to the y axis) is

$$P_1P_2 = y_2 - y_1$$

Similarly,

$$P_2P_1 = y_1 - y_2$$

The undirected distance $|P_1P_2|$ between P_1 and P_2 is

$$|P_1P_2| = |y_2 - y_1|$$

If two points $P_1(x_1, y_1)$ and $P_2(x_2, y_2)$ are not on the same vertical or horizontal line, the undirected distance d between them is

$$d = \sqrt{(x_2 - x_1)^2 + (y_2 - y_1)^2}$$

Point $P_0(x_0, y_0)$, on line $\overleftrightarrow{P_1P_2}$ for $P_1(x_1, y_1)$ and $P_2(x_2, y_2)$, is called a point of division when $P_1P_0/P_1P_2 = k$. The coordinates of P_0 are

$$x_0 = x_1 + k(x_2 - x_1) \qquad \text{and} \qquad y_0 = y_1 + k(y_2 - y_1)$$

A special case of the point of division is the midpoint P_0 of a segment $\overline{P_1P_2}$. Here $k = \frac{1}{2}$ and the coordinates of P_0 are

$$x_0 = \frac{x_1 + x_2}{2} \qquad \text{and} \qquad y_0 = \frac{y_1 + y_2}{2}$$

The measure of area of a triangular region $P_1P_2P_3$ that has vertices $P_1(x_1, y_1)$, $P_2(x_2, y_2)$, and $P_3(x_3, y_3)$ is the absolute value of the determinant

$$\frac{1}{2}\begin{vmatrix} x_1 & y_1 & 1 \\ x_2 & y_2 & 1 \\ x_3 & y_3 & 1 \end{vmatrix}$$

A second way of locating points in a plane is by means of polar coordinates. A ray $\overrightarrow{OA}$, called the polar axis, is drawn. Its end point O is called the pole. The position of any point P is specified by the number pair (ρ, θ), where ρ is the undirected distance OP and θ is the measure of angle AOP.

REVIEW PROBLEMS

1. Solve for x:

   (a) $|x + 3| = 7$ (b) $\left|\frac{1}{x - 3}\right| = 4$ (c) $|3x - 2| \leqslant 7$

2. Find the solution set for each of the following:

   (a) $\{x \mid |5x - 2| > 8\}$ (b) $\{r \mid |1 - 3r| = 3r - 1\}$
   (c) $\{y \mid |2 - 1/y| > 1\}$ (d) $\{z \mid |z| < -2\}$

3. Show that the triangle with vertices at $(-2, -3)$, $(5, -1)$, and $(3, 6)$ is a right triangle.
4. The ordinate of a certain point P is greater than its abscissa by 4 and the

undirected distances from P to $A(-2, 3)$ and from P to $B(6, -3)$ are equal. Find the coordinates of P.

5. The two coordinates of a point P are equal, and the point is equidistant from $A(-6, 4)$ and $B(2, -8)$. What are the coordinates of P?
6. Find the coordinates of the points P on $\overleftrightarrow{AB}$, with $AP/AB = k$ for $A(-6, 4)$ and $B(2, -8)$, if

   (a) $k = \frac{3}{2}$ (b) $k = n/3$ for n a nonnegative integer less than 4

7. The coordinates of point P in rectangular form are $(0, 6)$. Write the polar coordinates of P for each of the following conditions:

   (a) $\rho > 0$; $0° \leq \theta < 360°$ (b) $\rho < 0$; $-360° < \theta \leq 0$
   (c) $\rho > 0$; $\theta > 0°$ (d) $\rho > 0$; $\theta < 0°$

8. Find the area of polygonal region $ABCD$ where the coordinates of the vertices are $A(-4, 0)$, $B(0, -6)$, $C(8, -2)$, and $D(1, 5)$.
9. Show in two ways that points $P(0, 2)$, $Q(-3, 4)$, and $R(6, -2)$ are on the same line.
10. For what value or values of y will the triangular region ABC with vertices at $A(-3, 4)$, $B(6, 1)$, and $C(4, y)$ have an area of 25 square units?

FOR EXTENDED STUDY

1. Prove that if the coordinates of the vertices of a triangular region are even numbers, its area is an even number of square units.
2. If (ρ, θ) is one set of polar coordinates of point P, find general expressions for all possible polar coordinates of P if (a) θ is in degrees; (b) θ is in radians.
3. A regular hexagon has sides of length 8. It has a center at the origin and two vertices on the x axis. List the coordinates of all vertices (a) in rectangular form; (b) in polar form.
4. Solve $|2x|/|x - 3| > 1$ for x.
5. Given all points (ρ, θ), describe the set of points for which $\rho < 5$.

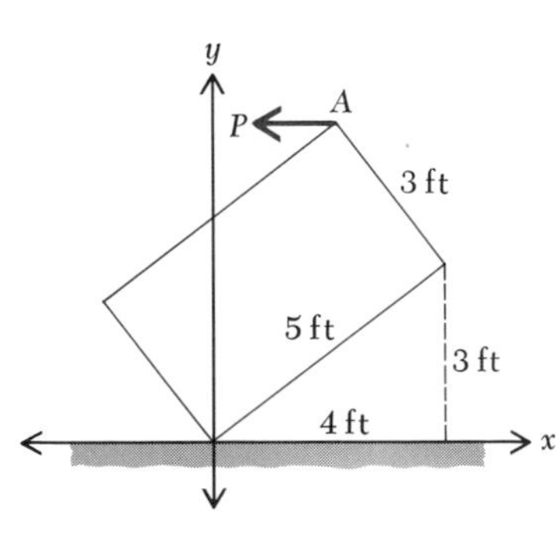

Fig. 1-22

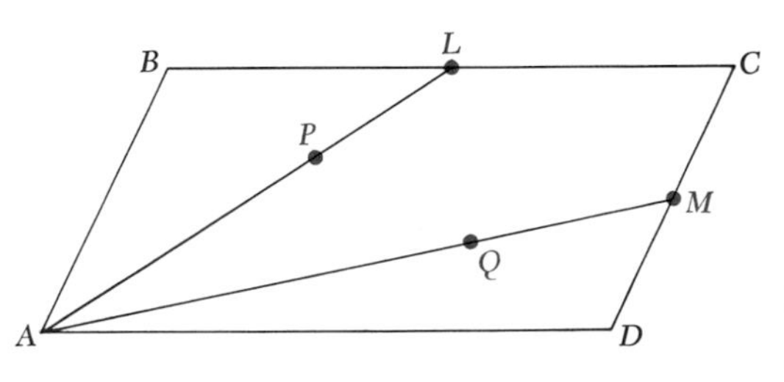

Fig. 1-23

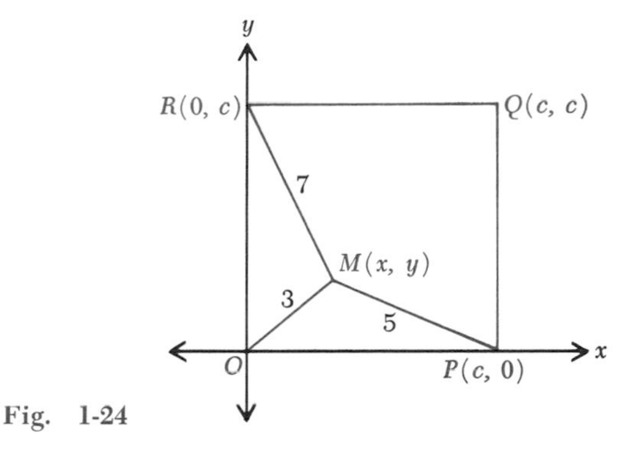

Fig. 1-24

6. Read about the life of Descartes in "The World of Mathematics," edited by James R. Newman.
7. Read Newton as the Originator of Polar Coordinates, in *American Mathematical Monthly,* vol. 56, p. 73, 1949.
8. For the system of rectangular coordinates established in this chapter the same units are used on the x and y axes. Suppose the ratio of the unit used on the x axis to that used on the y axis is a/b. Find the undirected distance d between $P_1(x_1, y_1)$ and $P_2(x_2, y_2)$ in units of (a) the x axis; (b) the y axis.
9. A 50-lb rectangular plate having the dimensions indicated in Fig. 1-22 is held in the position shown by a horizontal force P applied at A. To calculate the magnitude of the force P it would be necessary to know the coordinates of A with respect to the axes shown. Find the coordinates.
10. In Fig. 1-23 L and M are the midpoints of $\overline{BC}$ and $\overline{CD}$, respectively; the distance $AP = \frac{2}{3}AL$ and $AQ = \frac{2}{3}AM$. Show that P and Q lie on the diagonal $\overline{BD}$ and that they divide it into three congruent segments. $ABCD$ is any parallelogram. *Hint:* Take the origin at A and the x axis along $\overrightarrow{AD}$, thus letting the vertices of the parallelogram be $A(0, 0)$, $D(x_1, 0)$, $B(x_2, y_2)$, and $C(x_1 + x_2, y_2)$.
11. Point M is inside square $OPQR$ in Fig. 1-24. The undirected distances as shown are $|OM| = 3$, $|MP| = 5$, and $|MR| = 7$. Find the length c of the side of the square.

ANGLES AND DIRECTION | 2

2-1 TRIGONOMETRIC FUNCTIONS

If θ is any real number, the *trigonometric functions* of θ are defined as follows. Construct an angle of θ rad with vertex at the origin and initial side along the positive x axis (see Fig. 2-1). If θ is a positive number, the angle is generated by a counterclockwise rotation from the posi-

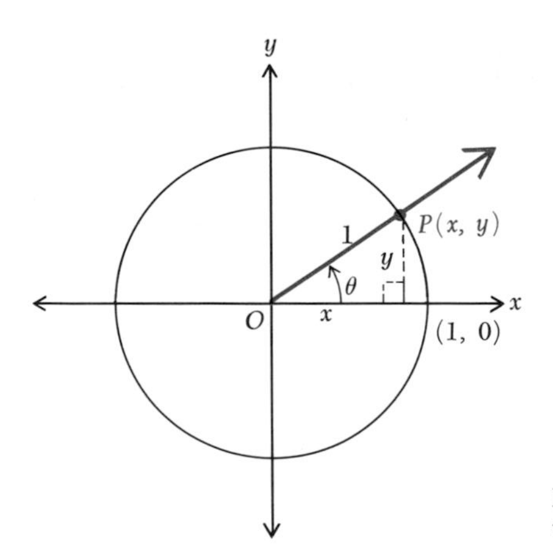

Fig. 2-1 Definitions of trigonometric functions in terms of a unit circle.

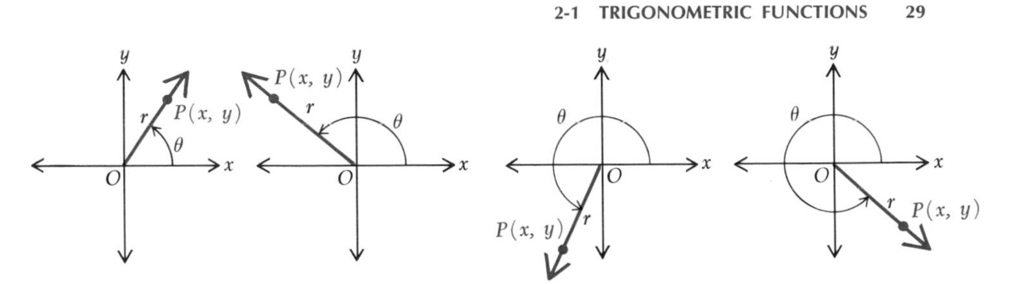

Fig. 2-2 An alternative definition of trigonometric functions.

tive portion of the x axis; if θ is a negative number, the angle is generated by a clockwise rotation. Now, a point P on the terminal side of the angle and on a unit circle with center at the origin has coordinates (x, y). The definitions of the trigonometric functions are then

$$\sin\theta = y \qquad \csc\theta = \frac{1}{y}$$

$$\cos\theta = x \qquad \sec\theta = \frac{1}{x}$$

$$\tan\theta = \frac{y}{x} \qquad \cot\theta = \frac{x}{y}$$

It is possible to define the trigonometric functions of an angle without reference to a point on a unit circle. Choose *any* point $P(x, y)$ (see Fig. 2-2) on the terminal side of any angle θ and denote the distance of P from the origin O by r, with the assumption that r is a positive number. The definitions are then

$$\sin\theta = \frac{y}{r} \qquad \csc\theta = \frac{r}{y}$$

$$\cos\theta = \frac{x}{r} \qquad \sec\theta = \frac{r}{x}$$

$$\tan\theta = \frac{y}{x} \qquad \cot\theta = \frac{x}{y}$$

The given number θ may be regarded as the number of degrees in an angle (rather than radians); an angle of $\theta°$ may be constructed, and the above procedure followed.

For many practical applications it is sufficient to consider the trigonometric functions of acute angles in a right triangle. As indicated by Fig. 2-3, the definitions derived from Fig. 2-1 can be modified to include the names of the sides of the right triangle PAB. For angle PAB, $\overline{AB}$

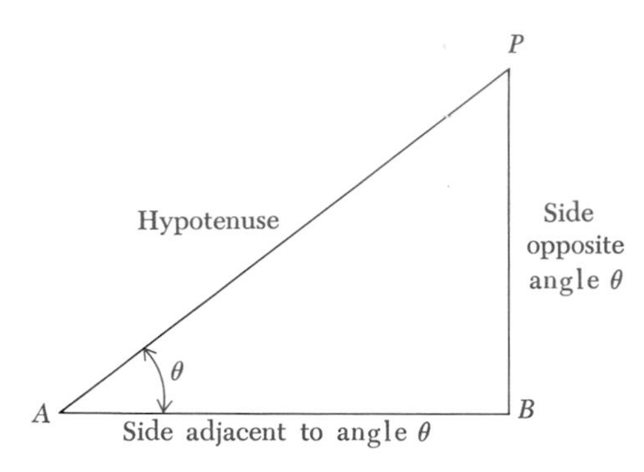

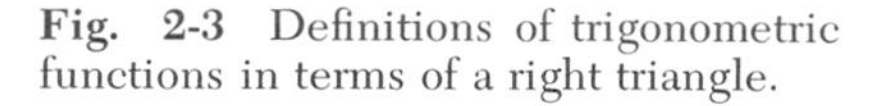
Fig. 2-3 Definitions of trigonometric functions in terms of a right triangle.

is the adjacent side and $\overline{BP}$ is the opposite side. Then

$$\sin \text{ angle } PAB = \frac{\text{measure of opposite side}}{\text{measure of hypotenuse}}$$

$$\cos \text{ angle } PAB = \frac{\text{measure of adjacent side}}{\text{measure of hypotenuse}}$$

$$\tan \text{ angle } PAB = \frac{\text{measure of opposite side}}{\text{measure of adjacent side}}$$

$$\csc \text{ angle } PAB = \frac{\text{measure of hypotenuse}}{\text{measure of opposite side}}$$

$$\sec \text{ angle } PAB = \frac{\text{measure of hypotenuse}}{\text{measure of adjacent side}}$$

$$\cot \text{ angle } PAB = \frac{\text{measure of adjacent side}}{\text{measure of opposite side}}$$

2-2 INCLINATION: SLOPE

A line l, not parallel to the x axis, intersects this axis at some point Q (see Fig. 2-4). The direction of the line in relation to the coordinate axes may be specified by giving the counterclockwise angle $\alpha < 180°$ through which $\overrightarrow{QX}$ would have to be rotated to bring it into coincidence with l. *This angle is called the inclination of l.* The inclination of a line parallel to the x axis is defined to be zero.

The *slope* of a line, usually denoted by the letter m, is defined as *the tangent of the inclination α.*

If $0 < \alpha < 90°$, m is a positive number.

If $90° < \alpha < 180°$, m is a negative number.
If $\alpha = 0$, $m = 0$.
If $\alpha = 90°$, the slope of the line is undefined. *Why?*

The slope of a line may be expressed in terms of the coordinates of two points $P_1(x_1, y_1)$ and $P_2(x_2, y_2)$ on it. It is apparent that for either of the cases shown in Fig. 2-5

$$m = \tan \alpha = \frac{MP_2}{P_1M} = \frac{y_2 - y_1}{x_2 - x_1} \qquad \text{for } x_2 \neq x_1$$

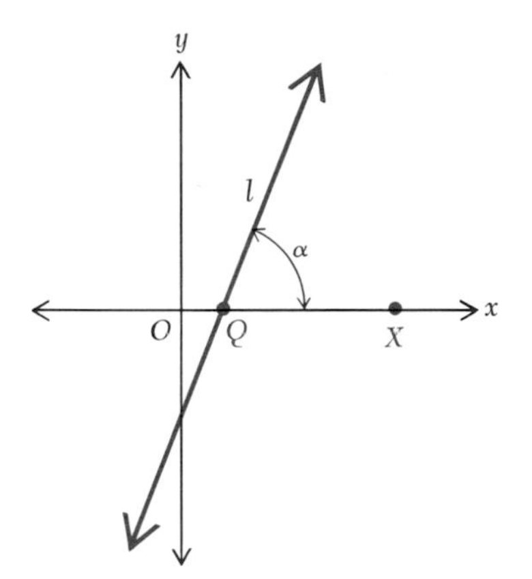

Fig. 2-4 Angle α is inclination of line l; $m = \tan \alpha$ is slope of line l.

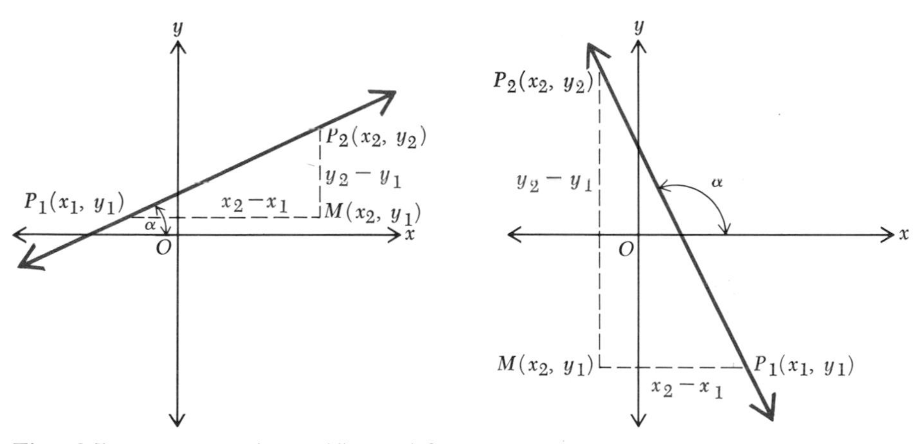

Fig. 2-5 $m = \tan \alpha = (y_2 - y_1)/(x_2 - x_1)$ for $x_2 \neq x_1$.

This formula holds for all positions of P_1 and P_2, provided $x_1 \neq x_2$. Therefore the following theorem may be stated.

Theorem 2-1

If $P_1(x_1, y_1)$ and $P_2(x_2, y_2)$ are any two points on a line l not perpendicular to the x axis, then the slope of l is

$$m = \frac{y_2 - y_1}{x_2 - x_1} \qquad x_2 \neq x_1$$

It makes no difference which of the two points is called P_1. The slope of the line through points $P(1, -3)$ and $Q(5, 7)$ (see Fig. 2-6) is, by Theorem 2-1,

$$\frac{7-(-3)}{5-1} = \frac{5}{2} \qquad \text{or} \qquad \frac{-3-7}{1-5} = \frac{5}{2}$$

The slope $\frac{5}{2}$ means that as x increases by 2, y increases by 5, or there is a change of 1 unit to the right for each rise of $2\frac{1}{2}$ units in the y direction. Similarly, a slope of $-\frac{3}{7}$ means that as x increases by 7 units, y decreases by 3 units, or as y increases by 3, x decreases by 7. The slope is thus a measure of steepness of a line.

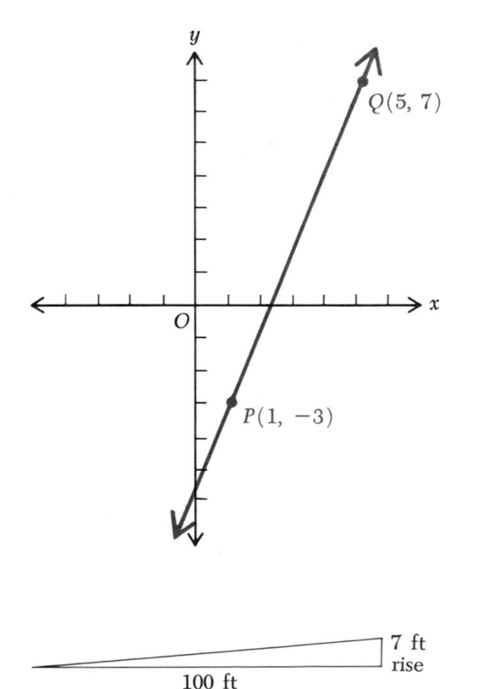

Fig. 2-6 $m = [7-(-3)]/(5-1) = \frac{5}{2}$ or $m = (-3-7)/(1-5) = \frac{5}{2}$.

7 ft rise

100 ft

Fig. 2-7 Meaning of percent grade: $m = 0.07$, or 7% grade.

The slope of a road is called its *grade*. A 7% grade means a slope of 0.07 or $\frac{7}{100}$; this indicates a rise of 7 ft for each 100 ft of horizontal distance (Fig. 2-7).

PROBLEMS

1. What is the slope of a line whose inclination is (*a*) 120°; (*b*) 90°; (*c*) 60°; (*d*) 30°; (*e*) $\pi/4$; (*f*) $5\pi/6$?
2. What is the inclination of a line whose slope is (*a*) 0; (*b*) -1; (*c*) 1; (*d*) $-1/\sqrt{3}$; (*e*) $\sqrt{3}/3$; (*f*) $-\sqrt{3}$?
3. Do the following sentences make sense? (*a*) The inclination of a line is $-60°$. (*b*) The line has a slope of zero. (*c*) The inclination of a line is 210°. (*d*) The inclination of a line is Arctan 3. Give reasons for your answers.
4. Find the slope of a line through a pair of points whose coordinates are: (*a*) (3, 7), (1, 2); (*b*) $(-4, -7)$, (1, 9); (*c*) $(-5, -3)$, $(8, -3)$; (*d*) $(3m^2 + n^2, n^2)$, $(-8m^2 - n^2, 4n^2)$; (*e*) (4, 7), $(4, -6)$; (*f*) $(\sqrt{5}, -5\sqrt{3})$, $(-6\sqrt{3}, 2\sqrt{5})$.
5. (*a*) Draw the line through the point (3, 4), with an inclination of 60°.
   (*b*) Draw the line through the point $(-2, 5)$, with slope of $\frac{3}{2}$.
   (*c*) Draw the line through the point $(-4, -1)$, with slope of $-\frac{5}{3}$.
6. List the coordinates of two other points on each of the lines in Prob. 5.
7. (*a*) Find the slope of each side of the triangle which has vertices $A(1, 6)$, $B(-2, -8)$, and $C(7, -2)$. (*b*) Find the slope of each of the three medians of the triangle.
8. Using slopes, show that the points $A(-2, 6)$, $B(1, 1)$, and $C(7, -9)$ lie on the same line.
9. If possible, determine p so that the line containing each pair of points has the given slope:

   (*a*) $(3, 2p)$, $(7, 14p)$; $m = \frac{1}{3}$ (*b*) $(5, 2/p)$, $(3, 1/p)$; $m = 4$
   (*c*) $(2, |p|)$, $(3, p)$; $m = -1$ (*d*) $(-3, p)$, $(5, p)$; $m = 2$

10. The point $(0, b)$ is on a line with a slope of m. (*a*) What is the x coordinate of the point where the line crosses the x axis? (*b*) Another point on the line has a y coordinate twice its x coordinate. What are the coordinates of this point?
11. Prove that the slope of any line parallel to the x axis is 0.
12. State and prove the converse of the statement in Prob. 11.

2-3 (OPTIONAL) DIRECTION COSINES

Intuitively it is possible to think of line $\overleftrightarrow{P_1P_2}$ in Fig. 2-8 as having two senses, or directions:

1. The sense, or direction, of $\overrightarrow{P_1P_2}$, which is down and to the right and
2. The sense, or direction, of $\overrightarrow{P_2P_1}$, which is up and to the left.

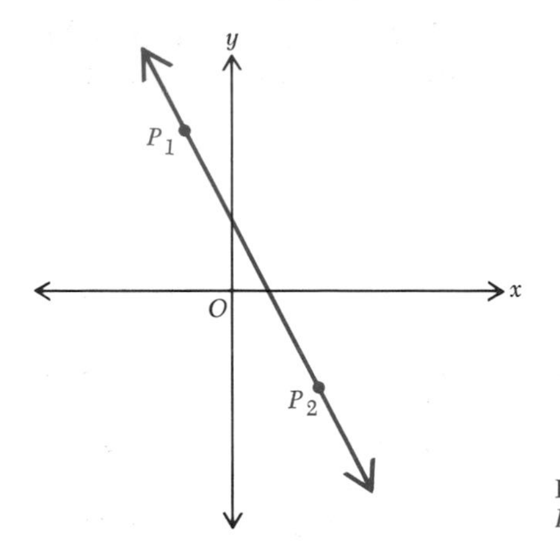

Fig. 2-8 Two directions on $\overleftrightarrow{P_1P_2}$: from P_1 to P_2 or from P_2 to P_1.

Either of these directions may be considered positive. That is, in Fig. 2-8 the direction from P_1 to P_2 may be thought of as positive; then the direction from P_2 to P_1 is negative. However, if the direction from P_2 to P_1 is considered positive, then the direction from P_1 to P_2 is negative.

The direction, or sense, of a line may be associated with two angles.

Definition

The undirected angles α and β formed by the positive sense of a line and the positive sense of the x and y axes, respectively, are the direction angles of a directed line and are expressed as the ordered pair $[\alpha, \beta]$.

The two numbers

$$\lambda = \cos \alpha \qquad \text{and} \qquad \mu = \cos \beta$$

are called the *direction cosines* of a directed line. They are expressed as the ordered pair $[\lambda, \mu]$. In Fig. 2-9*a*, for the line directed from P_1 to P_2, or for ray $\overrightarrow{P_1P_2}$, the direction cosines are

$$\lambda = \cos \alpha \qquad \text{and} \qquad \mu = \cos \beta$$

In Fig. 2-9*b*, for the same line directed from P_2 to P_1, or for ray $\overrightarrow{P_2P_1}$, the direction cosines are

$$\lambda = \cos \gamma \qquad \text{and} \qquad \mu = \cos \delta$$

Because the direction angles are considered nondirected angles, α and γ are supplementary, as are β and δ. Furthermore, as illustrated in Fig. 2-9, each of the direction angles of any directed line may measure any number from 0 through 180°.

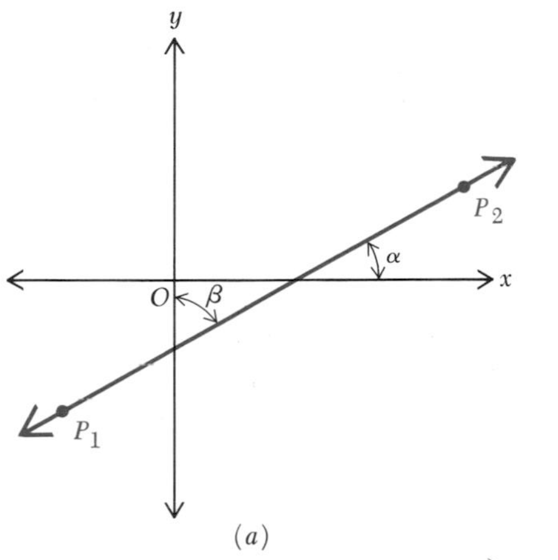

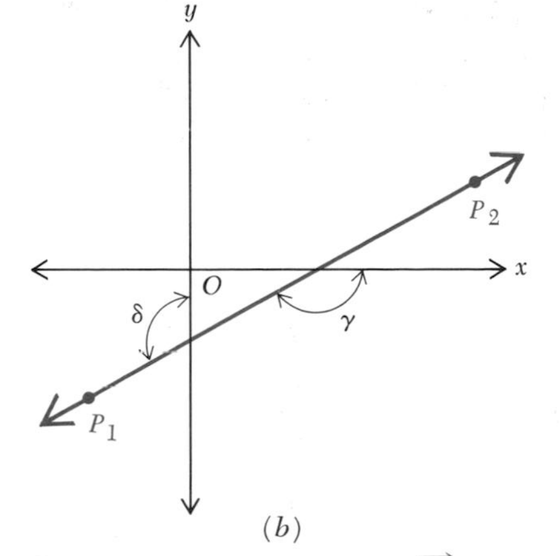

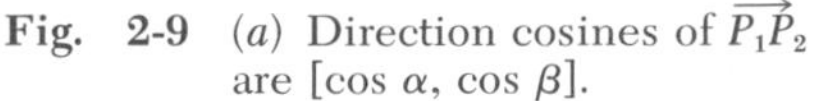
Fig. 2-9 (*a*) Direction cosines of $\overrightarrow{P_1P_2}$ are $[\cos \alpha, \cos \beta]$.

(*b*) Direction cosines of $\overrightarrow{P_2P_1}$ are $[\cos \gamma, \cos \delta]$.

Example

If $\lambda = \cos \alpha$ and $\mu = \cos \beta$ are the direction cosines of $\overrightarrow{P_1P_2}$, what are the direction cosines of $\overrightarrow{P_2P_1}$?

Solution The direction cosines of $\overrightarrow{P_2P_1}$ (Fig. 2-10) are

$$\cos (180 - \alpha) = -\cos \alpha = -\lambda$$

and

$$\cos (180 - \beta) = -\cos \beta = -\mu$$

The positive or negative signs for the direction cosines of a ray are readily determined. For ray $\overrightarrow{P_2P_1}$ in Fig. 2-10, directed downward from

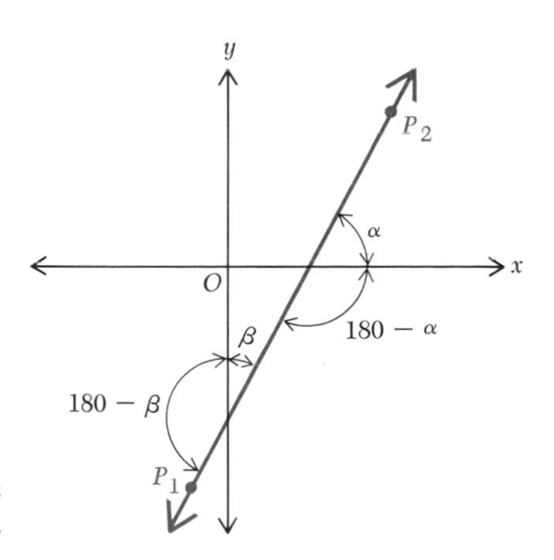

Fig. 2-10 $[\alpha, \beta]$ are direction angles of $\overrightarrow{P_1P_2}$; $[180 - \alpha, 180 - \beta]$ are direction angles of $\overrightarrow{P_2P_1}$.

upper right to lower left, the direction angles $180 - \alpha$ and $180 - \beta$ are each greater than 90° and less than 180°. Hence $\cos(180 - \alpha)$ and $\cos(180 - \beta)$ are both negative. The analysis of three other cases for rays that are not horizontal or vertical is left to the reader (see Prob. 8).

2-4 (OPTIONAL) DIRECTION NUMBERS

Any two numbers r and s which are proportional to the direction cosines of a line are called *direction numbers* of the line. Then $r = k\lambda = k \cos \alpha$ and $s = k\mu = k \cos \beta$, where k, the constant of proportionality, is a nonzero real number.

For $\overleftrightarrow{P_1P_2}$ in Fig. 2-11, $x_2 - x_1$ and $y_2 - y_1$ are a pair of direction numbers. This may be proved as follows. $d = |P_1P_2| = \sqrt{(x_2 - x_1)^2 + (y_2 - y_1)^2}$, angle α = angle $P_2P_1P_0$, and angle β = angle $P_1P_2P_0$. Then $\cos \alpha = (x_2 - x_1)/d$ and $\cos \beta = (y_2 - y_1)/d$, or $x_2 - x_1 = d \cos \alpha$ and $y_2 - y_1 = d \cos \beta$. With d the constant of proportionality, $x_2 - x_1$ and $y_2 - y_1$ are proportional to $\cos \alpha$ and $\cos \beta$, respectively. Then a pair of direction numbers for $\overleftrightarrow{P_1P_2}$ is

$$[x_2 - x_1, y_2 - y_1] \tag{2-1}$$

If $[a, b]$ is a pair of direction numbers for a line, then a and b can be interpreted as the directed distances in the direction of the x and y axes, respectively, for a distance $\sqrt{a^2 + b^2}$ along the line. It should also be pointed out that the direction cosines are the directed distances in the direction of the axes corresponding to a distance of 1 unit in the *positive direction* along the line.

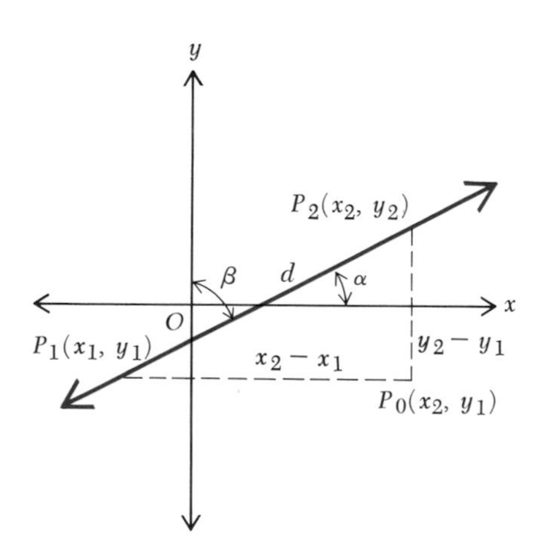

Fig. 2-11 Direction numbers of $\overrightarrow{P_1P_2}$ are $[x_2 - x_1, y_2 - y_1]$ and those of $\overrightarrow{P_2P_1}$ are $[-(x_2 - x_1), -(y_2 - y_1)]$.

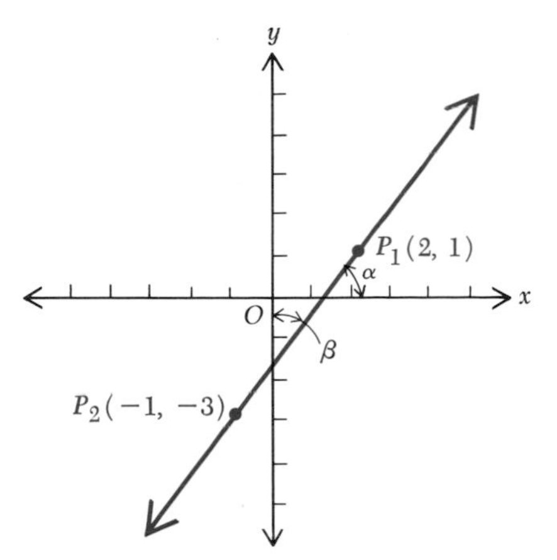

Fig. 2-12 For $\overrightarrow{P_2P_1}$ direction cosines are $[\frac{3}{5}, \frac{4}{5}]$ and for $\overrightarrow{P_1P_2}$ they are $[-\frac{3}{5}, -\frac{4}{5}]$.

Example 1

Two points on a line are $P_1(2, 1)$ and $P_2(-1, -3)$. What are the direction cosines of $\overleftrightarrow{P_2P_1}$ if the positive direction is from P_2 to P_1?

Solution (See Fig. 2-12.)

$$d = |P_2P_1| = \sqrt{(-1-2)^2 + (-3-1)^2} = 5$$

For $\overrightarrow{P_2P_1}$

$$\cos \alpha = \frac{x_2 - x_1}{d} = \frac{2-(-1)}{5} = \frac{3}{5}$$

and

$$\cos \beta = \frac{y_2 - y_1}{d} = \frac{1-(-3)}{5} = \frac{4}{5}$$

Thus for each unit along the line from P_2 to P_1, we move three-fifths of a unit in the positive x direction and four-fifths of a unit in the positive y direction.

An interesting relationship that holds for every pair of direction cosines $[\cos \alpha, \cos \beta]$ is

$$\cos^2 \alpha + \cos^2 \beta = 1 \qquad \text{or} \qquad \lambda^2 + \mu^2 = 1 \tag{2-2}$$

This is readily proved. In Fig. 2-11,

$$\cos \alpha = \frac{x_2 - x_1}{d}$$

$$\cos \beta = \frac{y_2 - y_1}{d}$$

and $$d = |P_1P_2| = \sqrt{(x_2 - x_1)^2 + (y_2 - y_1)^2}$$

Now
$$\begin{aligned}\cos^2\alpha + \cos^2\beta &= \left(\frac{x_2 - x_1}{d}\right)^2 + \left(\frac{y_2 - y_1}{d}\right)^2 \\ &= \frac{(x_2 - x_1)^2 + (y_2 - y_1)^2}{d^2} \\ &= \frac{(x_2 - x_1)^2 + (y_2 - y_1)^2}{(x_2 - x_1)^2 + (y_2 - y_1)^2} \\ &= 1\end{aligned}$$

Example 2
Is it possible for a line to have direction angles $\alpha = 135°$ and $\beta = 150°$?

Solution For any line, $\cos^2 \alpha + \cos^2 \beta = 1$. In this case $\cos \alpha = \cos 135° = -1/\sqrt{2}$ and $\cos \beta = \cos 150° = -\sqrt{3}/2$.

But
$$\left(\frac{1}{\sqrt{2}}\right)^2 + \left(-\frac{\sqrt{3}}{2}\right)^2 \neq 1$$

Hence it is impossible for the line to have a pair of direction angles [135°, 150°].

Example 3
If $[r, s]$ is a pair of direction numbers for a line, show that $\lambda = \pm r/\sqrt{r^2 + s^2}$ and $\mu = \pm s/\sqrt{r^2 + s^2}$.

Solution On the line with direction numbers $[r, s]$ (Fig. 2-13) are two

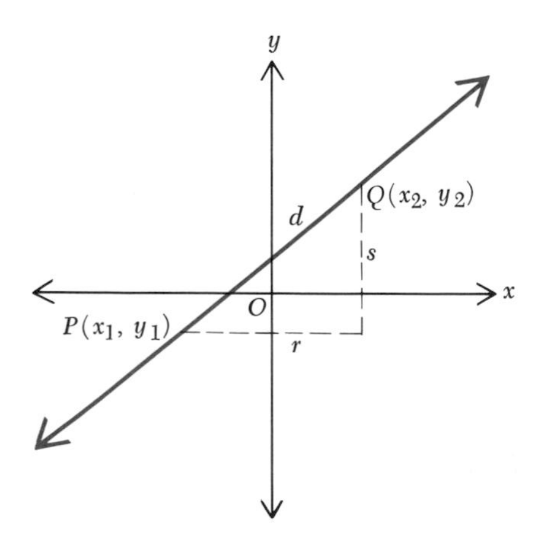

Fig. 2-13 A pair of direction numbers for $\overleftrightarrow{PQ}$ is $[r, s]$. Pairs of direction cosines are $[r/\sqrt{r^2 + s^2}, s/\sqrt{r^2 + s^2}]$ and $[-r/\sqrt{r^2 + s^2}, -s/\sqrt{r^2 + s^2}]$.

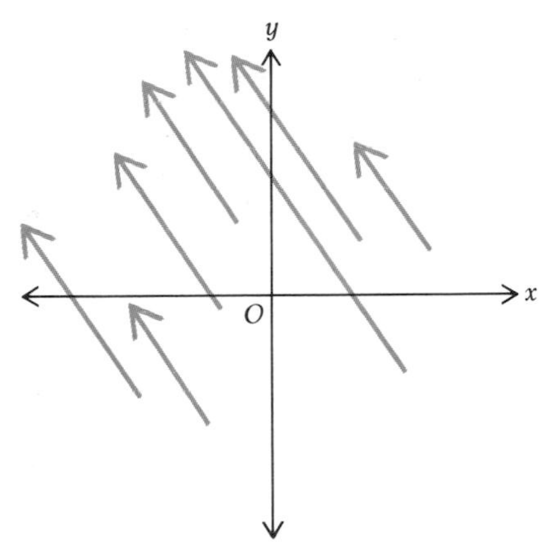

Fig. 2-14 Rays with the same pair of direction numbers.

points $P(x_1, y_1)$ and $Q(x_2, y_2)$. It follows that

$$r = \pm(x_2 - x_1) \qquad s = \pm(y_2 - y_1) \qquad d^2 = r^2 + s^2$$

with the choice of sign for r and s depending on the positive direction chosen for the line.

Now, $\lambda = r/(r^2 + s^2)$ for $\overrightarrow{PQ}$ or $-r/(r^2 + s^2)$ for $\overrightarrow{QP}$. For $\overleftrightarrow{PQ}$,

$$\lambda = \pm\frac{r}{r^2 + s^2} \qquad \text{and} \qquad \mu = \pm\frac{s}{r^2 + s^2} \tag{2-3}$$

The correct signs for λ and μ for a given ray are chosen by analyzing the conditions of the problem.

Example 4

A pair of direction numbers for a ray with positive direction upward and to the left is $r = -2$ and $s = 3$. What are its direction cosines?

Solution Any of the rays pictured in Fig. 2-14 meets the condition, as does any other parallel ray with the same direction. From Eqs. (2-3),

$$\lambda = \frac{-2}{\sqrt{(-2)^2 + (3)^2}} = -\frac{2}{\sqrt{13}} \qquad \text{and} \qquad \mu = \frac{3}{\sqrt{(-2)^2 + (3)^2}} = \frac{3}{\sqrt{13}}$$

PROBLEMS

1. Find a pair of direction numbers and the direction cosines of $\overrightarrow{P_1P_2}$ and $\overrightarrow{P_2P_1}$ for each of the following pairs of points:

(a) $P_1(3, 5)$, $P_2(6, 7)$ (b) $P_1(4, -3)$, $P_2(-6, -10)$
(c) $P_1(-7, 1)$, $P_2(-2, -5)$ (d) $P_1(-9, 4)$, $P_2(3, 8)$

2. Determine the direction angles of each ray in Prob. 1. Express answers in the form Arctan (a/b).
3. For each ordered pair of direction numbers given, find the slope of the line:

   (a) $[2, 3]$ (b) $[-4, 3]$ (c) $[\sqrt{7}, -3]$ (d) $[a/b, c/d]$

4. Draw the following rays:

   (a) End point $(0, 0)$; $\lambda = 1/\sqrt{2}$ and $\mu = -1/\sqrt{2}$
   (b) End point $(-1, 3)$; direction angles $\alpha = 120°$ and $\beta = 30°$
   (c) End point $(3, -4)$; a pair of direction numbers $r = 3 - 2$ and $s = -4 + 5$

5. The direction cosines of $\overrightarrow{AB}$ are $\lambda = -\frac{1}{2}$ and $\mu = \sqrt{3}/2$. (a) What is the slope of $\overleftrightarrow{AB}$? (b) What are the direction cosines of $\overrightarrow{BA}$? (c) Give three pairs of direction numbers for $\overrightarrow{AB}$.
6. (a) What are the direction angles of $\overrightarrow{CD}$ if the coordinates of C are $(4, 6)$ and those of D are $(-3, 6)$? (b) Repeat (a) for the line directed from D to C.
7. Which of these pairs of numbers can be direction cosines of a line?

   (a) $[\sqrt{2}/2, -\sqrt{2}/2]$ (b) $[\sqrt{5}, \sqrt{4}]$ (c) $[-\frac{4}{5}, \frac{3}{5}]$ (d) $[\frac{12}{13}, -\frac{5}{13}]$

8. For rays directed (a) upward to the left, (b) downward to the right, and (c) upward to the right, what are the signs for λ and μ? Make statements as in Sec. 2-3 for each of these rays.
9. Find the direction angles of (a) a line directed upward to the right with congruent direction angles; (b) a line directed upward to the left, with one direction cosine the negative of the other; (c) a line directed downward to the left with one direction cosine, $-\sqrt{2}/2$.
10. For two points $P_1(x_1, y_1)$ and $P_2(x_2, y_2)$ find the coordinates of $P(x, y)$ on $\overleftrightarrow{P_1P_2}$ such that $P_1P/PP_2 = c/d$.
11. A line is directed downward to the right. (a) If $\lambda = \frac{1}{2}$, what is μ? (b) If $\mu = -\frac{1}{4}$, what is λ?
12. For a horizontal line directed to the left, give (a) the direction angles; (b) the direction cosines.
13. What are the direction cosines of a line parallel to the y axis and (a) directed upward; (b) directed downward?
14. What are three pairs of direction numbers for a line parallel to the x axis and directed (a) to the right; (b) to the left?

2-5 PARALLEL AND PERPENDICULAR LINES

Two lines, l_1 and l_2, are parallel if and only if their inclinations are equal. If the lines have slopes m_1 and m_2, the condition for parallelism

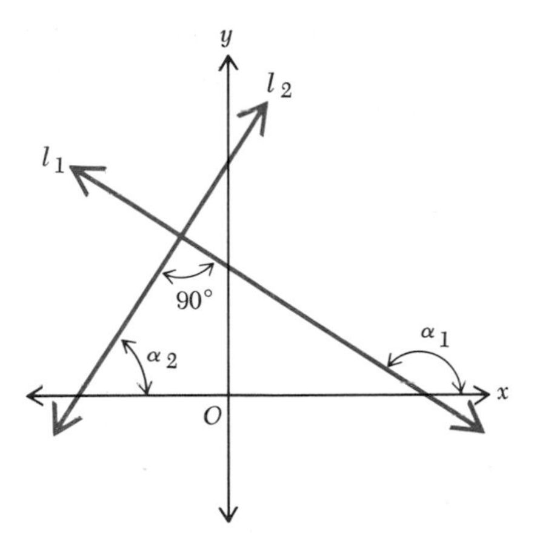

Fig. 2-15 Relationship of angles for perpendicular lines.

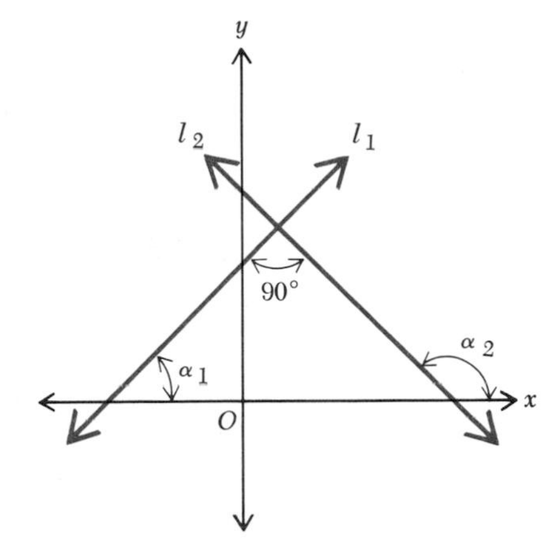

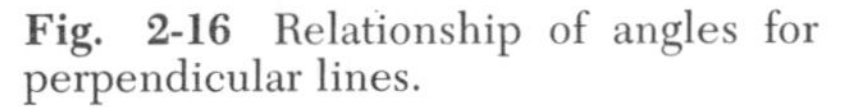

Fig. 2-16 Relationship of angles for perpendicular lines.

is that

$$m_1 = m_2$$

If the lines are perpendicular, the inclination of one of them exceeds that of the other by 90°; that is, either $\alpha_1 = \alpha_2 + 90°$ or $\alpha_2 = \alpha_1 + 90°$ (see Figs. 2-15 and 2-16).

In either case

$$\tan \alpha_1 = -\cot \alpha_2 = -\frac{1}{\tan \alpha_2} \qquad \text{for } \alpha_1 \neq 90° \qquad \text{and} \qquad \alpha_2 \neq 90°$$

or, since $\tan \alpha_1 = m_1$ and $\tan \alpha_2 = m_2$,

$$m_1 = -\frac{1}{m_2} \qquad \text{or} \qquad m_1 m_2 = -1$$

Conversely, if $m_1 = -1/m_2$, then

$$\tan \alpha_1 = -\cot \alpha_2 \qquad \text{and} \qquad \alpha_1 = \alpha_2 \pm 90°$$

From this it follows that the lines are perpendicular; we have proved the following theorem.

Theorem 2-2

Two lines, neither vertical nor horizontal, having slopes m_1 and m_2, are perpendicular if and only if

$$m_1 = -\frac{1}{m_2} \qquad \text{or} \qquad m_1 m_2 = -1$$

The only case not covered by the theorem is that of two lines parallel to the x and y axes, respectively. They are, of course, perpendicular; the slope of one of them is zero and the slope of the other is undefined.

2-6 THE ANGLE BETWEEN TWO LINES

Let l_1 and l_2 be two lines intersecting at P and having slopes $m_1 = \tan\alpha_1$ and $m_2 = \tan\alpha_2$, respectively ($\alpha_1 \neq \pi/2$ and $\alpha_2 \neq \pi/2$). (See Fig. 2-17.)

The smallest counterclockwise angle θ through which a ray with end point P on l_1 would have to be rotated about P to coincide with l_2 is called the *angle from* l_1 *to* l_2. This is a particular one of the two supplementary angles of intersection between l_1 and l_2. To find this angle (or its tangent) in terms of the slopes of the two lines, two cases must be considered.

Case 1 $\alpha_1 < \alpha_2$ Since angle α_2 is equal to $\alpha_1 + \theta$ (see Fig. 2-17*a*),

$$\begin{aligned}\tan\theta &= \tan(\alpha_2 - \alpha_1)\\ &= \frac{\tan\alpha_2 - \tan\alpha_1}{1 + \tan\alpha_2 \tan\alpha_1}\\ &= \frac{m_2 - m_1}{1 + m_2 m_1} \qquad \text{for } m_1 m_2 \neq -1\end{aligned}$$

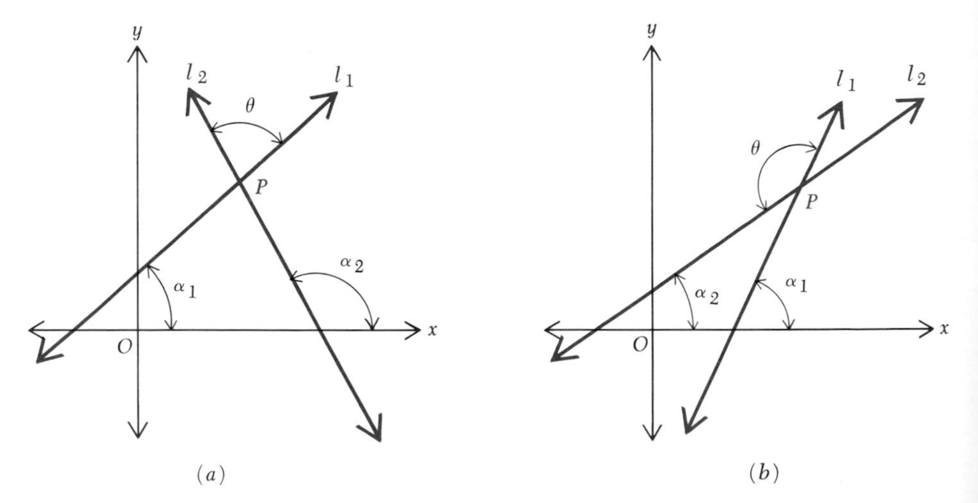

Fig. 2-17 θ is the angle between line l_1, with slope m, and line l_2, with slope m_2: $\tan\theta = (m_2 - m_1)/(1 + m_1 m_2)$.

Case 2 $\alpha_1 > \alpha_2$ (See Fig. 2-17*b*.) In this case

$$180° - \theta = \alpha_1 - \alpha_2$$

$$\tan(180° - \theta) = \frac{\tan \alpha_1 - \tan \alpha_2}{1 + \tan \alpha_1 \tan \alpha_2}$$

$$\tan \theta = \frac{\tan \alpha_2 - \tan \alpha_1}{1 + \tan \alpha_1 \tan \alpha_2}$$

$$= \frac{m_2 - m_1}{1 + m_1 m_2} \qquad \text{for } m_1 m_2 \neq -1$$

The result is the same in both cases, and therefore the following theorem may be stated.

Theorem 2-3
If two lines l_1 and l_2 have slopes m_1 and m_2, respectively, and if θ is the angle from l_1 to l_2, then

$$\tan \theta = \frac{m_2 - m_1}{1 + m_1 m_2} \qquad \text{for } m_1 m_2 \neq -1$$

In deriving the formula for tan θ it was assumed that neither line was parallel to the y axis. A case in which l_2 is parallel to the y axis is depicted in Fig. 2-18. In this case $\tan \theta = \cot \alpha_1 = 1/m_1$.

If the angle formed by two lines having slopes m_1 and m_2 is 90°, then $m_1 m_2 = -1$, and the denominator in the formula for tan θ becomes zero.

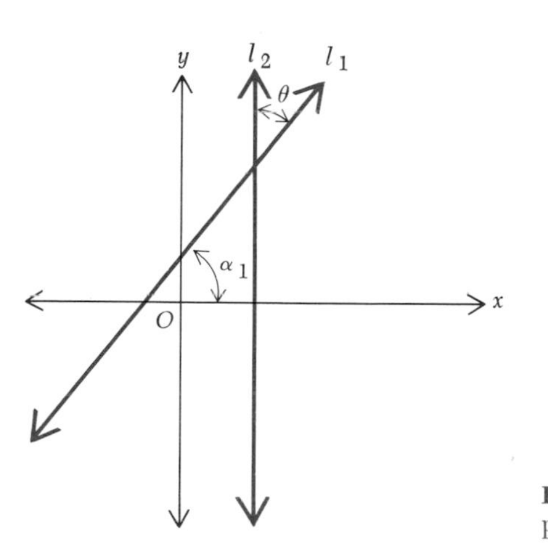

Fig. 2-18 Angle θ between two lines, one parallel to y axis.

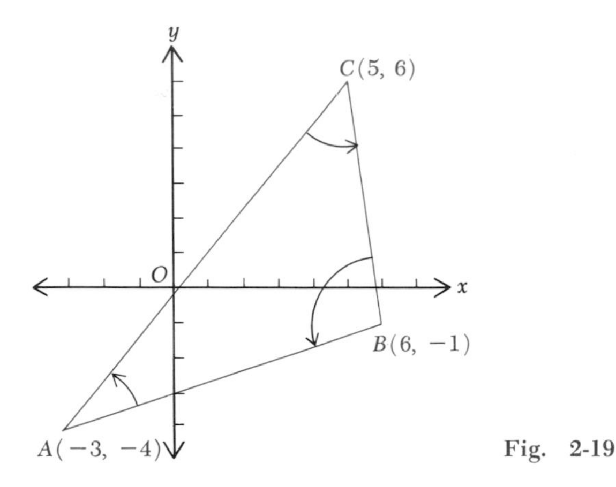

Fig. 2-19

Example
Find the tangent of angle A of the triangle with vertices $A(-3, -4)$, $B(6, -1)$, and $C(5, 6)$.

Solution Let the line determined by A and B be l_1 with slope m_1 and that determined by A and C be l_2 with slope m_2; then the required angle is the angle from l_1 to l_2 (see Fig. 2-19). The slopes are, by Theorem 2-1,

$$m_1 = \frac{-1-(-4)}{6-(-3)} = \frac{1}{3} \qquad m_2 = \frac{6-(-4)}{5-(-3)} = \frac{5}{4}$$

Then, by Theorem 2-3,

$$\tan A = \frac{m_2 - m_1}{1 + m_1 m_2} = \frac{\frac{5}{4} - \frac{1}{3}}{1 + \frac{5}{4}(\frac{1}{3})} = \frac{\frac{11}{12}}{\frac{17}{12}} = \frac{11}{17}$$

Tables may be used to find the angle whose tangent is $\frac{11}{17}$, or 0.6471 (to four decimal places). It is approximately 32°54′.

The other angles could be computed similarly.

PROBLEMS

1. Choose pairs of members of the following set that are slopes of parallel lines:

$$\frac{3}{\sqrt{3}}, \quad \frac{-\sqrt{3}}{3}, \quad \frac{-1}{\sqrt{3}}, \quad \sqrt{3}, \quad -\sqrt{3}, \quad \frac{-3}{\sqrt{3}}, \quad \log_3 3, \quad \tan\left(\text{Arctan}\,\frac{1}{\sqrt{3}}\right)$$

2. Repeat Prob. 1 for perpendicular lines.

3. Draw a line through $P(3, 4)$ with inclination 60°. (*a*) What is its slope? (*b*) What is the slope of a line through P which is perpendicular to this line?
4. Find the slope of the shortest segment that can be drawn from the origin to the line determined by $P(-4, 1)$ and $Q(6, 5)$.
5. A line is drawn through $A(-4, 5)$ perpendicular to a line connecting the origin and $P(6, -2)$. What is its slope?
6. Draw the triangle with vertices $A(-3, 6)$, $B(2, 8)$, and $C(4, -7)$. Show that the segment joining the midpoints of $\overline{AC}$ and $\overline{BC}$ is parallel to $\overline{AB}$ and its measure is half that of $\overline{AB}$.
7. Draw the triangle with vertices $A(-2, 7)$, $B(13, 1)$, and $C(0, -3)$. Find the slope of each of its altitudes and each of its medians.
8. Show that the triangle with vertices $P(-10, 4)$, $Q(7, 3)$, and $R(4, -3)$ is a right triangle.
9. Show that the quadrilateral with vertices $P(7, 9)$, $Q(8, -4)$, $R(-5, -5)$, and $S(-6, 8)$ is a square.
10. Show that the points $E(-5, -2)$, $F(1, 7)$, $G(2, 2)$, and $H(0, -1)$ are the vertices of an isosceles trapezoid.
11. In each of the following cases find the tangent of the acute angle formed by $\overleftrightarrow{AB}$ and $\overleftrightarrow{PQ}$. Then, from tables, find the measure of the angle:

    (*a*) $A(-2, -6)$, $B(8, 2)$; $P(2, 3)$, $Q(4, -2)$
    (*b*) $A(0, -3)$, $B(6, 1)$; $P(2, 0)$, $Q(5, -2)$

12. Find the tangents of the angles at B and D in the quadrilateral $ABCD$ with vertices $A(-2, -2)$, $B(-4, -7)$, $C(3, -5)$, and $D(5, 2)$.
13. In each of the following, find the tangent of each angle of the triangle:

    (*a*) $A(3, 6)$, $B(8, 1)$, $C(5, -2)$ (*b*) $P(-5, 2)$, $Q(0, 6)$, $R(-4, -4)$

14. A line is drawn through A perpendicular to $\overline{BC}$ in the triangle with vertices $A(2, 6)$, $B(8, 2)$, and $C(-2, -8)$. Find the tangent of the angle formed by this line and side $\overline{AB}$.
15. In the triangle of Prob. 14 a ray is drawn from A through the midpoint of $\overline{BC}$, and another is drawn from A perpendicular to $\overline{BC}$. Find the tangent of the acute angle formed by these rays.
16. A line l_1 is drawn through $P(4, 3)$ and the origin, and another line l_2 is drawn through P perpendicular to the line through $A(-2, 1)$ and $B(3, 7)$. Find the tangent of the acute angle formed by l_1 and l_2.
17. From Theorem 2-3 derive the formula

$$\cos\theta = \frac{1 + m_1 m_2}{\sqrt{1 + m_1^2}\,\sqrt{1 + m_2^2}}$$

    where θ, m_1, and m_2 have the same definitions as given for the theorem.
18. From the formula in Prob. 17 show that if two lines with slopes m_1 and m_2 are perpendicular, then $1 + m_1 m_2 = 0$.

2-7 APPLICATIONS TO ELEMENTARY GEOMETRY

The methods of analytic geometry can frequently be used to obtain simple proofs of theorems of elementary geometry. The procedure will be illustrated by two examples.

Example 1

Prove that the diagonals of a parallelogram bisect each other.

Solution After the parallelogram is drawn the coordinate system for it may be chosen in such a manner that the coordinates are as simple as possible. Hence, choose one vertex as the origin and let the x axis coincide with one side. The coordinates of three vertices are then $P_0(0, 0)$, $P_1(x_1, 0)$, and $P_2(x_2, y_2)$. Since the figure is a parallelogram, the abscissa of the fourth vertex must be $x_1 + x_2$ and its ordinate must be y_2. This point is then $P_3(x_1 + x_2, y_2)$ (see Fig. 2-20). The coordinates of the midpoint of $\overline{P_0P_3}$ are $\left(\frac{(x_1 + x_2)}{2}, \frac{y_2}{2}\right)$, and the coordinates of the midpoint of $\overline{P_1P_2}$ are $\left(\frac{(x_1 + x_2)}{2}, \frac{y_2}{2}\right)$. Since these coordinates are identical, the midpoints of the diagonals coincide and the theorem is proved.

Example 2

Prove that the midpoint of the hypotenuse of a right triangle is equidistant from the vertices of the triangle.

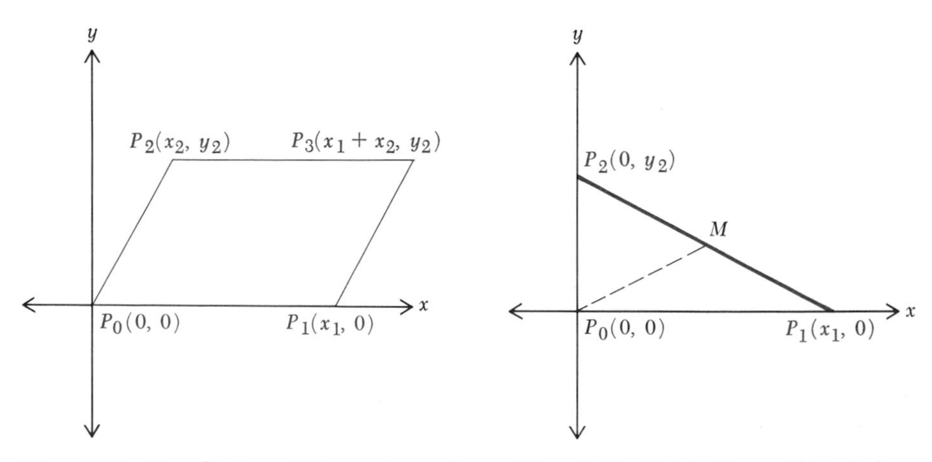

Fig. 2-20 Coordinates of the vertices of a parallelogram.

Fig. 2-21 Point M equidistant from the vertices of the right triangle $P_0P_1P_2$.

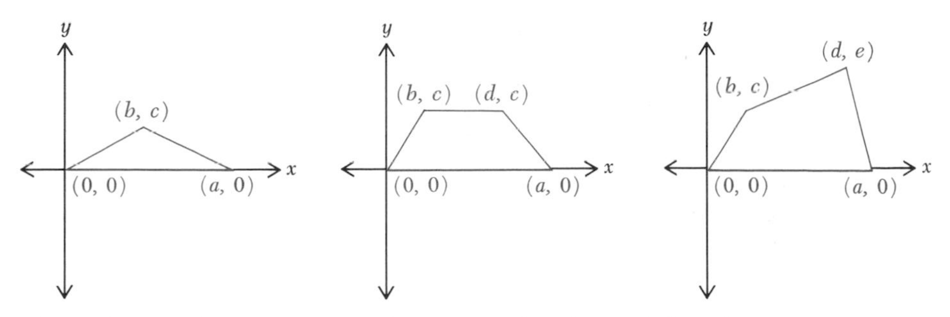

Fig. 2-22 Coordinates for a triangle, a trapezoid, and a quadrilateral.

Solution After drawing any right triangle, choose the coordinate system for greatest convenience (Fig. 2-21). Then if M is the midpoint of $\overline{P_1P_2}$, its coordinates are $(\frac{1}{2}x_1, \frac{1}{2}y_2)$. The undirected distance of M from P_1 or P_2 is $\frac{1}{2}P_1P_2$ or $\frac{1}{2}\sqrt{x_1^2 + y_2^2}$. Its distance from P_0 is also $\frac{1}{2}\sqrt{x_1^2 + y_2^2}$. Hence the theorem is proved.

The special placement of axes does not result in a loss in generality. The theorems concern properties of the geometric figures themselves and are independent of the coordinate system to which they are referred. In each case the figure is drawn first and then oriented to the coordinate system in a convenient position.

It must be emphasized for this analytic method, that although the proof is completed algebraically, it is based on the coordinates of the points. The figure that is drawn is used merely as an aid in visualizing the problem. The coordinates, then, rather than the figure, must express the given data. If in Example 1 x_3 and y_3 had been used as the coordinates of P_3, the figure would not have been a parallelogram, even though it looked like one. It is the use of $x_1 + x_2$ and y_2 as the coordinates of P_3 that makes the figure a parallelogram.

Suitable choices of axes and coordinates for proving theorems concerning triangles, trapezoids, or general quadrilaterals are shown in Fig. 2-22.

PROBLEMS

Prove the following theorems analytically.

1. The segment joining the midpoints of two sides of a triangle is parallel to the third side, and its measure is one-half the measure of the third side.
2. The measure of the segment joining the midpoints of the nonparallel sides of a trapezoid is equal to one-half the sum of the measures of the parallel sides.

3. The diagonals of a square are perpendicular to each other.
4. The diagonals of an isosceles trapezoid are congruent.
5. The union of the four line segments that join the midpoints of the successive sides of any rectangle is a rhombus.
6. The union of the four line segments that join the midpoints of the successive sides of any quadrilateral is a parallelogram.
7. If two medians of a triangle are congruent, the triangle is isosceles.
8. The sum of the squares of the measures of the four sides of any parallelogram equals the sum of the squares of the measures of the diagonals.
9. The line segments joining the midpoints of the opposite sides of any quadrilateral bisect each other.
10. In any triangle the sum of the squares of the measures of the medians is equal to three-fourths the sum of the squares of the measures of the three sides.
11. In any triangle the sum of the squares of the measures of two sides is equal to twice the square of one-half the measure of the third side plus twice the square of the measure of the median drawn to that side.
12. In any quadrilateral the sum of the squares of the measures of the four sides is equal to the sum of the squares of the measures of the diagonals plus four times the square of the measure of the line segment joining the midpoints of the diagonals.

2-8 INTRODUCTION TO VECTORS

A line segment $\overline{AB}$ is a set of points. Associated with $\overline{AB}$ is a real number called its *measure,* which indicates the length of the segment. For example, the number 3 may be associated with some particular segment 3 in. long. As discussed earlier in this chapter, another number associated with any segment is its slope. Furthermore, directed distances along a line have been defined. Thus the directed distance AB could be called $+3$; then the directed distance BA would be called -3. If a positive direction is specified from one end point to the other, the segment is called a *directed line segment.*

Directed line segments are commonly in use in physics because each segment can be thought of as representing a magnitude acting in a certain direction. Some examples are in representing velocity, acceleration, or force. For instance, velocity may be thought of as speed (a measure) in a particular direction. From a mathematical point of view directed line segments may be considered as elements of a "number" system on which operations can be performed. The name commonly given to a directed line segment regarded as an element of a mathematical system is *vector.*

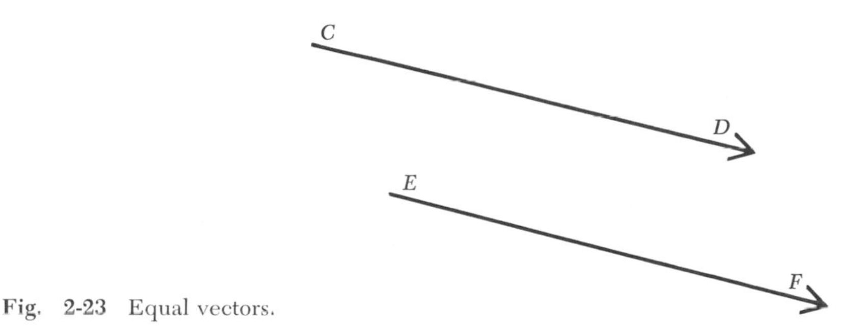

Fig. 2-23 Equal vectors.

Definition

A vector is a directed line segment considered as an element in a mathematical system.

The notation for a vector extending from A to B is **AB**. In this case A is called the *initial point* and B is the *terminal point*. The nonnegative real number representing the distance from A to B along the segment is called the *absolute value* of the vector or the *magnitude* of the vector, written $|\mathbf{AB}|$.

The fact that the magnitude $|\mathbf{AB}|$ of a vector may be any nonnegative real number means that $|\mathbf{AB}| = 0$ is a special case. This *null vector* has coincident initial and terminal points. It has no direction. While this perhaps seems a trivial case, it is necessary for completeness of the mathematical system of vectors.

Associated with $\overline{AB}$ are two vectors **AB** and **BA**, depending on the direction specified. These two vectors have the same magnitude, but they are not the same vector, nor are they equal, since their directions are opposite along $\overleftrightarrow{AB}$. Two vectors are considered equal if and only if they have both the same magnitude and the same direction; thus the two vectors pictured in Fig. 2-23 are equal. From a physical standpoint, they might represent a displacement in the same direction and with the same magnitude. It is not sufficient that two vectors have the same length and are parallel for them to be considered equal; they must also have the same sense or direction along the same line or parallel lines.

Definition

Two vectors are equal if and only if they have the same magnitude and the same direction.

As a consequence of the definition of equality of vectors, it can be concluded that, given any vector, we can find an infinite number of

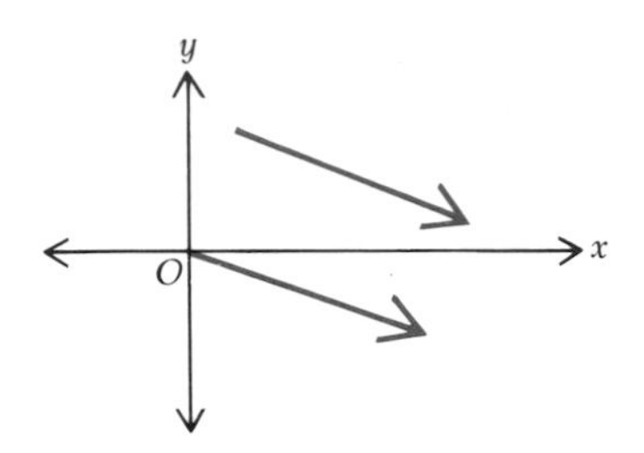

Fig. 2-24 Equal vectors, one with its initial point at the origin.

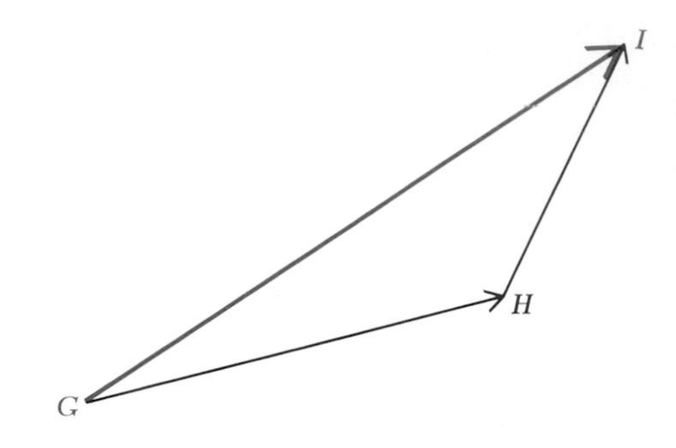

Fig. 2-25 Vector sum: **GH** + **HI** = **GI**.

other equal vectors. In every case exactly one of the equal vectors has the origin as its initial point (Fig. 2-24). In some applications it is necessary to substitute an equal vector for a given vector, and in many cases the substitute vector chosen is the one with its initial point at the origin.

The motivation for the choice of definitions for operations with vectors comes from physical applications. In Fig. 2-25 the net effect of a displacement from G to H and then from H to I is the same as if the displacement were directly from G to I. For vectors **GH** and **HI**, then, the vector sum should be **GI**. The sum of two vectors so positioned that the initial point of the second is the terminal point of the first is a vector whose initial point is the initial point of the first vector and whose terminal point is the terminal point of the second vector.

More generally, for **GH** and **AB** in a plane (Fig. 2-26*a*) the vector sum

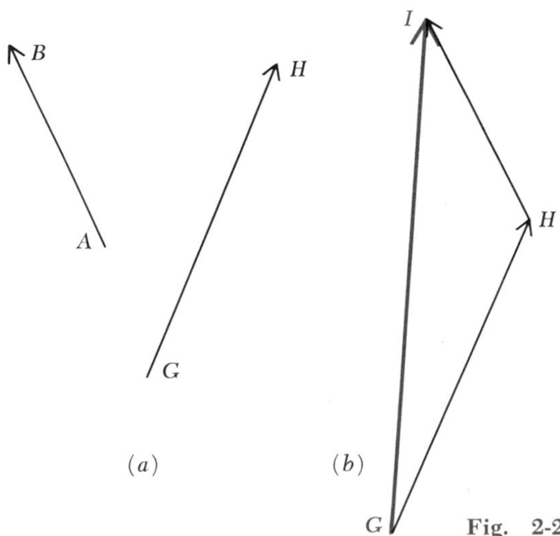

Fig. 2-26 Vector sum uniquely determined.

of **GH** and **AB** is defined to be the vector **GI** (Fig. 2-26*b*). Here **HI** is a vector equal to **AB**. It should be noted that the vector sum is a uniquely determined vector.

Theorem 2-4
For any two vectors **AB** *and* **CD**

$$\mathbf{AB} + \mathbf{CD} = \mathbf{CD} + \mathbf{AB}$$

That is, vector addition is commutative.

Proof In Fig. 2-27

$$\mathbf{AB} + \mathbf{BC} = \mathbf{AC}$$

and
$$\mathbf{AD} + \mathbf{DC} = \mathbf{AC}$$

hence
$$\mathbf{AB} + \mathbf{BC} = \mathbf{AD} + \mathbf{DC}$$

But
$$\mathbf{AD} = \mathbf{BC} \qquad \text{and} \qquad \mathbf{DC} = \mathbf{AB}$$

since they are on opposite sides of a parallelogram. Therefore

$$\mathbf{AB} + \mathbf{BC} = \mathbf{BC} + \mathbf{AB}$$

This completes the proof that vector addition is commutative.

Vector addition is also associative (see Prob. 5 in the next set).

As is true of subtraction in ordinary arithmetic, the definition of subtraction of vectors may be given in terms of addition. In Fig. 2-27

$$\mathbf{AB} + \mathbf{BC} = \mathbf{AC}$$

so
$$\mathbf{AC} - \mathbf{AB} = \mathbf{BC}$$

and
$$\mathbf{AC} - \mathbf{BC} = \mathbf{AB}$$

The difference **AC** − **AB** is equal to a vector **BC** whose initial point is the terminal point of the second vector **AB** and whose terminal point is the terminal point of the first vector **AC**. For the difference **AC** − **BC**, where the vectors do not have a common initial point, **BC** may be rep-

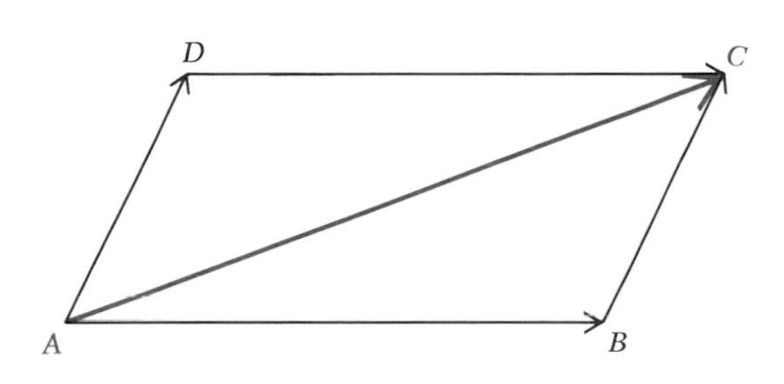

Fig. 2-27 Vector addition is commutative: **AB** + **BC** = **BC** + **AB**.

resented by an equal vector **AD**. Then

$$\begin{aligned}\mathbf{AC} - \mathbf{BC} &= \mathbf{AC} - \mathbf{AD}\\ &= \mathbf{DC}\\ &= \mathbf{AB}\end{aligned}$$

A special case occurs when the two given vectors are equal; in this event the difference is the null vector.

2-9 ELEMENTARY APPLICATIONS OF VECTORS

Vectors are used often in analytic geometry to prove geometric theorems. Sometimes the vector method is considerably shorter and less laborious than the procedure using coordinates.

Two additional concepts are required in order to prove elementary theorems by the vector approach. The first concept is *scalar multiplication* of a vector. A scalar is a real number used in connection with vectors. It differs from a vector in that it has magnitude but no direction.

Definition

Given any scalar k and any vector **a**, *the product k***a** *is a vector whose magnitude is* $|k|$ *times the magnitude of* **a**.

For nonzero vectors and scalars the product $k\mathbf{a}$ has the same direction as the original vector if k is positive and the opposite direction if k is negative. If $k = 0$, then $k\mathbf{a}$ is the zero vector.

Scalar multiplication of a vector makes possible the formulation of a second concept, that of parallel vectors. We see intuitively that parallel vectors lie along lines that are parallel (see Fig. 2-28).

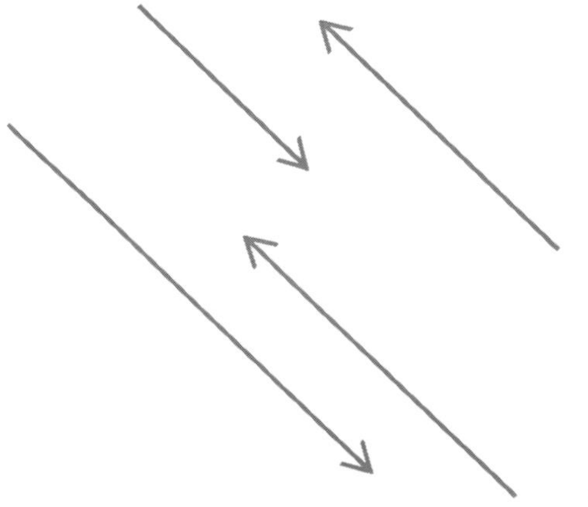

Fig. 2-28 Parallel vectors.

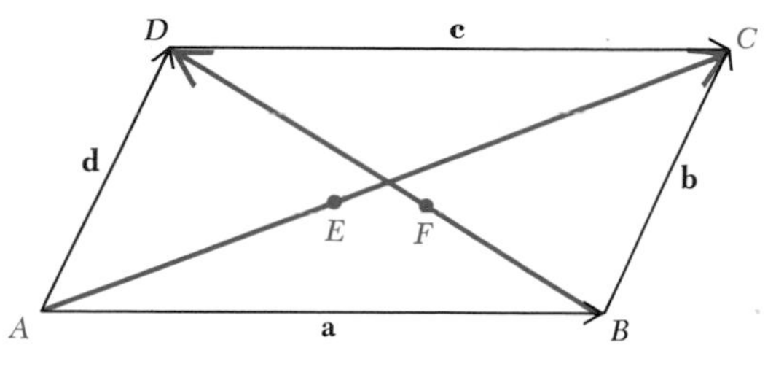

Fig. 2-29 Use of vectors in proving that the diagonals of a parallelogram bisect each other.

Definition

Two vectors are parallel if and only if one is a nonzero scalar multiple of the other.

Example

Prove that the diagonals of a parallelogram bisect each other.

Solution In Fig. 2-29

$$\mathbf{AC} = \mathbf{a} + \mathbf{d} \qquad \text{and} \qquad \mathbf{DB} = \mathbf{a} - \mathbf{d}$$

If E and F are the midpoints of $\overline{AC}$ and $\overline{DB}$, respectively, then

$$\mathbf{AE} = \tfrac{1}{2}(\mathbf{a} + \mathbf{d})$$

and

$$\mathbf{AF} = \mathbf{d} + \tfrac{1}{2}(\mathbf{a} - \mathbf{d}) = \tfrac{1}{2}(\mathbf{a} + \mathbf{d})$$

Thus $\mathbf{AE} = \mathbf{AF}$, and F and E coincide; hence the diagonals of a parallelogram bisect each other.

PROBLEMS

1. Find the magnitude and slope of the following vectors:

   (*a*) Initial point (1, 3) and terminal point (2, 7)
   (*b*) Initial point (4, −1) and terminal point (3, −2)
   (*c*) Initial point (2, 5) and terminal point (−2, −3)

2. Show the vector sum of each pair of vectors in Fig. 2-30.

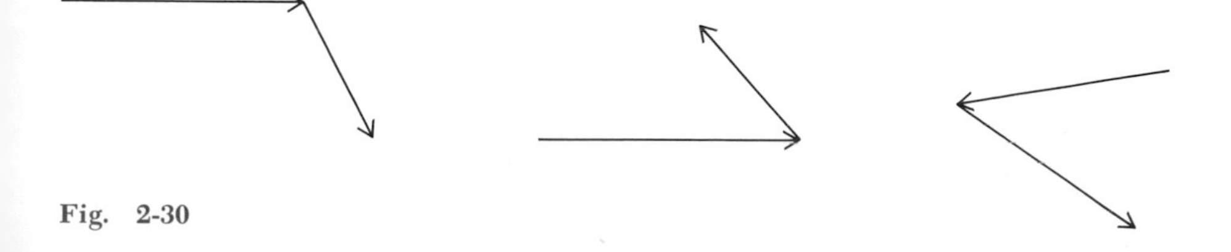

Fig. 2-30

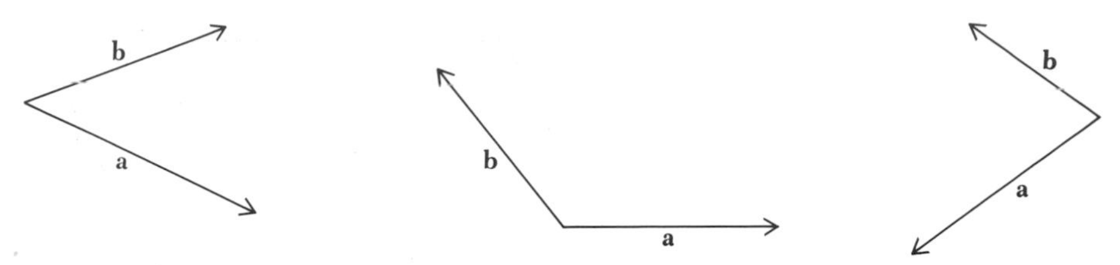

Fig. 2-31

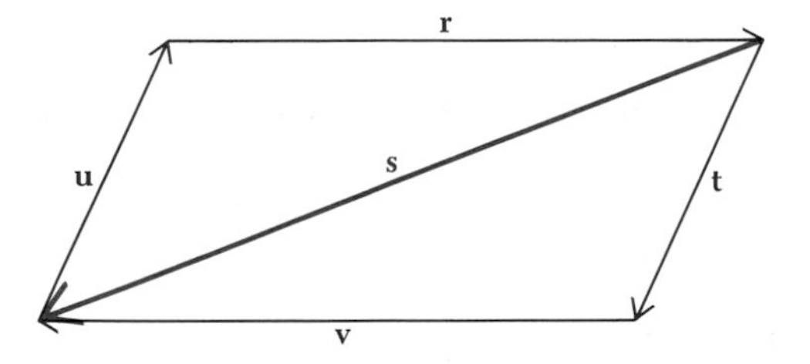

Fig. 2-32

3. Show the vector difference $\mathbf{a} - \mathbf{b}$ of each pair of vectors in Fig. 2-31.
4. In Fig. 2-32, find the following sum or difference:

   (*a*) $\mathbf{r} + \mathbf{t} + \mathbf{v}$ (*b*) $\mathbf{s} + \mathbf{u}$ (*c*) $\mathbf{r} + \mathbf{s} + \mathbf{u}$ (*d*) $\mathbf{s} - \mathbf{t}$

5. Prove that vector addition is associative. *Hint:* Consider a rectangular prism and its diagonal.

Prove the theorems in Probs. 6 to 10 with vectors.

6. The line segment joining the midpoints of two sides of a triangle is parallel to the third side, and its measure is equal to one-half the measure of the third side.
7. If a line divides two sides of a triangle proportionally, it is parallel to the third side.
8. The line segment joining the midpoints of the diagonals of a trapezoid is parallel to the bases.
9. If the diagonals of a quadrilateral bisect each other, the quadrilateral is a parallelogram.
10. The midpoints of the sides of a quadrilateral are the vertices of a parallelogram.

2-10 VECTOR MULTIPLICATION

So far the operations of addition, subtraction, and multiplication of a vector by a scalar have been defined. The *inner product* of two vectors, written $\mathbf{a} \cdot \mathbf{b}$, can now be defined.

Definition

$\mathbf{a} \cdot \mathbf{b} = |\mathbf{a}||\mathbf{b}|$ *cos* $(\mathbf{a}, \mathbf{b})$, *where* $(\mathbf{a}, \mathbf{b})$ *is the angle with the least positive measure between* **a** *and* **b** *when the vectors have a common initial point.*

For example, in Fig. 2-33

$$\begin{aligned} \mathbf{a} \cdot \mathbf{b} &= (4)(\sqrt{2}) \cos 45° \\ &= (4\sqrt{2})\left(\frac{1}{\sqrt{2}}\right) \\ &= 4 \end{aligned}$$

Other names often used for the inner product of two vectors are the *scalar product,* or *dot product.* The use of the term "scalar product" in this connection is justified by the observation that the result is a scalar, not a vector. The geometric significance of the inner product is illustrated in Fig. 2-34. There it may be seen that the inner product of two vectors is equal to the product of the projection [$|\mathbf{a}| \cos (\mathbf{a}, \mathbf{b})$] of the first vector onto the second vector and the magnitude of the second vector.

Theorem 2-5

The operation of finding the inner product of two vectors is commutative.

Proof

$$\begin{aligned} \mathbf{a} \cdot \mathbf{b} &= |\mathbf{a}||\mathbf{b}| \cos (\mathbf{a}, \mathbf{b}) \\ &= |\mathbf{b}||\mathbf{a}| \cos (\mathbf{a}, \mathbf{b}) \\ &= |\mathbf{b}||\mathbf{a}| \cos -(\mathbf{b}, \mathbf{a}) \\ &= |\mathbf{b}||\mathbf{a}| \cos (\mathbf{b}, \mathbf{a}) \\ &= \mathbf{b} \cdot \mathbf{a} \end{aligned}$$

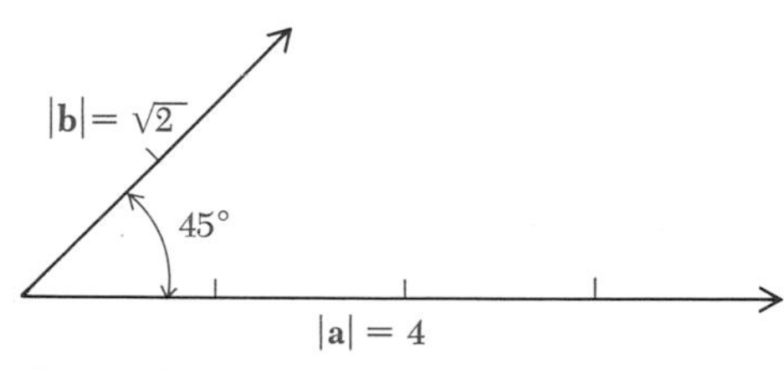

Fig. 2-33 Inner product of two vectors: $\mathbf{a} \cdot \mathbf{b} = |\mathbf{a}||\mathbf{b}| \cos (\mathbf{a}, \mathbf{b})$.

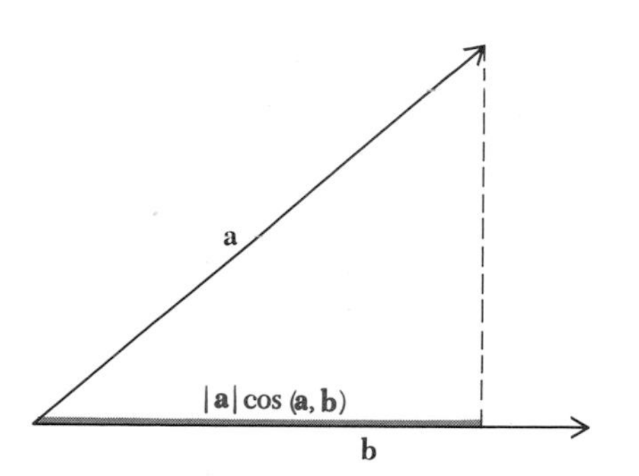

Fig. 2-34 Geometric interpretation of the inner product of two vectors.

Theorem 2-6

The inner product of two nonzero vectors is zero if and only if the two vectors are perpendicular.

Proof If $\mathbf{a} \cdot \mathbf{b} = 0$, where $a \neq 0$ and $b \neq 0$, then

$$|\mathbf{a}||\mathbf{b}| \cos(\mathbf{a}, \mathbf{b}) = 0$$
$$\cos(\mathbf{a}, \mathbf{b}) = 0$$

Hence
$$(\mathbf{a}, \mathbf{b}) = \frac{\pi}{2}$$

since $0 \leqslant (\mathbf{a}, \mathbf{b}) \leqslant \pi$.

On the other hand, if two nonzero vectors are perpendicular, then

$$\mathbf{a} \cdot \mathbf{b} = |\mathbf{a}||\mathbf{b}| \cos\frac{\pi}{2} = 0$$

The inner product is distributive with respect to the addition of vectors (see Prob. 9).

The definition of inner product extends significantly the applicability of vector notation to problem solving in analytic geometry. Two examples will illustrate the techniques used.

Example 1

Prove that the altitudes of any triangle are concurrent.

Solution The three altitudes of a right triangle are concurrent at the vertex of the right angle. Let triangle ABC in Fig. 2-35 be any triangle *except* a right triangle. Let $\overline{AD}$ and $\overline{BE}$ intersect at point H. Draw $\overline{CH}$ and extend it to $\overline{AB}$. Let $\overleftrightarrow{CH} \cap \overline{AB} = F$. The problem is to prove that $\overline{CF}$ is perpendicular to $\overline{AB}$.

$$\mathbf{HB} \cdot (\mathbf{HC} - \mathbf{HA}) = \mathbf{HB} \cdot \mathbf{AC} = 0$$

and
$$\mathbf{HA} \cdot (\mathbf{HB} - \mathbf{HC}) = \mathbf{HA} \cdot \mathbf{CB} = 0$$

so
$$\mathbf{HB} \cdot (\mathbf{HC} - \mathbf{HA}) + \mathbf{HA} \cdot (\mathbf{HB} - \mathbf{HC}) = 0$$
$$\mathbf{HB} \cdot \mathbf{HC} - \mathbf{HB} \cdot \mathbf{HA} + \mathbf{HA} \cdot \mathbf{HB} - \mathbf{HA} \cdot \mathbf{HC} = 0$$
$$\mathbf{HB} \cdot \mathbf{HC} - \mathbf{HA} \cdot \mathbf{HC} = 0$$
$$\mathbf{HC} \cdot (\mathbf{HB} - \mathbf{HA}) = 0$$
$$\mathbf{HC} \cdot \mathbf{AB} = 0$$

Hence **HC** is perpendicular to **AB**, and $\overline{CF}$ is perpendicular to $\overline{AB}$.

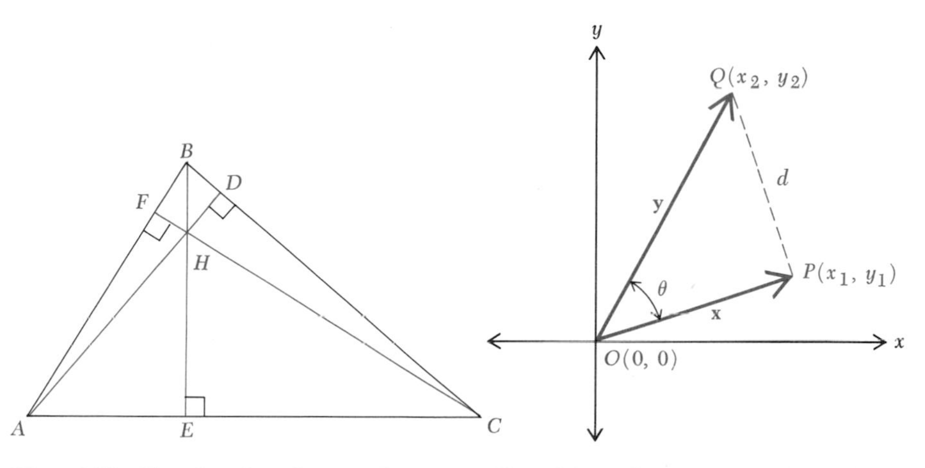

Fig. 2-35 The altitudes of a triangle are concurrent.

Fig. 2-36 The inner product of two vectors: $\mathbf{x} \cdot \mathbf{y} = x_1x_2 + y_1y_2$.

Example 2

Derive a formula for the inner product of two vectors with initial points at (0, 0) and with terminal points at (x_1, y_1) and (x_2, y_2) (see Fig. 2-36).

Solution Let **x** and **y** designate the two vectors. Then

$$\mathbf{x} \cdot \mathbf{y} = |\mathbf{x}||\mathbf{y}| \cos \theta$$

By the distance formula,

$$d = \sqrt{(x_1 - x_2)^2 + (y_1 - y_2)^2}$$

and

$$d^2 = |\mathbf{y}|^2 + |\mathbf{x}|^2 - 2|\mathbf{y}||\mathbf{x}| \cos \theta$$

Hence

$$\frac{d^2 - |\mathbf{y}|^2 - |\mathbf{x}|^2}{-2|\mathbf{y}||\mathbf{x}|} = \cos \theta$$

Therefore

$$\begin{aligned}
\mathbf{x} \cdot \mathbf{y} &= \frac{|\mathbf{x}||\mathbf{y}|(d^2 - |\mathbf{y}|^2 - |\mathbf{x}|^2)}{-2|\mathbf{y}||\mathbf{x}|} \\
&= \frac{d^2 - |\mathbf{y}|^2 - |\mathbf{x}|^2}{-2} \\
&= -\tfrac{1}{2}[(x_1 - x_2)^2 + (y_1 - y_2)^2 - (x_2^2 + y_2^2) - (x_1^2 + y_1^2)] \\
&= -\tfrac{1}{2}(x_1^2 - 2x_1x_2 + x_2^2 + y_1^2 - 2y_1y_2^2 + y_2^2 - x_2^2 - y_2^2 - x_1^2 - y_1^2) \\
&= x_1x_2 + y_1y_2
\end{aligned}$$

PROBLEMS

1. Prove that $\mathbf{a} \cdot \mathbf{a} = |\mathbf{a}|^2$.
2. If m is a real number, show that $m(\mathbf{a} \cdot \mathbf{b}) = (m\mathbf{a}) \cdot \mathbf{b}$.
3. If $|\mathbf{a}| = 5$, $|\mathbf{b}| = 3$, $|\mathbf{c}| = 7$, $\cos(\mathbf{a}, \mathbf{b}) = \frac{1}{2}$, $\cos(\mathbf{b}, \mathbf{c}) = \sqrt{3}/2$, and $\cos(\mathbf{a}, \mathbf{c}) = 0$, find the inner products:

   (*a*) $\mathbf{a} \cdot \mathbf{b}$ (*b*) $\mathbf{a} \cdot \mathbf{c}$ (*c*) $\mathbf{b} \cdot \mathbf{c}$

4. Two vectors have initial points at the origin. Find the inner product if their terminal points are as indicated:

   (*a*) (7, 2) and (−1, 5) (*b*) (5, 3) and (6, −4)
   (*c*) (2, 1) and (5, 3) (*d*) (1, 2) and (2, 1)

Use vectors to prove the following theorems.

5. The diagonals of a rhombus are perpendicular.
6. An angle inscribed in a semicircle is a right angle.
7. The median to the base of an isosceles triangle is perpendicular to the base.
8. The Pythagorean theorem.
9. The distributive property $\mathbf{a} \cdot (\mathbf{b} + \mathbf{c}) = \mathbf{a} \cdot \mathbf{b} + \mathbf{a} \cdot \mathbf{c}$ holds for the inner product of two vectors.

SUMMARY

The trigonometric functions of an angle in standard position are defined by choosing any point $P(x, y)$ on its terminal side and representing the distance of P from the origin by the positive number r. Then

$$\sin\theta = \frac{y}{r} \qquad \csc\theta = \frac{r}{y}$$

$$\cos\theta = \frac{x}{r} \qquad \sec\theta = \frac{r}{x}$$

$$\tan\theta = \frac{y}{x} \qquad \cot\theta = \frac{x}{y}$$

It is also possible to define the trigonometric functions by using a point on the unit circle or for special purposes as the ratios of the measures of the sides of a right triangle.

The inclination of a line is the angle θ, measured counterclockwise, for $0° \leqslant \theta < 180°$, through which the positive x axis would be rotated to bring it into coincidence with the line. The tangent of the inclination is the slope of the line. The slope m of a line through $P_1(x_1, y_1)$ and $P_2(x_2, y_2)$ is $m = (y_2 - y_1)/(x_2 - x_1)$, where $x_2 \neq x_1$.

Each line has two senses, or directions. The undirected angles formed by the positive sense of a line and the positive sense of the x and y axes are the direction angles of the directed line or ray. The cosines of these angles are the direction cosines of a directed line, and any two numbers proportional to the direction cosines are the direction numbers of the directed line. For line $\overleftrightarrow{P_1P_2}$, through points $P_1(x_1, y_1)$ and $P_2(x_2, y_2)$, one pair of direction numbers is

$$[x_2 - x_1, y_2 - y_1]$$

There are three important ideas associated with two lines l_1 and l_2, having slopes m_1 and m_2, respectively:

1. Lines l_1 and l_2 are parallel if and only if $m_1 = m_2$.
2. Lines l_1 and l_2 are perpendicular if and only if $m_1m_2 = -1$.
3. The tangent of the angle θ between l_1 and l_2 is

$$\tan \theta = \frac{(m_2 - m_1)}{(1 + m_1m_2)} \qquad \text{if } m_1m_2 \neq -1$$

A vector is a directed line segment considered as an element in a mathematical system. Some of the elementary definitions and concepts of vectors are as follows:

1. Associated with a vector are its direction and its magnitude.
2. The magnitude, or absolute value, of a vector is the nonnegative real number representing the distance from its initial point to its terminal point.
3. Two equal vectors have not only the same magnitude, but also the same direction.
4. The sum of two vectors **AB** and **BC** so positioned that the initial point of the second is the terminal point of the first is a vector whose initial point is the initial point of the first vector and whose terminal point is the terminal point of the second vector. Thus **AB** + **BC** = **AC**. In case the two vectors are not so positioned, an equal vector in the correct position can be used.
5. Vector subtraction is defined in terms of vector addition. Thus if **AB** + **BC** = **AC**, then **AB** = **AC** − **BC** or **BC** = **AC** − **AB**.
6. Vector addition is commutative and associative.
7. The product of a scalar k and a vector $\mathbf{a}$ is a vector with magnitude $|k||\mathbf{a}|$.

8. Two vectors are parallel if and only if one is a nonzero scalar multiple of the other.
9. The inner product (scalar, or dot, product) of two vectors **a** and **b** is defined as

$$\mathbf{a} \cdot \mathbf{b} = |\mathbf{a}||\mathbf{b}| \cos (\mathbf{a}, \mathbf{b})$$

where $(\mathbf{a}, \mathbf{b})$ is the angle with the least positive measure between **a** and **b** when the vectors have a common initial point.

The methods of analytic geometry are readily applied to obtain simple proofs of many theorems of elementary geometry. Figures are identified by assigning coordinates to points of the figure, and the proof is completed algebraically. Vectors are used in some cases to provide an alternative, and often shorter, proof for some theorems.

REVIEW PROBLEMS

1. Determine the slope of a line for each of the following conditions:

   (*a*) A line through points $P(3, 5)$ and $Q(-4, 1)$
   (*b*) A line parallel to the line through $A(-4, 2)$ and $B(12, -2)$
   (*c*) A line perpendicular to a line with slope $\frac{3}{2}$
   (*d*) A line that forms an angle of 45° with a line having a slope of 3
   (*e*) A line that has an inclination of 150°
   (*f*) A line through points $R(-6, 5)$ and $S(-6, -7)$
   (*g*) A line perpendicular to the y axis

2. Prove that the points $A(-2, 2)$, $B(2, 5)$, $C(5, 1)$, and $D(1, -2)$ are the vertices of a square.
3. Prove the converse of the theorem in Prob. 7, page 48. Do not use vectors.
4. Prove the following theorem analytically: The segment whose end points are the midpoints of two opposite sides of a quadrilateral and the segment whose end points are the midpoints of the diagonals bisect each other.
5. Using the vectors shown in Fig. 2-29, construct each of the following vectors:

   (*a*) $\mathbf{a} + \mathbf{c}$ (*b*) $\mathbf{c} - \mathbf{b}$ (*c*) $\mathbf{b} - \mathbf{c}$ (*d*) $\mathbf{a} + \mathbf{b} + \mathbf{c}$

6. Show by a drawing that if $\mathbf{a} + \mathbf{b} = \mathbf{c}$, then $\mathbf{b} = \mathbf{c} - \mathbf{a}$.
7. Two points on a line are $G(-2, 7)$ and $H(-3 - k, 3 + k)$. For what value of k does the line have a slope of $\frac{2}{3}$?
8. For points $P(5, 3)$, $Q(-3, 5)$, and $O(0, 0)$ determine each of the following:

   (*a*) $|\mathbf{OP}|$ (*b*) $|\mathbf{PQ}|$
   (*c*) The slope of $\overleftrightarrow{OP}$ (*d*) The slope of $\overleftrightarrow{OQ}$
   (*e*) The angle between $\overrightarrow{OP}$ and $\overrightarrow{OQ}$ (*f*) $\mathbf{OP} \cdot \mathbf{OQ}$

9. Find the tangent of each acute angle of triangle FEG whose vertices are $F(6, 7)$, $E(3, -4)$, and $G(-1, 0)$.
10. A line is directed from $P(4, -3)$ to $Q(10, 5)$. Find each of the following:
    (a) A pair of direction numbers for $\overrightarrow{PQ}$
    (b) A pair of direction cosines for $\overrightarrow{PQ}$
    (c) A pair of direction numbers for a ray perpendicular to $\overrightarrow{PQ}$ and directed upward and to the left
    (d) A pair of direction numbers for any ray on a line parallel to $\overleftrightarrow{PQ}$ and directed down to the left

FOR EXTENDED STUDY

1. Line l_1 has direction angles α_1 and β_1; its positive direction is determined by the pair of direction numbers $[a, b]$. Line l_2 has direction angles α_2 and β_2; its positive direction is determined by the direction numbers $[-a, -b]$. What is the relationship between α_1 and α_2 and between β_1 and β_2? Prove your answer.
2. To prove that the median to the base of an isosceles triangle is perpendicular to the base a student chose to draw the picture shown in Fig. 2-37. He used the fact that $\overline{AC} \cong \overline{BC}$ and showed that the product of the slope of $\overline{AB}$ and the slope of $\overline{CD}$ is -1. (a) Write his proof. (b) Prove the theorem a second way by choosing the best possible position in the coordinate plane for the isosceles triangle.
3. Prove analytically that the diagonals of an isosceles trapezoid have the same measure.
4. Graph a representative number of the set of all lines whose slope m may be any of the numbers described by $-1 \leq m \leq 1$ and which contain the origin. Describe by means of two inequalities the set of points you have graphed.

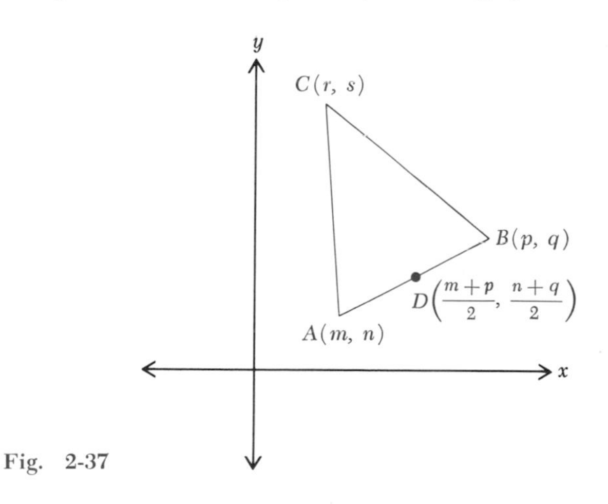

Fig. 2-37

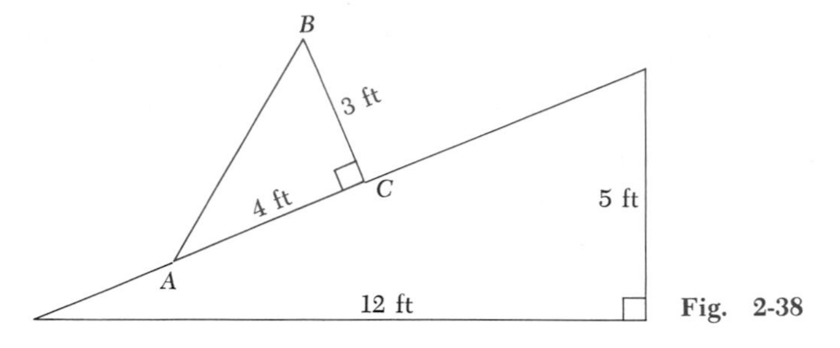

Fig. 2-38

5. A wedge ABC rests on an inclined plane as shown in Fig. 2-38. Find the slopes (to the horizontal) of faces $\overline{AB}$ and $\overline{BC}$.
6. Suppose that line l_1 with slope m_1 has direction cosines λ_1 and μ_1 and that line l_2 with slope m_2 has direction cosines λ_2 and μ_2. Remembering that $m = \tan\theta = \mu/\lambda$, show that the two lines are perpendicular if and only if $\lambda_1\lambda_2 + \mu_1\mu_2 = 0$. *Hint:* Start with the fact that if the lines are perpendicular, $m_1m_2 = -1$.
7. Demonstrate that the condition for perpendicularity of two lines l_1 and l_2 is $r_1r_2 + s_1s_2 = 0$, where r_1 and s_1 are the direction numbers of line l_1 and r_2 and s_2 are the direction numbers of line l_2.

RELATIONS, GRAPHS, AND LOCI | 3

3-1 CONSTANTS AND VARIABLES

As indicated in Chap. 1, when a letter represents a number, it is understood to be a real number unless the contrary is specifically stated. If the letter denotes a *constant,* it is understood to stand for a particular number, even though the number is not specified. If the letter signifies a *variable,* it may represent any one of some specified set of numbers. This set is called the *domain* of the variable.

Example

The total surface area of a right circular cylinder of radius x in. and height y in. is

$$2\pi x^2 + 2\pi xy \text{ sq in.}$$

The number 2 and the Greek letter π are constants. The letters x and y may represent any positive number and hence are variables in the above sense. If y is thought of as having some fixed value, then x is the only variable, and the expression gives the surface areas of cylinders having various radii with this fixed height.

The later letters of the alphabet, such as x, y, z, u, and v, will usually be employed for variables, and the early letters, such as a, b, c, and d, will represent constants. Symbols such as a_0, a_1, a_2, . . . are also used to represent constants.

3-2 RELATIONS

Let x denote a variable with some specified domain (the domain might be all real numbers, all negative numbers, all positive integers, all numbers between -2 and $+2$, and so on). Suppose now that some relation has been established between the value of x and the value of a second variable y such that to each value of x in its domain there correspond one or more values of y. Then the set of ordered pairs (x, y) is called a *relation*. If there is exactly one value of y corresponding to each admissible value of x, then the ordered pairs (x, y) specify a *function*, and y is said to be a *function of* x.

Definition

A relation is a set of ordered pairs (x, y). A function is a set of ordered pairs (x, y) such that exactly one value of y corresponds to each value of x. The set of values of x is the domain of the function, and the set of values of y is the range of the function.

For example, if x represents any real number and y depends upon x in accordance with the relation

$$y = \frac{6x}{x^2 + 3}$$

then y has been defined as a function of x; it is a function because for each number x there corresponds exactly one number y. The values of y constitute the *range* of the function.

Often expressions such as "the function x^2" or "the function $y = x^2$" are used to mean "the function consisting of all ordered pairs (x, y) such that $y = x^2$." When no domain is specified for a function it is to be understood that all real values of x for which the given relationship yields a value of y are to be taken as the domain.

There are many familiar examples of dependence of one variable upon another. The volume of a sphere depends upon, or is a function of, its radius. The function in this case is $V = \frac{4}{3}\pi r^3$. To every (positive) value of r there corresponds a definite value of V. The equation $3x - 4y = 12$ yields exactly one number for y for each number assigned to x, and therefore it defines y as a function of x. It also, of course, defines x as a function of y; that is, a number may be assigned to y, and the equation determines a corresponding and unique number for x. In a relation expressed with two variables, the variable to which values are assigned is called the *independent variable;* the other is called the *dependent variable.*

The statement "y is a function of x" is abbreviated as $y = f(x)$, which is read "y equals f of x." Symbols such as $f(x)$, $g(x)$, and $\phi(x)$ are used to express specific functions of x. Thus in a problem concerned with, say, the functions $x^3 + 14x^2 - 5$ and $3x^2 - 7x + 2$ it may be convenient to let $f(x)$ denote the first and $g(x)$ the second of these functions. Then the first function may be referred to as "the function $f(x)$," which is simpler than saying "the function $x^3 + 14x^2 - 5$."

If $f(x)$ represents a certain function of x, then $f(a)$ denotes the value of $f(x)$ when x has the value a; it is found by substituting a for x in the expression that defines $f(x)$. Thus if

$$f(x) = x^2 - 4x + 2$$

then

$$f(0) = 0^2 - 4(0) + 2 = 2$$
$$f(1) = 1^2 - 4(1) + 2 = -1$$
$$f(-2) = (-2)^2 - 4(-2) + 2 = 14$$
$$f(h) = h^2 - 4h + 2$$

3-3 GRAPH OF A RELATION

In the rectangular-coordinate system defined in Chap. 1 a point in the xy plane was associated with a pair of real numbers (x, y). It is possible to associate a line or curve or other geometric figure with a relation. Let each ordered pair of values of x and y (real numbers only) that satisfy the relation be regarded as the coordinates (x, y) of a point, and let all such points be plotted. The totality of these points is called the *graph of the relation.*

Example 1

The equation $2y = x + 2$ is satisfied by the pairs of values of x and y given in the following table:

x	-4	-3	-2	-1	0	1	2	3	4	5
y	-1	$-\frac{1}{2}$	0	$\frac{1}{2}$	1	$1\frac{1}{2}$	2	$2\frac{1}{2}$	3	$3\frac{1}{2}$

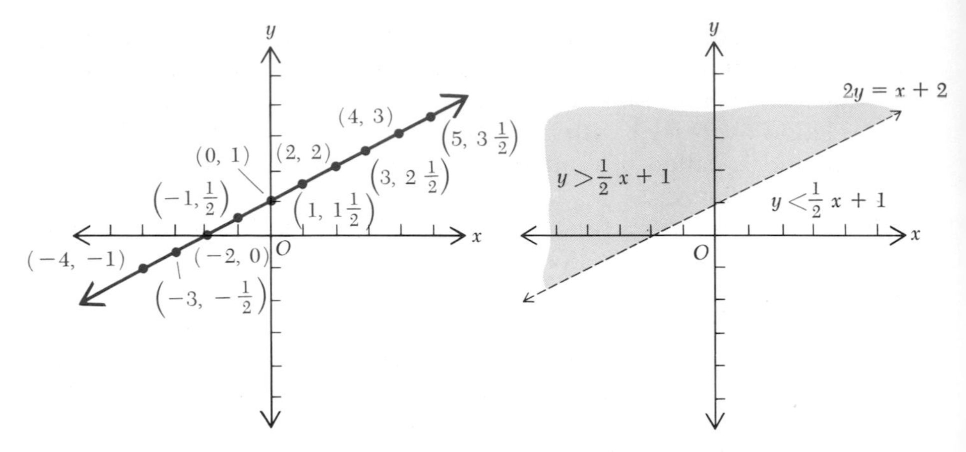

Fig. 3-1 Graph of $y = \frac{1}{2}x + 1$.

Fig. 3-2 Graph of $y > \frac{1}{2}x + 1$ (shaded region).

When these points are plotted, they appear to lie on a straight line (Fig. 3-1). We may suspect that the coordinates of every point on this line satisfy the equation and that the equation is not satisfied by the coordinates of any point not on the line. This is true, but a proof is not given here.

Figure 3-1 suggests that the line separates the xy plane into three disjoint sets of points:

1. *All points on the line*
2. *All points in the half-plane above the line*
3. *All points in the half-plane below the line*

Every point in the plane is in one and only one of these sets. This conclusion is verified by the following argument. For all ordered pairs (x, y) the y coordinate must be either equal to or not equal to $\frac{1}{2}x + 1$. The point with coordinates (x, y) is *on* the line if its coordinates satisfy $y = \frac{1}{2}x + 1$. A point with coordinates (x, y) is not on the line $y = \frac{1}{2}x + 1$ if its coordinates satisfy $y \neq \frac{1}{2}x + 1$. But all ordered pairs (x, y) that satisfy $y \neq \frac{1}{2}x + 1$ must satisfy either $y > \frac{1}{2}x + 1$ or $y < \frac{1}{2}x + 1$.

Example 2

Graph $y > \frac{1}{2}x + 1$.

Solution First the graph of $y = \frac{1}{2}x + 1$ is drawn as in Fig. 3-2. Now, either the region above the line or the region below the line represents

$y > \frac{1}{2}x + 1$. Choose points that are definitely above the line; two such points are (0, 2) and (2, 3).

If $(x, y) = (0, 2)$, then $2 > \frac{1}{2}(0) + 1$ is true.
If $(x, y) = (2, 3)$, then $3 > \frac{1}{2}(1) + 1$ is true.

It would seem that $y > \frac{1}{2}x + 1$ is a description of the set of points in the half-plane above $y = \frac{1}{2}x + 1$. Note that in Fig. 3-2 the graph of the line is dashed to show that it is not part of the graph of $y > \frac{1}{2}x + 1$. The graph of $y > \frac{1}{2}x + 1$ is shaded. The graph of $y \geqslant \frac{1}{2}x + 1$ includes points on the line, and its graph is like that for $y > \frac{1}{2}x + 1$, except that the graph of the line is solid, not dashed.

Example 3

The equation $x^2 + y^2 = 25$ is satisfied by the pairs of values of x and y given in the following table:

x	-5	-4	-3	-2	-1	0	1	2	3	4	5
y	0	± 3	± 4	$\pm\sqrt{21}$	$\pm\sqrt{24}$	± 5	$\pm\sqrt{24}$	$\pm\sqrt{21}$	± 4	± 3	0

When these points are plotted, they appear to lie on a circle with center at the origin and a radius of 5 (Fig. 3-3).

The proof that the equation is satisfied by the coordinates of every point on this circle and that it is not satisfied by the coordinates of any point not on the circle is as follows. The distance from the origin to any point (x, y) in the plane is equal to $\sqrt{x^2 + y^2}$ (Fig. 3-4). Hence for

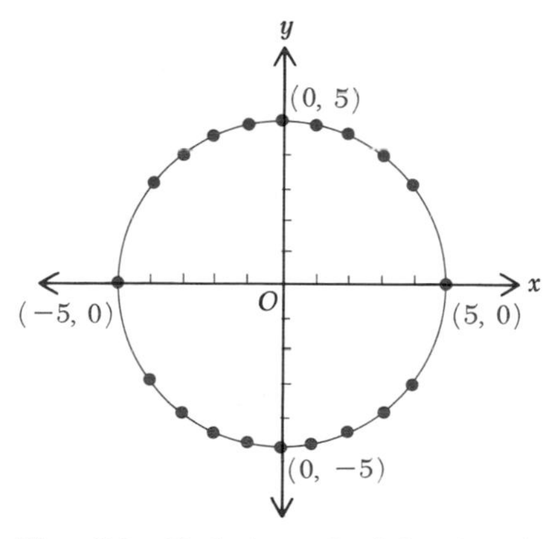

Fig. 3-3 Circle is graph of $x^2 + y^2 = 25$ for x and y real; dots represent the graph of $x^2 + y^2 = 25$ for x an integer.

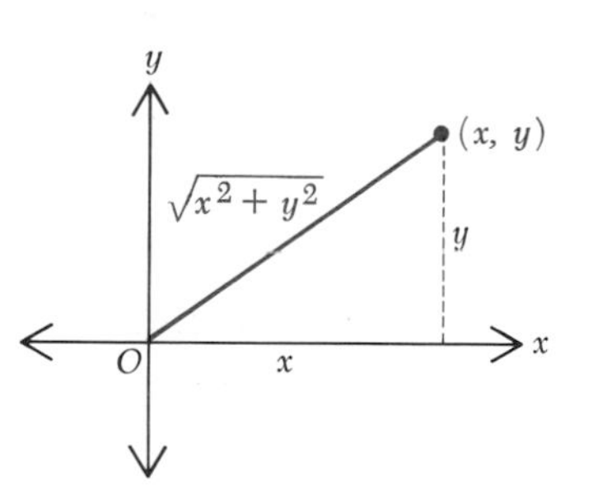

Fig. 3-4 Distance from origin to any point (x, y) is $\sqrt{x^2 + y^2}$.

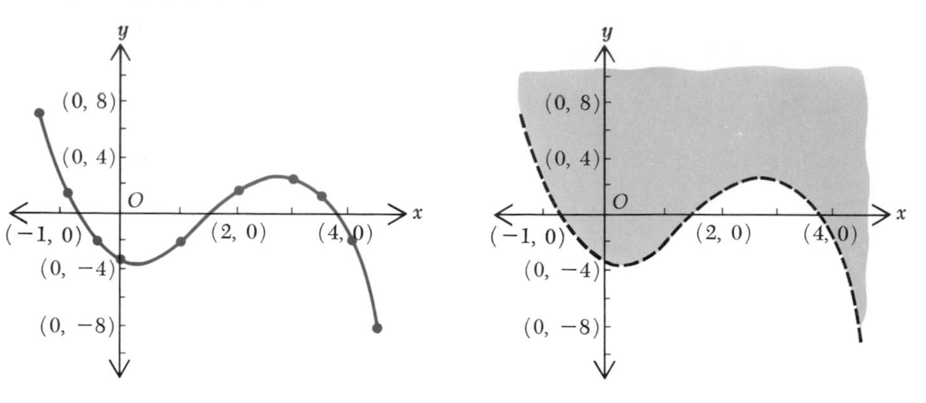

Fig. 3-5 Graph of $3y = -2x^3 + 9x^2 - 3x - 10$ for $-1.5 \leq x \leq 4.5$.

Fig. 3-6 $3y > -2x^3 + 9x^2 - 3x - 10$ for $-1.5 \leq x \leq 4.5$ is represented by the shaded region.

any point on the circle, $\sqrt{x^2 + y^2} = 5$, or $x^2 + y^2 = 25$. If (x, y) is a point of the plane that does not lie on this circle, the value of $\sqrt{x^2 + y^2}$ is either more than or less than 5, and hence $x^2 + y^2$ is not equal to 25.

The graph of the circle also separates the plane into three disjoint sets of points, points inside, points on, and points outside the circle. Substituting ordered pairs of numbers or using the reasoning of the last paragraph indicates that the three sets of points are described as

Interior of the circle: $\{(x, y) \mid x^2 + y^2 < 25\}$
The circle: $\{(x, y) \mid x^2 + y^2 = 25\}$
Exterior of the circle: $\{(x, y) \mid x^2 + y^2 > 25\}$

Note in Fig. 3-3 that the points represented by dots constitute the entire graph of the relation if the domain is specified as the set of integers. In this case the graph is no longer the circle, but a subset of points on it.

In graphing a given relation we cannot, of course, plot all the points. However, we can plot enough of them to indicate the shape of the graph and then draw a curve through the points. This curve is in general an approximation to the actual graph. Thus in Fig. 3-5 the graph of the equation

$$3y = -2x^3 + 9x^2 - 3x - 10$$

is drawn in the interval $-1.5 \leq x \leq 4.5$ through the following points (y coordinates are correct to one decimal place):

x	−1.5	−1	−0.5	0	1	2	3	3.5	4	4.5
y	7.2	1.3	−2	−3.3	−2	1.3	2.7	1.3	−2	−7.8

After the points corresponding to these values are plotted, a smooth curve is drawn through them.

It is also possible to find pairs of numbers such that $3y > -2x^3 + 9x^2 - 3x - 10$ (for the interval $-1.5 \leqslant x \leqslant 4.5$). For example at (0, 2) we have $3(2) > -2(0)^3 + 9(0)^2 - 3(0) - 10$. The graph of the inequality $3y > -2x^3 + 9x^2 - 3x - 10$ is the shaded region in Fig. 3-6. The unshaded region *below* the curve is the graph of $3y < -2x^3 + 9x^2 - 3x - 10$ (for $-1.5 \leqslant x \leqslant 4.5$).

It is possible, of course, that a given relation may have no graph, or that the graph may consist of one or more isolated points. Thus the equation $x^2 + y^2 = 0$ is not satisfied by any pair of real numbers except (0, 0); the graph consists of this one point. The equation $2x^2 + 3y^2 + 4 = 0$ has no graph because it is not satisfied by any pair of real numbers.

The expression "graph of a function $f(x)$" means "graph of the equation $y = f(x)$." Thus the phrases "the graph of the function $x^2 + 5$" and "the graph of the equation $y = x^2 + 5$" are used interchangeably. Figure 3-1 may be regarded as the graph of the function $\frac{1}{2}(x + 2)$. The top half of the circle in Fig. 3-3 is the graph of the function $\sqrt{25 - x^2}$; the bottom half is the graph of the function $-\sqrt{25 - x^2}$. The graph of the complete circle is not the graph of a function, since each value of x is associated with two values of y. Generally we can determine whether a graph is that of a function by drawing lines parallel to the y axis. If every such line in an interval intersects the graph at one and only one point, the graph denotes a function in that interval.

PROBLEMS

1. For what points (ordered pairs) is the abscissa equal to the ordinate? Write an equation that has these points for its graph.
2. For what points is the abscissa always equal to -3? Write an equation that has these points for its graph.
3. For what points is the ordinate always equal to 5? Write an equation that has these points for its graph.
4. For what points is the sum of the abscissa and ordinate equal to zero? Write an equation that has these points as its graph.
5. Using the definition of the graph of an equation, prove that the graph of the equation $xy = 0$ consists of all points on the x axis and all points on the y axis.
6. Show that the equations $xy = 0$ and $xy(x^2 + y^2) = 0$ have the same graph. *Hint:* Recall that a product is zero if and only if at least one of the factors is zero.

In Probs. 7 to 17 make a table of ordered pairs of numbers, plot the corresponding points, and draw a smooth curve through them for the specified interval on the x axis.

7. $y = x^2 + 4x$ for $-6 \leqslant x \leqslant +2$
8. $x - 3y = 6$ for $-2 \leqslant x \leqslant +8$
9. $2y = 2x^2 + 3x - 14$ for $-4 \leqslant x \leqslant +3$
10. $y^2 = 4x$ for $0 \leqslant x \leqslant +6$
11. $2y = x^3 - 4x^2$ for $-1 \leqslant x \leqslant +5$
12. $x^2 + y^2 = 36$ for $-6 \leqslant x \leqslant +6$
13. $y + 2x^2 = 5x + 12$ for $-3 \leqslant x \leqslant 6$
14. $8y = x^3 - 6x^2 + 4x - 24$ for $-3 \leqslant x \leqslant +7$
15. $y = \frac{1}{8}x^2(x - 6)^2$ for $-1 \leqslant x \leqslant +7$
16. $y = \dfrac{24}{x^2 + 4}$ for $-5 \leqslant x \leqslant +5$
17. $y = \dfrac{16x}{x^2 + 4}$ for $-6 \leqslant x \leqslant +6$

In Probs. 18 to 25 graph the given function for the specified interval:

18. $6 - \frac{1}{4}x^2$ for $0 \leqslant x \leqslant +6$
19. $\sqrt{x - 4}$ for $4 \leqslant x \leqslant 13$
20. $3 + 2x - x^2$ for $-3 \leqslant x \leqslant +5$
21. $\dfrac{4}{1 + x^2}$ for $-3 \leqslant x \leqslant +3$
22. $\sqrt[3]{2x}$ for $-10 \leqslant x \leqslant +10$
23. $\dfrac{x^3}{x^3 + 8}$ for $-5 \leqslant x \leqslant +5$
24. $-\frac{2}{3}\sqrt{36 - x^2}$ for $-6 \leqslant x \leqslant +6$
25. $\frac{1}{3}x\sqrt{36 - x^2}$ for $-6 \leqslant x \leqslant 6$

In Probs. 26 to 32 graph the region represented by the given inequality for the specified interval.

26. $y > x^2 + 4x$ for $-6 \leqslant x \leqslant +2$ (see Prob. 7)
27. $x - 3y < 6$ for $-2 \leqslant x \leqslant +8$ (see Prob. 8)
28. $y^2 > 4x$ for $0 \leqslant x \leqslant 6$ (see Prob. 10)
29. $2y < x^3 - 4x^2$ for $-1 \leqslant x \leqslant +5$ (see Prob. 11)
30. $y < 6 - \frac{1}{4}x^2$ for $0 \leqslant x \leqslant 6$ (see Prob. 18)
31. $y > \sqrt{x - 4}$ for $4 \leqslant x \leqslant 13$ (see Prob. 19)
32. $y < \dfrac{4}{1 + x^2}$ for $-3 \leqslant x \leqslant +3$ (see Prob. 21)
33. Express the volume V of a cube as a function of its surface area S. *Hint:* If the edge is x in., then $V = x^3$ and $S = 6x^2$. From these relations derive an expression for V in terms of S.
34. Express the volume of a sphere as a function of its surface area (see hint in Prob. 33).

35. The period T of a pendulum is expressed as a function of its length by the formula $T = 2\pi\sqrt{L/g}$, where g is a constant. Express L as a function of T.
36. Express the undirected distance d between the points $A(2, -3)$ and $P(x, 5)$ as a function of x.
37. A line is drawn parallel to the y axis and 3 units to the left of it. Express the directed distance from the line to the point $P(x, y)$ as a function of x. *Hint:* Observe that this is the same as the length of the directed segment AP, where A has the coordinates $(-3, y)$.

3-4 INTERSECTIONS OF GRAPHS

In Fig. 3-7 the graphs of the equations $3y - x = 5$ and $x^2 + y^2 = 25$ are drawn on the same coordinate axes. The first is a straight line* and the second is a circle. From the definition of the graph of a relation it is clear that the equation $3y - x = 5$ is satisfied by the coordinates of every point on the line and is not satisfied by the coordinates of any other point. Similarly, the equation $x^2 + y^2 = 25$ is satisfied by the coordinates of every point on the circle and is not satisfied by the coordinates of any other point. The two equations

$$3y - x = 5$$

and

$$x^2 + y^2 = 25$$

are *both* satisfied by the coordinates of those points that lie on both the line and the circle and are not *both* satisfied by any other pair of real numbers. Thus the coordinates of the points of intersection of the graphs are the pairs of real numbers that are *solutions* of the system of equations.

These coordinates can be found in simple cases by algebraic methods.

* It will be proved in Chap. 4 that the graph of any equation of first degree in x and y is a straight line.

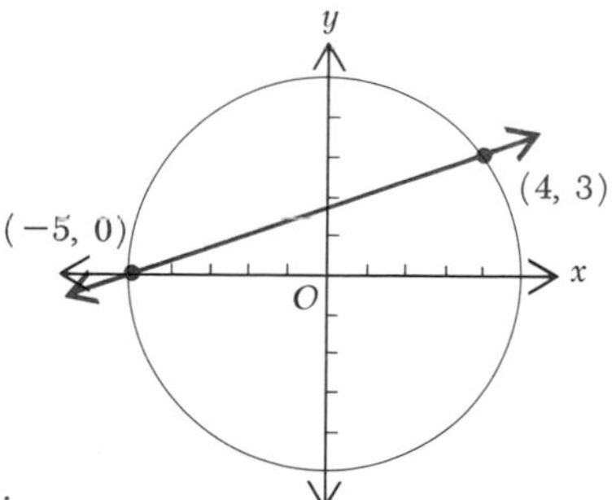

Fig. 3-7
$\{(-5, 0), (4, 3)\} = \{(x, y) \mid x^2 + y^2 = 25 \cap 3y - x = 5\}$.

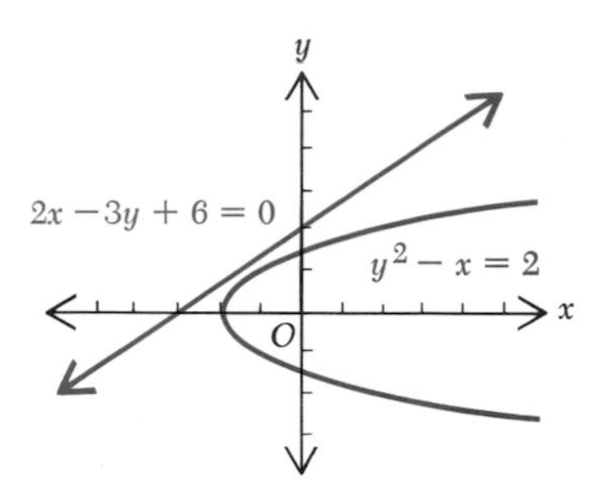

Fig. 3-8
$\{ \ \} = \{(x, y) \mid y^2 - x = 2 \cap 2x - 3y + 6 = 0\}$.

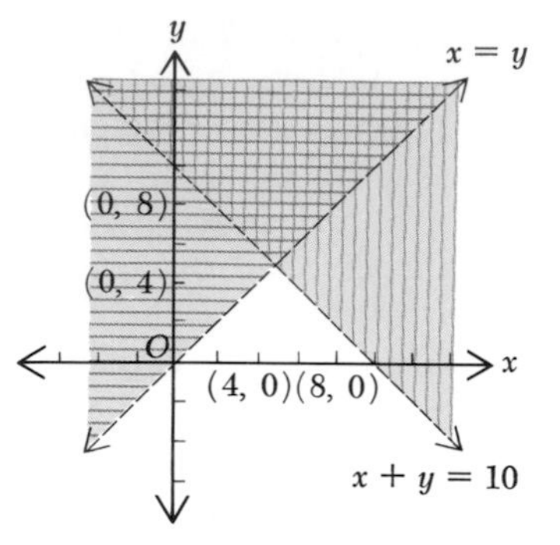

Fig. 3-9 The darker shaded region represents the set of points in $\{(x, y) \mid x + y > 10 \cap y > x\}$.

A suitable procedure is as follows. Given the equations

(3-1) $$3y - x = 5$$
(3-2) $$x^2 + y^2 = 25$$

solve (3-1) for x in terms of y to obtain $x = 3y - 5$. Now substitute this for x in (3-2) and then solve the resulting quadratic equation for y:

$$(3y - 5)^2 + y^2 = 25$$
$$9y^2 - 30y + 25 + y^2 = 25$$
$$y^2 - 3y = 0$$
$$y = 0 \quad \text{or} \quad y = 3$$

Substituting these values for y into Eq. (3-1) gives the corresponding values $x = -5$ or $x = 4$, respectively. The coordinates of the points of intersection are then $(-5, 0)$ and $(4, 3)$. This result may be expressed as

$$\{(x, y) \mid (3y - x = 5) \cap (x^2 + y^2 = 25)\} = \{(-5, 0), (4, 3)\}$$

If the graphs of the equations have no point in common, it can be concluded that the corresponding equations have no common solutions in the field of real numbers. Figure 3-8 shows the graphs of the equations

$$y^2 - x = 2$$
and
$$2x - 3y + 6 = 0$$

You may prove algebraically that these equations are not both satisfied by any pair of real numbers (x, y).

It seems plausible that there might be pairs of numbers that satisfy two inequalities. For example, the two dashed lines in Fig. 3-9 represent the equations $y = x$ and $x + y = 10$. The region shaded horizontally represents $y > x$ and that shaded vertically depicts $x + y > 10$.

The cross-hatched region is part of both regions. Coordinates of points in this common region satisfy both inequalities; the set of these points may be expressed analytically as

$$\{(x, y) \mid y > x \cap x + y > 10 \text{ for } x \text{ a real number}\}$$

3-5 APPLICATIONS OF GRAPHS

A problem frequently met is to determine the way some physical or geometrical quantity Q, which is known to be a function of a certain variable x, varies as the value of x changes. One procedure is to express the relation between Q and x in the form of an equation, $Q = f(x)$, and then draw the corresponding graph. This graph may be regarded as a picture of the value of Q for each value of x in a certain domain; it shows pictorially the way in which Q increases or decreases as x changes. The following example should make the idea clear.

Example 1

A manufacturer of an antifreeze compound packages his product in 1-gal cylindrical metal cans. He recognizes that such a can may be relatively small in radius and rather tall, or larger in radius and shorter. He knows that cans with different combinations of radius and height require different amounts of metal, and he suspects that there should be a particular radius that would result in a can with the least amount of surface area and therefore requiring the least amount of metal. What is this radius?

Solution The manufacturer knows that when the radius is specified the height is fixed by the requirement that the volume be 1 gal (231 cu in.). Consequently, the surface area S is determined if the radius x is specified; in other words, S is a function of x. To determine the relation between S and x observe that for any radius x (inches) there will be a bottom and a top each having an area of πx^2 sq in. In addition, there will be a lateral area of $2\pi x h$ sq in., where the height h is related to x by the requirement that $\pi x^2 h = 231$ cu in. Thus

$$S = 2\pi x^2 + 2\pi x h \qquad \text{where } h = \frac{231}{\pi x^2}$$

After substitution of the value of h in terms of x, the resulting relation is

$$S = 2\pi x^2 + \frac{462}{x}$$

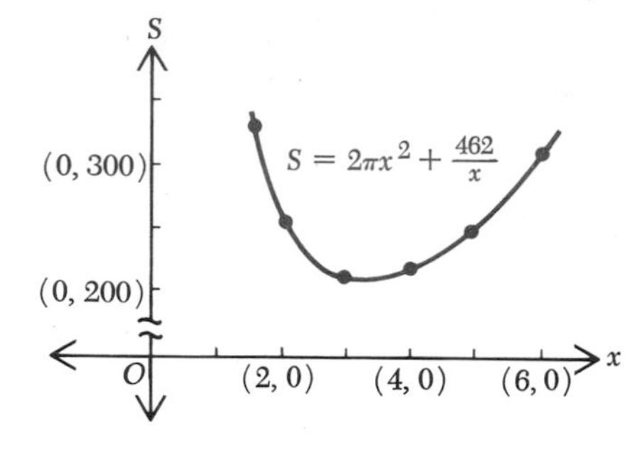

Fig. 3-10 Graph of surface area S for Example 1.

The graph of this equation is shown in Fig. 3-10. It is evident that S is great if x is small and that S decreases as x increases until x is approximately $3\frac{1}{4}$ in., where S appears to be smallest. For x greater than this the surface area increases again. With methods of calculus, we can show that the lowest point on the curve is actually at $x = \sqrt[3]{231/2\pi}$, or approximately 3.325 in.

Another situation in which a graph is used to estimate the value of x for which a certain function $f(x)$ has its greatest value is the following.

Example 2

A piece of tin is 20 in. square (Fig. 3-11a). A box is to be made from it by cutting a small square region from each corner and turning up the sides (Fig. 3-11b). Express the volume V of the resulting box as a function of the length x of the side of the square region cut off. From the graph of this equation, note the way in which V varies with x.

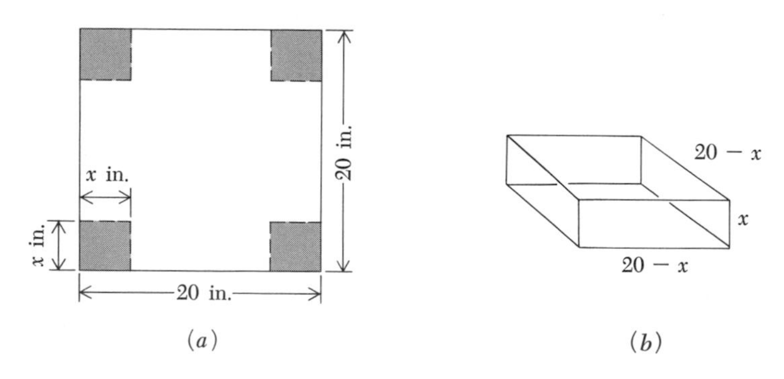

Fig. 3-11 Cutting the corners off a piece of tin and constructing a box.

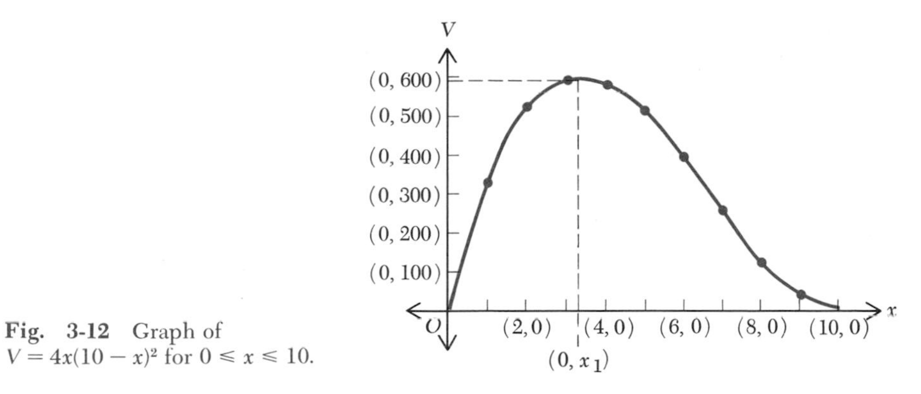

Fig. 3-12 Graph of $V = 4x(10 - x)^2$ for $0 \leq x \leq 10$.

Solution The length and width of the box are each $20 - 2x$, and the depth is x. Hence

$$V = x(20 - 2x)^2 \quad \text{or} \quad V = 4x(10 - x)^2$$

Physical considerations restrict x to the interval $0 < x < 10$. Substituting the integers in this interval for x gives the following table, from which the graph can be drawn:

x	1	2	3	4	5	6	7	8	9
V	324	512	588	576	500	384	252	128	36

The graph of Fig. 3-12 depicts the relation between V and x. It shows that V increases as x increases until x reaches a value x_1 between 3 and 4, for which V appears to be greatest. From here on an increase in x results in a smaller value of V; and finally, if $x = 10$, then $V = 0$. The exact value of x for which V is largest can be found by a method that requires the use of calculus. It amounts to finding the point at which the tangent line to the curve is parallel to the x axis.

PROBLEMS

In Probs. 1 to 12 two equations are given. Draw both graphs on the same axes and find or estimate the coordinates of the points of intersection. Then check by finding the solutions of the system by the methods of algebra.

1. $3x - y = 10$; $6x + 5y = 6$
2. $4x + 5y = 5$; $6x - y = -18$

3. $7x - 3y = 36; \quad 2x + 5y = 22$
4. $3x - y = 6; \quad 9x - \frac{1}{2}y = 13$
5. $\frac{3}{7}x + \frac{9}{4}y = 10; \quad \frac{5}{2}x - \frac{4}{3}y = \frac{1}{2}$
6. $x + 2y = 3; \quad 6x + 6y = 5$
7. $y = \frac{1}{4}x^2 - 1; \quad 2y = x + 2$
8. $4y + x^2 = 8x + 9; \quad x + 4y = 9$
9. $2y = 9 - x^2; \quad x + y = 5$
10. $2x^2 + y^2 = 36; \quad x^2 + y^2 = 20$
11. $y = \dfrac{12}{x^2 + 2}; \quad y = 2x^2 - 6$
12. $8x^2 + 3y^2 = 120; \quad 2x^2 = 3y + 6$

In Probs. 13 to 16 graph the region indicated.

13. $\{(x, y) \mid (x + 2y > 8) \cap (11x + 4y < -11)\}$
14. $\{(x, y) \mid (3x - 4y < 10) \cap (6x - 8y > 12)\}$
15. $\{(x, y) \mid (8y > x^2) \cap (y^2 < x)\}$
16. $\{(x, y) \mid (9x^2 + 25y^2 > 225) \cap (3x + 9 > y^2)\}$
17. Express the area A of a circular region as a function of its radius r. Make a graph of this function.
18. Express the surface area S and volume V of a cube as functions of the length x of its edge. Draw both graphs on the same axes. What is the significance of the point of intersection?
19. A rectangular box is made from a sheet of tin that is 18 in. long and 12 in. wide by cutting a square region from each corner and turning up the sides as in Example 2 on page 74. Express the volume V of the box obtained as a function of the edge x of the square cut out, and make a graph of the function. About what size square region would give the largest box?
20. A can in the form of a right circular cylinder is to contain 600 cu in. The material used for the top costs 3 cents/sq in. and that for the rest of the can costs 1 cent/sq in. Express the cost C of material as a function of the radius x of the can. Make a graph of this function. Estimate the value of x for which the cost is least.
21. A rectangle is inscribed in an isosceles triangle as shown in Fig. 3-13. Express the area A of the interior of the rectangle as a function of its height y. Draw the graph of this function. What value of y appears to give the largest

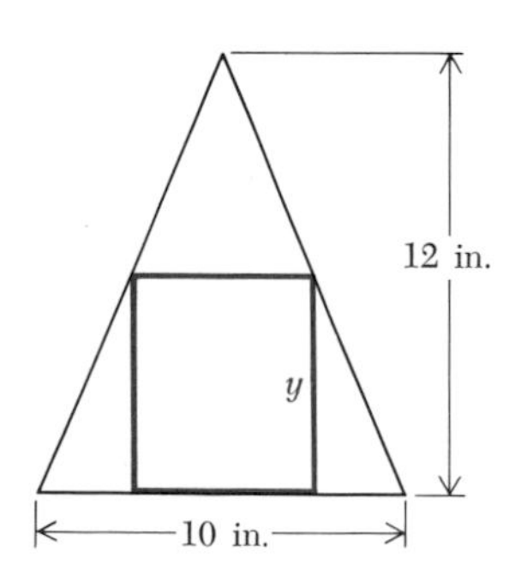

Fig. 3-13 Rectangle inscribed in an isosceles triangle.

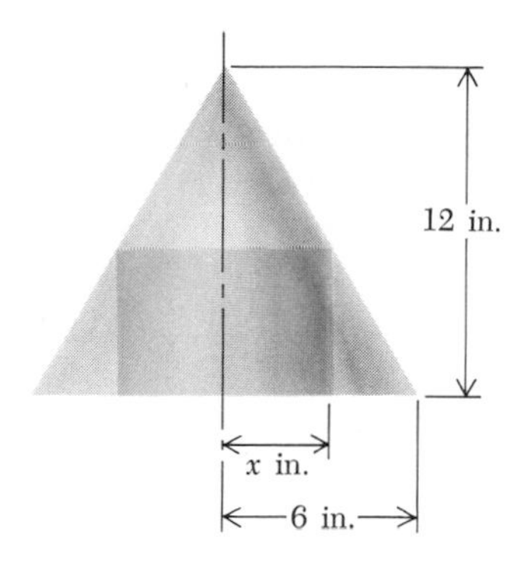

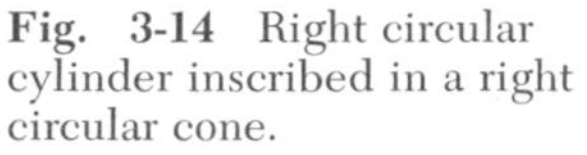

Fig. 3-14 Right circular cylinder inscribed in a right circular cone.

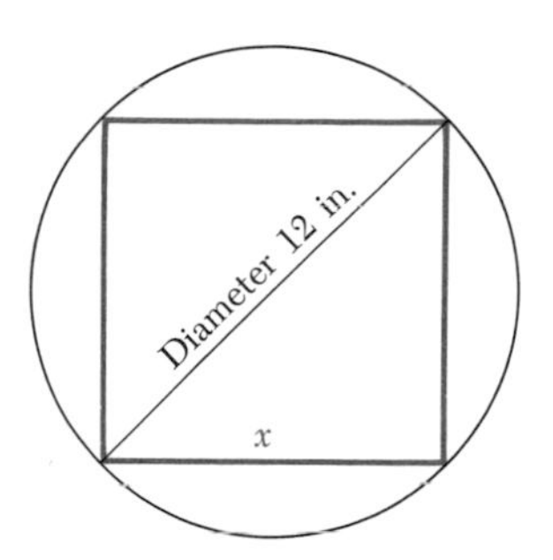

Fig. 3-15 Rectangle inscribed in a circle.

rectangular region? *Hint:* Find the other dimension of the rectangular region in terms of y by using similar triangles.

22. A chord is drawn in a circle of radius 6 in. Express the length L of the chord as a function of its distance x from the center of the circle. Make a graph of the function.
23. A right circular cylinder of radius x is inscribed in a right circular cone of radius 6 in. and height 12 in., as indicated in Fig. 3-14. Express the volume V of the cylinder as a function of x and make a graph of the function. What value of x appears to give the largest cylinder? *Hint:* In making the graph it is convenient to let a unit on the V axis represent π cu in.
24. A rectangle is inscribed in a circle of radius 6 in. Express the area A of the rectangular region as a function of the length x of its base (Fig. 3-15). Draw a graph of this function and estimate the value of x for which A is greatest.
25. The following experimentally determined data give the length L (inches) of a certain spring when it is subjected to a pull of x (pounds). Draw the corresponding graph showing how the length varies with the load.

x	0	1	2	3	4	5	6	7	8
L	13.10	14.32	15.56	16.82	18.04	19.30	20.58	21.88	23.18

26. The following data give the period T (seconds) of a simple pendulum of length L (feet). Draw the corresponding graph showing how the period varies with the length. The period is the time required for the pendulum to make one complete swing (over and back).

L	0.50	1.0	1.5	2.0	2.5	3.0	3.5	4.0
T	0.78	1.11	1.35	1.56	1.75	1.92	2.07	2.21

3-6 EQUATION OF A LOCUS

One of the principal problems solved by analytic geometry is that of finding the equation of a curve defined by some geometrical condition. In this context we shall consider both the static concept of sets of points and the dynamic concept of a point moving along a path. While the first is perhaps more sound as an abstract mathematical model for space, the second is extremely useful in many practical applications. For example, the problem may be to find the equation for the path of a point which moves in the xy plane in such a way that it is always equidistant from two given points. Or it may be to find the equation of the curve every point of which is the same distance from the y axis as from the point (4, 0). Or we may want the equation for the path in which a point moves if the sum of its distances from (−3, 0) and (3, 0) is always to equal 10. The curve, or path, is the *locus* of points satisfying the given condition. Thus the locus of points at a distance of 7 from $A(2, 3)$ is a circle with center at A and radius 7. The locus of points equidistant from $A(1, 1)$ and $B(6, 3)$ is a certain straight line, the perpendicular bisector of $\overline{AB}$.

Example 1

Find the equation of the locus of points equidistant from $A(1, 1)$ and $B(6, 3)$.

Solution Let $P(x, y)$ be the coordinates of a point in the xy plane (Fig. 3-16) on the perpendicular bisector of $\overline{AB}$. Let d_1 and d_2 be its (undirected) distances from A and B, respectively. Then

$$d_1 = \sqrt{(x-1)^2 + (y-1)^2} \qquad d_2 = \sqrt{(x-6)^2 + (y-3)^2}$$

Now, to meet the condition $d_1 = d_2$,

$$\sqrt{(x-1)^2 + (y-1)^2} = \sqrt{(x-6)^2 + (y-3)^2} \tag{3-3}$$

Squaring both sides and simplifying results in

$$x^2 - 2x + 1 + y^2 - 2y + 1 = x^2 - 12x + 36 + y^2 - 6y + 9 \tag{3-4}$$

$$-2x - 2y + 2 = -12x - 6y + 45 \tag{3-5}$$

and finally,

$$10x + 4y = 43 \tag{3-6}$$

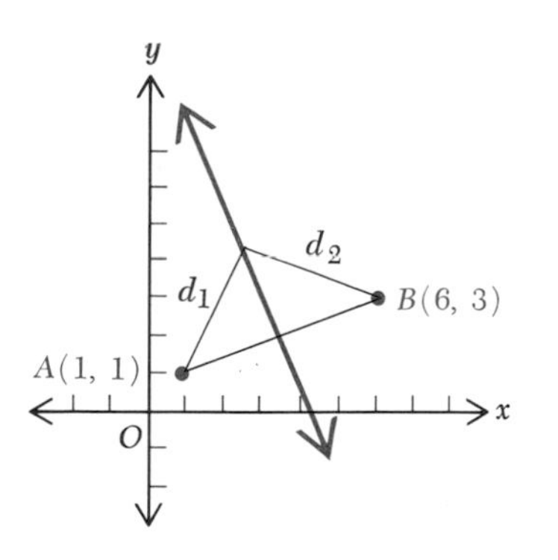

Fig. 3-16 Locus of points equidistant from points A and B: equation of locus $10x + 4y = 43$.

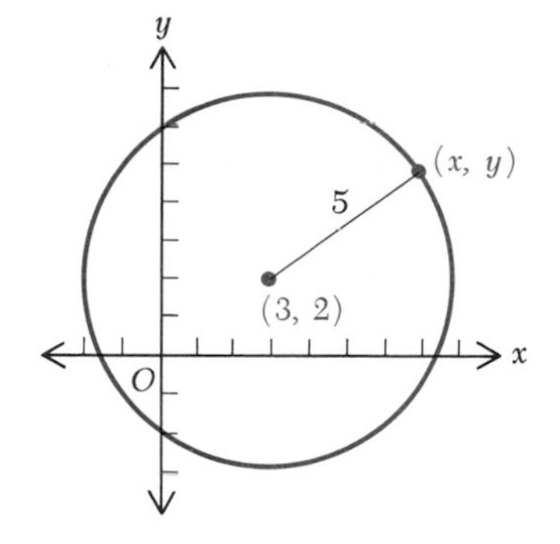

Fig. 3-17 Graph of points 5 units from (3, 2).

If Eq. (3-3) is to be true, then (3-6) must be true; $d_1 = d_2$ if the coordinates (x, y) satisfy (3-6). Conversely, Eq. (3-6) implies (3-5), (3-4), and finally (3-3), so that $d_1 = d_2$ for *every* point whose coordinates satisfy Eq. (3-6). Hence it has been proved that the points (x, y) whose coordinates satisfy (3-6) are those, and only those, for which $d_1 = d_2$.

Example 2
Find the equation of the locus of points whose undirected distance from (3, 2) is 5.

Solution Let (x, y) be the coordinates of a point in the xy plane (Fig. 3-17). Let d be its undirected distance from (3, 2). Then

$$d = \sqrt{(x-3)^2 + (y-2)^2}$$

To meet the condition $d = 5$,

$$\sqrt{(x-3)^2 + (y-2)^2} = 5 \tag{3-7}$$

Squaring both sides and simplifying results in

$$(x-3)^2 + (y-2)^2 = 25 \tag{3-8}$$

or

$$x^2 + y^2 - 6x - 4y - 12 = 0 \tag{3-9}$$

From the fact that Eq. (3-7) implies (3-9) we conclude that if $d = 5$ for a point (x, y), then its coordinates must satisfy Eq. (3-9). Thus the graph

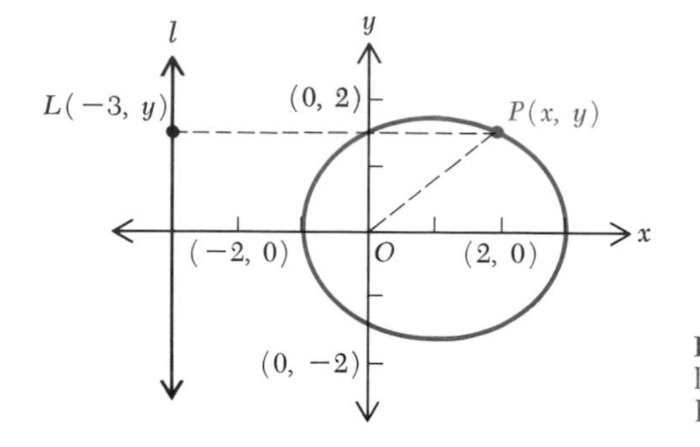

Fig. 3-18 Points twice as far from line l as from the origin: equation of locus $3x^2 + 4y^2 - 6x - 9 = 0$.

of Eq. (3-9) contains all points that are 5 units from (3, 2) and no other points. The locus is, of course, a circle.

Example 3
A line l is drawn parallel to the y axis and 3 units to the left of it. A point moves so that it is always twice as far from this line as from the origin. Find the equation of its path.

Solution Let (x, y) be the coordinates of a point in the plane. Let d_1 and d_2 be its undirected distances from l and from the origin, respectively (Fig. 3-18). Then

$$d_1 = \sqrt{(x+3)^2} \qquad (\text{or } |x+3|) \qquad d_2 = \sqrt{x^2+y^2}$$

To meet the condition $d_1 = 2d_2$, we have

$$\sqrt{(x+3)^2} = 2\sqrt{x^2+y^2} \tag{3-10}$$

Squaring both sides and simplifying yields

$$(x+3)^2 = 4(x^2+y^2) \tag{3-11}$$

$$x^2 + 6x + 9 = 4x^2 + 4y^2 \tag{3-12}$$

$$3x^2 + 4y^2 - 6x - 9 = 0 \tag{3-13}$$

Supply the reasoning to assure yourself that the points (x, y) whose coordinates satisfy Eq. (3-13), and no other points, have $d_1 = 2d_2$. The graph of the equation is a curve called an *ellipse*. Its properties will be studied later.

It is perhaps worthwhile to return briefly to Example 1 on page 78 and reemphasize the fact that before we can conclude that $10x + 4y = 43$

is the solution to the problem we must ascertain that the following two conditions are satisfied:

1. The coordinates of every point (x, y) having $d_1 = d_2$ satisfy the equation; briefly, if $d_1 = d_2$, then $10x + 4y = 43$.
2. Every point (x, y) such that $10x + 4y = 43$ has $d_1 = d_2$; briefly, if $10x + 4y = 43$, then $d_1 = d_2$.

The reasoning underlying the steps by which Eq. (3-6) was obtained from Eq. (3-3) takes care of the first part: If Eq. (3-3) is true, then (3-4) must be true, because if two numbers are equal, their squares are equal. If Eq. (3-4) is true, then (3-5) and (3-6) must be true because of the addition property of equality. It can therefore be concluded that

if $$\sqrt{(x-1)^2 + (y-1)^2} = \sqrt{(x-6)^2 + (y-3)^2}$$
then $$10x + 4y = 43$$

This means that if a point (x, y) in the plane is such that $d_1 = d_2$, then it lies on the graph of the equation $10x + 4y = 43$. However, to prove that $d_1 = d_2$ for every point on this graph we must prove the truth of the assertion that

if $$10x + 4y = 43$$
then $$\sqrt{(x-1)^2 + (y-1)^2} = \sqrt{(x-6)^2 + (y-3)^2}$$

This entails reversing the above steps—starting with Eq. (3-6) as the hypothesis and showing that (3-3) is a logical consequence. In this connection it is important to observe that if Eq. (3-4) is true, then (3-3) is true, because the positive square roots of two equal positive numbers are equal.

PROBLEMS

1. Find the equation of the locus of all points whose undirected distance from $A(-3, -1)$ is 3.
2. Find the equation of the circle with center at $A(6, -2)$ and radius 4.
3. Find the equation of the circle whose center is at $A(-5, 5)$, and which passes through the origin.
4. A point moves in the xy plane so as to be always equidistant from $A(-2, -6)$ and $B(4, -1)$. Find the equation of its path.
5. Find the equation of the perpendicular bisector of the segment $\overline{PQ}$ for $P(-2, 3)$ and $Q(4, -5)$.
6. A point moves so as to be always equidistant from two points A and B. Find the equation of its path, and draw the graph if

   (a) $A(6, 2)$, $B(-1, 5)$ (b) $A(-2, 4)$, $B(4, 0)$

7. Find the equation of the locus of points equidistant from the y axis and the point $A(4, 0)$.
8. Find the equation of the locus of points equidistant from the x axis and the point $A(3, 4)$.
9. A point moves so as to be always equidistant from the origin and a line drawn parallel to and 4 units above the x axis. Find the equation of its path.
10. A point moves so as to be always equidistant from the point $A(4, 0)$ and a line parallel to and 4 units to the left of the y axis. Find the equation of its path and sketch the graph.
11. Find the equation of the curve each of whose points is twice as far from the point $A(12, 0)$ as from the origin.
12. A point moves in the xy plane in such a way as to be always twice as far from $A(-8, 0)$ as from $B(4, 0)$. Find the equation of its path.
13. A point moves in the xy plane in such a way that the sum of the slopes of the lines joining it to $A(1, 4)$ and $B(1, -2)$ is always equal to 2. Find the equation of its path and plot it.
14. The slope of the line joining any point P of a certain curve to the origin is equal to one-third of the slope of the line joining P to $A(1, 1)$. What is the equation of this curve?
15. A point moves in the xy plane in such a way that its distance from the x axis is always equal to twice its distance from the point $A(0, 4)$. Find the equation of its path.
16. A point moves in the xy plane so that the sum of the squares of its distances from the points A and B is equal to the square of the undirected distance $|AB|$. Find the equation of its path and identify the curve if the points are

    (*a*) $A(-6, 0)$, $B(6, 0)$ (*b*) $A(-3, -4)$, $B(3, 0)$

17. From any point on a certain curve the sum of the undirected distances to $A(-3, 0)$ and $B(3, 0)$ is equal to 10. Find the equation of this curve.
18. From any point on a certain curve the undirected distance to $A(-5, 0)$ minus the undirected distance to $B(5, 0)$ is equal to 8. Find the equation of this curve.
19. A point moves in the xy plane in such a way that the sum of its distances from the origin and the point $A(8, 0)$ is equal to 12. Find the equation of its path.

In Probs. 20 to 25 state whether or not the given assertion is valid. Write the converse and state whether or not it is valid.

20. If $x = -4$, then $x^2 = 16$.
21. If $(x + 3)^2 = 1$, then $|x + 3| = 1$.
22. If $\sqrt{(x + 1)^2} = 5$, then $(x + 1)^2 = 25$.
23. If $|x + 3| = 5$, then $(x + 3)^2 = 25$.
24. If $\sqrt{(x - 4)^2 + (y + 1)^2} = 4 - x$, then $(x - 4)^2 + (y + 1)^2 = (4 - x)^2$.
25. If $\sqrt{(x - 1)^2 + (y - 2)^2} = 2\sqrt{(x + 4)^2 + (y + 2)^2}$, then $(x - 1)^2 + (y - 2)^2 = 4[(x + 4)^2 + (y + 2)^2]$.

26. Prove the following propositions:
    (*a*) If a point $P(x, y)$ is equidistant from $A(1, 6)$ and $B(4, -3)$, then its coordinates satisfy the equation $3y - x = 2$.
    (*b*) Every point whose coordinates satisfy the equation $3y - x = 2$ is equidistant from $A(1, 6)$ and $B(4, -3)$.

3-7 RELATIONS BETWEEN POLAR AND RECTANGULAR COORDINATES

It is possible to determine the relations that exist between the polar coordinates (ρ, θ) and the corresponding rectangular coordinates (x, y) of any point P when the polar axis coincides with the positive x axis. Let (ρ, θ) be the polar coordinates of any point P other than the origin, and consider first only the case in which $\rho > 0$. Then the definitions $\cos\theta = x/\rho$ and $\sin\theta = y/\rho$ (Fig. 3-19) immediately yield the following relations:

(3-14) $$x = \rho\cos\theta \qquad y = \rho\sin\theta$$

If P is at the origin, so that $\rho = 0$, these equations give $x = 0$ and $y = 0$ and thus apply. Also, if point P is associated with the number pair $(-\rho, \theta + \pi)$ the relations again hold, for $-\rho\cos(\theta + \pi) = \rho\cos\theta$ and $-\rho\sin(\theta + \pi) = \rho\sin\theta$. Finally, it can be asserted that if any pair of polar coordinates of a point is known, Eqs. (3-14) name the rectangular coordinates of that point.

If the coordinates (x, y) of P are given and the problem is to determine a corresponding pair of polar coordinates (ρ, θ), then

$$\rho^2 = x^2 + y^2$$

or

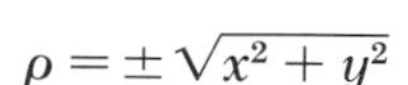

(3-15) $$\rho = \pm\sqrt{x^2 + y^2}$$

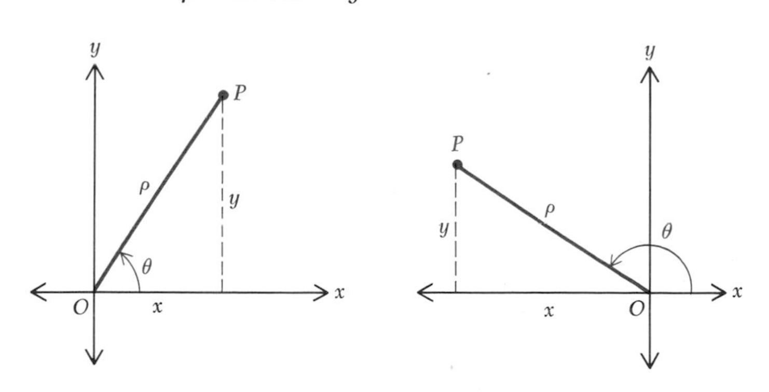

Fig. 3-19 Relations between polar and rectangular coordinates: $x = \rho\cos\theta$ and $y = \rho\sin\theta$.

We see from Fig. 3-19 that for any position of P

$$\tan \theta = \frac{y}{x} \qquad (x \neq 0)$$

or

$$\theta = \arctan \frac{y}{x} \tag{3-16}$$

Examination of Eqs. (3-15) and (3-16) indicates that ρ and θ are not uniquely defined. This, however, does not present special difficulties, as will be seen below. Equations (3-15) and (3-16) merely state that the ordered pair (x, y) does not have a unique representation using a single ordered pair (ρ, θ). Finally, for the point $x = 0$ and $y = 0$, θ in Eq. (3-16) is undefined and ρ in Eq. (3-15) is 0; but it has been agreed that if $\rho = 0$, then any value of θ will describe the origin in polar coordinates.

Example 1

Find the rectangular coordinates of the point whose polar coordinates are $(4, 7\pi/6)$.

Solution

$$x = 4 \cos \frac{7\pi}{6} = 4\left(-\frac{\sqrt{3}}{2}\right) = -2\sqrt{3}$$

$$y = 4 \sin \frac{7\pi}{6} = 4(-\tfrac{1}{2}) = -2$$

Example 2

Find the polar coordinates of the point whose rectangular coordinates are $(-\sqrt{3}, -1)$ (Fig. 3-20).

Solution

$$\rho^2 = (-\sqrt{3})^2 + (-1)^2$$

Hence

$$\rho = 2 \quad \text{or} \quad \rho = -2$$

$$\theta = \arctan \frac{-1}{-\sqrt{3}} = \frac{\pi}{6} + n\pi \qquad \text{for } n \text{ an integer}$$

We now need to find values for ρ and θ that meet the conditions of the

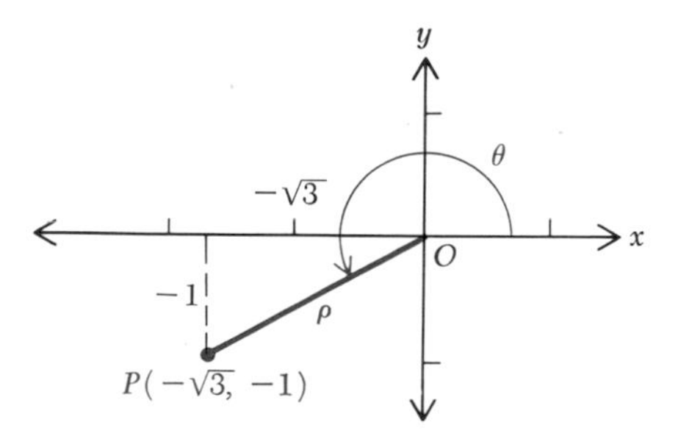

Fig. 3-20 Coordinates for P in rectangular form are $(-\sqrt{3}, -1)$ and in polar form are $(2, 7\pi/6)$.

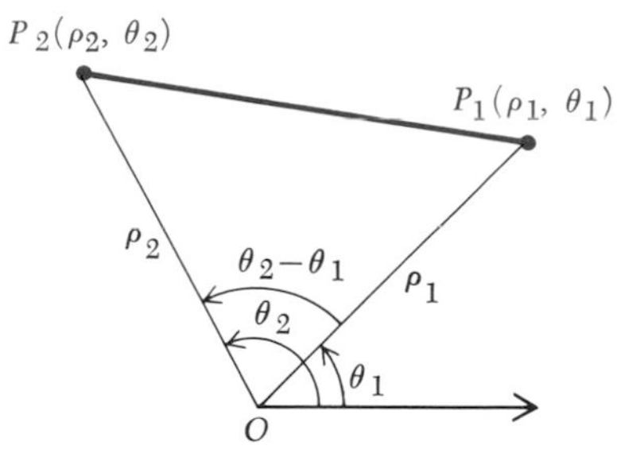

Fig. 3-21 Square of the distance between P_1 and P_2 in polar coordinates is $(P_1P_2)^2 = \rho_1{}^2 + \rho_2{}^2 - 2\rho_1\rho_2 \cos(\theta_2 - \theta_1)$.

problem. For example, $\rho = 2$ and $\theta = \pi/6$ do not comprise a solution because, as indicated in Fig. 3-20, the required point is not in quadrant I. However, $(2, 7\pi/6)$, $(-2, -5\pi/6)$, and $(2, 19\pi/6)$ are possible solutions.

The question may now be raised, "Is it possible to find the undirected distance between two points whose coordinates are expressed in polar form?" From Fig. 3-21 we see that in triangle OP_1P_2 the measure of two sides, $\overline{OP}_1$ and $\overline{OP}_2$, and the measure of angle P_1OP_2 are known, and the measure of the side opposite the known angle is to be determined. This is precisely what is obtained by applying the law of cosines:

$$(P_1P_2)^2 = (OP_1)^2 + (OP_2)^2 - 2(|OP_1|)(|OP_2|) \cos(\theta_2 - \theta_1)$$

$$= \rho_1{}^2 + \rho_2{}^2 - 2\rho_1\rho_2 \cos(\theta_2 - \theta_1) \tag{3-17}$$

While this formula has been developed for $\rho_1 > 0$ and $\rho_2 > 0$ and $0 \leq \theta_1 < 360°$ and $0 \leq \theta_2 < 360°$, it is quite general and may be used for all cases, including those in which either or both ρ_1 and ρ_2 are less than zero and θ_1 or θ_2 or both are less than zero.

Another approach to this problem would be to express the distance formula for rectangular coordinates in polar form. For two points $P_1(x_1, y_1)$ and $P_2(x_2, y_2)$

$$(P_1P_2)^2 = (x_2 - x_1)^2 + (y_2 - y_1)^2$$

If P_1 and P_2 are points determined by polar coordinates and represented by $P_1(\rho_1, \theta_1)$ and $P_2(\rho_2, \theta_2)$, the following relations hold:

$$x_1 = \rho_1 \cos\theta_1 \qquad x_2 = \rho_2 \cos\theta_2$$
$$y_1 = \rho_1 \sin\theta_1 \qquad y_2 = \rho_2 \sin\theta_2$$

$$\begin{aligned} \text{Then} \quad (P_1P_2)^2 &= (\rho_2 \cos \theta_2 - \rho_1 \cos \theta_1)^2 + (\rho_2 \sin \theta_2 - \rho_1 \sin \theta_1)^2 \\ &= \rho_2^2 \cos^2 \theta_2 - 2\rho_2\rho_1 \cos \theta_2 \cos \theta_1 + \rho_1^2 \cos^2 \theta_1 \\ &\qquad + \rho_2 \sin^2 \theta_2 - 2\rho_2\rho_1 \sin \theta_2 \sin \theta_1 + \rho_1^2 \sin^2 \theta_1 \\ &= \rho_2^2(\cos^2 \theta_2 + \sin^2 \theta_2) - 2\rho_2\rho_1 (\cos \theta_2 \cos \theta_1 + \sin \theta_2 \sin \theta_1) \\ &\qquad + \rho_1^2(\cos^2 \theta_1 + \sin^2 \theta_1) \end{aligned}$$

$$\text{Hence} \qquad (P_1P_2)^2 = \rho_2^2 + \rho_1^2 - 2\rho_1\rho_2 \cos (\theta_2 - \theta_1)$$

This is exactly the same as Eq. (3-17), found above by direct application of the law of cosines.

PROBLEMS

1. Find the rectangular coordinates of the points whose polar forms are as follows:

   $(a)\ \left(3, \frac{\pi}{3}\right) \qquad (b)\ \left(5, \frac{7\pi}{4}\right) \qquad (c)\ \left(-3, \frac{3\pi}{4}\right) \qquad (d)\ \left(6, -\frac{13\pi}{6}\right)$

2. Write the polar coordinates of each point whose rectangular coordinates are as given:

   (a) $(5, 5)$ $\qquad (b)$ $(2\sqrt{3}, -2)$ $\qquad (c)$ $(-\sqrt{3}/3, \frac{1}{3})$
   (d) $(1, 5)$ $\qquad (e)$ $(-3, -4)$ $\qquad (f)$ (p, q)

3. Find the distance between P and Q for each of the following. Use polar coordinates. Verify your result by expressing the coordinates of P and Q in rectangular form and finding the distance between them.

   $(a)\ P\left(3, \frac{\pi}{6}\right), Q\left(4, \frac{5\pi}{6}\right) \qquad (b)\ P(7, 135°), Q(-4, 225°)$

4. Find the distance between points J and K in each of the following:

   (a) $J(4, 0°), K(5, 90°)$ $\qquad (b)$ $J(3, 60°), K(5, 120°)$
   (c) $J(4, -60°), K(3, 180°)$ $\qquad (d)$ $J(4, 130°), K(5, 85°)$
   (e) $J(-2, -290°), K(4, 140°)$ $\qquad (f)$ $J(-5, 8\pi/9), K(1, 59\pi/72)$

3-8 (OPTIONAL) GRAPHING SETS OF POINTS ON A LINE

In previous sections points in a plane have been indicated by ordered pairs of numbers. There is also a one-to-one correspondence between the real numbers and the points on a number line. This makes it possible to graph an equation or inequality in one variable on a number line.

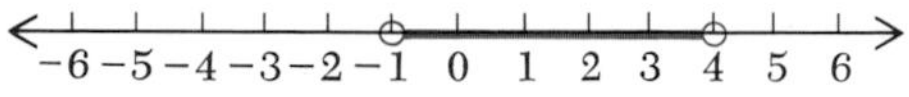

Fig. 3-22 Line graph of $-1 < x < 4$ for x a real number.

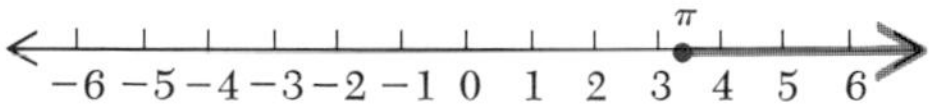

Fig. 3-23 Line graph of $\{x \mid x \geqslant \pi$ for x a real number$\}$.

Example 1
Graph $-1 < x < 4$ for x a real number.

Solution The solution set consists of all numbers greater than -1 and less than 4. The graph is shown in Fig. 3-22, with shading between -1 and 4; the unshaded circles at -1 and 4 indicate that these points are not members of the solution set.

Example 2
Graph on a number line members of the set $P = \{x \mid x \geqslant \pi$ for x a real number$\}$.

Solution $P = \{x \mid x \geqslant \pi\}$ is read "set P consists of all numbers x such that x is greater than or equal to π." The solution set consists of π and all numbers greater than π. The solution set is shaded in Fig. 3-23. The black dot indicates that π is a member of the solution set.

Example 3
Graph on a number line the inequality $|x + 2| > 3$ for x a real number.

Solution To graph this relation we need an equivalent inequality or inequalities which do not incorporate the absolute-value symbol. The solution set consists of numbers x that satisfy $x + 2 > 3$ when $x + 2 \geqslant 0$ or $-(x + 2) > 3$ when $x + 2 < 0$. This expression simplifies to

$$x > 1 \qquad \text{for } x \geqslant -2$$

or

$$x < -5 \qquad \text{for } x < -2$$

This is equivalent to

$$x > 1 \quad \text{or} \quad x < -5$$

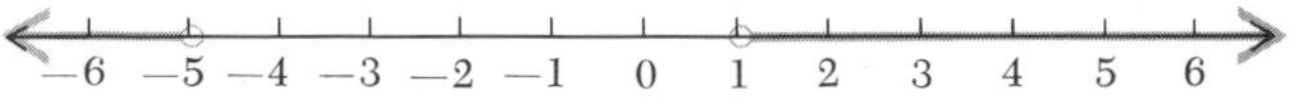

Fig. 3-24 Line graph of $\{x \mid x > 1 \text{ or } x < -5 \text{ for } x \text{ a real number}\}$.

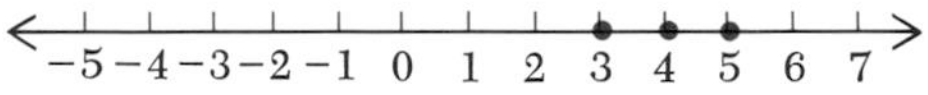

Fig. 3-25 Line graph of $x = 5$ or $x = 4$ or $x = 3$.

and the solution set may be written as

$$\{x \mid x > 1 \text{ or } x < -5 \text{ for } x \text{ a real number}\}$$

The set of points is pictured in Fig. 3-24.

Example 4

Graph on a number line the set $\{x \mid |x - 4| \leqslant 1 \text{ for } x \text{ an integer}\}$.

Solution $\{x \mid |x - 4| \leqslant 1 \text{ for } x \text{ an integer}\}$ is equivalent to $x - 4 \leqslant 1$ when $x - 4 \geqslant 0$ or $-(x - 4) \leqslant 1$ when $x - 4 < 0$. This is simplified by the following steps:

$x \leqslant 5$ when $x \geqslant 4$ or $x - 4 \geqslant -1$ when $x < 4$.
$x = 5$ or $x = 4$ or $x \geqslant 3$ and $x < 4$.
$x = 5$ or $x = 4$ or $x = 3$.

The graph is shown in Fig. 3-25.

PROBLEMS

Graph on a line the sets of points indicated in Probs. 1 to 14. The domain of the variable in each case is the set of real numbers.

1. $-4 < x < 9$
2. $-5 > u > -11$
3. $x \leqslant 5$ and $x > -1$
4. $y > 2$ or $y \leqslant -4$
5. $\{r \mid r^2 = 9\}$
6. $\{y \mid y^2 \leqslant 3\}$
7. $\{x \mid x(x - 1) = 0\}$
8. $\{y \mid y^2 - 8 < 2y\}$
9. $\{x \mid 2x - 3 > x\}$
10. $\{x \mid x^2 + 9 = 0\}$
11. $\{x \mid |x| > 0\}$
12. $\{r \mid |r| < 3\}$
13. $\{s \mid |2s| > 7\}$
14. $\{x \mid |3x - 5| < 2\}$

15. Graph Probs. 2, 4, 12, and 14 if the domain of the variable is the set of integers.

Write the inequalities describing the sets graphed in Probs. 16 to 20. State the domain of the variable.

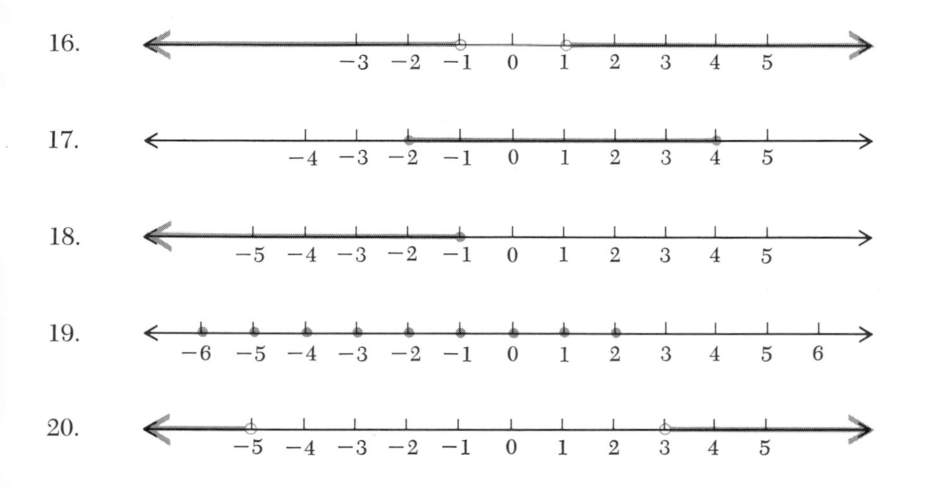

21. Repeat Probs. 16 to 20 using the symbol for absolute value.

SUMMARY

Analytic geometry may be briefly described as the study of the correspondence between an equation or inequality and a geometric figure or between an equation and the given set of conditions that determine a geometric figure. Hence two major problems are studied:

1. Given an equation or inequality, graph the associated geometric figure.
2. Given a geometric figure (or a word description of it), determine the equation or inequality.

In order to study these problems effectively the terms "relation" and "function" must be understood. For the cartesian system of coordinates a relation is a set of ordered pairs of numbers (x, y). A function is a set of ordered pairs (x, y) such that for each value of x there is exactly one corresponding value of y.

If the equation of the curve is given, one convenient method of graphing it (others are discussed later in the book) is to make a table. If the equation is in the form $y = f(x)$, then x may be replaced by certain numbers and corresponding numbers determined for y. If the domain, the set of specified replacements for x, is the set of integers, the graph of a relation will be different from that when the domain is the set of real numbers.

The other major aspect of analytic geometry, that of determining the equation or inequality from a set of conditions, is often called "finding the equation of a locus." A general procedure for solving this type of problem is to draw a picture indicating the given coordinates and then express the conditions algebraically.

The intersection of two curves is, in general, a set of points. The coordinates for these points are determined by methods of algebra. Three examples of intersections, given in set notation, are as follows. The first two are intersections of curves, and the last is the set of points common to two regions:

$\{(x, y) \mid x^2 + y^2 = 16 \cap x = y\}$ is a set of two points.
$\{(x, y) \mid y = x^2 \cap x + y = -4\}$ is an empty set.
$\{(x, y) \mid x^2 + y^2 < 16 \cap y < x\}$ is a region.

A relation in rectangular coordinates x and y may be expressed in polar coordinates ρ and θ by the substitutions

$$x = \rho \cos \theta \qquad \text{and} \qquad y = \rho \sin \theta$$

The equations for expressing a polar equation in rectangular form,

$$\rho = \pm \sqrt{x^2 + y^2} \qquad \text{and} \qquad \theta = \arctan \frac{y}{x}$$

must be applied carefully in order to choose the correct sign for ρ and the correct angle for θ, because for any pair (x, y) there is more than one pair (ρ, θ).

REVIEW PROBLEMS

For each of Probs. 1 to 3 make a table of ordered pairs of numbers, plot the corresponding points, and draw a smooth curve through them. Use the specified interval on the x axis.

1. $y = 6x - x^2$ for $-1 \leqslant x \leqslant 6$
2. $3y = 2x + 3$ for $-3 \leqslant x \leqslant 4$
3. $6y = x(x + 4)^2$ for $-6 \leqslant x \leqslant 2$
4. (*a*) Graph the function $x^3 - 4x$ in the interval $-3 \leqslant x \leqslant 3$. (*b*) Graph $y < x^3 - 4x$ in the interval $-3 \leqslant x \leqslant 3$.
5. On the same axes graph the equations $x^2 + y^2 = 25$ and $x - 7y = -25$. Esti-

mate the coordinates of the points of intersection. Solve for the points of intersection algebraically.

6. Graph the set of points P if $P = \{(x, y) \mid (y \geqslant x + 4) \cap (y \leqslant x^2 - 2)\}$.
7. A rectangular piece of tin is 4 by 8 in. A box is made from it by cutting a square region from each corner and turning up the sides. Express the volume V of the box obtained as a function of the length x of the side of the square region cut out. From the graph of this equation determine x so the volume of the box will be greatest.
8. Find the equation of the locus of a point $P(x, y)$ whose distance from point $Q(5, 0)$ is equal to the abscissa of P.
9. (*a*) Find the distance between points P and Q (given in polar coordinates) for $P(5, -60°)$ and $Q(10, -330°)$. (*b*) Express the coordinates of points P and Q in rectangular form and find $|PQ|$.
10. Graph the following sets of points on a number line; in each problem x is a real number:

    (*a*) $\{x \mid -2 \leqslant x < 5\}$ (*b*) $\{x \mid |x + 3| < 2\}$

FOR EXTENDED STUDY

In each of Probs. 1 to 7 describe the figure as a set of ordered pairs of numbers. For example, the positive "part" of the x axis in the xy plane is the set of points $P = \{(x, y) \mid y = 0 \text{ and } x > 0\}$.

1. The interior of a square with vertices $A(0, 0)$, $B(6, 0)$, $C(6, 6)$, and $D(0, 6)$
2. The square region with vertices listed in Prob. 1
3. A line which contains points $A(6, 1)$ and $B(6, -4)$
4. A segment with end points $A(-5, -7)$ and $B(-5, -2)$
5. The half-plane to the left of a line that goes through $A(3, 4)$ and $B(3, 8)$. (A half-plane consists of all the points on a plane on one side of a given line in the plane.) *Hint:* Draw a picture. The x coordinate can be what numbers? The y coordinate can be what numbers?
6. The region between a horizontal line through $A(0, 4)$ and one through $B(0, 7)$
7. A line perpendicular to the y axis through $A(0, a)$

Graph the sets of points in Probs. 8 to 14.

8. $P = \{(x, y) \mid (x + y + 2 \geqslant 0) \cap (x + y - 2 \leqslant 0)\}$
9. $Q = \{(x, y) \mid (x^2 + y^2 \leqslant 16) \cap (x^2 + y^2 \geqslant 9)\}$
10. $T = \{(\rho, \theta) \mid (\rho > 2) \cap (\rho < 4)\}$
11. $U = \{(x, y) \mid (x + y > 10) \cap (x + y \leqslant 6)\}$
12. $V = \{(\rho, \theta) \mid 0 \leqslant \theta \leqslant \frac{1}{4}\pi, \rho > 0\}$
13. $R = \{(x, y) \mid 1 = 0\}$
14. $S = \{(x, y) \mid 1 = 1\}$

4 THE LINE

4-1 INTRODUCTION

A line is completely determined if the coordinates of two points on it are specified. It may also be determined by one point and the slope or by certain other data, such as the length and direction of the shortest segment that can be drawn from the origin to the line.

In this chapter formulas are derived for writing the equation of a line in terms of the various data that determine it. The result, an equation of the *first degree* in x and y, will be discovered and proved. It should be remembered that the equation of a line has the following properties:

1. It shows a relationship among ordered pairs of numbers. For example, $y = 2x$ indicates that the second member of the ordered pair (x, y) is twice the first, or $x + y = 10$ indicates that the sum of the members of the ordered pair is 10.
2. It serves to partition the set of points in the plane, or acts as a "sorter" among all ordered pairs of numbers. Those that satisfy the equation may be graphed as points on the line; those that do not satisfy the equation may not be graphed on the line. Because $3(2) - 4(5) + 14 = 0$, the point $(2, 5)$ is on the line $3x - 4y + 14 = 0$. Similarly, because $3(-1) - 4(-4) + 14 \neq 0$, the point $(-1, -4)$ is not on the line $3x - 4y + 14 = 0$.

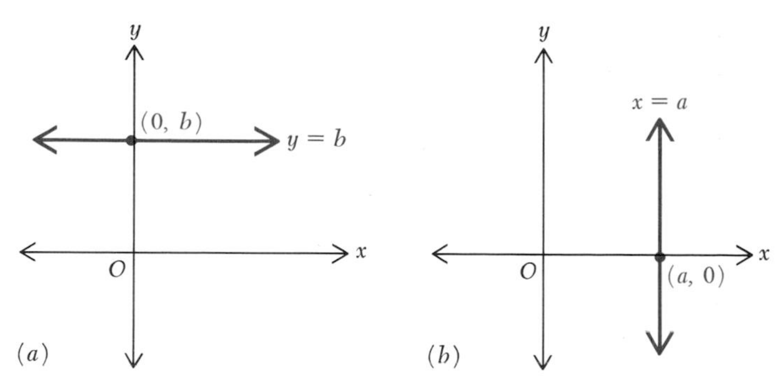

Fig. 4-1 (*a*) The line $y = b$ is parallel to the x axis. (*b*) The line $x = a$ is parallel to the y axis.

4-2 LINES PARALLEL TO THE COORDINATE AXES

Figure 4-1*a* shows a line with a y intercept of b, parallel to the x axis. Every point on this line is associated with an ordered pair of numbers whose second member is b. Thus every point on the line satisfies the equation $y = b$, and conversely, every point with an ordinate equal to b is on the line. The equation of the line is $y = b$.

Similarly, the equation of a line parallel to the y axis and intersecting the x axis at $(a, 0)$ corresponds to a set of ordered pairs whose first member is a. The equation of this line is $x = a$ (see Fig. 4-1*b*).

4-3 THE POINT-SLOPE FORM

Let a line be determined by specifying its slope m and the coordinates (x_1, y_1) of one point on it. The problem is to find the equation of the line, that is, to determine an equation in x and y which is satisfied by the coordinates (x, y) of every point on the line and which is not satisfied by any other pair of real numbers. We proceed as follows (see Fig. 4-2). If $P(x, y)$ is any point in the xy plane, then the slope of the line joining

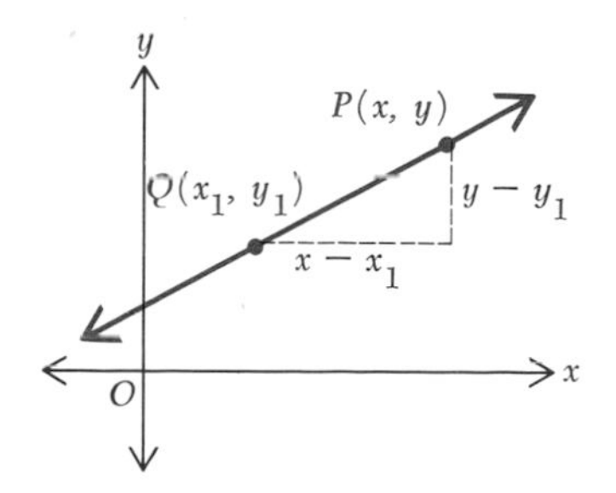

Fig. 4-2 Given the slope m and a point $Q(x_1, y_1)$ on a line, the equation of the line is $y - y_1 = (x - x_1)$.

P to the given point $Q(x_1, y_1)$ is $(y - y_1)/(x - x_1)$. This is equal to the given number m if and only if $P(x, y)$ lies on the specified line. The desired equation is then

$$\frac{y - y_1}{x - x_1} = m$$

or

$$y - y_1 = m(x - x_1) \tag{4-1}$$

This is called the *point-slope form* of the equation of a line. It may be used for writing the equation of a line when its slope and one point on it are known.

Example 1
Write the equation of a line with a slope of $\frac{3}{2}$ that passes through the point $Q(-4, 2)$.

Solution From Eq. (4-1),

$$y - 2 = \tfrac{3}{2}[x - (-4)]$$

or

$$3x - 2y + 16 = 0$$

If two different points $P_1(x_1, y_1)$ and $P_2(x_2, y_2)$ are given, the slope of the line containing P_1 and P_2 is $m = (y_2 - y_1)/(x_2 - x_1)$. Its equation is, from (4-1),

$$y - y_1 = \frac{y_2 - y_1}{x_2 - x_1}(x - x_1) \qquad \text{for } x_1 \neq x_2 \tag{4-2}$$

Let R be the set of points that are on this line; then, by Eq. (4-2),

$$R = \left\{(x, y) \mid y - y_1 = \frac{y_2 - y_1}{x_2 - x_1}(x - x_1)\right\}$$

Example 2
Find the equation of a line through $A(6, -2)$ and $B(-4, 3)$.

Solution The slope is $(-2 - 3)/[6 - (-4)]$; using the coordinates of B in Eq. (4-2) gives

$$y - 3 = \frac{-2 - 3}{6 - (-4)}(x + 4)$$

or

$$x + 2y - 2 = 0$$

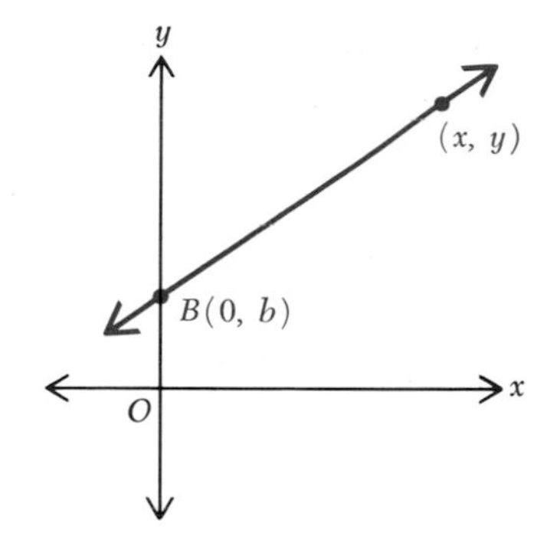

Fig. 4-3 Given the slope m and y intercept b, the equation of the line is $y = mx + b$.

4-4 THE SLOPE-INTERCEPT FORM

If a line intersects the y axis at $B(0, b)$, as in Fig. 4-3, then b is called the *y intercept* of the line. The equation of the line may be found in terms of its slope m and its y intercept b simply by substituting the point $B(0, b)$ for (x_1, y_1) in the point-slope formula. The result is

$$y - b = m(x - 0)$$

or

$$y = mx + b \tag{4-3}$$

This is called the *slope-intercept* form of the equation of a line. It may be used conveniently for writing the equation when the slope and y intercept are known.

Example 1
Find the equation of the line with a slope of $\frac{1}{2}$ and y intercept -3.

Solution From Eq. (4-3) we have

$$y = \tfrac{1}{2}x + (-3)$$

or

$$x - 2y - 6 = 0$$

Example 2
What is the x intercept of the line whose equation is $3x - 2y - 6 = 0$?

Solution The *x intercept* is the x coordinate of the point at which the line intersects the x axis. At this point the y coordinate is 0. Hence we substitute $y = 0$ in the equation $3x - 2y = 6$ to find $x = 2$. The x intercept is thus 2.

PROBLEMS

1. Find the equation of the line through the given point A with given slope m:

   (*a*) $A(-5, 3)$; $m = \frac{2}{3}$ (*b*) $A(3, -4)$; $m = -\frac{1}{2}$
   (*c*) $A(6, 1)$; $m = -2$ (*d*) $A(-2, 7)$; $m = \frac{1}{3}$

2. Find the equation of the line determined by the given points A and B:

   (*a*) $A(2, 3), B(6, -2)$ (*b*) $A(-5, -2), B(0, -1)$
   (*c*) $A(-3, 4), B(3, -2)$ (*d*) $A(5, 2), B(0, 3)$

3. Find the equations of the sets of points named by P, Q, R, and S:

   (*a*) $P = \{(x, y) \mid (x, y)$ is on a line through $A(-3, -5)$ with a slope of $0.3\}$
   (*b*) $Q = \{(x, y) \mid (x, y)$ is on a line through $A(-4, 0)$ with a slope of $-\frac{3}{4}\}$
   (*c*) $R = \{(x, y) \mid (x, y)$ is on a line containing $A(\frac{4}{3}, 0)$ and $B(\frac{5}{2}, -\frac{3}{2})\}$
   (*d*) $S = \{(x, y) \mid (x, y)$ is on a line containing $A(-2, -\frac{5}{3})$ and $B(0, \frac{2}{3})\}$

4. A line intersects the x axis at $(a, 0)$ and the y axis at $(0, b)$. The numbers a and b are thus the intercepts of the line. Show that the equation of this line is

$$\frac{x}{a} + \frac{y}{b} = 1 \qquad \text{for } a \neq 0 \quad \text{and} \quad b \neq 0$$

   Why are the restrictions placed on a and b?

5. Write the equation of the line for each of the following conditions:

   (*a*) y intercept of -4 and slope of $\frac{2}{3}$
   (*b*) y intercept of 6 and slope of $-\frac{1}{2}$
   (*c*) y intercept of -8 and slope of $-\frac{5}{2}$
   (*d*) x and y intercepts of $\frac{3}{2}$ and -2, respectively
   (*e*) x and y intercepts of -5 and $\frac{7}{2}$, respectively
   (*f*) x and y intercepts of $\frac{1}{2}$ and $-\frac{2}{3}$, respectively

6. The vertices of a rectangle are $A(0, 0)$, $B(-8, 0)$, $C(-8, -6)$, and $D(0, -6)$. Write equations for the lines which contain the sides and diagonals of the rectangle.
7. The vertices of a triangle are $A(2, 6)$, $B(8, -4)$, and $C(-3, -5)$. Two lines are drawn from A, one parallel to and the other perpendicular to $\overline{BC}$. Find the equation of these lines.
8. Write the equation of the line that is the perpendicular bisector of $\overline{AB}$:

   (*a*) $A(-2, 4), B(4, 0)$ (*b*) $A(0, -4), B(6, 0)$

9. In the triangle ABC find the equation of the altitude drawn from A perpendicular to side $\overline{BC}$:

   (a) $A(0, -4)$, $B(6, 2)$, $C(-3, 8)$ (b) $A(-4, -2)$, $B(0, -6)$, $C(6, 4)$

10. The equation of a straight line is $y = \frac{2}{3}x + 2$. A second line is drawn through $P(8, -3)$ perpendicular to the first. What is its equation?
11. What is the equation of the line drawn through the origin perpendicular to a line whose x and y intercepts are 5 and $-\frac{3}{2}$, respectively?
12. Find the equation of the line through the point $P(0, 6)$ and parallel to the line whose equation is $y = -3x + 2$.
13. Find the equation of the median drawn from B to the midpoint of $\overline{AC}$ in triangle ABC:

    (a) $A(0, 6)$, $B(8, 0)$, $C(3, -4)$ (b) $A(-7, 5)$, $B(1, 6)$, $C(5, -6)$

14. The equation of a line is $2x - 3y = 30$. What point on it is equidistant from $A(1, 3)$ and $B(7, 9)$? *Hint:* Find the point where the perpendicular bisector of $\overline{AB}$ intersects the given line.
15. Find the coordinates of the center of the circle that passes through $A(8, 7)$, $B(6, -7)$, and $C(-8, -5)$.
16. Graph the following equations:

    (a) $|y| = x$ (b) $|x| = y$ (c) $|x + y| = 7$
    (d) $|x| + y = 6$ (e) $|x| + |y| = 10$ (f) $|3x| + |5y| = 15$

17. The equation for a line through points $P_1(x_1, y_1)$ and $P_2(x_2, y_2)$ is

$$y - y_1 = \frac{y_2 - y_1}{x_2 - x_1}(x - x_1)$$

    Show that another form for this equation is

$$(y_1 - y_2)x + (x_2 - x_1)y + (x_1y_2 - x_2y_1) = 0$$

18. Give the x and y intercepts of the line whose equation is

$$y - y_1 = \frac{y_2 - y_1}{x_2 - x_1}(x - x_1)$$

19. Show that the equation of the line determined by the points $P_1(x , y_1)$ and $P_2(x_2, y_2)$ can be written in the form

$$\begin{vmatrix} x & y & 1 \\ x_1 & y_1 & 1 \\ x_2 & y_2 & 1 \end{vmatrix} = 0$$

20. Use the determinant of Prob. 19 to find the equation of the line determined by the given points A and B:

    (a) $A(-6, 2)$, $B(0, 4)$ (b) $A(4, 0)$, $B(-2, 2)$

4-5 THE GENERAL EQUATION OF FIRST DEGREE

Any equation of the first degree in the variables x and y can be written in the form

$$Ax + By + C = 0 \tag{4-4}$$

where A, B, and C are real numbers and A and B are not both zero. The graph of this linear equation in x and y is always a straight line. If $B = 0$, then the graph is a straight line because the equation becomes $Ax + C = 0$ or $x = -C/A$, and the locus of any equation of the form x equal to a constant is a straight line parallel to the y axis. If $B \neq 0$, then the equation can be solved for y in terms of x. Thus in this case the graph is the same as that of the equation

$$y = -\frac{A}{B}x - \frac{C}{B}$$

But this is an equation of the form $y = mx + b$, and it has been shown that its graph is a straight line with slope $-A/B$ and y intercept $-C/B$. If $A = 0$, the equation describes a straight line parallel to the x axis. Since B either is zero or is not zero, and since the graph is a straight line in both these cases, the proof is complete.

A general linear equation $Ax + By + C = 0$ in which $B \neq 0$ can be put into the slope-intercept form simply by solving it for y in terms of x. The coefficient of x in the resulting equation will be the slope, and the constant term will be the y intercept.

Example
What are the slope and y intercept of the line whose equation is $3x - 4y - 8 = 0$?

Solution Solving for y gives

$$y = \tfrac{3}{4}x - 2$$

Then the slope is $\frac{3}{4}$ and the y intercept is -2.

4-6 (OPTIONAL) DIRECTION COSINES OF $Ax + By + C = 0$

If the graph of $Ax + By + C = 0$ is to be a line, either $B \neq 0$ or $A \neq 0$. Let us assume $B \neq 0$ (a similar argument holds for $A \neq 0$). If $P_1(x_1, y_1)$ and $P_2(x_2, y_2)$ are members of the solution set of $Ax + By + C = 0$, then

$$Ax_1 + By_1 + C = 0 \qquad \text{and} \qquad Ax_2 + By_2 + C = 0$$

Then

$$y_1 = -\frac{C + Ax_1}{B} \qquad \text{and} \qquad y_2 = -\frac{C + Ax_2}{B}$$

so two points on the line $Ax + By + C = 0$ are

$$\left(x_1, -\frac{C + Ax_1}{B}\right) \qquad \text{and} \qquad \left(x_2, -\frac{C + Ax_2}{B}\right)$$

A pair of direction numbers r and s of the line are

$$r = x_2 - x_1 \qquad \text{and} \qquad s = -\frac{C + Ax_2}{B} - \left(-\frac{C + Ax_1}{B}\right) = -\frac{A}{B}(x_2 - x_1)$$

Another set of direction numbers, kr and ks, are found by using $k = B/(x_2 - x_1)$, so

$$kr = B \qquad \text{and} \qquad ks = -A$$

Then direction cosines (λ, μ) of $Ax + By + C = 0$ are

$$\lambda = \frac{B}{\sqrt{A^2 + B^2}} \qquad \text{and} \qquad \mu = \frac{-A}{\sqrt{A^2 + B^2}}$$

or

$$\lambda = \frac{-B}{\sqrt{A^2 + B^2}} \qquad \text{and} \qquad \mu = \frac{A}{\sqrt{A^2 + B^2}}$$

The slope is

$$m = \frac{\mu}{\lambda} = \frac{-A}{B} \qquad \text{for } B \neq 0$$

Example

A line has direction numbers $r = 4$ and $s = -7$ and passes through $P_1(-4, 1)$. What is its equation?

Solution The required equation is of the form $7x + 4y + k = 0$. Point $P_1(-4, 1)$ is on the line, so $7(-4) + 4(1) + k = 0$ and $k = 24$. The equation is $7x + 4y + 24 = 0$.

PROBLEMS

1. Find the slope and y intercept of the line whose equation is given and write the equation in the slope-intercept form:

   (*a*) $3x - 4y = 18$ (*b*) $0.4x - 2y = 6.4$
   (*c*) $2x - 3y + 4 = 0$ (*d*) $x/5 - y/3 = 1$

2. Find the intercepts of the line determined by the given equation or by the given points and write the equation of the line in the intercept form (see Prob. 4, page 96):

   (*a*) $3x + 4y = 24$ (*b*) $3x = y - 12$
   (*c*) $A(-3, 4)$, $B(1, -6)$ (*d*) $A(2.6, 1.4)$, $B(3.8, 2)$

3. The sum of the coefficients of x and y in the equation $Ax + By + C = 0$ is 6. The slope of the line represented by the equation is 7. Determine A, B, and C if the line passes through the point $P(1, -6)$.
4. The graph of

$$(A_1x + B_1y + C_1) + k(A_2x + B_2y + C_2) = 0$$

   is a straight line. What are its x and y intercepts? What is its slope?
5. Determine a and b such that $ax + by - 7 = 0$ is satisfied by the ordered pairs $(3, 1)$ and $(-3, -8)$.
6. Determine A and C such that the graph of $Ax - 4y + C = 0$ contains the points $P(3, -1)$ and $Q(15, 2)$.
7. Given the set of points $\{(x, y) \mid Ax + By + C = 0\}$, describe the graph when

   (*a*) $A = 0, B \neq 0, C \neq 0$ (*b*) $B = 0, A \neq 0, C \neq 0$
   (*c*) $C = 0, A \neq 0, B \neq 0$

8. The equation

$$\frac{x - x_1}{x_2 - x_1} = \frac{y - y_1}{y_2 - y_1}$$

   is called the symmetric form of the equation of a straight line containing the points $P(x, y)$, $P_1(x_1, y_1)$, and $P_2(x_2, y_2)$. Choose (x_1, y_1) as $(5, -4)$ and write $4x + 3y - 8 = 0$ in a symmetric form.
9. For each of the following determine a pair of direction numbers for $\overleftrightarrow{AB}$:

   (*a*) $A(3, 7)$, $B(5, 11)$ (*b*) $A(-5, -8)$, $B(-9, 4)$
   (*c*) $A(3, 2)$, $B(-1, -5)$ (*d*) $A(-1, 4)$, $B(3, -5)$

10. What is a pair of direction numbers for $\overrightarrow{BA}$ in Prob. 9?
11. What are the direction cosines of $\overleftrightarrow{AB}$ in Prob. 9?
12. A pair of direction numbers of a ray l are $[2, 3]$. With C a constant, which of the following could be equations of a line on which l is located?

    (*a*) $-3x + 2y + C = 0$ (*b*) $-3x - 2y + C = 0$
    (*c*) $3x - 2y + C = 0$ (*d*) $-12x + 6y - C = 0$
    (*e*) $2x - 3y + C = 0$ (*f*) $x - \frac{2}{3}y + C = 0$
    (*g*) $-6x - 4y + C = 0$ (*h*) $\frac{6}{4}x - y - C = 0$

13. Write the equations of the lines that meet the following conditions:

    (*a*) Direction numbers $r = -2$ and $s = -1$ and passing through $(0, 0)$
    (*b*) Direction numbers $r = -\frac{1}{2}$ and $s = -3$ and passing through $(-4, -7)$

(*c*) Direction cosines $\lambda = 4/\sqrt{17}$ and $\mu = 1/\sqrt{17}$ and passing through $(-3, 5)$

(*d*) Direction numbers $r = a_1$ and $s = b_1$ and passing through (x_1, y_1)

14. Find the direction cosines for lines whose equations are as follows:

(*a*) $12x - 5y + 4 = 0$ (*b*) $a_1x + b_1y + c_1 = 0$

(*c*) $-3x - y + 7 = 0$ (*d*) $y = mx + b$

(*e*) $6x + 4y = 0$ (*f*) $x/a + y/b = 1$

4-7 CONDITIONS FOR PARALLELISM AND PERPENDICULARITY

The condition for parallelism of two lines having slopes m_1 and m_2 is that $m_1 = m_2$. Let the equations of the lines l_1 and l_2 be, respectively,

$$A_1x + B_1y + C_1 = 0$$

and

$$A_2x + B_2y + C_2 = 0$$

The slope of l_1 is $m_1 = -A_1/B_1$ and that of l_2 is $m_2 = -A_2/B_2$. If the A's and B's are different from zero, the condition for parallelism is

$$-\frac{A_1}{B_1} = -\frac{A_2}{B_2} \quad \text{or} \quad \frac{A_1}{A_2} = \frac{B_1}{B_2} \tag{4-5}$$

This condition may be expressed as

$$\begin{vmatrix} A_1 & B_1 \\ A_2 & B_2 \end{vmatrix} = 0$$

Example 1

Are the lines $2x + 3y - 6 = 0$ and $4x + 6y + 5 = 0$ parallel?

Solution Since $A_1/A_2 = \frac{2}{4}$ and $B_1/B_2 = \frac{3}{6}$, then $A_1/A_2 = B_1/B_2$. Hence the lines are parallel, by Eq. (4-5).

The proof of Theorem 4-2 will require the following consequence of the condition for parallelism. If a line l has the equation $Ax + By + C = 0$, then the equation of any line l' parallel to l is $Ax + By + k = 0$, where k is a constant.

Example 2

Find the equation of the line that passes through the point $A(4, -7)$ and is parallel to the line $5x - 3y + 10 = 0$.

Solution The required line has the equation $5x - 3y + k = 0$, where k is to be determined from the fact that the line goes through the point $A(4, -7)$. Thus the value of k must be such that

$$5(4) - 3(-7) + k = 0$$

Then

$$k = -41$$

so the required equation is

$$5x - 3y - 41 = 0$$

The corresponding condition for perpendicularity of two lines with slopes m_1 and m_2 is that $m_1 = -1/m_2$ (see Theorem 2-2); this means that

$$-\frac{A_1}{B_1} = \frac{B_2}{A_2} \quad \text{or} \quad A_1A_2 + B_1B_2 = 0 \tag{4-6}$$

This condition may be expressed as

$$\begin{vmatrix} A_1 & -B_1 \\ B_2 & A_2 \end{vmatrix} = 0$$

If line l_1 has direction numbers r_1 and s_1 and line l_2 has direction numbers r_2 and s_2, then l_1 is perpendicular to l_2 if and only if

$$r_1r_2 + s_1s_2 = 0$$

This follows from Eq. (4-6) and the facts that $r_1 = B_1$, $s_1 = -A$, $r_2 = B_2$, and $s_2 = -A_2$.

From this conclusion we may deduce that if the direction cosines of l_1 are λ_1 and μ_1 and those of line l_2 are λ_2 and μ_2, the condition for perpendicularity may be expressed as

$$\lambda_1\lambda_2 + \mu_1\mu_2 = 0$$

Example 3
Are the lines $2x + 3y - 6 = 0$ and $9x - 6y + 1 = 0$ perpendicular?

Solution Since $A_1A_2 + B_1B_2 = 2(9) + 3(-6) = 0$, the lines are perpendicular, by Eq. (4-6).

Example 4
Find the equation of a line that passes through the point $C(4, 2)$ and is perpendicular to the line $3x + 5y - 7 = 0$.

Solution The required line has the equation $5x - 3y + k = 0$ in order to meet the condition for perpendicularity. Now, k must be determined from the fact that the line passes through the point $C(4, 2)$. Thus k must be such that

$$5(4) - 3(2) + k = 0$$

Then

$$k = -14$$

and the required equation is

$$5x - 3y - 14 = 0$$

4-8 LENGTH OF THE PERPENDICULAR SEGMENT FROM THE ORIGIN TO THE LINE $Ax + By + C = 0$

Theorem 4-1

The undirected length p of the perpendicular segment drawn from the origin to the line $Ax + By + C = 0$ is equal to the absolute value of C divided by $\sqrt{A^2 + B^2}$; that is,

$$p = \frac{|C|}{\sqrt{A^2 + B^2}} \tag{4-7}$$

Proof Let the line l in Fig. 4-4 be the graph of the equation $Ax + By + C = 0$, in which A, B, and C are all different from zero. Let α be the inclination of the perpendicular segment drawn from the origin to l, and let $p = |ON|$ be the required length. Then since p is a positive number,

$$|\cos \alpha| = \frac{p}{|OQ|}$$

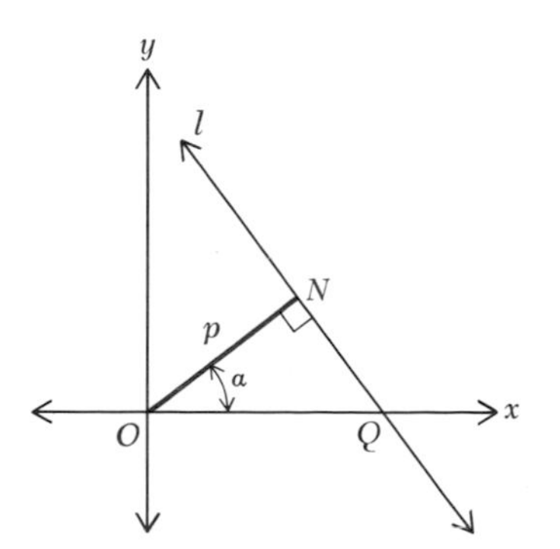

Fig. 4-4 Segment $\overline{ON}$ is perpendicular to the line $Ax + By + C = 0$: $ON = |C|/\sqrt{A^2 + B^2}$.

and $$p = |OQ||\cos\alpha|$$

Now, OQ is the x intercept of l; hence $OQ = -C/A$ and $|OQ| = |C/A|$. To find $|\cos\alpha|$ observe that the slope of l is $-A/B$, and since $\overline{ON}$ is perdicular to l, the slope of $\overline{ON}$ is B/A; thus

$$\tan\alpha = \frac{B}{A}$$

so that $$|\cos\alpha| = \frac{|A|}{\sqrt{A^2 + B^2}}$$

Hence $$p = |OQ||\cos\alpha| = \frac{|C|}{|A|}\frac{|A|}{\sqrt{A^2 + B^2}} = \frac{|C|}{\sqrt{A^2 + B^2}}$$

This completes the proof for the case in which A, B, and C are all different from zero. Suppose now that $C = 0$. In this case the line l goes through the origin, so $p = 0$, and the formula holds. If either $A = 0$ or $B = 0$, with $C \neq 0$, the formula again gives the correct result.

Example

Find the length of the segment drawn from the origin to the line $3x - 5y - 6 = 0$.

Solution From Eq. (4-7), the distance is

$$p = \frac{|-6|}{\sqrt{3^2 + (-5)^2}} = \frac{6}{\sqrt{34}}$$

4-9 LENGTH OF THE PERPENDICULAR SEGMENT DRAWN FROM A POINT $P(x_1, y_1)$ TO THE LINE $Ax + By + C = 0$

Theorem 4-2

The undirected length d of the perpendicular segment drawn from a point $P(x_1, y_1)$ to the line $Ax + By + C = 0$ is given by

$$d = \frac{|Ax_1 + By_1 + C|}{\sqrt{A^2 + B^2}} \tag{4-8}$$

Proof Let the equation of line l in Fig. 4-5 be

$$Ax + By + C = 0 \qquad \text{with } A \neq 0$$

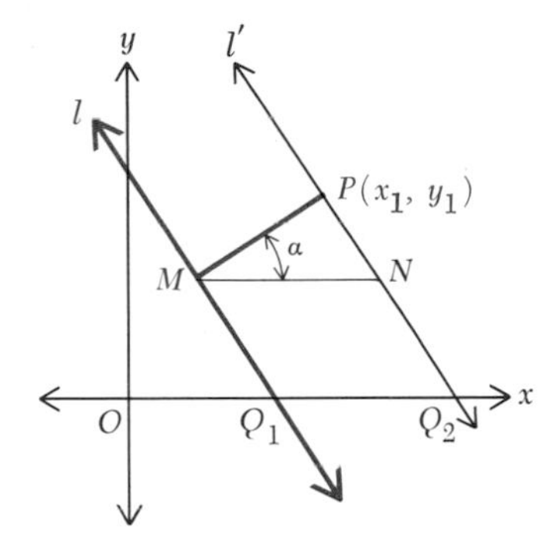

Fig. 4-5 $\overline{PM}$ is the segment perpendicular to $Ax + By + C = 0$: $PM = (Ax_1 + By_1 + C)/\sqrt{A^2 + B^2}$.

Let $P(x_1, y_1)$ be the given point, and let $d = |MP|$ be the required length. Now draw a line l' through P parallel to l and let OQ_1 and OQ_2 be the x intercepts of l and l', respectively. Draw $\overline{MN}$ parallel to $\overline{Q_1Q_2}$. Then

$$d = |MP| = |MN||\cos\alpha| = |Q_1Q_2||\cos\alpha| = |OQ_2 - OQ_1||\cos\alpha| \tag{4-9}$$

Now $OQ_1 = -C/A$. To find OQ_2 we first observe that the equation of l', since it is parallel to l, is $Ax + By + k = 0$, where k is determined by the fact that $P(x_1, y_1)$ is a point on the line. Thus $Ax_1 + By_1 + k = 0$; so $k = -Ax_1 - By_1$. Then

$$OQ_2 = \frac{-k}{A} = \frac{Ax_1 + By_1}{A}$$

and

$$OQ_2 - OQ_1 = \left(\frac{Ax_1 + By_1}{A}\right) - \left(-\frac{C}{A}\right) = \frac{Ax_1 + By_1 + C}{A}$$

As in the corresponding derivation of the preceding section,

$$|\cos\alpha| = \frac{|A|}{\sqrt{A^2 + B^2}}$$

Substitution in Eq. (4-9) gives

$$d = |OQ_2 - OQ_1||\cos\alpha| = \frac{|Ax_1 + By_1 + C|}{|A|}\frac{|A|}{\sqrt{A^2 + B^2}} = \frac{|Ax_1 + By_1 + C|}{\sqrt{A^2 + B^2}}$$

This completes the proof for the case in which $A \neq 0$. The reader may show that the formula is also correct if $A = 0$.

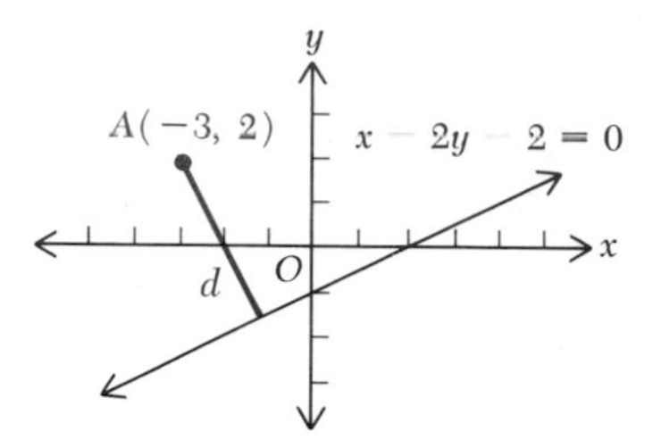

Fig. 4-6 d is the shortest distance from $A(-3, 2)$ to $x - 2y - 2 = 0$.

Example

Find the measure of the perpendicular segment from point $A(-3, 2)$ to the line $x - 2y - 2 = 0$.

Solution The measure d (see Fig. 4-6) is, from Eq. (4-8),

$$d = \frac{|(-3) - 2(2) - 2|}{\sqrt{(1)^2 + (-2)^2}}$$

$$= \frac{|-3 - 4 - 2|}{\sqrt{5}} = \frac{9}{\sqrt{5}}$$

PROBLEMS

1. Find the value that must be assigned to A or B so that the two lines whose equations are given will be parallel, and then find the value such that these lines are perpendicular:

   (*a*) $Ax - 2y = 4$; $x + 3y = 7$ (*b*) $4x + 5y = 20$; $Ax - 6y = 4$
   (*c*) $x + 3y + 7 = 0$; $5x + By + 4 = 0$ (*d*) $1.2x - 3.6y = 5$; $x + By = 2$

2. In each of the following find the equations of two lines through point A, one parallel to and the other perpendicular to the given line:

   (*a*) $A(3, -5)$; $3x + 2y = 10$ (*b*) $A(-2, -1)$; $x - 7y = 16$

3. Find the equation of a line that contains the point $A(6, -3)$ and whose intercepts add to 10.
4. Draw the two lines whose equations are given and find the tangent of the acute angle between them:

   (*a*) $3x - 4y = 12$; $x/2 + y/-6 = 1$
   (*b*) $x + y = 5$; $y = \frac{1}{2}x - 3$

5. Find the acute angle formed by the lines whose equations are $2x - 3y + 6 = 0$ and $x = 4$.

6. In the triangle with vertices $A(-2, 4)$, $B(6, 2)$, and $C(4, -3)$ the median is drawn from A to the midpoint of $\overline{BC}$. Find the tangent of the acute angle formed by this median and the line $y = 2x$.
7. In the triangle of Prob. 6 a line is drawn from B perpendicular to $\overline{AC}$. Find the tangent of the acute angle formed by this line and the line $x + y = 0$.
8. Find the length of the perpendicular segment from the origin to the line with the given equation:

   (a) $3x + y = 10$ (b) $y = \frac{3}{4}x + 5$ (c) $x/2 - y/6 = 1$

9. Find the length of the perpendicular segment from point A to the line:

   (a) $6x - 8y - 20 = 0$; $A(1, 2)$ (b) $x + y = 4$; $A(-1, -1)$

   (c) $5x + 12y = 30$; $A(-3, 1)$ (d) $\dfrac{x}{3} + \dfrac{y}{5} = 1$; $A(2, 1)$

10. A line is parallel to the line $3x - 4y = 6$, and the undirected length of the perpendicular segment drawn from the origin to it is 3. Find its equation.
11. A line is parallel to the line through $A(-6, 2)$ and $B(6, 7)$, and tangent to the circle with center at the origin and radius 5. Find its equation.
12. Find the area of the interior of the triangle with vertices $A(-2, 6)$, $B(5, 3)$, and $C(2, -3)$ by multiplying $\frac{1}{2}|BC|$ by the altitude. Check by using a determinant.
13. Find the distance between the parallel lines $x - 2y = 9$ and $2x - 4y + 12 = 0$.
14. Show that the equation of the line that passes through point $P(x_1, y_1)$ and is parallel to the line $Ax + By + C = 0$ is $Ax + By = Ax_1 + By_1$.
15. Show that the equation of the line that passes through point $P(x_1, y_1)$ and is perpendicular to the line $Ax + By + C = 0$ is $Bx - Ay = Bx_1 - Ay_1$.
16. For what value or values of A does the acute angle formed by the lines $3x + 2y = 10$ and $Ax - 2y = 3$ have a measurement of 45°?
17. The direction numbers of line l_1 are $r_1 = 2$ and $s_1 = -5$; line l_2 has the direction numbers r_2 and s_2. (a) Find r_2 if $s_2 = 6$ and l_1 is perpendicular to l_2. (b) Find s_2 if $r_2 = 3$ and l_1 is parallel to l_2.
18. A line l_1 has direction numbers $r = -4$ and $s = 5$. Write the equation of a line l_2 such that (a) l_2 is perpendicular to l_1 and passes through (2, 3); (b) l_2 is parallel to l_1 and passes through (−6, −5); (c) l_1 is perpendicular to l_2 at (0, 0); (d) l_2 is parallel to l_1 and passes through the midpoint of $\overline{AB}$, where $A(5, 3)$ and $B(-1, 7)$.
19. Consider line l_1 with the equation $x/3 + y/4 = 1$. (a) Line l_2 has as its equation $Ax = 6y = 9$. Determine A such that l_1 is perpendicular to l_2. (b) Line l_3 has as its equation $A_3x + B_3y + C_3 = 0$. What number must A_3/B_3 be for l_1 to be perpendicular to l_3?

4-10 (OPTIONAL) NUMBER OF ESSENTIAL CONSTANTS

Two of the forms in which the equation of a line can be written are

$$y = mx + b$$

and
$$\frac{x}{a} + \frac{y}{b} = 1$$

Each of these equations contains two constants which completely determine the line. The equation of a line written in the form

$$Ax + By + C = 0$$

contains *three* constants. It is possible, however, to divide both sides of the equation by any one of them not zero, thus reducing the number to two. Thus if $A \neq 0$, the equation $Ax + By + C = 0$ is equivalent to

$$x + dy + e = 0$$

where $d = B/A$ and $e = C/A$. The number of essential constants in this equation is therefore also two. This is, of course, connected with the fact that two conditions—such as two points, or one point and the slope—suffice to determine a line. Each such condition will yield an equation in the constants to be determined. For example, the condition that the line $y = mx + b$ passes through the point $A(6, 2)$ means that m and b must satisfy the equation

$$6m + b = 2$$

The additional condition that the x intercept shall be 5 means that

$$5m + b = 0$$

This system of two equations in m and b has the unique solution $m = 2$ and $b = -10$. The equation of the line satisfying the conditions is then

$$y = 2x - 10$$

4-11 PARAMETRIC EQUATIONS OF A LINE

A line in the xy plane is defined by an equation in two variables x and y. Sometimes it is more convenient to define a line by means of two equations which express the coordinates of points on it as functions of a third variable. These equations may have the form

$$x = g(t)$$
$$y = h(t)$$

The third variable (t in this case) is called a *parameter*, and the equations are called *parametric equations* of a line.

Parametric equations defining the path of a point are often used to simplify complex problems in physics or engineering. For example, the tracking of satellites is simplified with parametric representation. In many practical problems the parameter represents time elapsed from a given instant, as, for example, the parametric equations defining the path of a projectile (see page 206). The motion of an object moving with constant speed on a straight line is best studied by expressing the coordinates (x, y) of the object in terms of elapsed time. If an object moves around a circle at constant speed, a convenient parameter to use is θ, the measure of the angle of rotation.

The coordinates of any point $P(x, y)$ on a line passing through $P_1(x_1, y_1)$ and $P_2(x_2, y_2)$, as indicated by Eq. (1-2), are

$$x = x_1 + k(x_2 - x_1)$$
$$y = y_1 + k(y_2 - y_1)$$

where k is any real number. The expressions $x_2 - x_1$ and $y_2 - y_1$ are, of course, a pair of direction numbers for the line.

Example 1

Find parametric equations for the line through (2, 6) and (−3, −4).

Solution If (2, 6) is chosen as P_1, then

$$x_2 - x_1 = -3 - 2 = -5$$
$$y_2 - y_1 = -4 - 6 = -10$$

Then, with parameter t,

$$x = 2 - 5t$$
$$y = 6 - 10t$$

If (−3, −4) is chosen as P_1, then the parametric equations, with parameter s, are

$$x = -3 + 5s$$
$$y = -4 + 10s$$

These are only two of an infinite number of parametric representations of the same line. The graph of the line is shown in Fig. 4-7, along with the table of computed values. The line may be plotted by first making a table, assigning values to t or s, and determining x and y for each value of t or s.

It should be observed that a certain number for t determines a different pair of numbers (x, y) than the same number for s. However, the table

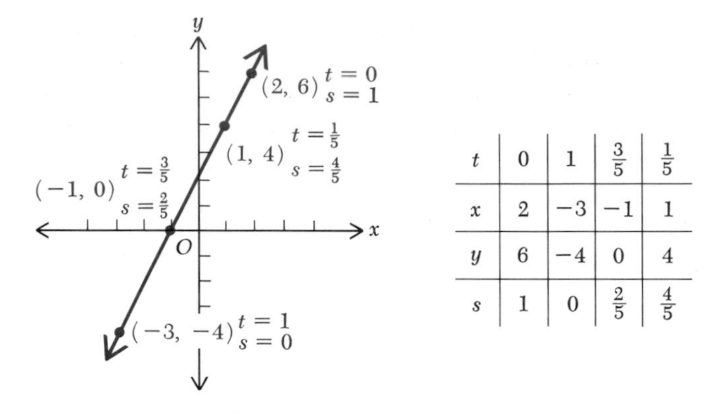

t	0	1	$\frac{3}{5}$	$\frac{1}{5}$
x	2	−3	−1	1
y	6	−4	0	4
s	1	0	$\frac{2}{5}$	$\frac{4}{5}$

Fig. 4-7 Points on a line determined from the given parametric equations.

indicates that for a given replacement for t there is a real number s which generates the same pair (x, y).

It is evident that for every real number t there exists a pair of numbers (x, y), and for every pair (x, y) there exists one real number t. Thus the parametric representation of a line establishes a one-to-one correspondence between the real numbers and points on a line.

Example 2
Write the equation in rectangular form of the line represented parametrically by

$$x = 3 - 2t$$
$$y = 1 + 4t$$

Solution From the first equation, $t = (3 - x)/2$. If this is substituted in the second equation the result is

$$y = 1 + 4\,\frac{(3 - x)}{2} \qquad \text{or} \qquad y = -2x + 7$$

Example 3
Find the intersection of line l_1 through $(3, -1)$ and $(1, 3)$ and line l_2 through $(5, 1)$ and $(7, -2)$. Use parametric representations for the lines.

Solution Parametric equations of the lines are

$$l_1:\quad x = 3 - 2t \qquad \text{and} \qquad l_2:\quad x = 5 + 2u$$
$$y = -1 + 4t \qquad\qquad\qquad y = 1 - 3u$$

Where the lines intersect, the x coordinates are equal and the y coordinates are equal. Hence

$$3 - 2t = 5 + 2u \qquad \text{and} \qquad -1 + 4t = 1 - 3u$$

The simultaneous solution is $t = 5$ and $u = -6$. By substituting in either pair of parametric equations, the coordinates of the point of intersection are found to be $(-7, 19)$.

Example 4

Find a parametric representation for the line whose equation is $3x + y - 7 = 0$.

Solution The coordinates of two points on the line, found by choosing values for x and substituting, are $(2, 1)$ and $(3, -2)$. A pair of direction numbers are $x_2 - x_1 = 3 - 2 = 1$ and $y_2 - y_1 = -2 - 1 = -3$. Hence one parametric form is

$$x = 2 + t$$
$$y = 1 - 3t$$

Another solution may be found by replacing x by t in $3x + y - 7 = 0$. Then $y = 7 - 3t$, and the parametric form is

$$x = t$$
$$y = 7 - 3t$$

A third solution is obtained if x is replaced by $1 - 2t$. Then $y = 4 + 6t$, and another set of parametric equations is

$$x = 1 - 2t$$
$$y = 4 + 6t$$

All these pairs of equations when graphed will be $\overleftrightarrow{P_1P_2}$ in Fig. 4-8. However, care must be exercised in assigning a parameter. If, for example, x is replaced by t^2 in $3x + y - 7 = 0$, then $y = 7 - 3t^2$. The parametric form is

$$x = t^2$$
$$y = 7 - 3t^2$$

The graph of this system is not $\overleftrightarrow{P_1P_2}$, but $\overrightarrow{P_1P_2}$ (Fig. 4-8), because $x = t^2$ allows only nonnegative values of x. Similarly, if x is replaced by $\sin\theta$, then

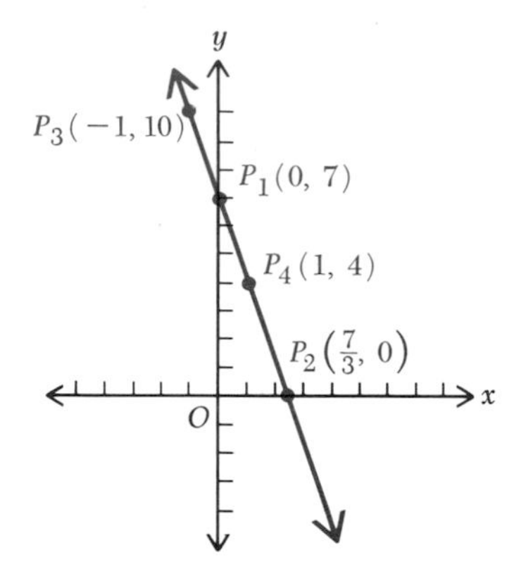

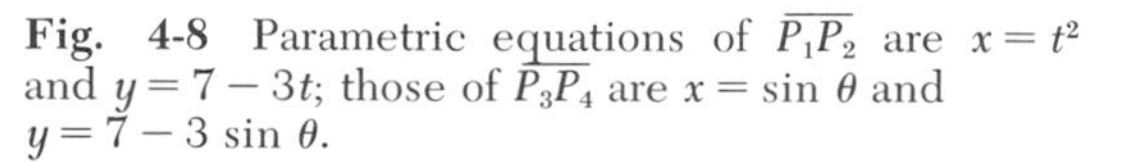
Fig. 4-8 Parametric equations of $\overline{P_1P_2}$ are $x = t^2$ and $y = 7 - 3t$; those of $\overline{P_3P_4}$ are $x = \sin\theta$ and $y = 7 - 3\sin\theta$.

$$x = \sin\theta$$
$$y = 7 - 3\sin\theta$$

and in this parametric representation the domain for x is satisfied by the inequality $-1 \leqslant x \leqslant 1$ for x a real number. These parametric equations represent $\overline{P_3P_4}$ in Fig. 4-8.

4-12 (OPTIONAL) APPLICATIONS OF PARAMETRIC EQUATIONS

As suggested in Sec. 4-11, it is often practical to study the motion of an object by means of parametric equations. For example, if a train is moving on a straight track at a constant speed, it may be thought of as at a particular point at a given time t. Then its actual position may be expressed by a pair of numbers (x, y), where $x = g(t)$ and $y = h(t)$. It will then be possible to study components of the motion parallel to the x axis and parallel to the y axis. This may be best understood by studying some examples.

Example 1

Analyze the motion of the point described in the following table:

t	0	1	2	3
x	2	5	8	11
y	6	10	14	18

Here t is given in seconds, and x and y are given in the same linear units.

Solution The table is read "at $t = 0$, the point is at (2, 6)," "at $t = 1$ the point is at (5, 10)," and so on. As t increases the x coordinate increases and the y coordinate increases; hence the motion is up and to the right.

The motion is uniform (at a constant rate) because in each 1-sec interval the x coordinate increases by 3 and the y coordinate increases by 4. In any 1 sec, then, the particle moves 3 units to the right, parallel to the x axis, and 4 units up, parallel to the y axis (Fig. 4-9). It follows that $|P_1P_2| = \sqrt{3^2 + 4^2} = 5$. Then in 1 sec the particle moves 5 units along the line.

Because $x_2 - x_1 = 3$ and $y_2 - y_1 = 4$, the slope of the line on which the particle is moving is $\frac{4}{3}$. Its equation is

$$y = \tfrac{4}{3}x + b$$

or, substituting the point (2, 6) and calculating b

$$y = \tfrac{4}{3}x + \tfrac{10}{3}$$

Pairs of direction numbers for the line (from the table) are as follows:

$P_1(2, 6), P_2(5, 10)$	$P_1(2, 6), P_2'(8, 14)$	$P_1(2, 6), P_2''(11, 18)$	$\cdots$
$x_2 - x_1 = 3$	$x_2 - x_1 = 6$	$x_2 - x_1 = 9$	$\cdots$
$y_2 - y_1 = 4$	$y_2 - y_1 = 8$	$y_2 - y_1 = 12$	$\cdots$

In writing parametric equations of the motion a pair of direction numbers must be chosen. It is convenient to choose the direction numbers for a unit interval of t, in this case $x_2 - x_1 = 3$ and $y_2 - y_1 = 4$. Now the

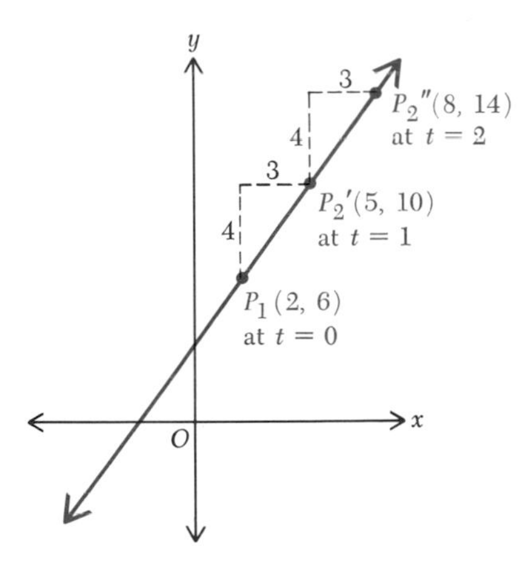

Fig. 4-9 Positions of a point moving along the line whose parametric representation is $x = 2 + 3t$ and $y = 6 + 4t$.

point $P_1(2, 6)$ is chosen for $t = 0$, and the parametric equations of motion are

$$x = 2 + 3t$$
$$y = 6 + 4t$$

where t is the elapsed time. The conclusions above may be verified readily from these parametric representations.

Example 2

A point travels left to right as time increases along the line whose equation is $3x + 2y - 2 = 0$ at a rate of 5 units/sec and passes through $P_2(4, -5)$ when $t = 2$. Write the parametric equations for its position at any time t.

Solution It is known that in 1 sec the point moves 5 units (Fig. 4-10), so $\overline{P_1P_2}$ and $\overline{P_2P_3}$ are each 5 units in length. The point is moving from P_1 to P_2 to P_3, because a pair of direction numbers for $\overrightarrow{P_1P_2}$ are $x_2 - x_1 = 2$ and $y_2 - y_1 = -3$. The direction cosines of $\overrightarrow{P_1P_2}$ are

$$\lambda = \frac{2}{\sqrt{2^2 + (-3)^2}} = \frac{2}{\sqrt{13}} \qquad \text{and} \qquad \mu = \frac{-3}{\sqrt{13}}$$

The displacement in the x direction per unit of time is 5 (displacement along the line) times λ, and in the y direction is 5 times μ. Hence a pair of direction numbers for the parametric representation of the motion are

$$x_2 - x_1 = 5\left(\frac{2}{\sqrt{13}}\right) \qquad \text{and} \qquad y_2 - y_1 = 5\left(-\frac{3}{\sqrt{13}}\right)$$

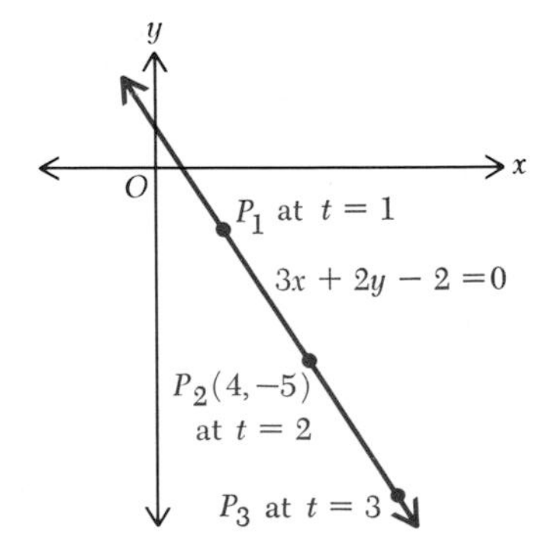

Fig. 4-10 The position of a point moving uniformly on $3x + 2y - 2 = 0$ at a rate of 5 units/sec.

Since the point goes through $(4, -5)$ at $t = 2$, the parametric equations are

$$x = 4 + \frac{10}{\sqrt{13}}(t-2)$$
$$y = -5 - \frac{15}{\sqrt{13}}(t-2)$$

PROBLEMS

1. Which of these pairs of parametric equations represent the same line?

$$l_1:\ x = 4 + 2t,\ y = 1 - 3t \qquad l_2:\ x = 6 - 2t,\ y = 2 + 3t \qquad l_3:\ x = 2 + 4t,\ y = 4 - 6t$$

2. Write parametric equations of a line through the pairs of points:

   (*a*) $P_1(3, 7)$, $P_2(-2, 4)$ (*b*) $P_1(0, 0)$, $P_2(2, -5)$
   (*c*) $P_1(-1, -3)$, $P_2(-2, 8)$ (*d*) $P_1(4, 6)$, $P_2(4, -7)$

3. Write another pair of parametric equations for each line of Prob. 2.
4. Each of the following pairs of parametric equations represents a line. Write the equation of each line in the form $Ax + By + C = 0$:

   (*a*) $x = 3 - 4t$, $y = 2 + t$ (*b*) $x = -2 - 5s$, $y = -4 + 8s$ (*c*) $x = 3(2 - t)$, $y = -4(3t + 1)$

5. Write a set of parametric equations representing the same line as each of the following:

   (*a*) $3x - y = 5$ (*b*) $7x + 6y = -4$
   (*c*) $-4x + 3y = 12$ (*d*) $a_1x + b_1y + c_1 = 0$
   (*e*) $x + y = 10$ (*f*) $y = -4$

6. Write a pair of parametric equations for each of the following lines:

   (*a*) l_1 has x and y intercepts of -3 and 2, respectively.
   (*b*) l_2 has slope of $-\frac{3}{2}$ and passes through $P(-4, 1)$.
   (*c*) l_3 has direction cosines of $\lambda = -\frac{1}{2}$ and $\mu = \frac{3}{2}$ and passes through $(3, -5)$.
   (*d*) l_4 is parallel to the line $3x - 4y + 7 = 0$ and passes through $(1, -4)$.
   (*e*) l_5 has direction numbers $r = -3$ and $s = 4$ and passes through $(-7, 2)$.

7. Draw the graph of each of the following parametric equations. Name the graph as a line, line segment, or ray if possible.

   (*a*) $x = 4$, $y = t^2$ (*b*) $x = \sin^2 t$, $y = \cos^2 t$ (*c*) $x = \sec^2 \theta$, $y = \tan^2 \theta$

(*d*) $x = 1 - 3t$, $y = 4 + 5t$ (*e*) $x = \log t$, $y = -\log t$ (*f*) $x = 2^t$, $y = -2^t$

8. Parametric equations for a line l_1 are $x = 5 + 2t$ and $y = -4 - t$ and those of l_2 are $x = -3 + 2t$ and $y = 5 + 3t$. (*a*) Find the coordinates of points P_1 on l_1 and P_2 on l_2 when $t = 3$. (*b*) Determine $|P_1P_2|$.
9. Point P_1 is on line l_1 whose parametric equations are $x = a + bt$ and $y = c + dt$ and point P_2 is on line l_2 whose parametric equations are $x = e + ft$ and $y = g + ht$. Express $|P_1P_2|$ in terms of the constants $a, b, c, \ldots, h$ in these equations for $t = t_1$.
10. Line l is represented by the parametric equations $x = a + bt$ and $y = c + dt$. Points P_1 and P_2 are on l; $P_1(x_1, y_1)$ is the point where $t = t_1$, and $P_2(x_2, y_2)$ is the point where $t = t_2$. Find $|P_1P_2|$.
11. Each of the following tables describes the motion of a point:

Point *A*

t	0	1	2	3
x	4	1	−2	−5
y	8	12	16	20

Point *B*

t	0	1	2	3
x	0	12	24	36
y	−3	2	7	12

For each table answer the following questions: (*a*) In what direction is the point moving? (*b*) In a unit time, how many units does the point move parallel to the x axis and parallel to the y axis? (*c*) How do you know that this is a uniform motion along a straight line? (*d*) How many units does the point move along the line in a unit of time? (*e*) What is the equation, in rectangular form, of the line along which the point is moving? (*f*) Name a set of parametric equations for the position of the point at any time t.
12. Write the set of parametric equations for the position of the point at the origin at $t = 0$ and moving uniformly up and to the right at a rate of 4 units/sec when the angle of inclination of its path is 45°.
13. Parametric equations of the paths of three different particles are

(*a*) $x = -2 + 7t$, $y = 5 + 24t$ (*b*) $x = 2$, $y = 6t$ (*c*) $x = 1 - 3t$, $y = -4 + t$

where t represents time and the unit is 1 sec. Discuss each motion, including its direction, speed along the line, and where each particle is at time $t = 5$.
14. Point $P(x, y)$ moves uniformly to the right at 6 units/sec. It starts on the y axis and is always 3 units above the x axis. Write the parametric representation of its position at any time t.
15. A point $P(x, y)$ moves uniformly in the xy plane in such a way that its displacement parallel to the x axis is 5 units/sec and its displacement parallel to the y axis is −4 units/sec. If it is at A(−2, 3) when $t = 0$, what are the parametric equations of its position at any time?
16. A point $P(x, y)$ moves uniformly in such a way that the absolute values of the

direction numbers of the equation of its path are equal. Draw a picture to show all possible directions in which the point may be moving.

17. Write the parametric equations of the position of a point $P(x, y)$ which is at $A(6, 0)$ at $t = 0$ and moves uniformly so that 10 sec later it passes through $B(6, 8)$.
18. A point $P(x, y)$ is moving upward to the right along the line whose equation is $4x - 3y + 6 = 0$ at a uniform rate of 8 units/sec, and it passes through $A(0, 2)$ at $t = 2$. Write the parametric representation of its position at any time t.
19. A point $P(x, y)$ is moving uniformly in a straight line from $P_1(2, -18)$ toward $P_2(23, 54)$. Its speed is 25 units/sec. Write the parametric equations of its path if it is at $P_1(2, -18)$ when $t = 0$.
20. A point $P(x, y)$ moves linearly from $P_1(4, -5)$ toward $P_2(7, 0)$ at a rate of 6 linear units/sec. If it is at P_2 when $t = -3$, what are the parametric equations of its path?
21. A point $P(x, y)$ moving at a uniform rate passes through $A(-5, 5)$ at $t = 0$ and through $B(4, -7)$ at $t = 3$. With t as the number of minutes elapsed, write the parametric equations of the position of the point at any time.
22. Write the parametric equations of the position of a point that moves along the line $x + 3y - 6 = 0$ at a uniform rate of 6 units/sec and passes through $A(3, 1)$ at $t = 3$.

4-13 POLAR EQUATIONS REPRESENTING STRAIGHT LINES

The general equation of first degree in x and y is $ax + by = c$. To find the corresponding equation in polar coordinates x may be replaced by $\rho \cos \theta$ and y by $\rho \sin \theta$. Thus

$$a \rho \cos \theta + b \rho \sin \theta = c$$

or

$$\rho = \frac{c}{a \cos \theta + b \sin \theta} \tag{4-10}$$

is its polar equation.

There are three special cases. In terms of $ax + by = c$, they are as follows:

1. $c = 0$ and both a and b are not equal to 0. The graph of this equation is a line passing through the origin. In polar form its equation is $\theta = k$, where k is a constant.
2. $a \neq 0$ and $b = 0$. The equation becomes $ax = c$ and is graphed as a line parallel to the y axis. If the x intercept of such a line is d, its equation is $x = d$; then the polar form is $\rho \cos \theta = d$ (Fig. 4-11).
3. $a = 0$ and $b \neq 0$. The equation becomes $by = c$ and is graphed as a line parallel to the x axis. If such a line has a y intercept e, its equation is $y = e$ and in polar form is $\rho \sin \theta = e$.

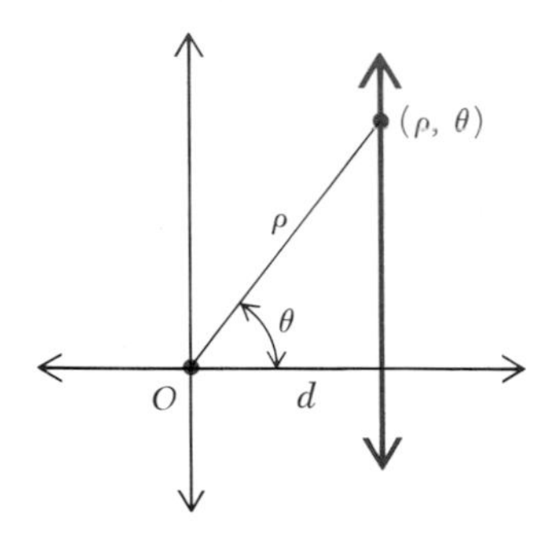

Fig. 4-11 Equation of a line perpendicular to the polar axis and d units to the right of O is $\rho \cos \theta = d$.

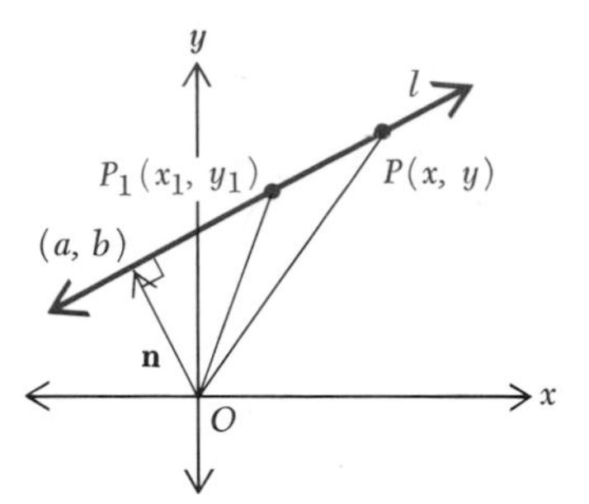

Fig. 4-12 Vector equation of a line passing through $P_1(x_1, y_1)$ and perpendicular to $\mathbf{n}$ is $(\mathbf{OP} - \mathbf{OP}_1) \cdot \mathbf{n} = 0$.

4-14 VECTOR FORM OF THE EQUATION OF A LINE

In addition to the many forms of the equation of the straight line already developed, one more is significant, since it employs the notation of vectors. Let $P(x, y)$ be a point on line l, with $P_1(x_1, y_1)$ as a specific point and $\mathbf{n}$ a vector with initial point at the origin and normal to line l (Fig. 4-12). The vector $\mathbf{OP} - \mathbf{OP}_1$ is perpendicular to $\mathbf{n}$ for $\mathbf{n} \neq 0$; hence

$$(\mathbf{OP} - \mathbf{OP}_1) \cdot \mathbf{n} = 0 \tag{4-11}$$

This is a vector form of the equation of line l.

A second form of the vector equation of a line may be derived using the concept of *unit vectors*. A unit vector has a magnitude of 1. A vector may be written as a linear combination of unit vectors along the coordinate axes. In Fig. 4-13 $\mathbf{OP} = \mathbf{r} + \mathbf{s}$. But $\mathbf{r}$ may be expressed as the product $x\mathbf{i}$, where $\mathbf{i}$ is a unit vector along the x axis. Similarly, $\mathbf{s} = y\mathbf{j}$, where $\mathbf{j}$ is a unit vector along the y axis. Then $\mathbf{OP} = x\mathbf{i} + y\mathbf{j}$.

In Eq. (4-11) $\mathbf{OP}_1$ may also be expressed in unit vectors as $x_1\mathbf{i} + y_1\mathbf{j}$, and $\mathbf{n}$ may be expressed as $a\mathbf{i} + b\mathbf{j}$. Thus a second form for the vector equation of a line is

$$[(x\mathbf{i} + y\mathbf{j}) - (x_1\mathbf{i} + y_1\mathbf{j})] \cdot (a\mathbf{i} + b\mathbf{j}) = 0 \tag{4-12}$$

It is interesting to see how this expression may be transformed into an equation of a line in rectangular coordinates. Since $\mathbf{i} \cdot \mathbf{i} = \mathbf{j} \cdot \mathbf{j} = 1$ and $\mathbf{i} \cdot \mathbf{j} = 0$ (the dot product of two perpendicular vectors is zero),

$$[(x\mathbf{i} + y\mathbf{j}) - (x_1\mathbf{i} + y_1\mathbf{j})] \cdot (a\mathbf{i} + b\mathbf{j}) = ax - ax_1 + by - by_1 = 0$$

or

$$a(x - x_1) + b(y - y_1) = 0$$

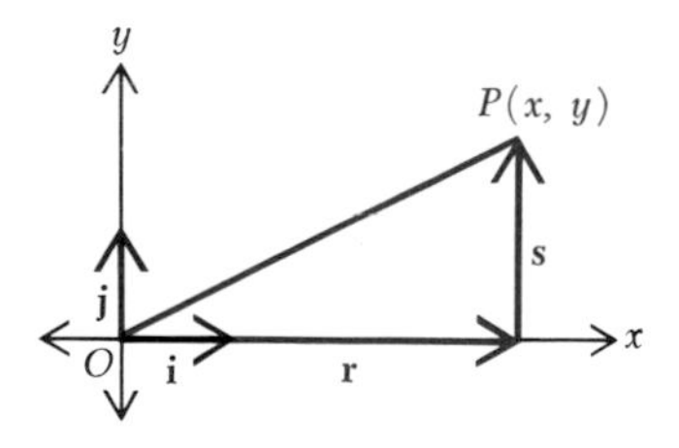

Fig. 4-13 **i** and **j** are unit vectors; $\mathbf{r} = x\mathbf{i}$ and $\mathbf{s} = y\mathbf{j}$: $\mathbf{OP} = \mathbf{r} + \mathbf{s} = x\mathbf{i} + y\mathbf{j}$.

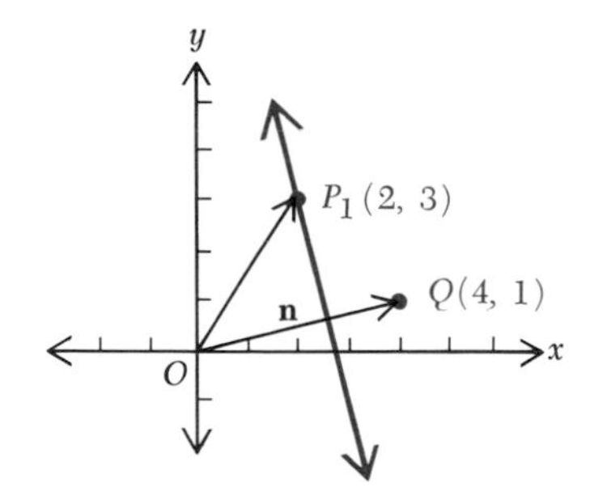

Fig. 4-14 Line passing through $P_1(2, 3)$ perpendicular to vector **OQ** has an equation $(x - 2)\mathbf{i} + (y - 3)\mathbf{j} \cdot (4\mathbf{i} + \mathbf{j}) = \mathbf{0}$.

Example

Find the equation in vector form of the line through $P_1(2, 3)$ and perpendicular to a vector with initial point at the origin and terminal point $Q(4, 1)$.

Solution In Fig. 4-14, $\mathbf{OP}_1 = 2\mathbf{i} + 3\mathbf{j}$ and $\mathbf{n} = 4\mathbf{i} + \mathbf{j}$; so one form for the equation of the line is, from Eq. (4-12),

$$[(x\mathbf{i} + y\mathbf{j}) - (2\mathbf{i} + 3\mathbf{j})] \cdot (4\mathbf{i} + \mathbf{j}) = 0$$

or

$$[(x - 2)\mathbf{i} + (y - 3)\mathbf{j}] \cdot (4\mathbf{i} + \mathbf{j}) = 0$$

PROBLEMS

1. Derive the polar equations $\theta = k$ and $\rho \sin \theta = e$ for lines passing through the origin and parallel to the x axis, respectively.
2. Which of the following equations graph as straight lines?

(a) $\rho(\cos \theta + \sin \theta) = 3$ (b) $\rho = \dfrac{12}{2 \cos \theta - 3 \sin \theta}$

(c) $\rho \cos \theta + 4 = 0$ (d) $\rho + \rho(3 \sin \theta - \cos \theta) = 0$

(e) $\rho + 3 \sin \theta = 0$ (f) $\rho + 2 \cos \theta = 3 \sin \theta$

(g) $\rho \left[\cos \left(\theta - \dfrac{\pi}{4}\right)\right] = 0$ (h) $\rho = \dfrac{8}{2 + \sin \theta}$

3. Graph all the equations that represent straight lines in Prob. 2.
4. Every point on a line is at a distance of 6 sec θ from the origin, where θ is the angle formed by the positive x axis and the ray connecting the origin and the point on the line. Write the equation of the line and graph it.

5. For each point P express **OP** in terms of unit vectors:

   (*a*) $P(3, 5)$ (*b*) $P(3, 8)$ (*c*) $P(2, 9)$ (*d*) $P(7, 10)$

6. Write a vector form for the equation of each of the following lines:

   (*a*) l_1 through (4, 3) and perpendicular to a vector from the origin to (3, −1)
   (*b*) l_2 through (−1, 4) and perpendicular to a vector from the origin to (−3, −4)
   (*c*) l_3 through (7, 6) and perpendicular to a vector from the origin to (−1, −2)
   (*d*) l_4 through (3, 9) and perpendicular to a vector from the origin to (2, 5)

4-15 SYSTEMS OF LINES

In the equation $y = mx + b$ let the value of m be fixed, say at $m = 2$, but leave the value of b unspecified. The equation

$$y = 2x + b$$

defines a line with slope 2 for each value assigned to b. Furthermore, every line with slope 2 is represented by this equation. It is therefore said to represent the *family of lines*, or *system of lines*, with slope 2. Several are shown in Fig. 4-15.

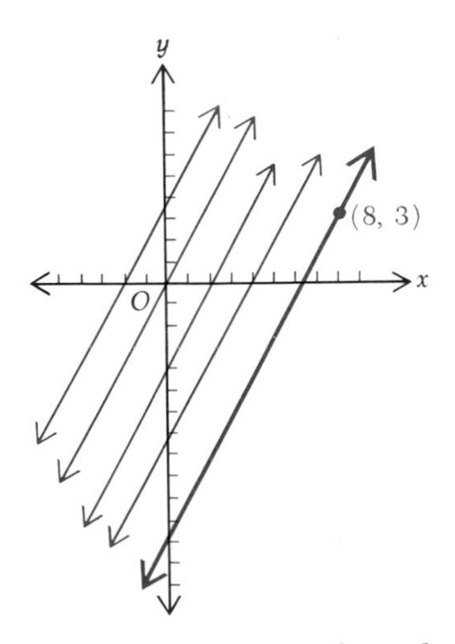

Fig. 4-15 Some members of a system of lines with a slope of 2: equation of the system is $y = 2x + b$.

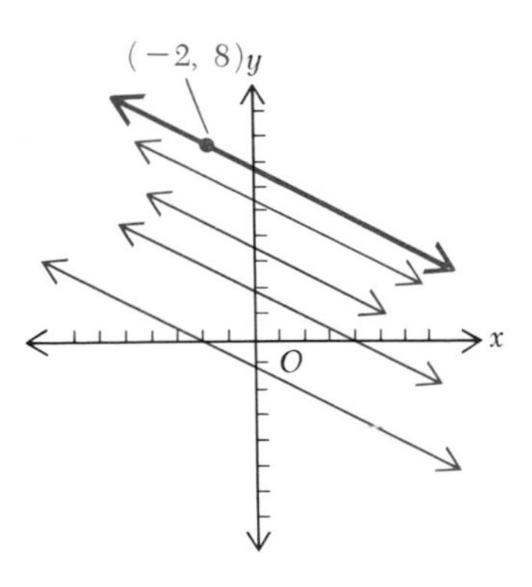

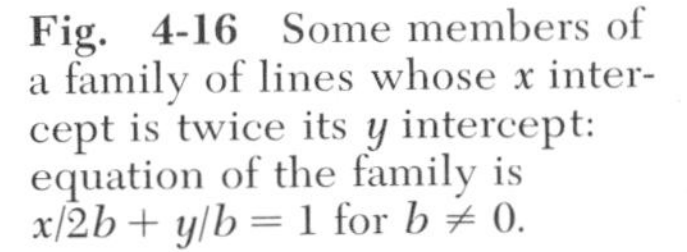

Fig. 4-16 Some members of a family of lines whose x intercept is twice its y intercept: equation of the family is $x/2b + y/b = 1$ for $b \neq 0$.

It is possible, by determining the proper value of b, to find the equation of the line with slope 2 which also satisfies one other condition.

Example 1
Find the equation of the line with slope 2 which passes through the point $A(8, 3)$.

Solution Since the slope of the line is 2, the equation of the line takes the form

$$y = 2x + b$$

where b is to be determined. Now, if the line passes through $A(8, 3)$ (Fig. 4-15), the following must hold:

$$3 = 2(8) + b \qquad \text{or} \qquad b = -13$$

The required equation is then

$$y = 2x - 13$$

In general, to find the equation of the line that satisfies two given conditions, start by writing the equation of the system of lines which satisfy one of these conditions. This equation will contain one arbitrary constant, called the *parameter of the system.* Then proceed to determine the value of this parameter which satisfies the second condition.

Example 2
Find the equation of the line with an x intercept equal to twice the y intercept and which passes through the point $A(-2, 8)$.

Solution An obvious solution is, of course, the line through $A(-2, 8)$ and the origin, since both intercepts are then zero. Aside from this case, all lines having x intercept equal to twice the y intercept are defined by the equation

$$\frac{x}{2b} + \frac{y}{b} = 1 \qquad \text{for } b \neq 0$$

It is required to find the member of this family (see Fig. 4-16) for which $y = 8$ when $x = -2$; this means that

$$\frac{-2}{2b} + \frac{8}{b} = 1 \qquad \text{or} \qquad b = 7$$

The desired equation is then

$$\frac{x}{14} + \frac{y}{7} = 1 \qquad \text{or} \qquad x + 2y = 14$$

4-16 EQUATIONS REPRESENTING TWO OR MORE LINES

The graphs of the equations

$$x - y = 0$$

and

$$x + 6y - 6 = 0$$

are, respectively, the lines l_1 and l_2 in Fig. 4-17. If the left-hand members of the equations are multiplied, and the right-hand members are also multiplied, the result is

$$(x - y)(x + 6y - 6) = 0 \tag{4-13}$$

or

$$x^2 + 5xy - 6y^2 - 6x + 6y = 0 \tag{4-14}$$

The left-hand product in (4-13) is equal to zero for all pairs (x, y) that make either factor zero and for no other values. Consequently, the graph of Eq. (4-13), which is the same as that of Eq. (4-14), consists of the two lines l_1 and l_2 in Fig. 4-17.

More generally, if the graph of an equation $g(x, y) = 0$ is a curve C_1 and the graph of an equation $h(x, y) = 0$ is a curve C_2, then the graph

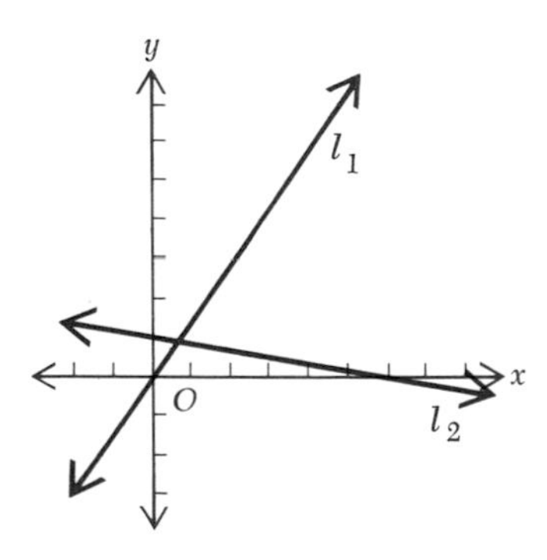

Fig. 4-17 The graph of $x^2 + 5xy - 6y^2 - 6x + 6y = 0$ or an equivalent equation $(x - y)(x + 6y - 6) = 0$.

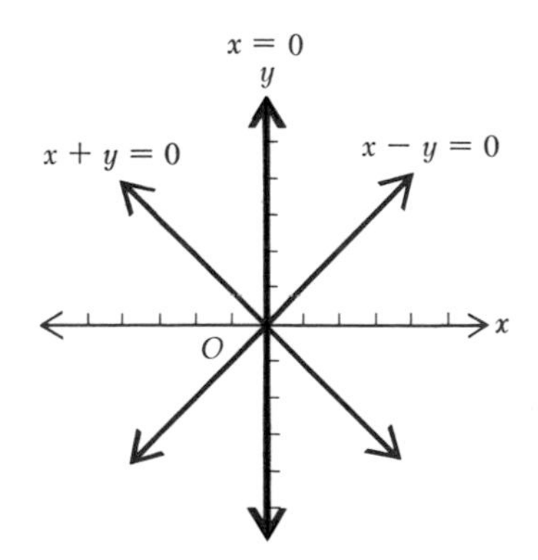

Fig. 4-18 The graph of $x^3 - xy^2 = 0$ or an equivalent equation $x(x - y)(x + y) = 0$.

of the equation

$$g(x, y)h(x, y) = 0$$

consists of the curves C_1 and C_2. This statement may be generalized to any number of factors. It follows that if the left-hand member of an equation of the form

$$f(x, y) = 0$$

can be factored into real linear factors, the graph consists of the lines represented by each factor.

Example 1

Graph the equation $x^3 - xy^2 = 0$.

Solution The equation $x^3 - xy^2 = 0$ can be written in the form

$$x(x - y)(x + y) = 0$$

The graph then consists of the lines $x = 0$, $x - y = 0$, and $x + y = 0$ (Fig. 4-18).

Example 2

Graph the equation $x^3 - 2x^2y + xy^2 - 2y^3 = 0$.

Solution The equation $x^3 - 2x^2y + xy^2 - 2y^3 = 0$ can be written in the form

$$(x^2 + y^2)(x - 2y) = 0$$

Its graph, therefore, is precisely that of the linear equation $x - 2y = 0$. Why?

We should perhaps comment at this point that instead of speaking of *the* equation of a certain line or curve, it would be more appropriate to speak of *an* equation. As indicated in Example 2, the line which has the equation $x - 2y = 0$ also has the equation $x^3 - 2x^2y + xy^2 - 2y^3 = 0$ and an unlimited number of others. The circle that has the equation $x^2 + y^2 = 25$ has also the equation $(x^2 + y^2 - 25)^3 = 0$, and so on. We shall continue to speak of "the equation" of a given curve even though it is understood that other equations could be devised that would have precisely the same locus when graphed.

PROBLEMS

1. Draw several members of each of the following systems of lines and determine what property is possessed by all members of the system:

   (a) $2y = 3x + b$ (b) $4x + by = 12$
   (c) $(y - 3) = a(x + 2)$ (d) $y = k(x - 3)$
   (e) $\frac{x}{a} + \frac{y}{2a} = 1$ (f) $\frac{x}{b+4} + \frac{y}{b} = 1$

2. Write the equation of the family of lines perpendicular to the line that passes through $A(-2, -3)$ and $B(6, 3)$; then find the member of this system that has y intercept equal to 4.
3. Write the equation of the family of lines passing through $P(6, 4)$; then find the member of this system that passes through the point $A(2, -5)$.
4. Write the equation of the system of lines through the point $P(-3, 6)$; then find the member of this system that has x intercept equal to 9.
5. Write the equation of the system of lines each of which has y intercept equal to the negative of its x intercept; then find members of this system for which the length of the perpendicular drawn from the origin to the line is equal to $\sqrt{50}$.
6. Write the equation of the system of lines through $P(6, 4)$, then find the member (or members) of the system having the sum of its two intercepts equal to 20.
7. Write the equation of the system of lines with y intercept equal to -4. Sketch the curve $y = x^2$ and find the members of this system that are tangent to the curve. *Hint:* Solving $y = mx - 4$ and $y = x^2$ simultaneously leads to the equation $x^2 - mx + 4 = 0$. The two roots of this quadratic equation must be equal for tangency.
8. Write the equation of the system of lines with slope -2. Sketch the curve $y = x^2 - 8x + 18$ and find the members of this system that are tangent to the curve (see the hint in Prob. 7).
9. Show that if the lines $A_1x + B_1y + C_1 = 0$ and $A_2x + B_2y + C_2 = 0$ intersect at a point P, then the equation

$$(A_1x + B_1y + C_1) + k(A_2x + B_2y + C_2) = 0$$

   where k is any real number, represents a line through P. *Hint:* Observe that the equation is of first degree in x and y and is satisfied by the coordinates of P.

10. Write the equation of the system of lines through the point of intersection of the lines $3x + 4y - 6 = 0$ and $x - 2y + 10 = 0$. Find the member of the system that passes through the point $A(-2, 1)$ (use the result of Prob. 9).
11. Find the intersection of each of the following pairs of lines:

   (a) $x = 3 + 2t$, $y = 6 - 4t$ and $x = 8 - 3s$, $y = -3 + 5s$
   (b) $x = 5 + 2t$, $y = -1 + t$ and $x = 6 - 3s$, $y = -8 + 6s$

12. Show that the graphs of the two equations $x + y = 0$ and $x^2 + 2xy + y^2 = 0$

are identical. *Hint:* Prove that the second equation is satisfied by every pair of real numbers (x, y) that satisfies the first and is not satisfied by any other pair of real numbers.

13. In what way does the graph of the equation $(x^2 + y^2)(x^2 + y^2 - 25) = 0$ differ from that of the equation $x^2 + y^2 = 25$?

For Probs. 14 to 18 draw the graph of the given equation after factoring the left-hand member.

14. $x^2 - 6x = 0$
15. $y^2 - 9 = 0$
16. $9x^2 - 4y^2 = 0$
17. $x^2 - y^2 + 2x - 2y = 0$
18. $xy + 2x - y - 2 = 0$
19. A function of the form $ax + b$ is called a *linear function of* x, and if some quantity Q varies with respect to x in accordance with the formula

$$Q = ax + b$$

then Q is said to *vary linearly with* x. Show that a and b can be interpreted such that b is the value of Q when $x = 0$ and a is the number of units by which the value of Q changes for each unit of change in x. (Thus, if $Q = -\frac{1}{2}x + 6$, then $Q = 6$ when $x = 0$, and Q decreases by $\frac{1}{2}$ unit for each unit of increase in x.)

20. The coefficient of expansion (change in length per unit length per degree change in temperature) for steel is 0.0000065 for the Fahrenheit scale. A steel rod is 30 ft long at 50°F. Express its length L at t°F as a function of t and draw the graph of length versus temperature. What are the physical interpretations of its slope and its L intercept (see Prob. 19)?

4-17 SIMULTANEOUS LINEAR INEQUALITIES

In Chap. 3 the locus of the inequality $3x + 4y > 12$ was graphed as the open half-plane shown shaded in Fig. 4-19*a*. Furthermore, the locus

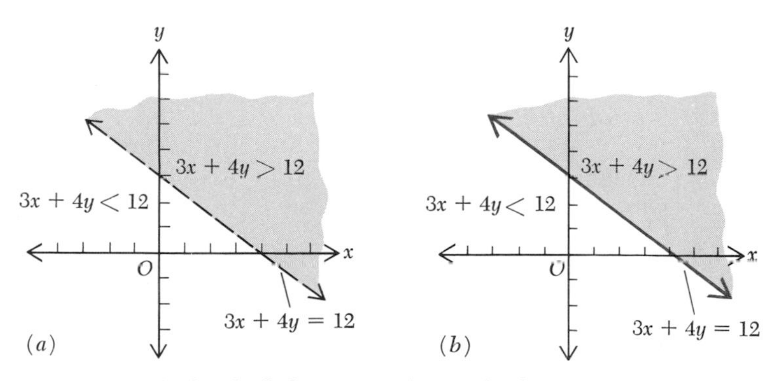

Fig. 4-19 (*a*) The shaded region is the graph of $3x + 4y > 12$. (*b*) The line and the shaded region constitute the graph of $3x + 4y \geqslant 12$.

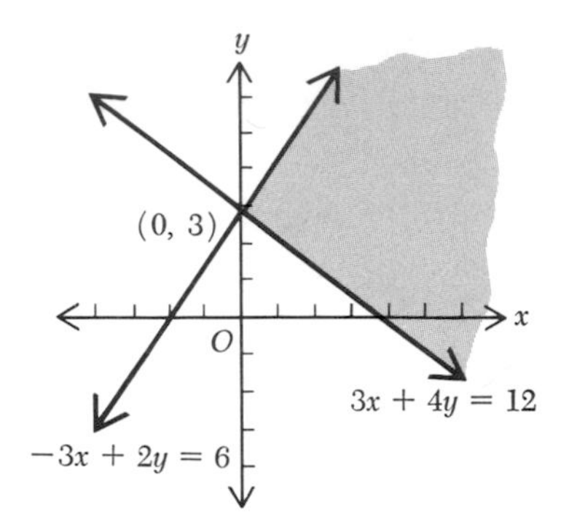

Fig. 4-20 The graph of $3x + 4y \geqslant 12$ and $-3x + 2y \leqslant 6$ is the shaded region and the rays that bound the region.

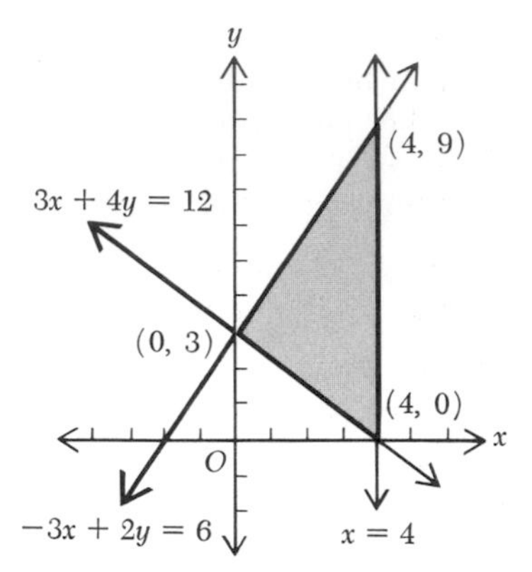

Fig. 4-21 The shaded triangular region is the graph of $\{(x, y) \mid (3x + 4y \geqslant 12) \cap (-3x + 2y \leqslant 6) \cap (x \leqslant 4)\}$.

of $3x + 4y \geqslant 12$ was determined as the union of the same open half-plane and the line whose equation is $3x + 4y = 12$ (Fig. 4-19*b*). This region is called a *closed half-plane.* The solution of the system

$$\begin{aligned} 3x + 4y &\geqslant 12 \\ -3x + 2y &\leqslant 6 \end{aligned} \tag{4-15}$$

is the set of ordered pairs in the intersection of the sets of ordered pairs in the solution for each inequality. The locus of the solution of this system is the shaded region and its boundaries in Fig. 4-20.

The graph of the simultaneous solution of three or more inequalities may be a *closed region.* Such a region is shaded in Fig. 4-21 for the system

$$\begin{aligned} 3x + 4y &\geqslant 12 \\ -3x + 2y &\leqslant 6 \\ x &\leqslant 4 \end{aligned}$$

This region is the intersection of three closed half-planes. The set of points P, depicted as a triangular region, may be expressed as

$$P = \{(x, y) \mid (3x + 4y \geqslant 12) \cap (-3x + 2y \leqslant 6) \cap (x \leqslant 4)\}$$

4-18 (OPTIONAL) CONVEX POLYGONAL REGIONS

The set of points P in Fig. 4-21 is often referred to as a *convex polygonal region.* Other examples of convex polygonal regions are shaded in Fig.

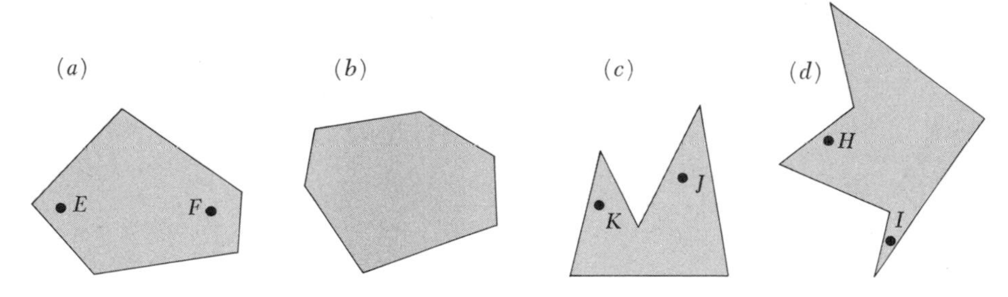

Fig. 4-22 (*a*) and (*b*) Convex polygonal regions. (*c*) and (*d*) Not convex polygonal regions.

4-22*a* and *b*. The shaded parts in Fig. 4-22*c* and *d* are not convex polygonal regions. The basic difference between regions that are convex polygonal and those that are not is explained by the following definition.

Definition

A convex polygonal region is a region such that for any two of its points, every point on the line segment joining these two points is in the region.

In Fig. 4-22*c* all points on $\overline{JK}$ are not in the shaded region. Similarly, in Fig. 4-22*d* all points on $\overline{HI}$ are not in the shaded region. However, for any choice of two points E and F in the region shaded in Fig. 4-22*a* every point on $\overline{EF}$ will be in the shaded region.

PROBLEMS

Graph the regions whose points are the solution sets in real numbers for the systems of simultaneous inequalities in Probs. 1 to 9.

1. $x + y < 3$
   $x - y > -1$
2. $y \leq 4x$
   $y > -2$
3. $2x - y \leq 3$
   $x + 2y \geq 8$
4. $x + 4y \geq 8$
   $3x - y \leq 6$
5. $3x + 2y \leq 12$
   $2x - y \geq 3$
   $y \geq -3$
6. $x \leq 0$
   $y \leq 0$
   $x + y \geq -4$
7. $x + y \geq 6$
   $x - y \leq 6$
   $x - y \geq -6$
8. $3x + 5y \leq 21$
   $x - 3y \geq -3$
   $x + 7y \geq 7$
9. $3x - 2y \geq -4$
   $x - 2y \leq -2$
   $x - 4y \geq -8$
10. Which of the regions in Probs. 1 to 9 are convex polygonal regions?

11. Graph the following inequalities, where the domain is the set of integers:

$$x + y < 4$$
$$x > 0$$
$$y > 0$$

In Probs. 12 to 16 graph each set of points, where

$$P = \text{locus of } x - 2y \leqslant 8$$
$$Q = \text{locus of } 2x - y \geqslant 4$$
$$R = \text{locus of } y + x \geqslant 0$$
$$S = \text{locus of } -4x + 8y \geqslant 8$$
$$T = \text{locus of } 3x - y \leqslant 4$$

for the given subsets of the xy plane.

12. $P \cup S$
13. $Q \cap R$
14. $Q \cap R \cap S$
15. $(Q \cap R) \cap (S \cap T)$
16. $(P \cup S) \cap (Q \cap R)$

Graph the systems of inequalities in Probs. 17 to 19 and name each geometric figure.

17. $x + y = 100$, $x \geqslant 50$, $y \geqslant 25$

18. $x - y = 8$, $x \geqslant 8$, $y \geqslant 0$

19. $x > 0$, $y \ngtr 0$, $y \nless 0$

20. Write the system of inequalities whose graph is described by a triangular region ABC whose vertices are $A(3, 5)$, $B(2, -7)$, and $C(-4, 1)$.
21. Write the system of inequalities for a square region $ABCD$ whose vertices are $A(3, 5)$, $B(7, 9)$, $C(3, 13)$, and $D(-1, 9)$.
22. Write the system of inequalities for a right triangular region in quadrant I with one vertex at $A(1, 2)$, another at $B(5, 2)$, and a point on the hypotenuse at $C(4, \frac{17}{4})$.
23. A club wants to buy two kinds of favors for a party. One type costs \$2 each and the other costs \$1 each. They want to buy at least one of each kind but must spend less than \$8. How many of each kind of favor can they buy?
24. Answer Prob. 23 if the club wishes to spend at most \$8.

4-19 (OPTIONAL) LINEAR PROGRAMMING

The most useful and interesting application of a system of inequalities that defines a convex polygonal set is linear *programming*. As an example of this, suppose a manufacturing firm wants to plan its production for maximum profit. It produces two items, J and K, with a profit of \$2 and \$1, respectively, per item. But there are a number of practical considerations which limit the number of each item that can be manufactured, such as the number of hours required to produce each item and the availability of machines for producing or packaging. How many of each item should be manufactured for the profit to be a maximum?

Stated in mathematical terminology, the problem is: Given a linear function (an expression for profit) such as $5x + 2y + 5$, what pairs of numbers for x and y make the linear function a maximum and a minimum? Certain constraints or restrictions in the problem limit the values that can be substituted for x and y.

The method of solution is best illustrated by a number of examples. The first concerns the maximum and minimum values of a function defined for points on a line segment.

Example 1

The function $5x + 2y + 5$ is defined for all points on the segment of the line $y = 2x - 4$ between and including $(1, -2)$ and $(5, 6)$ (see Fig. 4-23). What are the maximum and minimum values of the function for this set of points?

Solution As a start, $5x + 2y + 5$ may be evaluated for points whose number pairs are integers. Then:

At $(1, -2)$, $5x + 2y + 5 = 5(1) + 2(-2) + 5 = 6$
At $(2, 0)$, $5x + 2y + 5 = 5(2) + 2(0) + 5 = 15$
At $(3, 2)$, $5x + 2y + 5 = 5(3) + 2(2) + 5 = 24$
At $(4, 4)$, $5x + 2y + 5 = 5(4) + 2(4) + 5 = 33$
At $(5, 6)$, $5x + 2y + 5 = 5(5) + 2(6) + 5 = 42$

Here the maximum and minimum values of $5x + 2y + 5$ occur at the end points $(5, 6)$ and $(1, -2)$ of the segment.

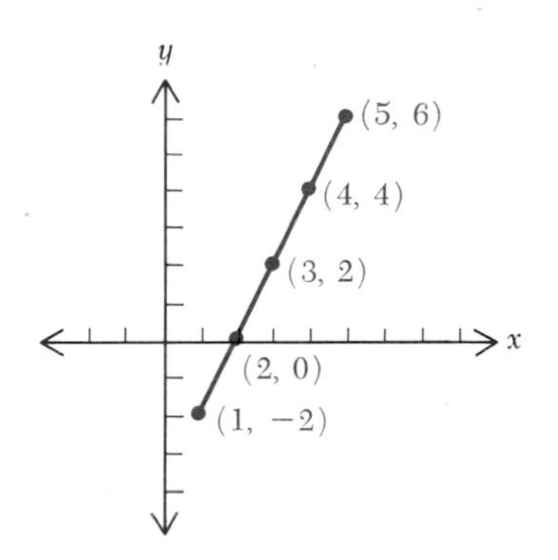

Fig. 4-23 Points with integers as coordinates on a line segment of $y = 2x - 4$.

Even if coordinates that are not integers are chosen for points on the line segment, the maximum and minimum values of $5x + 2y + 5$ will be obtained from coordinates for the end points. As a demonstration, consider a line segment on the line whose equation is $y = 2x - 4$. We wish to find number pairs (x, y) on this line which make the function $5x + 2y + 5$ a maximum. If $2x - 4$ is substituted for y in this function, then

$$5x + 2y + 5 = 5x + 2(2x - 4) + 5 = 9x - 3$$

This function increases as x increases and will be greatest for the greatest x and least for the least x. This agrees with the computation above, in which (5, 6), the number pair with the greatest x, produced a maximum and (1, −2), the number pair with the least x, produced a minimum.

These results can be generalized in the following theorem.

Theorem 4-3

If a linear function $ax + by + c$ is evaluated for coordinates of all points on a segment of $y = mx + b$, the maximum value of the function will be found for the coordinates of one end point and the minimum for the other.

The proof of this theorem may be completed by generalizing the argument for the special case in the preceding paragraph (see Prob. 15).

Example 2

If a given linear function is defined for all points on a convex polygon, the coordinates of what points on the polygon determine maximum and minimum values of the function?

Solution The question will be answered first for polygon $ABCDE$ in Fig. 4-24 and the linear function $-3x + 2y - 4$. With respect to line segment $\overline{AE}$ of the polygon, the maximum and minimum values of the function $-3x + 2y - 4$ are determined when the coordinates of the end points of segment $\overline{AE}$ are substituted in $-3x + 2y - 4$. This was established by Theorem 4.3. A similar argument may be made concerning maximum and minimum values of $-3x + 2y - 4$ for points on $\overline{AB}$, $\overline{BC}$, $\overline{CD}$, and $\overline{DE}$. Therefore the function is evaluated for coordinates of the vertices of polygon $ABCDE$:

At $A(4, -4)$, $-3x + 2y - 4 = -3(4) + 2(-4) - 4 = -24$
At $B(7, -4)$, $-3x + 2y - 4 = -3(7) + 2(-4) - 4 = -33$

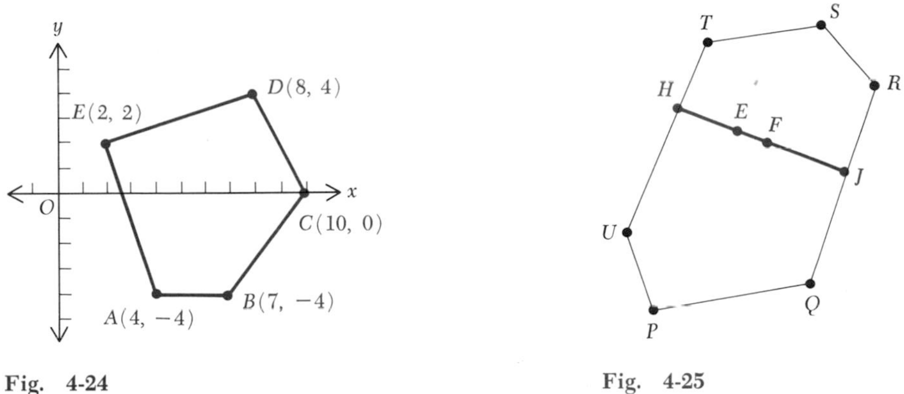

Fig. 4-24

Fig. 4-25

At $C(10, 0)$, $-3x + 2y - 4 = -3(10) + 2(0) - 4 = -34$
At $D(8, 4)$, $-3x + 2y - 4 = -3(8) + 2(4) - 4 = -20$
At $E(2, 2)$, $-3x + 2y - 4 = -3(2) + 2(2) - 4 = -6$

Hence, considering all points on the convex polygon $ABCDE$, the linear function $-3x + 2y - 4$ has a maximum of -6 at $E(2, 2)$ and a minimum of -34 at $C(10, 0)$.

The solution to this special case may be generalized by the following theorem.

Theorem 4-4
The maximum and minimum values of a linear function defined for all points on a convex polygon occur at vertices of the polygon.

The final aspect of the linear-programming problem is an investigation of whether points in the interior of the convex polygon determine maximum or minimum values of a linear function. The answer may be obtained by considering any two points E and F in the interior of convex polygon $PQRSTU$ in Fig. 4-25. $\overleftrightarrow{EF}$ intersects the polygon in two points, H and J. Now, every point on $\overline{HJ}$ is in the convex polygonal region $PQRSTU$. If a linear function is evaluated for points on $\overline{HJ}$, its maximum is determined by the coordinates of one end point and the minimum is determined by the coordinates of the other. Hence for any line segment with end points on the convex polygon, the maximum and minimum values of the linear function occur at points on the polygon. It has been shown, however, that for all points on a convex polygon the maximum and minimum values of a linear function occur at vertices of the polygon. This completes an analysis of the following basic theorem of linear programming.

Theorem 4-5

If a linear function $ax + by + c$ is defined for all points of a convex polygonal region, the maximum value of the function is determined by coordinates of one vertex and its minimum value by the coordinates of another vertex.

Example 3

In order to make two kinds of items, P and Q, two machines, I and II, are required. To make item P machine I runs for 1 hr and machine II runs for $\frac{8}{5}$ hr. To make item Q machine I runs 1 hr and machine II runs $\frac{4}{5}$ hr. Each machine can run 16 hr or less per day. On item P the company realizes a profit of \$30 and on item Q a profit of \$20. How many of each item should be produced per day for a maximum profit?

Solution For the conditions concerning number of items per day, let x be the number of P items per day for maximum profit, and y the number of Q items per day for maximum profit. Obviously, $x \geqslant 0$ and $y \geqslant 0$.

For the conditions concerning machines, machine I works $x(1)$ hours on item P and machine I works $y(1)$ hours on item Q; the machine works 16 hours per day or less. Hence $x + y \leqslant 16$.

A similar analysis shows that the restriction on machine II is

$$\tfrac{8}{5}x + \tfrac{4}{5}y \leqslant 16 \qquad \text{or} \qquad 2x + y \leqslant 20$$

The next step is to draw the convex polygonal region. The conditions

$$\begin{aligned} x &\geqslant 0 \\ y &\geqslant 0 \\ x + y &\leqslant 16 \\ 2x + y &\leqslant 20 \end{aligned}$$

determine the shaded region in Fig. 4-26.

The function to be maximized is $30x + 20y$. This is evaluated for the coordinates of vertices of the polygon:

At $(0, 0)$, $30x + 20y = 0 + 0 = 0$
At $(10, 0)$, $30x + 20y = 300 + 0 = 300$
At $(4, 12)$, $30x + 20y = 120 + 240 = 360$
At $(0, 16)$, $30x + 20y = 0 + 320 = 320$

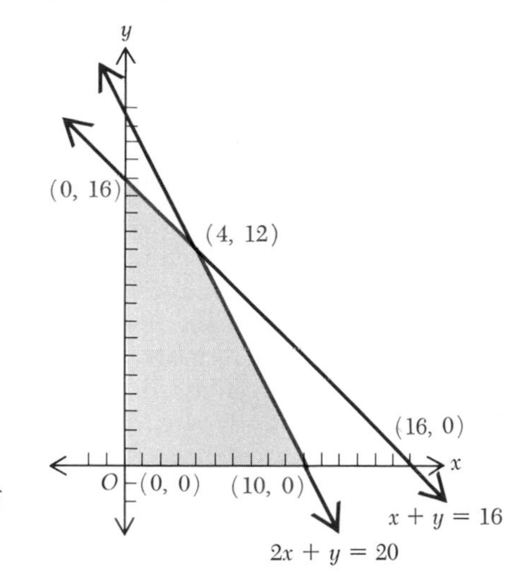

Fig. 4-26 The shaded region is the set of points $\{(x, y) \mid (x \geqslant 0) \cap (y \geqslant 0) \cap (x + y \leqslant 16) \cap (2x + y \leqslant 20)\}$.

The maximum profit is realized when 4 P items and 12 Q items are manufactured.

PROBLEMS

1. Construct the graph for the system of inequalities $x \geqslant 0, y \geqslant 0, x \leqslant 4, y \leqslant 4$, and $x + y \leqslant 6$.
2. Find the coordinates of vertices of the polygonal region defined in Prob. 1.
3. Evaluate the function $3x - 5y + 4$ at the vertices of the polygon in Prob. 3. What are the maximum and minimum values of the function?
4. For the region defined in Prob. 1 list points with integers as coordinates and evaluate the function $3x - 5y + 4$ at these points. Are the values so determined between the maximum and minimum found in Prob. 3?
5. A farmer has facilities on his ranch for 500 cows and horses. He wants no more than 400 cows or 200 horses. He can make a profit of \$50 on a cow and \$80 on a horse. How many of each should he raise for a maximum profit?
6. Find x and y to maximize the function $2x + 5y$ and find the maximum if x and y are subject to the restrictions $x \geqslant 0, y \geqslant 0, x \leqslant 8, y \leqslant 6$, and $x - y + 4 \geqslant 0$.
7. Find x and y to minimize the function $2x - 5y$ if x and y are subject to the restrictions $x \leqslant 2, y \geqslant -8, x + y \leqslant 5$, and $y - x \leqslant 5$.
8. A manufacturer produces models A and B of a device. To manufacture each model requires two machines, I and II. It takes 2 hr on machine I and 1 hr on machine II to manufacture model A. It requires 1 hr on machine I and 2 hr on machine II to manufacture model B. Each machine can be used no more than 12 hr/day. The profit on model A is \$30 each and on model B

is $50 each. How many of each model should be manufactured for a maximum profit?

9. A candy manufacturer produces two kinds of candy, Quality and Grade A. It takes 5 min of time on the preparing machine and 2 min on the packing machine to process a box of Quality candy. A box of Grade A candy is processed in 2 min on the preparing machine and 1 min on the packing machine. The preparing machine can work 12 hr or less per day, and while the packing machine is being repaired, it can run only 5 hr/day. What number of boxes of each kind of candy should be produced per day for a maximum profit if the profit per box of Quality candy is 36 cents and that per box of Grade A candy is 15 cents?
10. If the profit on Grade A candy in Prob. 9 were increased to 20 cents/lb, how many pounds of each kind of candy would be needed for a maximum profit?
11. In Prob. 9 if the profit on Quality candy were 40 cents/lb, how many pounds of each kind of candy would provide a maximum profit? How many hours would each machine be running under these conditions?
12. Mr. Hancock's physician prescribed a daily dosage of at most 15 units of vitamin P and at most 16 units of vitamin Q. Vitamin tablets are available at drug stores. Max tablets contain 1 unit of P and 4 units of Q, and Min tablets contain 3 units of P and 1 unit of Q. Max tablets cost 5 cents each and Min tablets cost 6 cents each. How many tablets of each should Mr. Hancock buy for a minimum cost each day if he decides to take the maximum number of the required vitamins?
13. For a final examination a teacher gave the class fifty 3-min problems (problems that supposedly could be solved in 3 min each), each worth 5 points, and fifty 2-min problems, each worth 4 points. Each pupil was to solve a maximum of 30 problems. The examination was to be for a maximum of 84 min. If a student were certain he could solve every problem he tried, how many of each should he solve for a maximum score?
14. Answer Prob. 13 if 60 min are allotted for the test.
15. Prove Theorem 4.3.

SUMMARY

A straight line is uniquely determined, and hence its equation may be found from certain conditions. Some of these are

1. Two points
2. One point and the slope of the line
3. One point and the direction cosines of the line

Five most often used equations of the line are summarized in the following table.

Conditions	Equation
Parallel to x axis through $Q(x_1, y_1)$	$y = y_1$
Parallel to y axis through $Q(x_1, y_1)$	$x = x_1$
Through $Q(x_1, y_1)$ with a slope of m	$y - y_1 = m(x - x_1)$
Through $Q(x_1, y_1)$ and $R(x_2, y_2)$	$y - y_1 = \dfrac{y_2 - y_1}{x_2 - x_1}(x - x_1)$
Having y intercept b and slope m	$y = mx + b$

The general form of the straight line is

$$Ax + By + C = 0$$

where A and B are not both 0. From this form considerable information about a line can be obtained:

1. Its slope is $-A/B$.
2. Its undirected distance from the origin is $|C|/\sqrt{A^2 + B^2}$.
3. Its undirected distance from $S(x_1, y_1)$ is $|Ax_1 + By_1 + C|/\sqrt{A^2 + B^2}$.
4. A line parallel to $Ax + By + C = 0$ is of the form $Ax + By + k = 0$, where k is determined from a further condition.
5. A line perpendicular to $Ax + By + C = 0$ is of the form $Bx - Ay + j = 0$, where j is determined from a further condition.
6. If the line passes through $P(x_1, y_1)$ and $Q(x_2, y_2)$, the direction numbers of $\overleftrightarrow{PQ}$ are $[x_2 - x_1, y_2 - y_1]$ or $[-(x_2 - x_1), -(y_2 - y_1)]$.

The parametric form of a line through $P(x_1, y_1)$ and $Q(x_2, y_2)$ is

$$x = x_1 + t(x_2 - x_1)$$
$$y = y_1 + t(x_2 - x_1)$$

where t is a variable called a parameter. By assigning numbers to t, coordinates (x, y) are obtained for graphing the line.

The polar form of $Ax + By + C = 0$,

$$p = \frac{-C}{A \cos \theta + B \sin \theta}$$

is seldom used but may be convenient when the distance of a point (on a line) from the origin is of major consideration.

A family of lines or a system of lines is a set of lines that meet a specific condition. For example, the family of lines passing through the point

(2, −9) is given by the equation

$$y + 9 = m(x - 2)$$

and the family of lines with a slope of −3 is given by the equation

$$y = -3x + b$$

While the graph of every equation linear in x and y is a straight line, a straight line may result from an equation that is not linear in x and y.

When a linear function in x and y is greater than or less than zero, the set of points is graphed as a region. When three or more inequalities are considered simultaneously, a polygon may bound the region. This concept is of major interest because of a recent application called linear programming.

REVIEW PROBLEMS

1. Write the equations of the following lines:

   (*a*) Passing through $A(-6, -1)$ and $B(4, -1)$
   (*b*) Intercepts at $A(5, 0)$ and $B(0, -3)$
   (*c*) The perpendicular bisector of $\overline{AB}$, with $A(3, -2)$ and $B(5, 6)$
   (*d*) Parallel to $7x - 4y = 3$ and passing through $A(2, -5)$
   (*e*) Passing through $P(2, -4)$, with x intercept a and y intercept $-a$
   (*f*) Containing the midpoint of $\overline{AB}$, with $A(-7, 2)$ and $B(3, 4)$, and having the same slope as $\overleftrightarrow{OA}$
   (*g*) Containing $A(-3, 8)$, with an inclination of $3\pi/4$

2. (*a*) Write the equation of a system of lines that passes through the intersection of the lines $5x + 3y + 2 = 0$ and $x - y - 2 = 0$. (*b*) Write the equation of the member of the system that has a slope of $-\frac{2}{3}$.
3. Find the undirected distance from the line $3x - 2y + 5 = 0$ to (*a*) the origin; (*b*) the point $Q(-3, 7)$.
4. (*a*) Write the equation of a line having a y intercept of 5 and a slope of −3. (*b*) Determine r if point $T(r, 4)$ is on the line.
5. Graph each of the following sets of points:

   (*a*) $\{(x, y) \mid 5x + 4y + 20 = 0\}$
   (*b*) $\{(x, y) \mid x = t^2 \text{ and } y = -t^2\}$
   (*c*) $\{(x, y) \mid \frac{3}{\sqrt{2}}x - \frac{5}{\sqrt{2}}y - 15\sqrt{2} = 0\}$
   (*d*) $\{(x, y) \mid 2x - 5y + 9 > 0\}$
   (*e*) $\{(x, y) \mid (4x - y + 11 \geq 0) \cap (x + 3y - 9 = 0)\}$
   (*f*) $\{(x, y) \mid x^2 = 16y^2\}$

6. Find two values of r such that the undirected distance from the line $4x - 3y - 24 = 0$ to the point $Q(r, 2)$ is 6.
7. The line $3x - y + 4 = 0$ is midway between two parallel lines which are 10 units apart. Find the equations of the two lines.
8. Express the following equations of straight lines in polar and parametric form:

   (*a*) $x + 4 = 0$ (*b*) $3x + 5y - 7 = 0$

9. Express in rectangular form and vector form the equation of a line which passes through point $P(2, -6)$ and is perpendicular to a vector determined by the origin and point $Q(-4, 1)$.
10. Graph the simultaneous solution of the system of inequalities

$$x \leqslant 3$$
$$6x - 7y + 17 \geqslant 0$$
$$3x + 7y + 19 \geqslant 0$$

FOR EXTENDED STUDY

1. Suppose line l_1 has the equation $a_1x + b_1y + c_1 = 0$ and line l_2 has the equation $a_2x + b_2y + c_2 = 0$. Neither line is horizontal or vertical. (*a*) Write the equation of l_3 so l_3 is perpendicular to l_1 and that of l_4 so l_4 is perpendicular to l_2. (*b*) If θ is the acute angle formed by l_1 and l_2 and ϕ the acute angle formed by l_3 and l_4, what is the relation between θ and ϕ? State your discovery as a theorem and prove it.
2. Write the equation of the line parallel to $ax + by + c = 0$ containing the point $P(x_1, y_1)$.
3. Use a determinant to show that if $P_1(x_1, y_1)$, $P_2(y_1, x_1)$, and $P(x, y)$ are three distinct points on a line, with $x_1 \neq y_1$, then the relations among x, y, x_1, and y_1 are given in the equation $x + y = x_1 + y_1$.
4. Describe the graph of each of these sets (in the xy plane):

   (*a*) $\{(x, y) \mid 1 = 0\}$ (*b*) $\{(x, y) \mid 1 = 1\}$

5. Describe the graph of the following polar equations:

   (*a*) $\rho = 5$ (*b*) $\rho > 5$ (*c*) $\rho < 5$

6. A line l is completely determined if the length and direction of the perpendicular segment from the origin to l are specified. In Fig. 4-27 let the length of the directed segment $OP = p$ be positive, and let the direction $\overrightarrow{OP}$ be given by specifying the counterclockwise angle $\omega < 360°$ from $\overrightarrow{OA}$ to $\overrightarrow{OP}$. Show that the coordinates of P are

$$x_1 = p \cos \omega$$
$$y_1 = p \sin \omega$$

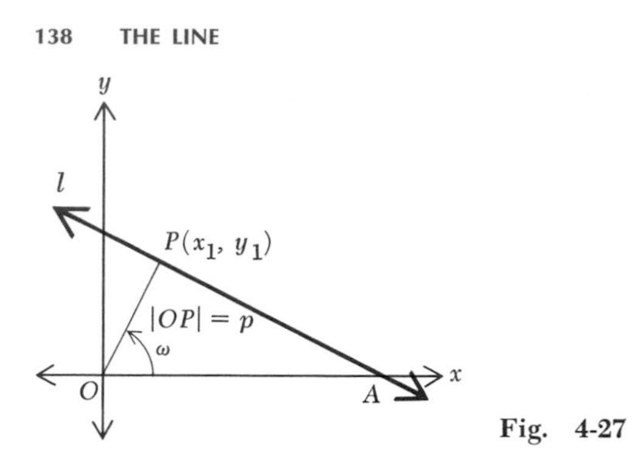

Fig. 4-27

and that the slope of l is $-\cot \omega$. Then, using the point-slope formula, show that the equation of l, in terms of p and ω, is

$$x \cos \omega + y \sin \omega = p$$

This is called the *normal form* of the equation of the line.

7. In each of the following cases write the equation of the line in the normal form, using the result of Prob. 6:

   (*a*) $\omega = 135°$; $p = 8$ (*b*) $\omega = 90°$; $p = 3$
   (*c*) $\omega = 240°$; $p = 6$ (*d*) $\omega = 180°$; $p = 5$

8. Prove that the three lines

$$(a_1 - a_2)x + (b_1 - b_2)y + (c_1 - c_2) = 0$$
$$(a_2 - a_3)x + (b_2 - b_3)y + (c_2 - c_3) = 0$$
$$(a_1 - a_3)x + (b_1 - b_3)y + (c_1 - c_3) = 0$$

   are concurrent or parallel.

9. Write equations for the bisectors of the angles formed by two lines whose equations are $3x - 9y = 10$ and $x + 3y = 6$.
10. A triangle has vertices at $P(4, 3)$, $Q(0, 5)$, and $R(-4, 1)$. Prove that the intersection of the medians, the intersection of the perpendicular bisectors of the sides, and the intersection of the altitudes lie on a straight line. Find the equation of the line.
11. Prove that the lines containing the bisectors of the adjacent angles formed by two intersecting lines are perpendicular.

TRANSFORMATION OF COORDINATES | 5

5-1 INTRODUCTION

The equation of a curve is a statement of the relation between the pair of coordinates (x, y) of any point on the curve. The equation then depends also upon the location of the x and y axes relative to the curve. For example, the line in Fig. 5-1 has the equation $2x - y = 2$ with respect to the x and y axes shown. It is also possible to determine the equation of the line with respect to the x' and y' axes in this figure. Since the line goes through the origin of the new coordinate system and has slope 2, its equation is $y' = 2x'$.

It is often necessary to solve the following problem. Given the equation of a curve with respect to an initial set of x and y axes, find the equation of this same curve with respect to another specified set of axes. If the new axes are parallel to the original ones and have the same positive directions, the transformation from which the new equation is obtained is called a *translation* of axes. If the new origin is identical with the original one, and the new axes are obtained by revolving the original axes about this origin through a specified angle, the transformation is called a *rotation* of axes.

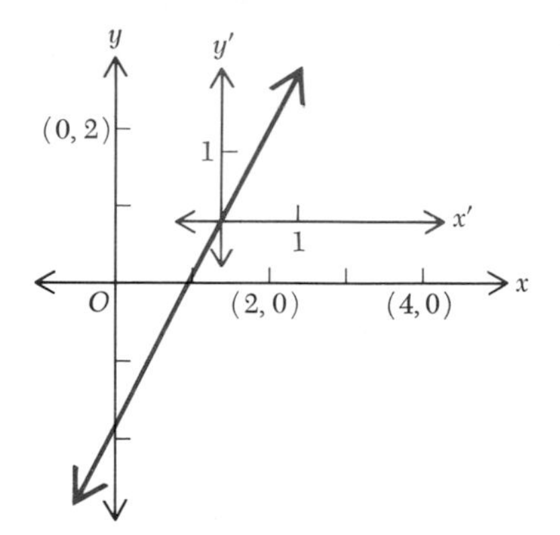

Fig. 5-1 Equation of line is $2x - y = 2$ with respect to xy axes and $y' = 2x'$ with respect to $x'y'$ axes.

5-2 TRANSLATION OF AXES

Suppose that the equation of a curve is given with respect to an initial set of x and y axes. A translation of the axes is made such that the point whose coordinates were (h, k) becomes the new origin. The equation of the curve relative to this new coordinate system is to be determined (Fig. 5-2).

Let P be any point on the curve (or, in fact, any point in the plane). Let its coordinates be (x, y) relative to the original axes and (x', y') relative to the new axes. Then the following relations are derived from the basic properties of directed lines:

$$(5\text{-}1) \qquad \begin{aligned} OM &= OA + AM \qquad \text{or} \qquad x = h + x' \\ ON &= OB + BN \qquad \text{or} \qquad y = k + y' \end{aligned}$$

where h and k are real numbers. Thus if in the original equation of the curve x is replaced by $x' + h$ and y is replaced by $y' + k$, the resulting relation between x' and y' is the equation of the curve relative to the new set of axes.

Example 1

The equation of the curve shown in Fig. 5-3 is $y = x^2 - 6x + 5$. The "lowest" point is at $R(3, -4)$. What is the equation of the curve relative to axes x' and y' that pass through $R(3, -4)$ as shown?

Solution In this case substitute $h = 3$ and $k = -4$ in Eqs. (5-1). In $y = x^2 - 6x + 5$ replace x by $x' + 3$ and y by $y' - 4$. Then the result is

$$y' - 4 = (x' + 3)^2 - 6(x' + 3) + 5$$
$$y' - 4 = x'^2 + 6x' + 9 - 6x' - 18 + 5$$
$$y' - 4 = x'^2 - 4$$

or

$$y' = x'^2$$

The system of equations relating coordinates of the old and new axes in the above example was

$$x = h + x'$$
$$y = k + y'$$

In the next example the *inverse transformation* is to be performed. An inverse transformation is one that "undoes" the original transformation. The result of first performing a transformation and then performing the inverse transformation is the *identity transformation*. In other words, the net effect is to make no change.

The system of equations for the inverse transformation in this case is found by solving Eqs. (5-1) for x' and y'; hence the system is

(5-2)
$$x' = x - h$$
$$y' = y - k$$

Example 2

It is known that the equation of the curve relative to the x' and y' axes in Fig. 5-3 is $y' = x'^2$. Find the equation of the curve relative to the x and y axes.

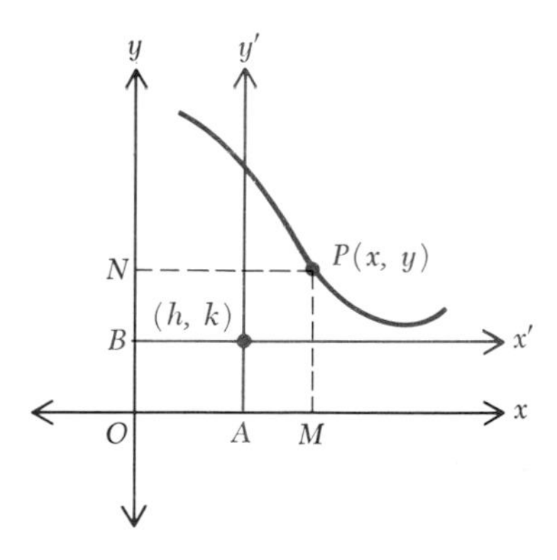

Fig. 5-2 If the equation of a curve is $y = f(x)$ with respect to xy axes, it is $y' + k = f(x' + h)$ with respect to $x'y'$ axes.

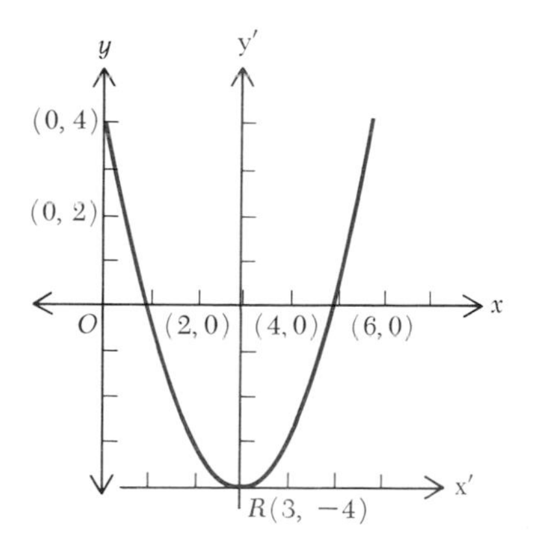

Fig. 5-3 Equation of curve for xy axes is $y = x^2 - 6x + 5$ and for $x'y'$ axes is $y' = x'^2$.

Solution In the equation $y' = x'^2$ replace x' with $x - 3$ and y' with $y + 4$. Then the result is

$$\begin{aligned} y + 4 &= (x - 3)^2 \\ y + 4 &= x^2 - 6x + 9 \\ y &= x^2 - 6x + 5 \end{aligned}$$

Notice that the result is the equation in its original form from Example 1.

5-3 ROTATION OF AXES

Consider the equation of a curve with respect to a set of x and y axes. A new set of axes, denoted by x' and y', is obtained by rotating the original axes through an angle θ (see Fig. 5-4) (the measure of θ is considered positive when measured counterclockwise and is considered negative when measured clockwise, as in trigonometry). It is required to find the equation of the curve corresponding to the new set of axes.

Let $P(x, y)$ be any point on the curve; then $x = OL$ and $y = LP$. Let the coordinates of P relative to the new axes be x' and y'; then $x' = OL'$ and $y' = L'P$. If $\overline{OP}$ is drawn, if the undirected distance $|OP| = r$, where r is a positive real number, and if the angle $L'OP$ is denoted by ϕ, then

$$x = OL = r \cos (\theta + \phi) \qquad y = LP = r \sin (\theta + \phi)$$
$$x' = OL' = r \cos \phi \qquad y' = L'P = r \sin \phi$$

By the formulas for the cosine and sine of the sum of the measures of two angles,

$$\begin{aligned} x = r \cos (\theta + \phi) &= r(\cos \theta \cos \phi - \sin \theta \sin \phi) \\ &= (r \cos \phi) \cos \theta - (r \sin \phi) \sin \theta \\ &= x' \cos \theta - y' \sin \theta \end{aligned}$$

Similarly,

$$\begin{aligned} y = r \sin (\theta + \phi) &= r(\sin \theta \cos \phi + \cos \theta \sin \phi) \\ &= (r \cos \phi) \sin \theta + (r \sin \phi) \cos \theta \\ &= x' \sin \theta + y' \cos \theta \end{aligned}$$

In order to find the new equation of the curve (in x' and y'), replace x by $x' \cos \theta - y' \sin \theta$ and replace y by $x' \sin \theta + y' \cos \theta$ in the original equation in x and y.

Thus the set of rotation equations is

$$\begin{aligned} x &= x' \cos \theta - y' \sin \theta \\ y &= x' \sin \theta + y' \cos \theta \end{aligned} \qquad (5\text{-}3)$$

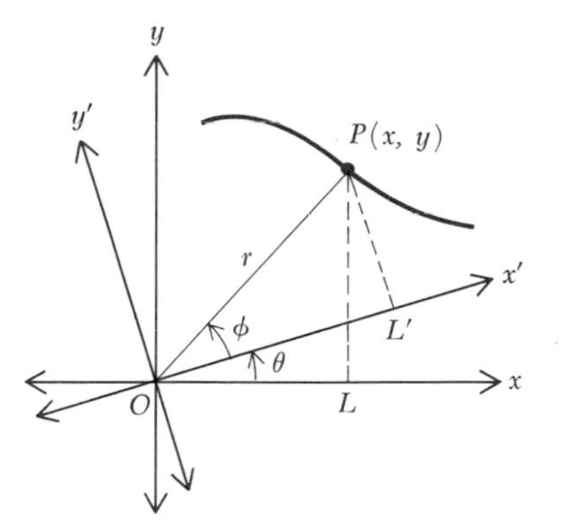

Fig. 5-4 x and y axes rotated an angle θ to obtain new x' and y' axes.

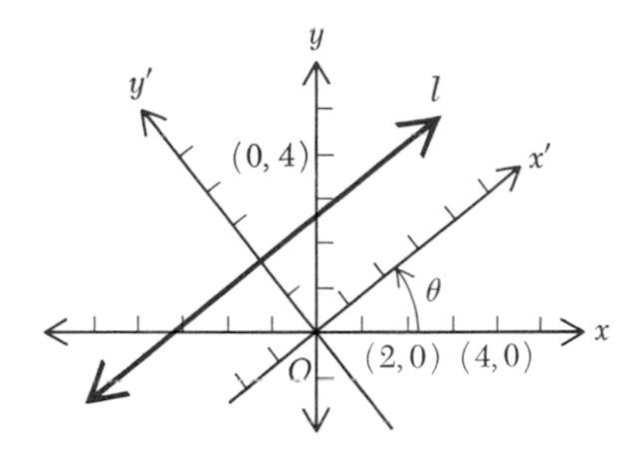

Fig. 5-5 Equation of line l is $4y - 3x = 10$ with respect to xy axes and $y' = 2$ with respect to $x'y'$ axes.

The set of equations for the inverse transformation can be obtained from Eqs. (5-3) by solving for x' and y' (see Prob. 21, page 145).

Example 1

The line l shown in Fig. 5-5 has the equation $4y - 3x = 10$. What is its equation with respect to axes x' and y' which have been obtained by rotating the original axes through the acute angle θ whose tangent is $\frac{3}{4}$?

Solution The sine and cosine of the acute angle whose tangent is $\frac{3}{4}$ are $\sin\theta = \frac{3}{5}$ and $\cos\theta = \frac{4}{5}$. In the original equation $4y - 3x = 10$, replace x with $\frac{4}{5}x' - \frac{3}{5}y'$ and replace y with $\frac{3}{5}x' + \frac{4}{5}y'$. The result is

$$4(\tfrac{3}{5}x' + \tfrac{4}{5}y') - 3(\tfrac{4}{5}x' - \tfrac{3}{5}y') = 10$$
$$\tfrac{12}{5}x' + \tfrac{16}{5}y' - \tfrac{12}{5}x' + \tfrac{9}{5}y' = 10$$
$$5y' = 10$$
$$y' = 2$$

This final result was to be expected, for the slope of the given line is $\frac{3}{4}$, and its distance from the origin is 2. Thus the line is parallel to the new x' axis and 2 units from it in the positive direction.

Example 2

The circle shown in Fig. 5-6 has the equation $x^2 + y^2 = 25$. Show that for any value of the angle θ of rotation, the new equation of the circle is $x'^2 + y'^2 = 25$.

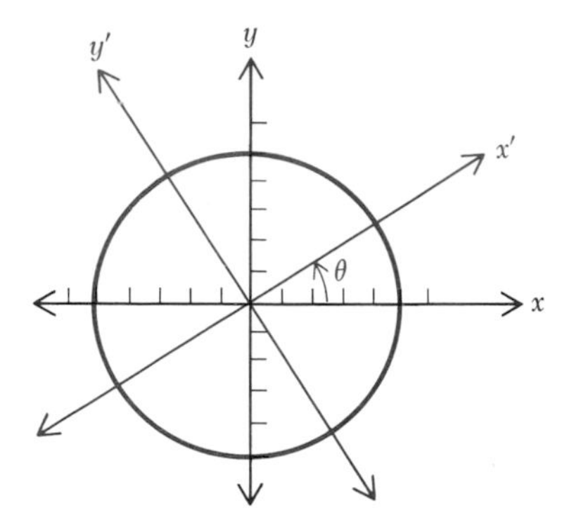

Fig. 5-6 The equation of a circle with center at O is invariant under rotation of axes through an angle θ.

Solution In $x^2 + y^2 = 25$ replace x by $x' \cos \theta - y' \sin \theta$ and replace y by $x' \sin \theta + y' \cos \theta$, without specifying the value of θ. The result is

$$(x' \cos \theta - y' \sin \theta)^2 + (x' \sin \theta + y' \cos \theta)^2 = 25$$

This can be simplified to

$$x'^2 \cos^2 \theta - 2x'y' \sin \theta \cos \theta + y'^2 \sin^2 \theta + x'^2 \sin^2 \theta + 2x'y' \sin \theta \cos \theta + y'^2 \cos^2 \theta = 25$$

or

$$x'^2(\cos^2 \theta + \sin^2 \theta) + y'^2(\sin^2 \theta + \cos^2 \theta) = 25$$

Now, for any value of θ, $\sin^2 \theta + \cos^2 \theta = 1$. The above equation therefore becomes

$$x'^2 + y'^2 = 25$$

Translation and rotation are the only two types of transformations studied in plane analytic geometry. For both there is a one-to-one correspondence between the members of the solution set of the original equation and those of the equation resulting from the transformation. Many geometric properties remain *invariant* (do not change) under either of these transformations, and it is precisely those properties that constitute much of the material to be studied in analytic geometry. For example, the distance between two points is an invariant under the transformations of translation and rotation.

PROBLEMS

1. Write the system of equations for the inverse transformation for each translation:

(a) $x = x' + 4$ (b) $x' = x - 7$
$y = y' - 2$ $y' = y + 3$

In Probs. 2 to 13 sketch the graph of the given equation and plot the given point A. Then draw new axes x' and y' through A parallel to the original axes and having the same positive directions. Find the equation of the curve with respect to these new axes:

2. $2y = x + 4$; A(2, 3)
3. $x - y = 5$; A(1, −1)
4. $\frac{x}{6} + \frac{y}{4} = 1$; A(0, 4)
5. $y - 3 = \frac{2}{3}(x - 5)$; A(5, 3)
6. $y = 8 + 2x - x^2$; A(1, 9)
7. $y = 2x^2 - 7x - 15$; A(5, 0)
8. $y = \frac{1}{2}(x^2 + 8x)$; A(−4, 0)
9. $y - 8 = x(x^2 + 6x + 12)$; A(−2, 0)
10. $x + y = x^3 - 3x^2 + 3$; A(1, 0)
11. $y = x(x^3 - 4x^2 + 2x + 4)$; A(1, 3)
12. $y = \frac{2x + 1}{x - 1}$; A(1, 2)
13. $y = \frac{-2x^2}{x^2 + 4}$; A(0, −2)
14. The graph of the equation $y = Ax^2 + Bx + C$ passes through the point $\left(-\frac{B}{2A}, -\frac{B^2 - 4AC}{4A}\right)$. Show that if new x' and y' axes, parallel to the original ones and having the same positive directions, are drawn through this point, the equation of the curve with respect to these axes is $y' = Ax'^2$. Hence infer that the graphs of the equations $y = Ax^2 + Bx + C$ and $y' = Ax'^2$ are precisely the same curve.
15. Plot each of the points $A(8, -6)$, $B(-4, -4)$, and $C(0, 10)$. Then obtain new axes by rotating the original ones through an angle $\theta = 30°$ and find the new coordinates of each point.
16. Repeat Prob. 15 for $\theta = -45°$.
17. Draw the line whose equation is $2y + x = 4$. Rotate the axes through the acute angle θ whose tangent is 2 and find the new equation of the line.
18. Draw the line whose equation is $5x - 12y = 52$. Rotate the axes through the acute angle whose tangent is $\frac{5}{12}$ and find the new equation of the line.
19. Sketch the curve whose equation is $y = \sqrt{2}x^2$. Rotate the axes through an angle $\theta = 45°$ and find the new equation of the curve.
20. Sketch the curve whose equation is $y = 1/x$. Rotate the axes through an angle $\theta = 45°$ and find the new equation of the curve.
21. Solve the system of equations $x = x' \cos \theta - y' \sin \theta$ and $y = x' \sin \theta + y' \cos \theta$ for x' and y' in terms of x and y.
22. Show that the equation $5x^2 + 4xy + 8y^2 = 9$ is transformed into $9x'^2 + 4y'^2 = 9$ when the axes are rotated through the acute angle whose tangent is 2.

SUMMARY

The equation of a curve depends on the location of the x and y axes relative to the curve. Given an equation with respect to a set of axes, it is possible to find the equation for the same curve with respect to a different set of axes. In analytic geometry this second equation may be obtained from the first one by transformations of translation and rotation. For each of these transformations there is an inverse transformation that "undoes" the original transformation.

A transformation may be indicated by a set of equations. The equations and their inverses for translation and rotation are summarized in the following table:

Transformation	Equations for Transformation	Equations for Inverse of Transformation
Translation of axes to origin (h, k)	$x = h + x'$ $y = k + y'$	$x' = x - h$ $y' = y - k$
Rotation of axes through angle θ	$x = x' \cos \theta - y' \sin \theta$ $y = x' \sin \theta + y' \cos \theta$	$x' = x \cos \theta + y \sin \theta$ $y' = -x \sin \theta + y \cos \theta$

REVIEW PROBLEMS

1. Graph the equation $y = x^2 - 2x - 8$. Replace x by $x' + 1$ and y by $y' - 9$ and write the new equation in the form $y' = f(x')$.
2. Determine the equation in terms of x' and y' of $x^2 + y^2 - 6x + y + 3 = 0$ after making the substitutions $x = x' + 3$ and $y = y' - \frac{1}{2}$.
3. The equation of a straight line is $3x + y = 7$, and the equation of the line relative to the x' and y' axes is $3x' + y' = 0$. What are the coordinates of the origin of this $x'y'$ coordinate system with respect to the x and y axes?
4. The graphs of $y = x^2 - 4x$ and $y = 16 - x^2$ intersect at $(4, 0)$ and $(-2, 12)$. What are the coordinates of the points of intersection after the transformation $x = x' - 2$ and $y = y' + 4$?
5. Write the equations of the lines in terms of x' and y' in each of the following cases after rotating the x and y axes through the angle indicated:

   (*a*) $y = x$; $\theta = 45°$ (*b*) $x = 4$; $\theta = 90°$
   (*c*) $y = -7$; $\theta = 180°$ (*d*) $x = 0$; $\theta = \arctan 1$; $0 < \theta < 90$

6. Plot and label points $A(4, 0)$, $B(6, 2)$, $C(-3, 7)$, $D(-5, -1)$, and $E(2, -\sqrt{3})$.

Rotate the axes through an angle of $-30°$ and find the coordinates of the given points with respect to the new axes. Draw both sets of axes.

7. Draw the line $2x - 3y + 7 = 0$. Rotate the axes through an acute angle whose tangent is $\frac{2}{3}$ and find the new equation of the line.
8. Show that the equation $4x^2 + 24xy - 3y^2 = 312$ is transformed into $x'^2/24 - y'^2/26 = 1$ by rotating the axes through an acute angle $\theta = \arctan \frac{3}{4}$.

FOR EXTENDED STUDY

1. By translation of axes remove the terms of first degree from the equation $x^2 - 4y^2 + x - 4y = 0$.
2. Consider the set of equations for a transformation as $x' = x$ and $y' = ky$, where k is a positive number. (*a*) With $k = 3$, what is the new equation for $4 - x - y = 0$ after the transformation is performed? (*b*) Suppose the transformation equations were $x' = kx$ and $y' = y$. Answer part (*a*). Compare the new graph to the graph of the original equation and to that in part (*a*).
3. Describe the transformation $x' = x$ and $y' = ky$ for each of these conditions:

   (*a*) $0 < k < 1$ (*b*) $k = 1$ (*c*) $k > 1$

4. Describe how the graph of $y = f(x)$ changes under the transformation $x' = kx$ and $y' = ky$ for k a positive number.
5. Rotate the x and y axes through a 45° angle and find the equation of $x^{\frac{1}{2}} + y^{\frac{1}{2}} = a^{\frac{1}{2}}$ relative to the new axes.
6. A set of axes is rotated through an angle of measure θ. The equations of rotation are

$$y = x' \cos \theta - y' \sin \theta$$
$$y = x' \sin \theta + y' \cos \theta$$

This is followed by a second rotation through an angle of measure β with the equations of rotation

$$x' = x'' \cos \beta - y'' \sin \beta$$
$$y' = x'' \sin \beta - y'' \cos \beta$$

Prove that

$$x = x'' \cos (\theta + \beta) - y'' \sin (\theta + \beta)$$
$$y = y'' \sin (\theta + \beta) + y'' \cos (\theta + \beta)$$

THE CIRCLE AND THE ELLIPSE | 6

6-1 STANDARD FORM FOR THE EQUATION OF A CIRCLE

A circle may be defined as the locus of all points in a plane at a given distance from a fixed point of the plane. The fixed point is called the *center,* and the given distance is called the *radius* of the circle. The equation may be derived directly from this definition.

Let the center be the point $Q(h, k)$, and let the radius be r (Fig. 6-1). Then if $P(x, y)$ is any point of the plane, its undirected distance from

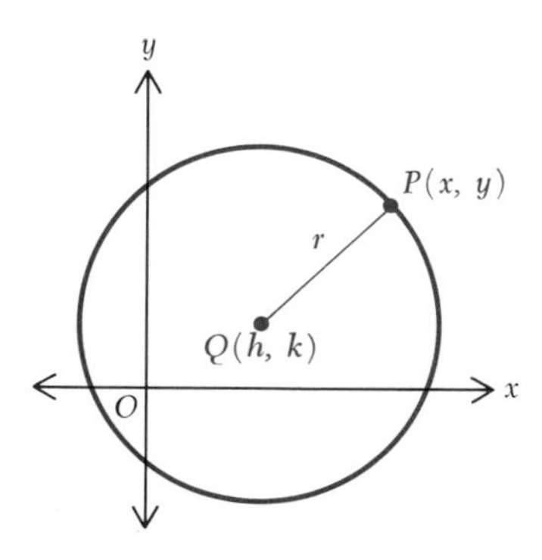

Fig. 6-1 Circle with center $Q(h, k)$ and radius r: $(x-h)^2 + (y-k)^2 = r^2$.

the point $Q(h, k)$ is

$$d = \sqrt{(x-h)^2 + (y-k)^2}$$

Every point P which lies on the circle must have $d = r$; hence its coordinates must satisfy the equation

$$\sqrt{(x-h)^2 + (y-k)^2} = r \tag{6-1}$$

or

$$(x-h)^2 + (y-k)^2 = r^2 \tag{6-2}$$

Conversely, every point in the plane whose coordinates (x, y) satisfy Eq. (6-2) lies on the circle, for if (6-2) is true for a point $P(x, y)$, then (6-1) is also true, and the point is therefore at a distance r from the point $Q(h, k)$. Equation (6-2) is thus the equation of a circle with center at the point $Q(h, k)$ and radius r; it is called the *standard form* of the equation of the circle. If the circle has its center at the origin, then $h = k = 0$, and the equation becomes

$$x^2 + y^2 = r^2$$

Example
The equation of the circle with center at $(2, -1)$ and radius 3 is

$$(x-2)^2 + (y+1)^2 = 9$$

By performing the indicated operations and simplifying, the equation may be expressed as

$$x^2 + y^2 - 4x + 2y - 4 = 0$$

This is the general form to be discussed in the next section. The set of points inside the circle is $(x-2)^2 + (y+1)^2 < 9$, and the set of points outside the circle is $(x-2)^2 + (y+1)^2 > 9$.

6-2 GENERAL FORM FOR THE EQUATION OF A CIRCLE

If the indicated operations in Eq. (6-2) are performed, the result can be written in the form

$$x^2 + y^2 - 2hx - 2ky + h^2 + k^2 - r^2 = 0$$

This equation is of the general type

$$x^2 + y^2 + Dx + Ey + F = 0 \tag{6-3}$$

It is evident that any equation of form (6-2) can be expressed in form (6-3), so that every circle has an equation of this type. Now consider the converse question of whether every equation of form (6-3) represents a circle. An example indicates intuitively that, with suitable restrictions on the values of D, E, and F, it does.

Example

Write the equation $x^2 + y^2 - 6x + 2y + 4 = 0$ in standard form.

Solution First write the equation in the following way:

$$x^2 - 6x + \quad + y^2 + 2y + \quad = -4$$

In order to complete the square in x add 9 to both members, and in order to complete the square in y add 1 to both members. Thus

$$x^2 - 6x + 9 + y^2 + 2y + 1 = -4 + 9 + 1$$

or

$$(x-3)^2 + (y+1)^2 = 6$$

It is now clear that the given equation is that of a circle with center at $(3, -1)$ and radius $\sqrt{6}$. Observe, however, that if the constant term in the given equation had been 10 instead of 4, the result would have been

$$(x-3)^2 + (y+1)^2 = 0$$

The graph in this case is a circle of radius zero with center at $(3, -1)$, or a *point circle*. Finally, if the constant term had been greater than 10, say 12, the result would have been

$$(x-3)^2 + (y+1)^2 = -2$$

In this case there is no graph; that is, the equation is not satisfied by any pair of real numbers.

By similarly completing the square in the general case it can be proved that the equation

$$x^2 + y^2 + Dx + Ey + F = 0$$

either invariably represents a circle if

$$-F + \frac{D^2}{4} + \frac{E^2}{4} \geqslant 0$$

or has no graph at all if

$$-F + \frac{D^2}{4} + \frac{E^2}{4} < 0$$

The word "circle" is to be understood to include a point circle (circle of radius zero). Hence the equation $x^2 + y^2 + Dx + Ey + F = 0$, where the coefficients of the x^2 and y^2 terms are 1, is called the *general form* of the equation of a circle. The result of completing the square shows that the center of the circle is at $(-\frac{1}{2}D, -\frac{1}{2}E)$ (see Prob. 4, page 153).

6-3 CIRCLE SATISFYING THREE CONDITIONS

Both the standard form and the general form of the equation of a circle contain three arbitrary constants: h, k, and r in the standard form and D, E, and F in the general form. Given three conditions which determine a circle, its equation can usually be found by expressing three equations in terms of these constants and solving for them. In most cases either form may be used, but depending upon the nature of the given conditions, one form may be more convenient than the other.

Example 1

Find the equation of the circle that passes through the points $A(1, 1)$, $B(4, 0)$, and $C(2, -1)$.

Solution The circle (Fig. 6-2) has an equation of the form

$$x^2 + y^2 + Dx + Ey + F = 0$$

The constants D, E, and F must be determined so that the equation is satisfied by the coordinates of each of the given points, since each is to be on the circle. Thus the following conditions, derived by substituting the coordinates of each point into the general equation, must be met:

$$1 + 1 + D + E + F = 0$$
$$16 + 0 + 4D + 0 + F = 0$$
$$4 + 1 + 2D - E + F = 0$$

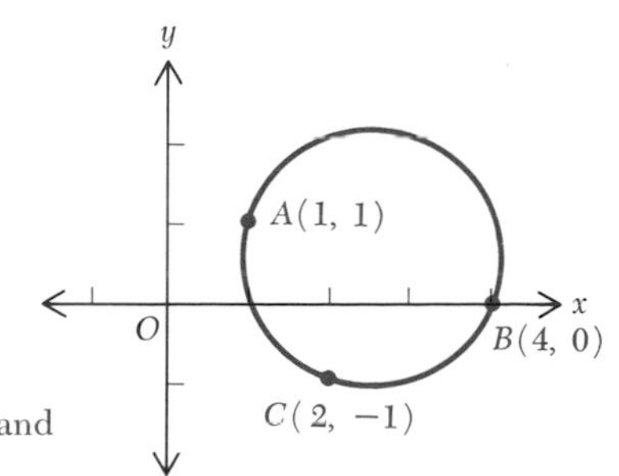

Fig. 6-2 Circle passing through $A(1, 1)$, $B(4, 0)$, and $C(2, -1)$.

This system of equations has the solution $D = -5$, $E = -1$, and $F = 4$ and no other solution. The required equation is then

$$x^2 + y^2 - 5x - y + 4 = 0$$

To find the center and radius of the circle complete the squares in x and y. The result is the standard form

$$(x - \tfrac{5}{2})^2 + (y - \tfrac{1}{2})^2 = \tfrac{10}{4}$$

The center is at the point $(\frac{5}{2}, \frac{1}{2})$, and the radius is $\frac{1}{2}\sqrt{10}$.

Perhaps you can see other ways of solving the problem.

Example 2

Find the equation of a circle that is tangent to the x axis, has its center on the line $y = 2x + 1$, and passes through the point $A(-2, 8)$.

Solution Assume an equation of the form

$$(x - h)^2 + (y - k)^2 = r^2$$

and determine h, k, and r to satisfy the specified conditions. The fact that the center is on the line $y = 2x + 1$ means that its coordinates (h, k) must satisfy this equation; that is,

$$k = 2h + 1$$

The condition that the circle be tangent to the x axis means that the y coordinate of the center is equal to r or $-r$:

$$k = \pm r$$

Finally, the condition that the circle passes through the point $A(-2, 8)$ requires that

$$(-2 - h)^2 + (8 - k)^2 = r^2$$

This system of equations has two (and only two) solutions, as follows:

$$h = 2 \qquad k = 5 \qquad r = 5$$

or

$$h = 26 \qquad k = 53 \qquad r = 53$$

Each of these solutions leads to the equation for a circle. There are then two circles satisfying the given conditions,

$$(x - 2)^2 + (y - 5)^2 = 25 \qquad \text{or} \qquad (x - 26)^2 + (y - 53)^2 = 2{,}809$$

The first of these is drawn in Fig. 6-3; the second cannot easily be shown with the same scale.

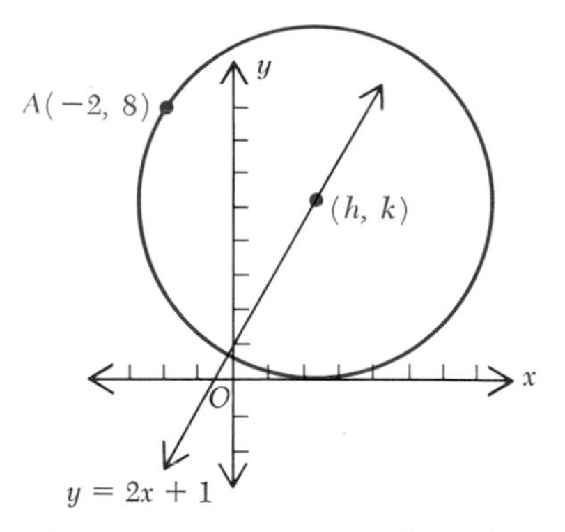

Fig. 6-3 Circle passing through $A(-2, 8)$, with center on $y = 2x + 1$ and tangent to x axis.

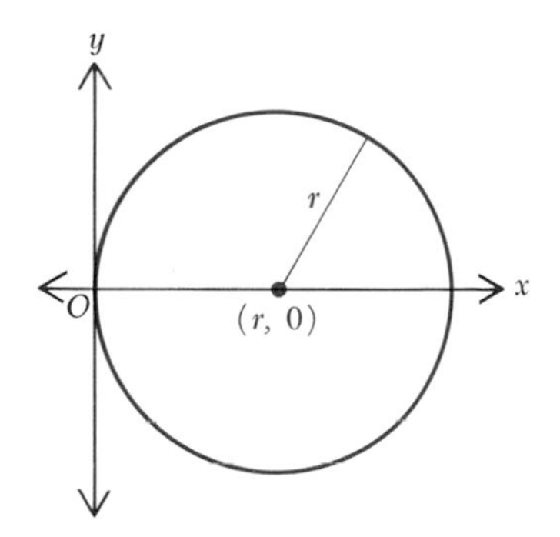

Fig. 6-4 Circle with center at $(r, 0)$ and radius r: $x^2 + y^2 = 2rx$.

PROBLEMS

1. In each of the following cases write the equation of the circle with center at C and radius r in both the standard and general form, and then write the inequalities representing the sets of points inside and outside the circle:

   (*a*) $C(4, -2)$; $r = 3$
   (*b*) $C(-2, -6)$; $r = 2$
   (*c*) $C(0, -5)$; $r = 5$
   (*d*) $C(2, 0)$; $r = 4$
   (*e*) $C(4, 1)$; $r = 4$
   (*f*) $C(-3, 4)$; $r = 5$
   (*g*) $C(3, 3)$; $r = 3$
   (*h*) $C(-5, 2)$; $r = 7$

2. In each of the following cases complete the squares in x and y and thus determine the center and radius of the circle represented by the equation, or determine that the equation has no graph. Sketch the graph if there is one.

   (*a*) $x^2 + y^2 - 4x + 6y = 12$
   (*b*) $x^2 + y^2 + 2x + 10y + 1 = 0$
   (*c*) $x^2 + y^2 - 3x + 5y + 7 = 0$
   (*d*) $x^2 + y^2 = x + y$
   (*e*) $2x^2 + 2y^2 = 5y - 4x - 2$
   (*f*) $x^2 + y^2 + 4x + 4y + 10 = 0$
   (*g*) $x^2 + y^2 - 2x - 3y + 4 = 0$
   (*h*) $x^2 + y^2 - 10x + 10y + 1 = 0$

3. (*a*) Show that the equation of the circle with center at $(r, 0)$ and radius r (Fig. 6-4) is $x^2 + y^2 = 2rx$. (*b*) What is the equation of a circle with center at $(0, r)$ and radius r?
4. Show that the equation $x^2 + y^2 + Dx + Ey + F = 0$ represents a circle with center at $(-\frac{1}{2}D, -\frac{1}{2}E)$ and radius $\sqrt{D^2/4 + E^2/4 - F}$ if $D^2/4 + E^2/4 - F$ is a positive number, and that it represents a point circle if $D^2/4 + E^2/4 - F = 0$.
5. Show that the equation $x^2 + y^2 = ax + by$ always represents a circle and that this is a point circle if and only if $a = b = 0$.

6. Find the equations of all circles that

   (*a*) Have the segment $\overline{AB}$ with $A(-2, -4)$ and $B(6, 4)$ as a diameter
   (*b*) Pass through $A(3, 0)$ and $B(12, 0)$ and are tangent to the y axis
   (*c*) Have centers on the line $y = \frac{1}{2}x + 3$ and are tangent to both axes
   (*d*) Pass through $A(3, 6)$ and $B(-4, -1)$ and have centers on the y axis
   (*e*) Have centers at $P(8, 4)$ and are tangent to the line whose intercepts are $A(6, 0)$ and $B(0, 3)$.

7. A point moves so as to be always twice as far from the origin as from the point $R(12, 0)$. Find the equation of its path and sketch it.
8. Find the equation of the locus of all points that are twice as far from the point $P(-3, 0)$ as from the point $Q(6, 6)$.
9. Find the equation of the locus of all points which are three times as far from the point $H(-4, 2)$ as from the point $J(8, 6)$ and sketch the locus.
10. What equation represents the family of all circles with center on the x axis that pass through the origin?
11. What equation represents the family of all circles with center on the line $y = x$ that pass through the origin?
12. What equation represents the family of all circles with center on the line $x + y = 0$ that are tangent to both axes?
13. Write an equation which represents the family of all circles with center on the line $y = 2x$ that pass through the origin. Find the particular member of the family that passes through the point $S(6, 2)$.
14. Draw a square with center at the origin and with each side 6 units long, the sides being parallel to the coordinate axes. A point moves in the plane so that the sum of the squares of its distances from the four sides of the square is 100. Find the equation of its path.
15. Does the circle whose equation is $4x^2 + 4y^2 - 32x - 48y + 127 = 0$ intersect the line $x - 2y = 2$ at two points, at one point, or not at all? *Hint:* Compare the radius of the circle with the distance from its center to the line.
16. In each of the following cases find the equation of the circle that passes through the three given points:

   (*a*) $E(6, 10), F(0, 2), G(2, 8)$
   (*b*) $H(10, 2), J(3, 9), K(-2, 10)$
   (*c*) $M(-2, -2), N(-1, 6), P(0, 1)$
   (*d*) $Q(0, 6), R(2, 2), S(3, 5)$

17. Find the equation of the circle with center at $C(5, -3)$ that is tangent to the line $3x - 4y = 12$.
18. Find the equation of the circle that has its center on the y axis and is tangent to the line $x + 4y = 10$ at the point $B(2, 2)$.
19. Find the equation of the circle that is tangent to the line $4x - 3y = 2$ at the point $D(-1, -2)$ and passes through the point $E(6, -1)$.
20. Find the equation of the circle that is tangent to the line $x + 2y = 4$ at the point $E(6, -1)$ and passes through the point $F(5, 2)$.
21. Find the equations of all circles that pass through the origin and the point $G(-1, -3)$ and are tangent to the line $2x + y + 10 = 0$.

6-4 POLAR EQUATIONS REPRESENTING CIRCLES

The standard equation of a circle with center at $P(h, k)$ and radius r is

$$(x-h)^2 + (y-k)^2 = r^2$$

If x is replaced by $\rho \cos \theta$ and y is replaced by $\rho \sin \theta$, the result is

$$(\rho \cos \theta - h)^2 + (\rho \sin \theta - k)^2 = r^2 \tag{6-4}$$

This equation can be written in the form

$$\rho^2 - 2\rho(h \cos \theta + k \sin \theta) = c \tag{6-5}$$

where $c = r^2 - h^2 - k^2$. The polar equation (6-5) of a circle is not used very often, but certain special cases arise frequently. These are as follows:

Circle with center at origin and radius r:

$$\rho = r \tag{6-6}$$

Circle with center at $C(r, 0)$ and radius r:

$$\rho = 2r \cos \theta \tag{6-7}$$

Circle with center at $B(r, \frac{1}{2}\pi)$ and radius r:

$$\rho = 2r \sin \theta \tag{6-8}$$

Equations (6-7) and (6-8) can be derived directly from Figs. 6-5 and 6-6. They are of course special cases of the general equation (6-5) in polar coordinates.

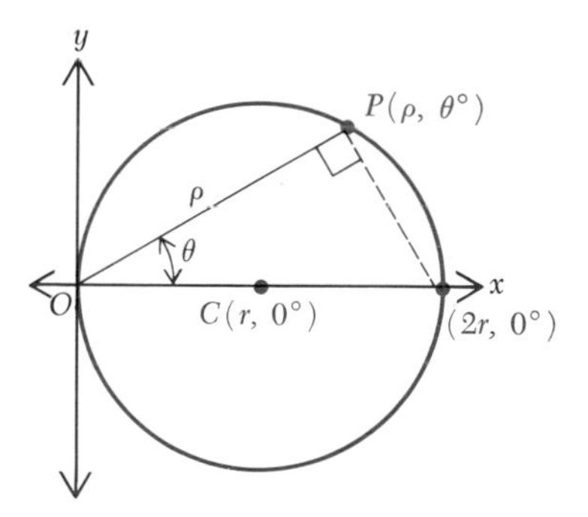

Fig. 6-5 Circle with center $C(r, 0°)$ and radius r: $\rho = 2r \cos \theta$.

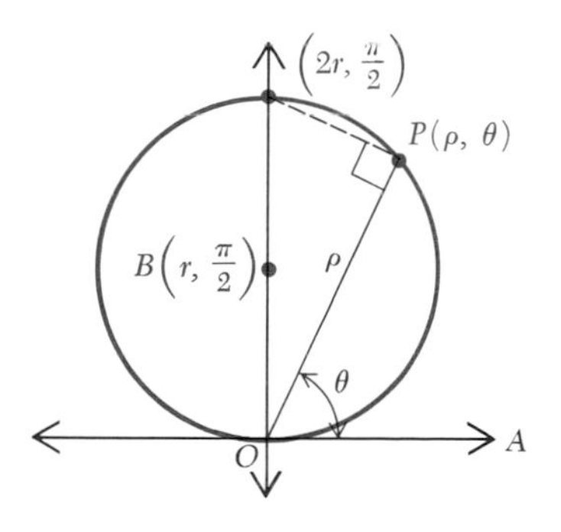

Fig. 6-6 Circle with center $B(r, \pi/2)$ and radius r: $\rho = 2r \sin \theta$.

6-5 VECTOR FORM OF THE EQUATION OF A CIRCLE

In Fig. 6-7 the circle with radius r has center $A(x_1, y_1)$. Then

$$|\mathbf{AP}| = r$$

and
$$\mathbf{AP} \cdot \mathbf{AP} = |\mathbf{AP}||\mathbf{AP}| \cos(\mathbf{AP}, \mathbf{AP})$$

But
$$\cos(\mathbf{AP}, \mathbf{AP}) = \cos 0° = 1$$

hence

$$\mathbf{AP} \cdot \mathbf{AP} = r^2 \tag{6-9}$$

Equation (6-9) is a vector form of the equation of the circle.

Since $\mathbf{AP} = \mathbf{OP} - \mathbf{OA}$, a second vector form of the equation of the circle is

$$(\mathbf{OP} - \mathbf{OA}) \cdot (\mathbf{OP} - \mathbf{OA}) = r^2 \tag{6-10}$$

To find the equation of the circle in cartesian coordinates from Eq. (6-10) observe that $\mathbf{OP} = x\mathbf{i} + y\mathbf{j}$ and $\mathbf{OA} = x_1\mathbf{i} + y_1\mathbf{j}$. Then

$$\begin{aligned} \mathbf{OP} - \mathbf{OA} &= x\mathbf{i} + y\mathbf{j} - (x_1\mathbf{i} + y_1\mathbf{j}) \\ &= (x - x_1)\mathbf{i} + (y - y_1)\mathbf{j} \end{aligned}$$

Hence, by Eq. (6-9),

$$\mathbf{AP} \cdot \mathbf{AP} = (x - x_1)^2 + (y - y_1)^2 = r^2$$

which is the equation of a circle in rectangular coordinates.

Example

Use Eqs. (6-9) and (6-10) to find the equation in rectangular coordinates of a circle with radius 2 and center at $A(1, 3)$.

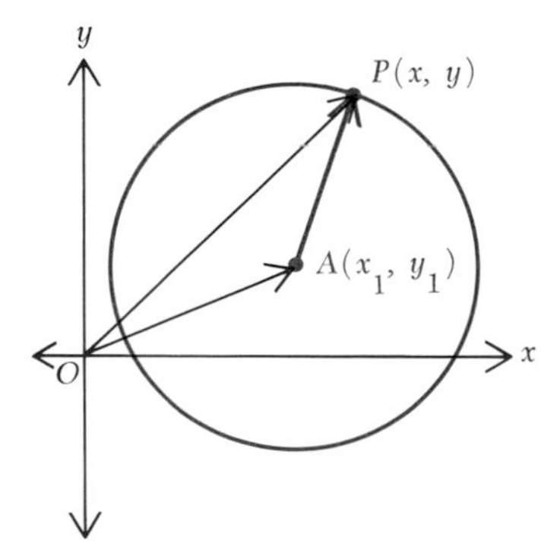

Fig. 6-7 Vector equation of circle with center at $A(x_1, y_1)$ is $\mathbf{AP} \cdot \mathbf{AP} = r^2$.

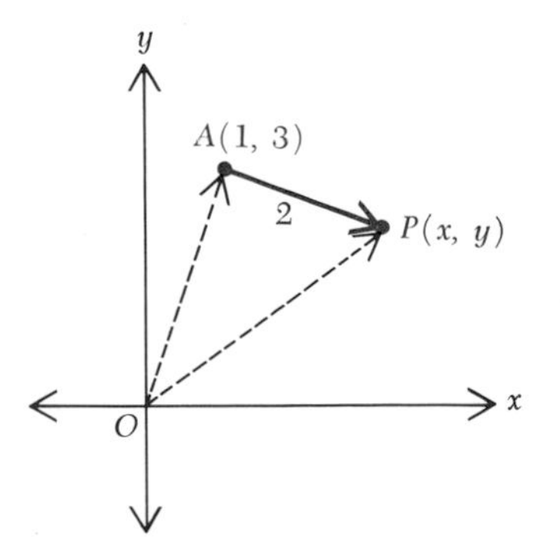

Fig. 6-8 For a circle with center at $A(1, 3)$ and radius 2, $\mathbf{AP} \cdot \mathbf{AP} = 4$.

Solution We start with $|\mathbf{AP}| = 2$ (see Fig. 6-8). Then, from Eq. (6-9),

$$\mathbf{AP} \cdot \mathbf{AP} = 4$$

and from Eq. (6-10),

$$(\mathbf{OP} - \mathbf{OA}) \cdot (\mathbf{OP} - \mathbf{OA}) = 4$$

Here $\mathbf{OP} = x\mathbf{i} + y\mathbf{j}$

and $\mathbf{OA} = \mathbf{i} + 3\mathbf{j}$

Then
$$(\mathbf{OP} - \mathbf{OA}) = (x\mathbf{i} + y\mathbf{j}) - (\mathbf{i} + 3\mathbf{j})$$
$$= (x-1)\mathbf{i} + (y-3)\mathbf{j}$$

Now, $\mathbf{AP} \cdot \mathbf{AP} = |\mathbf{AP}||\mathbf{AP}| \cos (\mathbf{AP}, \mathbf{AP})$

and $|\mathbf{AP}| = \sqrt{(x-1)^2 + (y-3)^2}$

Substituting in $\mathbf{AP} \cdot \mathbf{AP} = 4$ results in

$$\sqrt{(x-1)^2 + (y-3)^2}\ \sqrt{(x-1)^2 + (y-3)^2} \cos (\mathbf{AP}, \mathbf{AP}) = 4$$

Since $\cos (\mathbf{AP}, \mathbf{AP}) = 1$, we have

$$(x-1)^2 + (y-3)^2 = 4$$

the equation of a circle with center at $A(1, 3)$ and radius of 2 in rectangular form.

PROBLEMS

In Probs. 1 to 10 sketch the graph of the given equation; and then find the corresponding equation for the curve in rectangular coordinates.

1. $\rho = 3$
2. $\rho = 5$
3. $\rho = 4 \cos \theta$
4. $\rho = \cos \theta$
5. $\rho = 5 \sin \theta$
6. $\rho = 3 \sin \theta$
7. $\rho^2 - 6\rho \cos \theta + 4\rho \sin \theta = 0$
8. $\rho^2 = 6\rho \sin \theta + 16$
9. $\rho^2 = 4\rho \sin \theta + 12$
10. $\rho = 5 \cos (\theta - \frac{1}{4}\pi)$
11. Derive Eq. (6-7).
12. Derive Eq. (6-8).

In Probs. 13 to 16 use Eqs. (6-9) and (6-10) to find the equation in rectangular coordinates.

13. Circle with center $A(2, 0)$ and radius 4
14. Circle with center $A(5, 3)$ and radius 2
15. Circle with center $A(-1, 1)$ and radius 3
16. Circle with center $A(-3, -2)$ and radius 5

6-6 THE ELLIPSE

An *ellipse* is defined as the locus of all points each of which has the sum of its undirected distances from two fixed points equal to a particular nonnegative real number. In Fig. 6-9 the points F and F' are 8 units apart. If P moves so that the sum of its undirected distances from F and F' is equal to 10 units, it will trace the ellipse shown. F and F' are called the *foci* of the ellipse, and the point midway between them is called its *center*.

To obtain a simple equation for this ellipse choose axes as shown in Fig. 6-10, with the x axis through the foci and the origin at the center. If the distance between the foci is denoted by $2c$, the coordinates of F' and F are $F'(-c, 0)$ and $F(c, 0)$.

Now let $P(x, y)$ be a point of the plane, and let its undirected distances from F' and F be d_1 and d_2, respectively; then

$$d_1 = \sqrt{(x+c)^2 + y^2} \qquad \text{and} \qquad d_2 = \sqrt{(x-c)^2 + y^2}$$

Now $d_1 + d_2$ must be equal to some specified constant, which must, of course, be greater than $2c$. Denote this constant by $2a$, with the understanding that $a > c$. Then the coordinates of every point for which $d_1 + d_2 = 2a$ must satisfy the relation

$$\sqrt{(x+c)^2 + y^2} + \sqrt{(x-c)^2 + y^2} = 2a$$

Subtract the second radical from both sides, square both sides, and simplify to obtain

$$a\sqrt{(x-c)^2 + y^2} = a^2 - cx$$

Squaring again, obtain

$$a^2(x^2 - 2cx + c^2 + y^2) = a^4 - 2a^2cx + c^2x^2$$

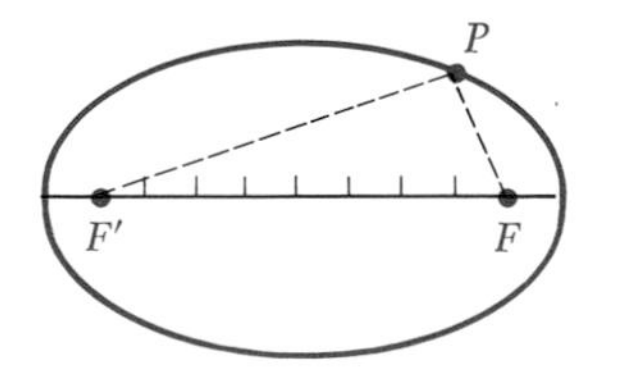

Fig. 6-9 $FP + F'P = 10$; point P traces an ellipse.

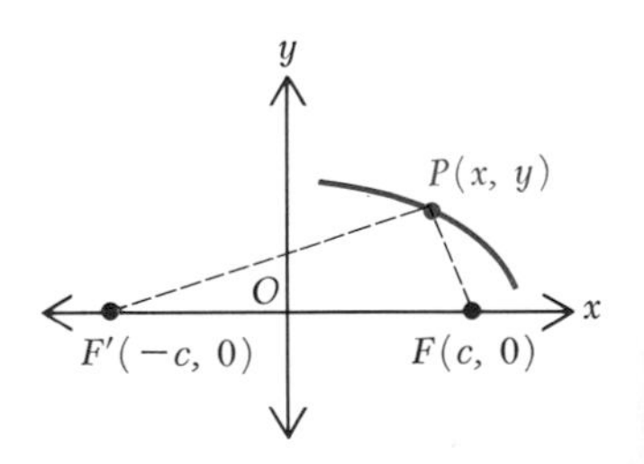

Fig. 6-10 Point P is on ellipse with center at O and foci at $F(c, 0)$ and $F'(-c, 0)$, and $PF + PF' = 2a$.

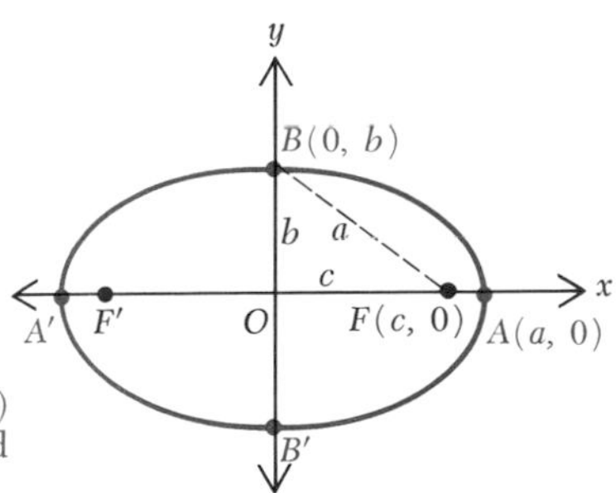

Fig. 6-11 Ellipse with center at O, foci at $F(c, 0)$ and $F'(-c, 0)$, vertices $A(a, 0)$, $A'(-a, 0)$, $B(0, b)$, and $B'(0, -b)$, and equation $x^2/a^2 + y^2/b^2 = 1$.

or
$$(a^2 - c^2)x^2 + a^2y^2 = a^2(a^2 - c^2)$$

Divide both sides by $a^2(a^2 - c^2)$ to get

$$\frac{x^2}{a^2} + \frac{y^2}{a^2 - c^2} = 1$$

Now, $a^2 - c^2$ is positive, because $a > c$, and we denote it by the new symbol b^2. The equation then takes the form

$$\frac{x^2}{a^2} + \frac{y^2}{b^2} = 1 \tag{6-11}$$

where a, b, and c are connected by the relation

$$a^2 = b^2 + c^2 \tag{6-12}$$

We have proved that the coordinates of every point in the plane that has $d_1 + d_2 = 2a$ must satisfy Eq. (6-11). It can be proved, conversely, that the locus of Eq. (6-11) does not contain any points other than those for which $d_1 + d_2 = 2a$. Equation (6-11) is called the *standard equation* of the ellipse with center at the origin and foci on the x axis.

The intercepts of the ellipse on the x and y axes are readily found to be $(\pm a, 0)$ and $(0, \pm b)$. The points A' and A (Fig. 6-11) are called the *vertices* of the ellipse; the line segment $\overline{A'A}$, whose length is $2a$, is called its *major axis;* and the segment $\overline{B'B}$, whose length is $2b$, is called its *minor axis.* Thus the graphical interpretation of the constants a and b in the equation $x^2/a^2 + y^2/b^2 = 1$ and the constant c is:

$c =$ distance from center to focus $= |OF|$
$a =$ distance from center to end of major axis $= |OA|$
$b =$ distance from center to end of minor axis $= |OB|$

The relation of a, b, and c can be seen in the triangle in Fig. 6-11; note that $|BF| = |OA| = a$.

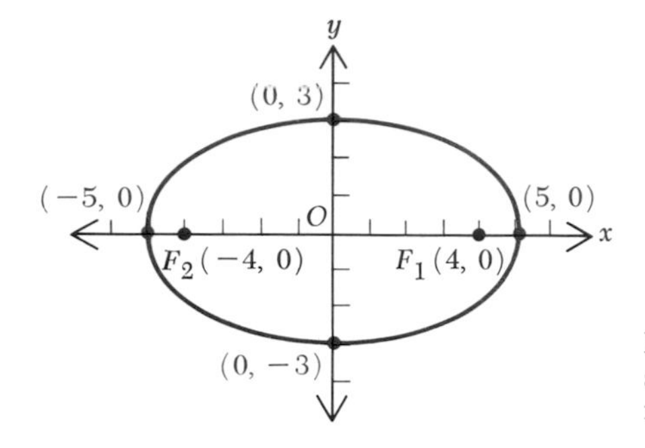

Fig. 6-12 Ellipse with center at O and semimajor and semiminor axes 5 and 3 units, respectively.

Example 1

In the equation $x^2/25 + y^2/9 = 1$, $a = 5$ and $b = 3$; $c = \sqrt{a^2 - b^2} = 4$ and the foci are at $F_1(4, 0)$ and $F_2(-4, 0)$. This is the equation of the ellipse shown in Fig. 6-12. All points outside this ellipse are represented by $x^2/25 + y^2/9 > 1$. Similarly, $x^2/25 + y^2/9 < 1$ represents all points inside the ellipse.

If the y axis (rather than the x axis) passes through the foci of an ellipse, coordinates of the foci are $(0, \pm c)$ and the equation of the ellipse is

$$\frac{x^2}{b^2} + \frac{y^2}{a^2} = 1 \tag{6-13}$$

Observe that since $a^2 = b^2 + c^2$, the larger of the two denominators in the equation of the ellipse is always a^2. The foci are on the x axis if this larger number is the denominator of the x^2 term and on the y axis if it is the denominator of the y^2 term.

Example 2

In the equation $x^2/9 + y^2/16 = 1$, $a = 4$ and $b = 3$; $c = \sqrt{a^2 - b^2} = \sqrt{7}$, and the foci are at $(0, \pm\sqrt{7})$.

6-7 ECCENTRICITY

The shape of the ellipse is determined by the ratio of c to a. This ratio, which is a number between 0 and 1, because $a > c > 0$, is called the *eccentricity* of the ellipse, and is denoted by the letter e:

$$e = \frac{c}{a}$$

Example

Find the eccentricity of the ellipse with equation $x^2/64 + y^2/16 = 1$.

Solution The equation $x^2/64 + y^2/16 = 1$ represents an ellipse in which $a = 8$ and $b = 4$; $c = \sqrt{64 - 16} = 4\sqrt{3}$, and the eccentricity of the ellipse is

$$\frac{4\sqrt{3}}{8} = \frac{\sqrt{3}}{2} \approx 0.866$$

It can easily be shown that if c/a, which is the same as $\sqrt{(a^2 - b^2)}/a$, is near 1, then b is small compared with a and the ellipse is long and narrow. On the other hand, if c/a is near zero, then b is nearly equal to a and the ellipse is nearly a circle. In fact, as c approaches zero, and the foci consequently approach the center of the ellipse, the value of b approaches that of a. If $b = a$, then $e = 0$, and the equation $x^2/a^2 + y^2/b^2 = 1$ becomes $x^2 + y^2 = a^2$. The circle may therefore be regarded as an ellipse with eccentricity zero.

PROBLEMS

In Probs. 1 to 8, sketch the ellipse, find the coordinates of its foci, compute its eccentricity, and write the inequality representing the set of points inside it.

1. $\dfrac{x^2}{9} + \dfrac{y^2}{25} = 1$
2. $\dfrac{x^2}{27} + \dfrac{y^2}{4} = 1$
3. $\dfrac{x^2}{1} + \dfrac{y^2}{16} = 1$
4. $\dfrac{x^2}{49} + \dfrac{y^2}{81} = 1$
5. $4x^2 + 3y^2 = 36$
6. $x^2 + \dfrac{3y^2}{4} = 24$
7. $\dfrac{2x^2}{3} + \dfrac{3y^2}{2} = 24$
8. $\frac{3}{4}x^2 + 4y^2 = 108$

In Probs. 9 to 14 find the equation of the ellipse with center at the origin that satisfies each of the given conditions:

9. One focus at $F(0, 6)$ and $e = \frac{1}{2}$
10. One vertex at $V(9, 0)$ and $e = \frac{2}{3}$
11. One end of the minor axis at $A(0, 4)$ and $e = 0.6$
12. One focus at $F(0, \frac{3}{2})$ and $e = \frac{1}{2}$
13. One vertex at $V(\frac{7}{2}, 0)$ and $e = \frac{2}{3}$
14. One end of the minor axis at $A(0, 1.4)$ and $e = 0.4$
15. What is the eccentricity of an ellipse whose major axis is (*a*) twice as long as its minor axis? (*b*) Ten times as long?
16. If an ellipse has its center at the origin, one focus at $F(3, 0)$, and one end of the major axis at $A(5, 0)$, can the end of the minor axis be at $B(0, 3)$? Why?

17. The chord drawn through either focus of an ellipse perpendicular to its major axis is called the *latus rectum* of the ellipse. Show that its length is $2b^2/a$.
18. Find the length of the latus rectum of the ellipse in Probs. 1 to 4.

6-8 ELLIPSE WITH CENTER AT $C(h, k)$

Let an ellipse have its center at $C(h, k)$ and its major and minor axes parallel to the coordinate axes (Fig. 6-13). If axes x' and y' are drawn through the center as shown, the equation of the curve relative to these new axes is

$$\frac{x'^2}{a^2} + \frac{y'^2}{b^2} = 1$$

In order to find the equation of the ellipse relative to the x and y axes let $x' = x - h$ and $y' = y - k$. The result is

$$\frac{(x-h)^2}{a^2} + \frac{(y-k)^2}{b^2} = 1 \tag{6-14}$$

This is the standard equation of the ellipse with center at $C(h, k)$ and major axis parallel to the x axis.

If the major axis is parallel to the y axis, the equation is

$$\frac{(x-h)^2}{b^2} + \frac{(y-k)^2}{a^2} = 1 \tag{6-15}$$

Either of these equations may be rewritten in the form

$$Ax^2 + Cy^2 + Dx + Ey + F = 0 \tag{6-16}$$

in which A and C have the same sign. Conversely, any equation of

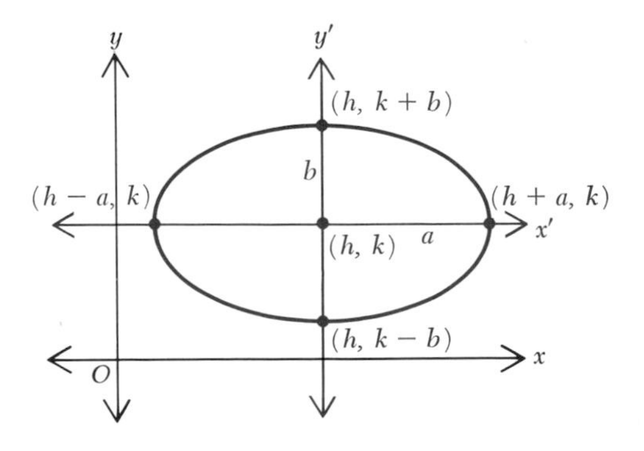

Fig. 6-13 Ellipse with center $C(h, k)$, semimajor axis a units, and semiminor axis b units: $(x-h)^2/a^2 + (y-k)^2/b^2 = 1$.

form (6-16) can be put into form (6-14) or (6-15) by completing the squares in x and y, provided that A and C have the same sign and provided that the constant $-F + D^2/4A + E^2/4C$, which appears on the right side of Eq. (6-16) when the left side is expressed as a sum of squares, is positive.

Example 1
Write the equation $x^2 + 4y^2 - 16x + 16y + 76 = 0$ in standard form.

Solution First write the equation in the form

$$(x^2 - 16x) + 4(y^2 + 4y) = -76$$

Now add to both members the numbers necessary to complete the squares:

$$(x^2 - 16x + 64) + 4(y^2 + 4y + 4) = -76 + 64 + 16$$

Note that the value of the right-hand member is now positive. The equation may be written as

$$(x - 8)^2 + 4(y + 2)^2 = 4$$

or, in standard form, as

$$\frac{(x-8)^2}{4} + \frac{(y+2)^2}{1} = 1 \tag{6-17}$$

The graph is an ellipse with center at $(8, -2)$ and with $a = 2$ and $b = 1$; the major axis is parallel to the x axis, and $c = \sqrt{3}$. The coordinates of the foci are $(8 \pm \sqrt{3}, -2)$.

If the right-hand member of Eq. (6-17) had been zero, the graph would have been a *point ellipse*, that is, the point $(8, -2)$. If it had been negative, there would have been no locus. It is evident then that every equation of form (6-16) in which A and C have the same sign represents an ellipse, a single point, or no locus.

Example 2
The equation

$$x^2 + 2y^2 - 2x + 16y + 45 = 0$$

becomes

$$(x - 1)^2 + 2(y + 4)^2 = -12$$

in standard form. There is no locus. If the constant 45 had been 33, the locus would have been the point $(1, -4)$.

6-9 PARAMETRIC EQUATIONS OF THE CIRCLE AND ELLIPSE

Consider a circle with center at the origin and radius a as shown in Fig. 6-14. Let $P(x, y)$ be a point of the plane, and let θ be the angle LOP. It is clear that if and only if P lies on the circle,

$$\begin{aligned} x &= a \cos \theta \\ y &= a \sin \theta \end{aligned} \qquad (6\text{-}18)$$

Equations (6-18) are then parametric equations of the circle, with θ the parameter. The direct relation between x and y can, of course, be obtained by eliminating θ between these equations. This can readily be done by squaring both sides of each equation in (6-18) and adding:

$$\begin{aligned} x^2 &= a^2 \cos^2 \theta \\ y^2 &= a^2 \sin^2 \theta \end{aligned}$$

Then

$$x^2 + y^2 = a^2 (\cos^2 \theta + \sin^2 \theta)$$

Now consider the equations

$$\begin{aligned} x &= a \cos \theta \\ y &= b \sin \theta \end{aligned} \qquad (6\text{-}19)$$

It is easy to show that they represent an ellipse whose semiaxes are a and b, for if they are written in the equivalent form

$$\frac{x}{a} = \cos \theta$$

$$\frac{y}{b} = \sin \theta$$

and then both sides of each equation are squared and the results added, the equation is

$$\frac{x^2}{a^2} + \frac{y^2}{b^2} = 1$$

The ellipse can be constructed point by point from the parametric equations (6-19). Draw two concentric circles having radii a and b, where $a > b$, as shown in Fig. 6-15. These are called the *major* and *minor auxiliary circles,* respectively. To determine a point on the ellipse, draw a line through the origin at any angle θ with the positive x axis and cutting the major and minor auxiliary circles at A and B, re-

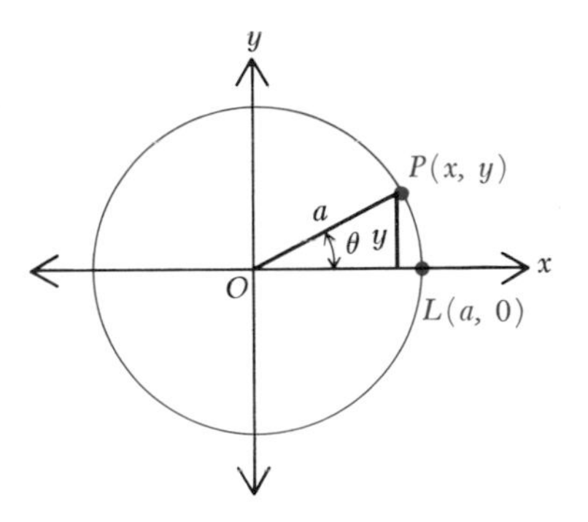

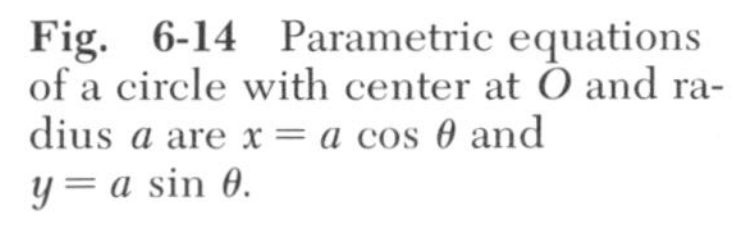
Fig. 6-14 Parametric equations of a circle with center at O and radius a are $x = a \cos \theta$ and $y = a \sin \theta$.

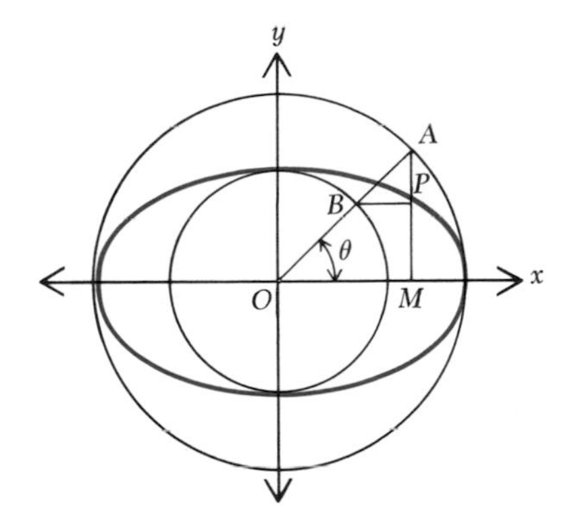

Fig. 6-15 Constructing an ellipse from its parametric equations.

spectively. Through A draw a line parallel to the y axis, and through B draw a line parallel to the x axis. These lines intersect at a point P which is a point on the ellipse; for if the coordinates of P are (x, y), then

$$x = OM = OA \cos \theta = a \cos \theta$$
$$y = MP = OB \sin \theta = b \sin \theta$$

The ellipse has many scientific applications. Elliptical gears are used in certain kinds of machinery to give a slow powerful stroke and a quick return; arches in the form of semiellipses are often employed in the construction of stone and concrete bridges. The orbits in which the planets revolve about the sun are ellipses, with the sun at one focus (the orbit of the earth has an eccentricity of about $\frac{1}{60}$). The projection of a circle on a plane not parallel to the plane of the circle is an ellipse (you may enjoy trying this).

PROBLEMS

For each ellipse in Probs. 1 to 6 find the center, give the length of the major and minor axes, find the eccentricity, and sketch the graph.

1. $\dfrac{(x-2)^2}{9} + \dfrac{(y-1)^2}{1} = 1$

2. $\dfrac{(x-3)^2}{4} + \dfrac{(y-2)^2}{1} = 1$

3. $\dfrac{(x+2)^2}{16} + \dfrac{(y-2)^2}{9} = 1$

4. $\dfrac{(x+2)^2}{25} + \dfrac{(y+2)^2}{16} = 1$

5. $\dfrac{(x+5)^2}{100} + \dfrac{(y+3)^2}{64} = 1$

6. $\dfrac{(x-\frac{3}{2})^2}{64} + \dfrac{(y+\frac{1}{2})^2}{25} = 1$

In Probs. 7 to 10 find the equation of the ellipse that satisfies each set of conditions:

7. Center at $A(6, -4)$, one vertex at $B(10, -4)$, and $e = \frac{1}{2}$
8. Vertices at $A(-5, -2)$ and $B(7, -2)$ and one focus at $C(5, -2)$
9. Center at $A(4, -4)$, one end of minor axis at $B(0, -4)$, and one focus at $C(4, 0)$
10. Foci at $F_1(-2, 3)$ and $F_2(6, 3)$ and one vertex at $V(8, 3)$
11. Let the undirected distances of a point $P(x, y)$ from $A(1, 0)$ and $B(9, 0)$ be d_1 and d_2, respectively. Show that if $d_1 + d_2 = 10$, then the coordinates of P must satisfy the equation $9x^2 + 25y^2 = 90x$. Have you proved that every point whose coordinates satisfy this equation has $d_1 + d_2 = 10$?
12. A point moves so that the sum of its undirected distances from $A(0, -3)$ and $B(0, -9)$ is equal to 10. Find the equation of its path and sketch it.
13. A point moves so that its undirected distance from the origin is equal to one-half its undirected distance from the line $x + 9 = 0$. Find the equation of its path and sketch it.
14. A point moves so that its undirected distance from the point $A(0, -2)$ is equal to one-half its undirected distance from the line $y = 4$. Find the equation of its path and sketch it.
15. Find the equation of the locus of points that are two-thirds as far from the point $A(6, 0)$ as from the y axis.
16. A semielliptic arch in a stone bridge has a span of 20 ft and a height of 6 ft, as shown in Fig. 6-16. To construct the arch it is necessary to know its height at distances of 2, 4, 6, 8, and 9 ft from its center, as indicated by the dashed segments. Compute these heights to the nearest $\frac{1}{10}$ ft.

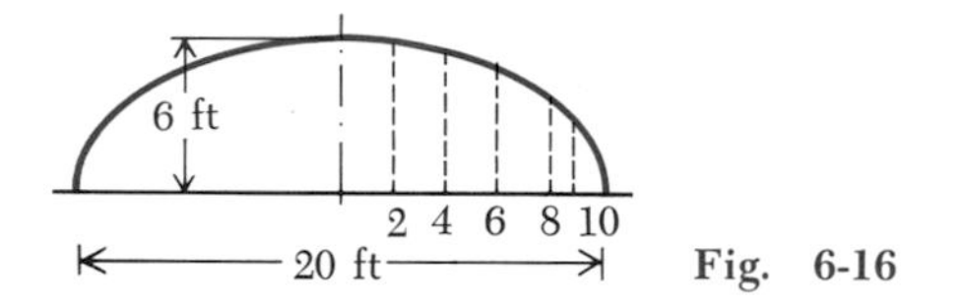

Fig. 6-16

17. A segment $\overline{PQ}$ of length 24 units moves so that P is always on the y axis and Q is on the x axis. A point M is on $\overline{PQ}$ two-thirds of the way from P to Q. Find the equation of the path traveled by M.
18. The orbit in which the earth travels about the sun is an ellipse with the sun at one focus. The semimajor axis of the ellipse is approximately 92.9 million miles, and its eccentricity is approximately 0.0168. Determine the greatest and least distances of the earth from the sun, correct to the nearest 100,000 miles.

Write the equations in Probs. 19 to 26 in standard form and draw the corresponding graph if there is one.

19. $9x^2 + 4y^2 - 54x + 8y + 49 = 0$
20. $4x^2 + 9y^2 + 24x - 36y + 36 = 0$
21. $x^2 + 4y^2 - 32y + 48 = 0$

22. $4x^2 + y^2 - 16x = 0$
23. $x^2 + y^2 + 3x - 4y = 0$
24. $x^2 + 2y^2 + 4x + 6y = 0$
25. $2x^2 + 3y^2 = 8x + 18y - 29$
26. $9x^2 + 4y^2 + 36x + 24y + 72 = 0$
27. Find the points of intersection of the line $3x - 5y + 3 = 0$ and the ellipse $9x^2 + 25y^2 = 225$ and draw the graph.
28. Find the points of intersection of the line $9x - 5y = 60$ and the ellipse $9x^2 + 25y^2 = 90x$ and draw the graph.
29. Show that the equations

$$x = h + a \cos \theta$$
$$y = k + a \sin \theta$$

represent a circle of radius a with center at the point $D(h, k)$.

30. Show that the equations

$$x = h + a \cos \theta$$
$$y = k + b \sin \theta$$

represent an ellipse with center at the point $C(h, k)$ and with semiaxes a and b.

In Probs. 31 to 36 determine if each set of parametric equations represents a circle or an ellipse and graph each.

31. $x = 5 \cos \theta$, $y = 5 \sin \theta$
32. $x = 6 \cos \theta$, $y = 4 \sin \theta$
33. $x = 3 \cos \theta - 2$, $y = 3 \sin \theta + 4$
34. $x = 4(2 + \cos \theta)$, $y = 2 \sin \theta$
35. $x = 3 - 3 \cos \theta$, $y = 3 \sin \theta$
36. $x = 3 \sin \theta$, $y = 1 + \cos \theta$

6-10 (OPTIONAL) APPLICATIONS OF PARAMETRIC EQUATIONS OF A CIRCLE

It was found convenient to use parametric equations to represent the path of an object moving along a straight line with constant speed. In this section a parameter which is the measure of the angle of rotation will be used to study an object moving with constant speed around a circle. We shall analyze several simplified versions of very practical problems concerned with flywheels, armatures, and other objects which rotate with constant speed.

If a point on a circle of radius r moves counterclockwise from point A in Fig. 6-17, its position P at any time is determined by the parametric equations

$$x = r \cos \theta$$
$$y = r \sin \theta$$

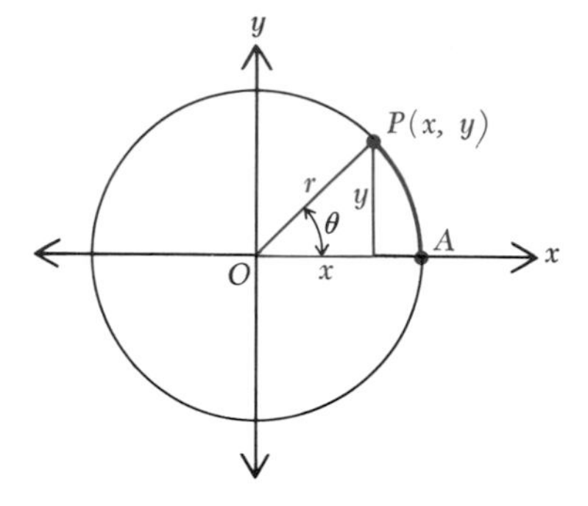

Fig. 6-17 Point P moving counterclockwise on circle, starting at point A.

If the point moves around the circle at a constant speed of so many inches per second, it is often convenient to express its speed in units of rotations per second. More often this *angular velocity* is given as a number of degrees per unit of time or a number of radians per unit of time and will be represented by the Greek letter ω. Thus if a wheel turns at a rate of 100 rpm, its angular velocity ω is 100(360) deg/min or $100(2\pi)$ rad/min.

Now, if point P is moving at a constant angular velocity ω, then the number of radians through which it moves in a time t is $\theta = \omega t$. The parametric equations of the path of a point P starting at A at $t = 0$ (Fig. 6-17) and moving counterclockwise with a constant angular velocity are therefore

$$x = r \cos \omega t \qquad (6\text{-}20)$$
$$y = r \sin \omega t$$

In problems relating to constant speed on a circle, ω will be given in radians per unit of time.

Example 1

Describe the motions of points Q and R which move according to the following parametric equations, in which t is a number of minutes:

For point Q:
$$x = 10 \cos 16\pi t$$
$$y = 10 \sin 16\pi t$$

For point R:
$$x = 10 \cos -20\pi t$$
$$y = 10 \sin -20\pi t$$

Solution Points Q and R both move on the circle with equation $x^2 + y^2 = 100$. For point Q, for example,

$$\begin{aligned} x^2 + y^2 &= 100 \cos^2 16\pi t + 100 \sin^2 16\pi t \\ &= 100 (\cos^2 16\pi t + \sin^2 16\pi t) \\ &= 100 \end{aligned}$$

Much additional information about the motion of the points can be derived by a careful study of the parametric equations.

1. Both points are at point A at $t = 0$ (Fig. 6-18).
2. For point Q, $16\pi = 8(2\pi)$. Hence it makes 8 rpm. For point R, $-20 = -10(2\pi)$. Hence it makes 10 rpm. However, point Q is moving counterclockwise and point R is moving clockwise. This is indicated by the fact that the angle for the motion of point Q increases and the angle for the motion of point R decreases as t increases from zero.
3. In 1 min point Q has swept out an angle 16π starting at zero. Its angular displacement is then 16π rad/min. The angular displacement for point R is -20π rad/min.
4. Point Q is at A (Fig. 6-18) when t equals any member of the set

$$\{0, \tfrac{1}{8}, \tfrac{2}{8}, \tfrac{3}{8}, \ldots, n/8\} \qquad \text{for } n \text{ a nonnegative integer}$$

Furthermore, its position may be located at any time t. For example, for its first revolution starting at $t = 0$, number pairs (x, y) may be determined from the following table:

t	0	$\frac{1}{96}$	$\frac{1}{32}$	$\frac{1}{16}$	$\frac{1}{12}$	$\frac{7}{64}$
x	10	$\frac{10\sqrt{3}}{2}$	0	-10	$-10(\frac{1}{2})$	$\frac{10\sqrt{2}}{2}$
y	0	$10(\frac{1}{2})$	10	0	$-\frac{10\sqrt{3}}{2}$	$-\frac{10\sqrt{2}}{2}$

The ordered pairs are graphed in Fig. 6-18.

Positions for point R may be graphed in a similar manner. From the

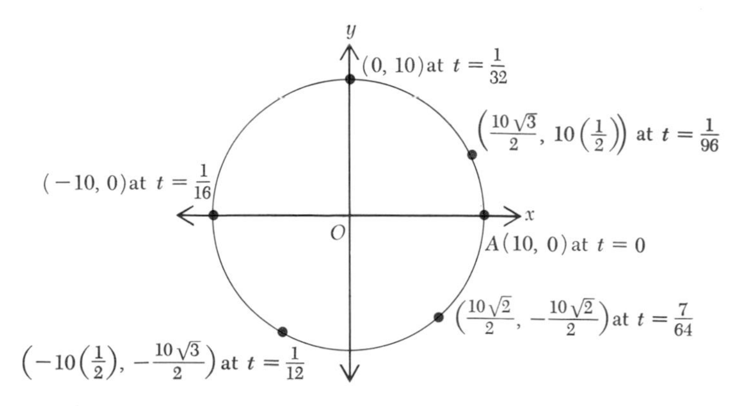

Fig. 6-18 Positions for a point moving counterclockwise around a circle.

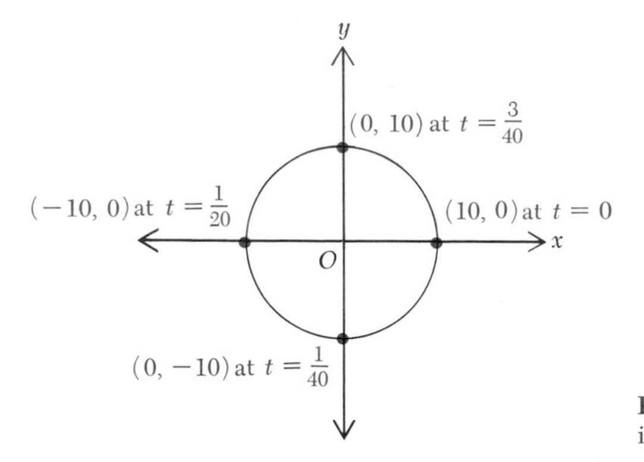

Fig. 6-19 Positions for a point moving clockwise around a circle.

following table, positions during the first revolution are listed as number pairs and graphed in Fig. 6-19.

t	0	$\frac{1}{40}$	$\frac{1}{20}$	$\frac{3}{40}$	$\frac{1}{10}$
x	10	0	−10	0	10
y	0	−10	0	10	0

The directions of the motion, counterclockwise for Q and clockwise for R, are verified by the above tables and Figs. 6-18 and 6-19.

Example 2

A point P starts at B in Fig. 6-20 when $t = 0$ and moves counterclockwise at a constant speed. At $t = \frac{1}{6}$ sec the point is at A. Write the parametric equations of its motion.

Solution It takes $\frac{1}{6}$ sec for half a revolution, or $\frac{1}{3}$ sec for one revolution; hence it makes 3 rps. The angular velocity is

$$3(2\pi) \text{ rad/sec}$$

The point starts (when $t = 0$) at $B(-5, 0)$; hence the equations of motion are

$$x = 5 \cos (\pi + 6\pi t)$$
$$y = 5 \sin (\pi + 6\pi t)$$

It should be noted that the π in $\pi + 6\pi t$ represents the displacement of the point P from A at the time $t = 0$, and that 6π, the coefficient of t, is the angular velocity.

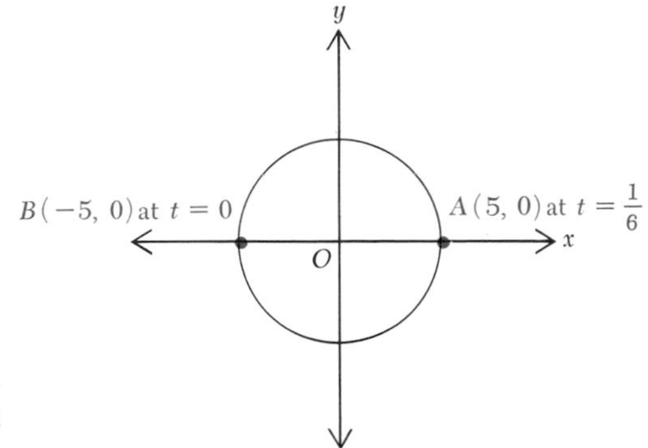

Fig. 6-20 Position for a point moving clockwise around a circle, starting at B.

Example 3

Write the equations of the path of point P in Example 2 if it moves clockwise and all other conditions are the same.

Solution In Example 2 the angle swept out by point P increased as t increased. In this example the angle decreases as t increases. Since the angular displacement is -6π rad/sec and the point is at $B(-5, 0)$ when $t = 0$, the parametric equations of the motion are

$$x = 5 \cos (\pi - 6\pi t)$$
$$y = 5 \sin (\pi - 6\pi t)$$

PROBLEMS

1. Write the parametric equations for a circle of radius 20 with center at the origin.
2. Write the parametric equations for the path of a point moving around the circle in Prob. 1 if it rotates counterclockwise at the rate of 3 rps and is at the three o'clock position (point A in Fig. 6-17) when $t = 0$.
3. Solve Prob. 2 if the rate of motion is 25 rps.
4. Solve Prob. 2 if the point moves clockwise instead of counterclockwise.

Describe in words the motions of the points whose paths are described by the parametric equations in Probs. 5 to 8.

5. $x = 8 \cos 6\pi t$
   $y = 8 \sin 6\pi t$
6. $x = 10 \cos - 12\pi t$
   $y = 10 \sin - 12\pi t$
7. $x = 4 \cos \left(\dfrac{\pi}{2} + 4\pi t\right)$
   $y = 4 \sin \left(\dfrac{\pi}{2} + 4\pi t\right)$
8. $x = 12 \cos \left(\dfrac{3\pi}{2} - 6\pi t\right)$
   $y = 12 \sin \left(\dfrac{3\pi}{2} - 6\pi t\right)$

Write the parametric equations for the path of a point that moves around a circle of radius 10 with center at the origin according to the conditions in each of Probs. 9 to 12.

9. The point moves counterclockwise, starting at its lowest point, at a rate of 25 rps.
10. The point moves clockwise at a rate of 16 rps and is at $A(-10, 0)$ at the time $t = 0$.
11. The point moves counterclockwise and is at $A(5\sqrt{2}, -5\sqrt{2})$ at the time $t = 0$.
12. The point moves clockwise and is at $A(0, 10)$ at time $t = \frac{1}{8}$ and at $B(10, 0)$ at time $t = \frac{3}{8}$. The times $t = \frac{1}{8}$ and $t = \frac{3}{8}$ are in the same revolution.

SUMMARY

Equations for the circle include the standard form, the general form, the polar form, the vector form, and the parametric form.

The standard form, $(x - h)^2 + (y - k)^2 = r^2$, represents a circle with center (h, k) and radius r.

The general form is $x^2 + y^2 + Dx + Ey + F = 0$. Both the general and the standard forms have three arbitrary constants, indicating that the circle satisfies three conditions.

Some special cases of the polar form for a circle are:

Center at origin and radius r:	$\rho = r$
Center at $(r, 0)$ and radius r:	$\rho = 2r \cos \theta$
Center at $(r, \pi/2)$ and radius r:	$\rho = 2r \sin \theta$

The simplest vector form for the equation of a circle is $\mathbf{AP} \cdot \mathbf{AP} = r^2$, where A is the center of the circle, P is a point on the circle, and r is the radius.

The parametric equations of the circle with center at the origin and radius a are $x = a \cos \theta$ and $y = a \sin \theta$. The parametric equations are particularly useful in studying an object moving with constant speed around a circle.

An ellipse is the locus of all points each of which has the sum of its undirected distances from two fixed points equal to a nonnegative real number. A circle may be considered as a special case of an ellipse. The following forms of the equation of an ellipse each display certain information about the ellipse:

1. The standard form for an ellipse with center at the origin is $x^2/a^2 + y^2/b^2 = 1$. The vertices are at $(\pm a, 0)$, with the length of the major

axis $2a$ and the length of the minor axis $2b$. The foci are $(\pm c, 0)$, where $a^2 = b^2 + c^2$. The shape of the ellipse is determined by the eccentricity, which is defined as the ratio c/a.

2. An ellipse with center at (h, k) and major axis parallel to the x axis has the equation $(x-h)^2/a^2 + (y-k)^2/b^2 = 1$. This equation may be written in the general form $Ax^2 + Cy^2 + Dx + Ey + F = 0$.
3. The parametric equations of an ellipse whose semiaxes are a and b, with center at the origin, is $x = a \cos \theta$ and $y = b \sin \theta$.

REVIEW PROBLEMS

1. Identify each of the following equations as representing a circle, an ellipse, or a point or as having no graph. If the equation has a graph, sketch it.

   (*a*) $(x-3)^2 + (y+2)^2 = 16$ (*b*) $x^2/16 + y^2/16 - 1 = 0$
   (*c*) $25x^2 + 36y^2 - 1 = 0$ (*d*) $x^2 + 4y^2 + 4x - 24y + 24 = 0$
   (*e*) $x^2 + y^2 - 7y = 0$ (*f*) $4x^2 + 3y^2 = 9$
   (*g*) $3x^2 + 4y^2 + 12x - 24y + 60 = 0$ (*h*) $x^2 + y^2 - 4x - 6y + 14 = 0$

2. Find the equation of each of the following circles:

   (*a*) Center at $C(0, 0)$ and passing through $P(-4, 7)$
   (*b*) Center at $C(5, -3)$ and circumference 18π
   (*c*) Tangent to the x axis, with center at $C(-3, 5)$
   (*d*) Passing through $A(-2, 3)$ and concentric to $x^2 + y^2 + 2x + 10y + 25 = 0$

3. The height of a semicircular arch 2 ft from the end is 6 ft. Find the maximum height of the arch.
4. A triangle has vertices at $A(3, 0)$, $B(4, 2)$, and $C(0, 1)$. Determine the equation of a circle that passes through each vertex of the triangle.
5. Prove that the locus of a point which moves so that the sum of the squares of the distances from two fixed points is a positive constant is a circle. *Hint:* Use $A(a, 0)$ and $B(-a, 0)$ as the fixed points.
6. Sketch the following curves:

   (*a*) $\rho = 3 \sin \theta$ (*b*) $\rho = 5 \sin (\theta + \frac{1}{4}\pi)$
   (*c*) $x = 8 \cos \theta$, $y = 8 \sin \theta$ (*d*) $x = 3 \cos \theta$, $y = 5 \sin \theta$
   (*e*) $2x^2 + 2y^2 - 6x + 2y - 5 = 0$ (*f*) $4y^2 + 9x^2 - 24y - 72x + 144 = 0$

7. Translate the axes for each equation so the new origin will be at the point indicated:

   (*a*) $9x^2 + 16y^2 + 72x - 96y + 144 = 0$ and origin at $P(-4, 3)$
   (*b*) $x^2 + y^2 - 5x + 2y - 5 = 0$ and origin at $P(\frac{5}{2}, -1)$

8. Find the locus of a point $P(x, y)$ that moves so that its distance from $Q(6, 0)$ is one-half its distance from the line $x = 24$.

9. Determine the equations of the following ellipses, each of which has axes parallel to the coordinate axes:

   (*a*) One end of the major axis at $V(6, 0)$, one focus at $F(-1, 0)$, and center at $Q(0, 0)$
   (*b*) Foci at $A(-2, 1)$ and $B(-2, 0)$ and $e = \frac{1}{2}$
   (*c*) Center at $C(0, -\frac{1}{2})$, one focus at $F(0, 1)$, and passing through $G(2, 2)$

10. For each ellipse find the coordinates for the center, the ends of the major and minor axes, and the foci and determine the eccentricity:

    (*a*) $16x^2 + 25y^2 - 160x + 50y - 1{,}175 = 0$
    (*b*) $2x^2 + 3y^2 - 8x - 18y + 29 = 0$

11. An arch of an underpass is a semiellipse 50 ft wide and 20 ft high. Find the clearance at the right edge of a lane if the edge is 18 ft from the middle.

FOR EXTENDED STUDY

1. Find the equation of the line tangent to $x^2 + y^2 - 10x - 6y + 29 = 0$ at the point $P(3, 4)$.
2. Show that the line tangent to the circle $x^2 + y^2 = r^2$ at any point $R(x_1, y_1)$ on the circle has the equation $x_1x + y_1y = r^2$.
3. Let the equations of two intersecting circles C_1 and C_2 (Fig. 6-21) be

$$x^2 + y^2 + D_1x + E_1y + F_1 = 0$$

and

$$x^2 + y^2 + D_2x + E_2y + F_2 = 0$$

Show that for every value of k except -1 the equation

$$(x^2 + y^2 + D_1x + E_1y + F_1) + k(x^2 + y^2 + D_2x + E_2y + F_2) = 0$$

represents a circle that passes through the points of intersection of C_1 and C_2. What does the equation represent for $k = -1$? *Hint:* If $k \neq -1$ the equation represents a circle. Furthermore, it is satisfied by the coordinates of the points of intersection of C_1 and C_2. Why?

4. Let $A(x_1, y_1)$, $B(x_2, y_2)$, and $C(x_3, y_3)$ be three distinct points, not all on the

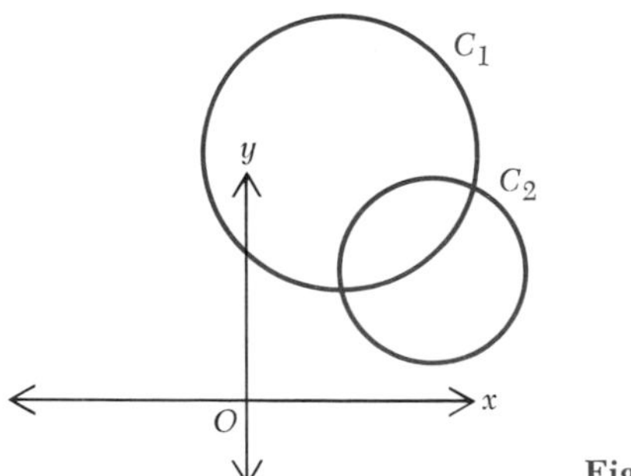

Fig. 6-21

same straight line, having rational numbers as coordinates. Is it possible that the coordinates of the center of the circle passing through A, B, and C may be irrational numbers? Is it possible that the radius may be an irrational number?

5. A point $P(x, y)$ moves so that its undirected distance from the point $A(-2, 0)$ remains equal to one-half its undirected distance from the line $x + 5 = 0$. Find the coordinates of the point or points on its path that are farthest from the x axis.
6. Use a reference book to find the meaning of the circle of Apollonius with respect to two points. Write a definition and draw an example of the circle.
7. Find the points of intersection of the two circles whose equations are

$$x^2 + y^2 - 16x - 10y + 4 = 0$$

and

$$x^2 + y^2 + 4x + 2y - 12 = 0$$

by the following two methods: (*a*) Solve the first equation for y and substitute in the second equation. (*b*) Multiply both sides of the first equation by -1 and add. The result is a linear equation. What is its relation to the circles? Continue the solution.

8. Prove that in an ellipse the length of the major axis is the mean proportional of the distance between the foci and the distance between the directrices.

THE PARABOLA AND THE HYPERBOLA 7

7-1 INTRODUCTION

In Chap. 6 definitions and equations for circles and ellipses were given. In this chapter we shall study the definitions of two closely related curves, the *parabola* and the *hyperbola*. These definitions will be used to derive standard equations for the curves and to form a basis for extracting their most important properties. The standard equations of circles, ellipses, parabolas, and hyperbolas are special cases of the general equation of second degree in x and y,

$$Ax^2 + Bxy + Cy^2 + Dx + Ey + F = 0$$

in which some of the coefficients may be zero (in Chap. 6 we saw that if $B = 0$ and $A = C$ the equation represents a circle or has no locus). Finally, we shall see that every equation of the second degree in x and y has for its graph (if it has a graph) a circle, a parabola, an ellipse, or a hyperbola, or a limiting case thereof; these loci are the various intersections of a right circular cone and a plane. For this reason they are called *conic sections*.

HISTORICAL NOTE

The Greek mathematician Apollonius, who died about 200 B.C., made a thorough investigation of conic sections. His work, "Conic Sections," included about 400 propositions on conics. Apollonius is credited with providing the names "ellipse," "parabola," and "hyperbola" and for discovering that all the conic sections result from intersections of a cone and certain planes, as shown in Fig. 7-1. The conics continued to interest mathematicians, and the theory associated with them was essentially completed by Descartes and Pascal during the seventeenth century.

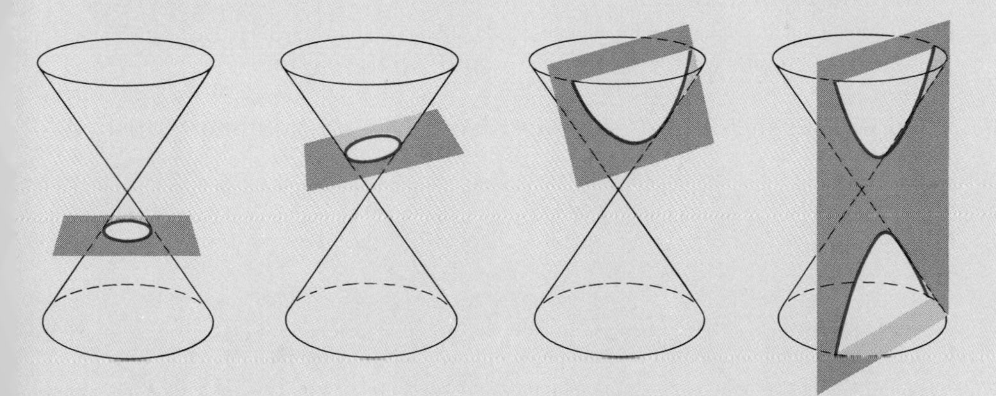

Fig. 7-1 The intersection of a cone and a plane is a circle, an ellipse, a parabola, or a hyperbola.

7-2 THE PARABOLA

A *parabola* is defined as the locus of all points equidistant from a fixed point and a fixed line (the distances referred to here are undirected). Thus if a point P (Fig. 7-2) moves so that the undirected distances $|FP|$ and $|LP|$ remain equal, it will trace out a parabola. The fixed point F is called the *focus*, and the fixed line l is called the *directrix* of the parabola.

Naturally, the equation of the curve depends on the choice of coordinate axes relative to the focus and directrix. The simplest equation results when axes are chosen as indicated in Fig. 7-3. Here a segment $\overline{MF}$ is drawn such that point F is the focus, point M is on the directrix l, and $\overline{MF}$ is perpendicular to l. The origin is chosen as the midpoint of $\overline{MF}$, with $\overleftrightarrow{MF}$ as the x axis. If p is the directed distance from the directrix to the focus, then the coordinates of the focus are $(\frac{1}{2}p, 0)$, and the equation of the directrix is $x = -p/2$. The equation of the curve with respect to these axes can be found as follows. Let $P(x, y)$ be a point of the plane and let d_1 and d_2 be the undirected lengths of $\overline{FP}$ and $\overline{LP}$, respectively; then

$$d_1 = \sqrt{\left(x - \frac{p}{2}\right)^2 + y^2} \qquad \text{and} \qquad d_2 = \sqrt{\left(x + \frac{p}{2}\right)^2}$$

Every point such that $d_1 = d_2$ must have $d_1{}^2 = d_2{}^2$, so it must satisfy the equation

$$\left(x - \frac{p}{2}\right)^2 + y^2 = \left(x + \frac{p}{2}\right)^2$$

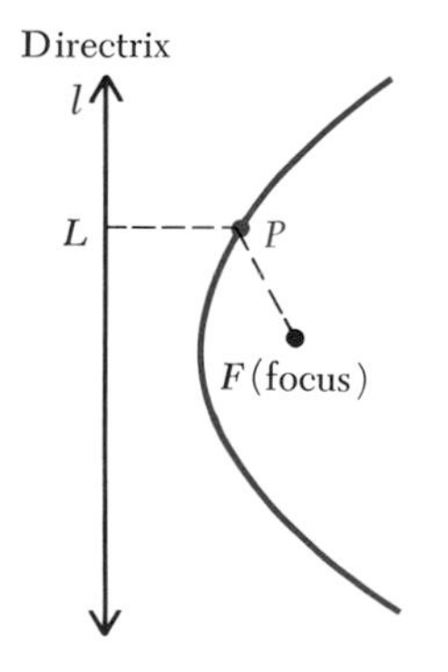

Fig. 7-2 Point P on a parabola is equidistant from the focus and the directrix.

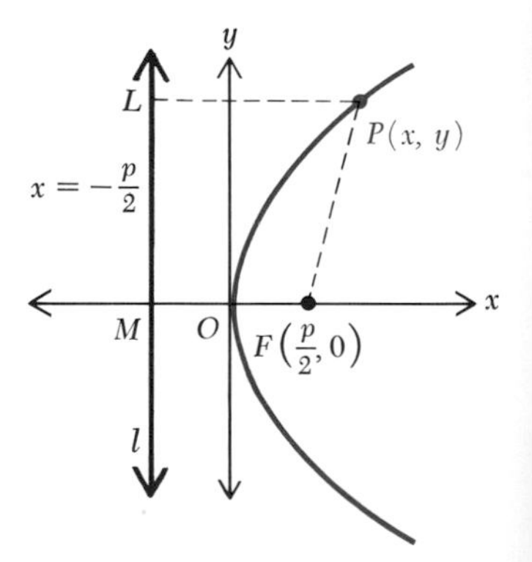

Fig. 7-3 Parabola with vertex at O, focus at $(p/2, 0)$, and directrix $x = -p/2$: $y^2 = 2px$.

This equation may be expressed as

$$x^2 - px + \frac{p^2}{4} + y^2 = x^2 + px + \frac{p^2}{4}$$

or

$$y^2 = 2px \tag{7-1}$$

Every point in the plane for which $d_1 = d_2$ must satisfy Eq. (7-1). Conversely, any point $P(x, y)$ which satisfies (7-1) has $d_1 = d_2$. Any point $P(x, y)$ for which $y^2 = 2px$ will also satisfy that equation with $(x - p/2)^2$ added to each member. Hence

$$\begin{aligned} y^2 + \left(x - \frac{p}{2}\right)^2 &= 2px + \left(x - \frac{p}{2}\right)^2 \\ &= 2px + x^2 - px + \frac{p^2}{4} \\ &= x^2 + px + \frac{p^2}{4} \\ &= \left(x + \frac{p}{2}\right)^2 \end{aligned}$$

It will thus have $d_1{}^2 = d_2{}^2$, and since the positive square roots of two equal positive numbers are equal, it will be such that $d_1 = d_2$. Thus the graph of the equation $y^2 = 2px$ includes all points for which $d_1 = d_2$ and no points for which $d_1 \neq d_2$.

Equation (7-1) is the equation of the parabola with respect to axes taken as shown in Fig. 7-3. The line of symmetry of the curve, which in this case is the x axis, is called the *axis* of the parabola. The point at which the axis intersects the curve is called the *vertex* of the parabola. The vertex in this case is at the origin.

The case illustrated is that in which p is positive. If p is negative, the focus is to the left of the directrix, and the parabola opens to the left.

Example 1

The equation $y^2 = -8x$ is in the form $y^2 = 2px$, with $p = -4$. Thus the focus is at $F(-2, 0)$, and the directrix is the line $x = 2$ (see Fig. 7-4).

If the axes are chosen such that the focus is on the y axis at the point $(0, \frac{1}{2}p)$ and the directrix is the line $y = -\frac{1}{2}p$, the corresponding equation is

$$x^2 = 2py \tag{7-2}$$

In this case if p is positive, the focus is above the directrix, and the parabola opens upward; if p is negative, the focus is below the directrix and the parabola opens downward.

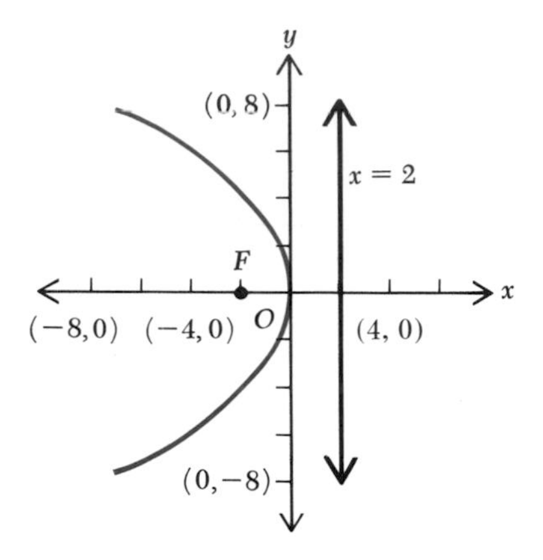

Fig. 7-4 Parabola with focus at $(-2, 0)$ and directrix $x = 2$: $y^2 = -8x$.

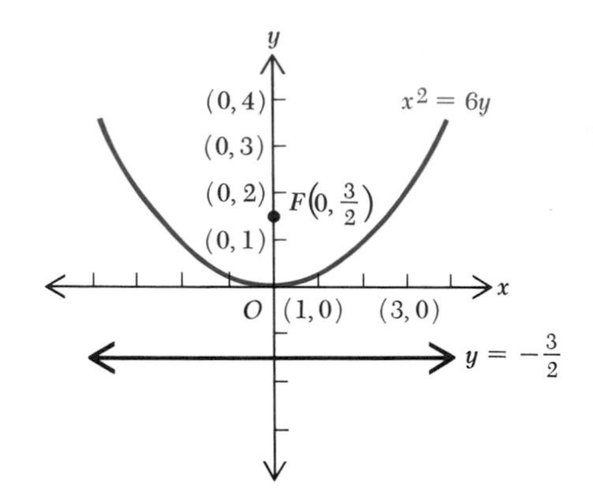

Fig. 7-5 Parabola with focus at $(0, \frac{3}{2})$ and directrix $y = -\frac{3}{2}$: $x^2 = 6y$.

Example 2

In the equation $x^2 = 6y$, or $y = \frac{1}{6}x^2$, $p = 3$. Thus the focus is at $F(0, \frac{3}{2})$, and the directrix is the line $y = -\frac{3}{2}$ (see Fig. 7-5).

7-3 PARABOLA WITH VERTEX AT $V(h, k)$ AND AXIS PARALLEL TO A COORDINATE AXIS

Let a parabola have its vertex at $V(h, k)$ and its axis parallel to the x axis (Fig. 7-6). New axes x' and y' may be drawn through the vertex as shown; the equation of the curve relative to these axes is

$$y'^2 = 2px'$$

In order to find the equation of the parabola relative to the x and y axes, use the transformation for translation of axes, $y' = y - k$ and $x' = x - h$. Then

$$(y - k)^2 = 2p(x - h) \tag{7-3}$$

This is standard form of the equation of a parabola with vertex at $V(h, k)$ and axis parallel to the x axis. The parabola opens to the right if p is positive and to the left if p is negative. Equation (7-1) is, of course, the special case of Eq. (7-3) in which $h = k = 0$.

If Eq. (7-3) is expanded, the result is an equation of the form

$$y^2 + Dx + Ey + F = 0 \tag{7-4}$$

Conversely, any equation of form (7-4) can be expressed in form (7-3) by completing the square in y if $D \neq 0$. It is therefore evident that any equation of form (7-4) with $D \neq 0$ represents a parabola with its axis parallel to the x axis.

If $D = 0$, the equation contains only one variable and represents a pair of parallel or coincident lines or has no locus. Thus the graph of $y^2 - 7y + 10 = 0$ is two parallel lines, $y = 2$ and $y = 5$; the equation $y^2 + 2y + 3 = 0$ has no graph.

Example 1

Express the equation $y^2 - 6y - 6x + 39 = 0$ in standard form and draw the graph.

Solution First write the equation in the form

$$y^2 - 6y = 6x - 39$$

To complete the square in y , add 9 to both sides:

$$y^2 - 6y + 9 = 6x - 30$$

This equation may now be written in the standard form (7-3) as

$$(y - 3)^2 = 6(x - 5)$$

The graph is a parabola with vertex at $V(5, 3)$. The value of p is 3; so the focus is $\frac{3}{2}$ units to the right of the vertex. The focus is at $F(\frac{13}{2}, 3)$, and the equation of the directrix is $x = \frac{7}{2}$. The graph is shown in Fig. 7-7.

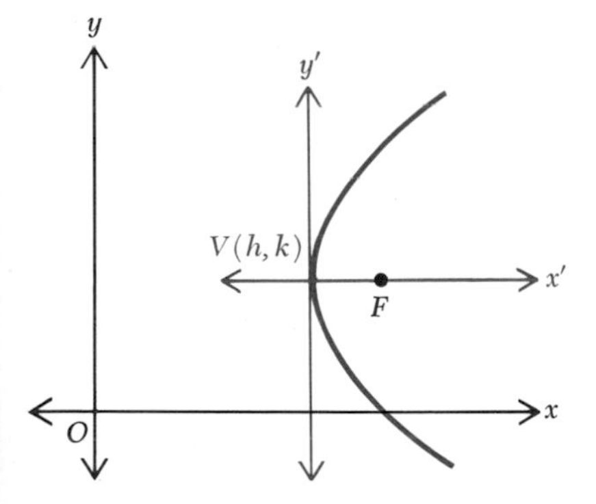

Fig. 7-6 Parabola with vertex $V(h, k)$ and horizontal axis: $(y - k^2) = 2p(x - h)$.

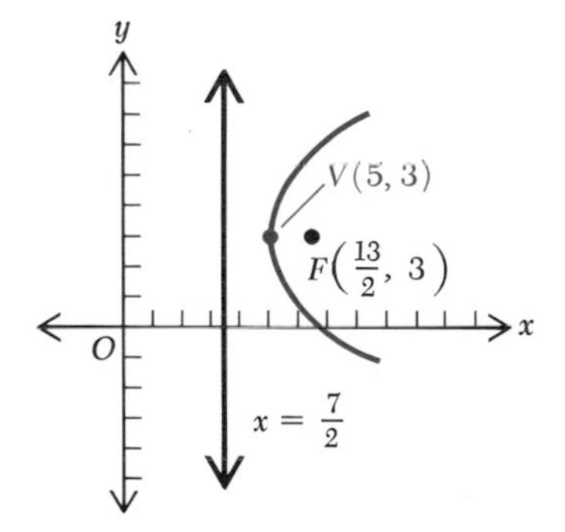

Fig. 7-7 Parabola with vertex $V(5, 3)$, focus $F(\frac{13}{2}, 3)$, and $x = \frac{7}{2}$. $(y - 3)^2 = 6(x - 5)$.

By simply completing the square in x, any equation of the form $x^2 + Dx + Ey + F = 0$ can be expressed in the form

$$(x - h)^2 = 2p(y - k) \qquad \text{if } E \neq 0 \tag{7-5}$$

This is the standard form of the equation of a parabola with vertex at $V(h, k)$ and axis parallel to the y axis. It opens upward if p is positive and downward if p is negative.

Example 2
Put the equation $y = x^2 + 4x + 1$ into the standard form (7-5).

Solution Complete the square in x:

$$y + 3 = x^2 + 4x + 4$$
$$(x + 2)^2 = y + 3$$

The graph is a parabola with vertex at $(-2, -3)$ and axis parallel to the y axis. In this case $2p = 1$, so the parabola opens upward, and the focus is one-fourth of a unit above the vertex.

7-4 APPLICATIONS

The parabola has a multitude of scientific applications. For example, when a projectile such as a ball or stone is thrown into the air, its path is a parabola except for deviations due to factors such as air resistance and revolution of the object; this deviation is appreciable in the case of a projectile fired at high velocity from a gun. Certain types of bridge construction employ parabolic arches; the curve in which a suspension-bridge cable hangs is approximately a parabola if the load is distributed uniformly along the horizontal roadbed.

If a parabola is rotated about its axis, a surface called a *paraboloid* is generated. This surface is used as a reflector in automobile headlights and in searchlights because of the following property of the parabola. From any point P on a parabola draw a segment $\overline{PF}$ to the focus and a ray PL parallel to the axis (Fig. 7-8). It can be proved that $\overrightarrow{PF}$ and $\overrightarrow{PL}$ make congruent angles with the tangent to the curve at P. This means that if a source of light is placed at F, light rays striking the reflecting surface will be reflected in rays parallel to the axis, thus throwing a cylindrical beam of light in this direction. This same principle is used in the reverse sense in the reflecting telescope: if the axis of a parabolic mirror is pointed toward a star, the rays from the star will, upon striking the mirror, be reflected to the focus.

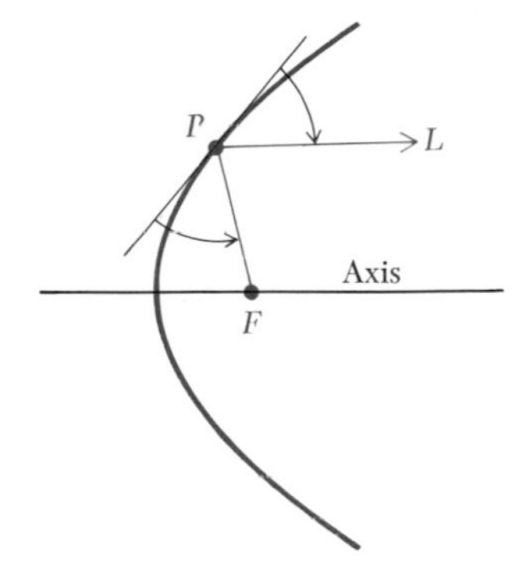

Fig. 7-8 Source of light at F reflects off parabolic surface in rays parallel to axis.

PROBLEMS

1. Write the coordinates of the focus and the equation of the directrix for each parabola whose equation is given below and sketch each graph:

   (*a*) $y^2 = 6x$ (*b*) $y^2 = -4x$ (*c*) $x^2 = -8y$
   (*d*) $y = 4x^2$ (*e*) $x = 0.6y^2$ (*f*) $x^2 = -0.4y$
   (*g*) $4y^2 = x$ (*h*) $3x = -8y^2$ (*i*) $9y = 4x^2$

2. Write the equation of the parabola satisfying each set of conditions:

   (*a*) Vertex at the origin, axis along the x axis, and passing through $A(3, 6)$ (*Hint:* The equation can be written as $y^2 = 2px$ or as $y^2 = cx$; then p or c can be found from the condition that $y = 6$ when $x = 3$.)
   (*b*) Vertex at the origin, axis along the y axis, and passing through $B(-2, -1)$
   (*c*) Vertex at the origin, axis along the x axis, and passing through $C(-3, 4)$

3. Write the equation of the parabola satisfying each set of conditions:

   (*a*) Vertex at $(4, 2)$, axis parallel to the x axis, and passing through $A(8, 7)$
   (*b*) Vertex at $(5, -3)$, axis parallel to the y axis, and passing through $B(1, 2)$
   (*c*) Vertex at $(0, -6)$, axis along the y axis, and passing through $(8, 0)$

4. The focus of parabola is at $F(3, 4)$ and its vertex at $W(3, -2)$. What is its equation?
5. A segment drawn through the focus of a parabola perpendicular to its axis with end points on the parabola is called the *latus rectum* of the parabola. Show that its length is $|2p|$. [This means that the "width" of a parabola at its focus is $|2p|$; for example, the parabola $y^2 = 8x$ is 8 units wide at its focus, which is the point $F(2, 0)$.]

6. Write the equation of the parabola whose axis is parallel to the y axis and which passes through each set of three points. *Hint:* The standard equation of a parabola can be written in the form $x^2 + Dx + Ey + F = 0$ or $y = ax^2 + bx + c$. Solve for D, E, and F or for a, b, and c.

   (*a*) (1, 3), (−2, 15), (3, 5) (*b*) (2, 3), (0, 2), (4, 12)
   (*c*) (1, −5), (−2, 6), (2, −2)

7. Write the equation of the parabola whose axis is parallel to the x axis and which passes through each set of three points:

   (*a*) (6, 2), (−6, −2), (−2, 0) (*b*) (0, 4), (12, −2), (4, 0)
   (*c*) (−1, 3), (−4, 0), (8, −6)

8. The diameter of a parabolic reflector is 12 in. and its depth is 4 in. Locate the focus.
9. A parabola is 16 in. wide at a distance of 6 in. from the vertex. How wide is it at the focus?
10. A parabolic arch has the dimensions shown in Fig. 7-9. Find the equation of the parabola with respect to the axes shown. Compute the coordinate y at the points where x equals 5, 10, and 15 ft.
11. Draw a graph that shows the way in which $\cos 2\theta$ varies with $\cos\theta$; that is, use $\cos\theta$ as the abscissa and $\cos 2\theta$ as the ordinate of the curve. Show that the curve is a parabola.
12. If a ball is thrown vertically upward from the ground with initial velocity v_0 ft/sec, its distance above the ground at the end of t sec is given approximately by the formula

$$y = v_0 t - 16t^2$$

   Draw this "distance-time graph," taking $v_0 = 40$ ft/sec. Show that the graph is a parabola. Do the focus, vertex, and directrix have any meaning in the context of the problem?
13. Prove that if a point $P(x, y)$ is equidistant from the y axis and the point $F(6, 0)$, then its coordinates must satisfy the equation $y^2 = 12(x - 3)$. Prove, con-

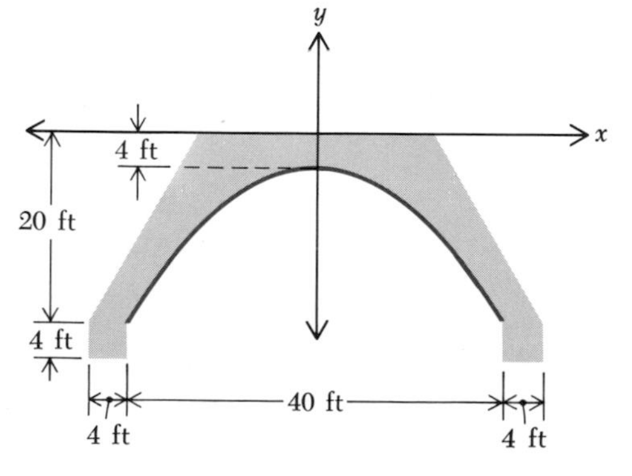

Fig. 7-9 Parabolic arch.

versely, that every point $P(x, y)$ having $y^2 = 12(x - 3)$ is equidistant from the point $F(6, 0)$ and the y axis. The distances involved are undirected distances.

14. Prove that if a point $P(x, y)$ is equidistant from $A(8, 0)$ and the line $x = 4$, then its coordinates must satisfy the equation $y^2 = 8(x - 6)$. Prove, conversely, that every point $P(x, y)$ having $y^2 = 8(x - 6)$ is equidistant from the point $A(8, 0)$ and the line $x = 4$.
15. Let d_1 and d_2 be the undirected distances of a point $P(x, y)$ from the point $A(10, 0)$ and from the line $x = 4$, respectively. Show that if $d_1 = d_2$, then $y^2 = 12(x - 7)$. Which, if any, of the following assertions is thereby proved?

    (*a*) Every point whose coordinates satisfy the equation $y^2 = 12(x - 7)$ is equidistant from the given line and the given point.
    (*b*) There are no points on the graph of the equation $y^2 = 12(x - 7)$ which are not equidistant from the given line and the given point.
    (*c*) Every point (of the plane) that is equidistant from the given line and the given point is on the graph of the equation $y^2 = 12(x - 7)$.

16. By completing the square, express the equation $y = ax^2 + bx + c$ in the standard form (7-5); find the coordinates of the vertex of the parabola and the value of p in terms of a, b, and c, and show that the directed distance from the vertex to the focus is $\frac{1}{4}a$.
17. In each of the following cases express the given equation in standard form (7-3) or (7-5) and sketch the curve. In making the sketch it may be convenient to use the fact that the width of the parabola at its focus is $2p$ (see Prob. 5).

    (*a*) $y^2 - 8y - 3x + 22 = 0$ (*b*) $y^2 - 4y - 6x - 10 = 0$
    (*c*) $x^2 + 2x - 4y = 11$ (*d*) $x^2 - 8x - 4y + 8 = 0$
    (*e*) $y = 12x - x^2$ (*f*) $y^2 = 4(x + y)$
    (*g*) $4y^2 - 12y - 12x + 33 = 0$ (*h*) $2x^2 - 10x - 3y + 8 = 0$

18. The length of fencing available to enclose a rectangular plot of ground is 100 ft. Let x (feet) be the width of the plot, and express the area that may be enclosed as a function of x. Sketch the graph. For what value of x is the area greatest?
19. A rectangular field is to be closed and separated into four lots by fences parallel to one of the sides. A total of 1,200 yd of fencing is available. Find the dimensions of the largest field that can be enclosed.

7-5 THE HYPERBOLA

A *hyperbola* is defined as the locus of points P such that the difference between the undirected distances from two fixed points to P is a constant. The fixed points are called the *foci* of the hyperbola, and the point midway between them is its *center*. The derivation of the standard equation for this curve parallels that for the ellipse. Choose axes as

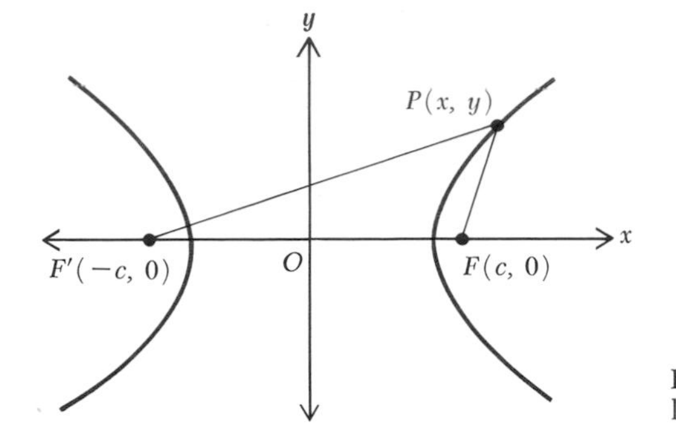

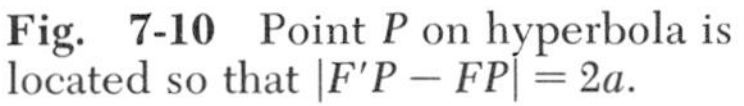
Fig. 7-10 Point P on hyperbola is located so that $|F'P - FP| = 2a$.

shown in Fig. 7-10, and let the undirected distance between the foci F and F' be $2c$, just as in the case of the ellipse. The foci are then the points $F(c, 0)$ and $F'(-c, 0)$, as shown. Let $P(x, y)$ be a point of the plane, and let d_1 and d_2 be its undirected distances from F' and F, respectively; then

$$d_1 = \sqrt{(x+c)^2 + y^2} \qquad \text{and} \qquad d_2 = \sqrt{(x-c)^2 + y^2}$$

Let $2a$ be the absolute value of the difference between d_1 and d_2. The definition then implies that if P is to be a point of the locus, its coordinates must be such that either $d_1 - d_2 = 2a$ or $d_2 - d_1 = 2a$ for $d_1 > d_2$ or $d_2 > d_1$, respectively. Thus

$$|\sqrt{(x+c)^2 + y^2} - \sqrt{(x-c)^2 + y^2}| = 2a$$

As a result of algebraic operations similar to those used in deriving the equation of the ellipse (Sec. 6-6), this equation may be expressed as

$$\frac{x^2}{a^2} - \frac{y^2}{c^2 - a^2} = 1$$

Observe now that $a < c$, because the difference in the measures of two sides of a triangle is less than the measure of the third side. Therefore $c^2 - a^2$ is positive and may be denoted as b^2. The equation then becomes

$$\frac{x^2}{a^2} - \frac{y^2}{b^2} = 1 \tag{7-6}$$

where a, b, and c are connected by the relation

$$c^2 = b^2 + a^2 \tag{7-7}$$

The graph has the general shape shown in Fig. 7-11. Deduced from Eq. (7-6) are the facts that its intercepts on the x axis are the points $I(a, 0)$ and $I'(-a, 0)$ and that it does not intersect the y axis. In fact, Eq. (7-6)

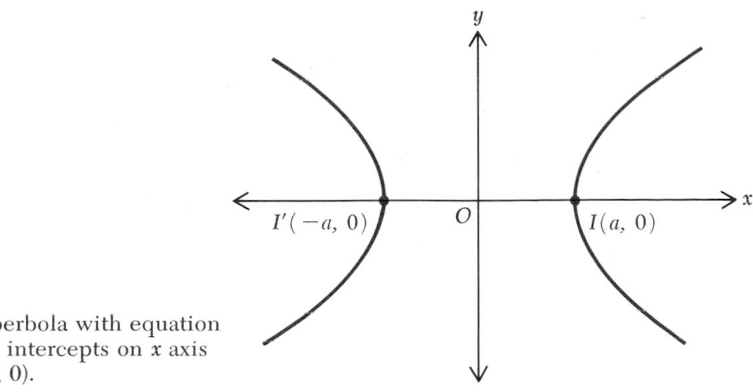

Fig. 7-11 Hyperbola with equation $x^2/a^2 - y^2/b^2 = 1$; intercepts on x axis at $(a, 0)$ and $(-a, 0)$.

solved explicitly for y in terms of x is

$$y = \pm \frac{b}{a} \sqrt{x^2 - a^2} \tag{7-8}$$

For a domain between $-a$ and a, the range is the empty set. It is also evident that $|y|$ increases indefinitely as x increases. Equation (7-8) may be written in the form

$$y = \pm \frac{bx}{a} \sqrt{1 - \frac{a^2}{x^2}} \tag{7-9}$$

From (7-9) we may deduce that for very large values of x, values of y in quadrant I are nearly equal to $(b/a)x$ (similar conclusions are possible for the other quadrants). We may also suspect that for sufficiently large absolute values of x the graph of the hyperbola nearly coincides with that of the lines

$$y = \frac{b}{a} x \qquad \text{and} \qquad y = -\frac{b}{a} x$$

This conclusion may be made more convincing by examining the difference d (Fig. 7-12) between the ordinates of the line $y = (b/a)x$ and the branches of the hyperbola. Since $y = (b/a)\sqrt{x^2 - a^2}$ in quadrant I,

$$\begin{aligned} d &= \frac{b}{a} x - \frac{b}{a} \sqrt{x^2 - a^2} \\ &= \frac{b}{a} (x - \sqrt{x^2 - a^2}) \end{aligned}$$

If the numerator and denominator are multiplied by the nonzero number

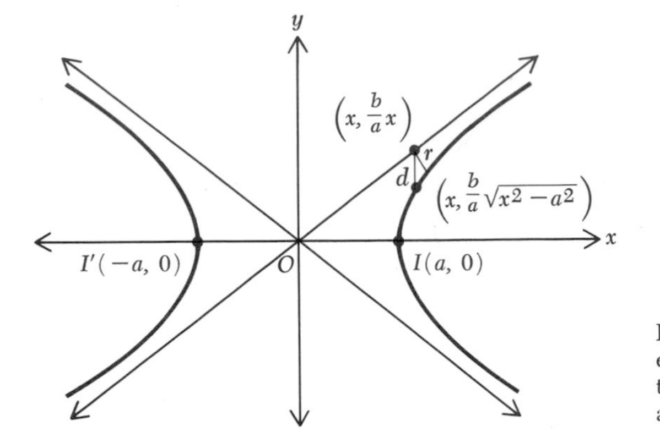

Fig. 7-12 Hyperbola with equation $x^2/a^2 - y^2/b^2 = 1$; equations of asymptotes are $x = (b/a)x$ and $x = -(b/a)x$.

$x + \sqrt{x^2 - a^2}$, then

$$d = \frac{b}{a} \frac{(x - \sqrt{x^2 - a^2})(x + \sqrt{x^2 - a^2})}{x + \sqrt{x^2 - a^2}}$$

$$= \frac{b}{a} \frac{a^2}{x + \sqrt{x^2 - a^2}}$$

$$= \frac{ab}{x + \sqrt{x^2 - a^2}}$$

The numerator of this fraction is a constant. The denominator increases as x increases. Actually, the denominator can be made arbitrarily large for a sufficiently large value of x. This means that the difference between the ordinates of the line and the hyperbola gets closer and closer to zero as x gets larger and larger. It can also be shown that the perpendicular distance r (Fig. 7-12) from a point on the line to the hyperbola is less than the fraction representing d (see Prob. 10 in For Extended Study). When a line and a curve are related in this manner, the line is called an *asymptote* to the curve. More details on asymptotes are discussed in Chap. 8.

The segment $\overline{A'A}$ in Fig. 7-13 is called the *transverse axis* of the hyperbola and has length $2a$; segment $\overline{B'B}$ is called *conjugate axis* and has length $2b$. The diagonals of the rectangle extended have the equations $y = \pm(b/a)x$; they are therefore the asymptotes of the hyperbola.

The ratio of c to a is called the *eccentricity* of the hyperbola, denoted by the letter e:

$$e = \frac{c}{a}$$

Since $a < c$, the eccentricity of a hyperbola is greater than 1.

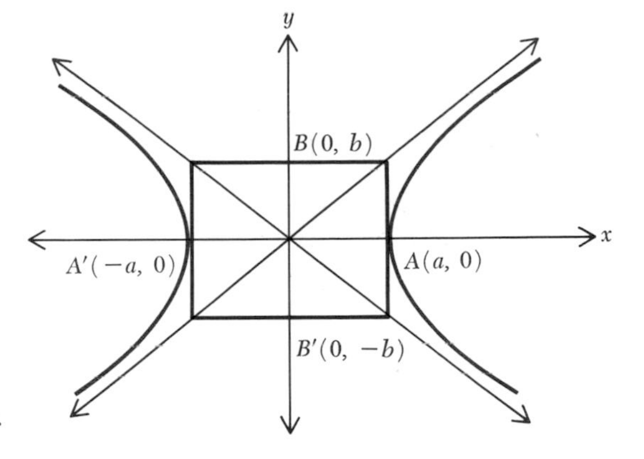

Fig. 7-13 Hyperbola with equation $x^2/a^2 - y^2/b^2 = 1$; $\overline{A'A}$ is transverse axis of length $2a$ and $\overline{B'B}$ is conjugate axis of length $2b$.

If the foci $(0, c)$ and $(0, -c)$ are on the y axis, the equation of the hyperbola is

$$\frac{y^2}{a^2} - \frac{x^2}{b^2} = 1 \tag{7-10}$$

In this equation a may be less than, equal to, or greater than b. We determine whether the foci are on the x or y axis by putting the equation in standard form (7-6) or (7-10); if the coefficient of x^2 is positive the foci are on the x axis, and if the coefficient of y^2 is positive they are on the y axis.

The standard form of the equation of a hyperbola with center at $C(h, k)$ and transverse axis parallel to the x axis is

$$\frac{(x-h)^2}{a^2} - \frac{(y-k)^2}{b^2} = 1 \tag{7-11}$$

If the transverse axis is parallel to the y axis, the corresponding equation is

$$\frac{(y-k)^2}{a^2} - \frac{(x-h)^2}{b^2} = 1 \tag{7-12}$$

Either (7-11) or (7-12) may be expressed in the general form

$$Ax^2 + Cy^2 + Dx + Ey + F = 0 \tag{7-13}$$

in which A and C have opposite signs. Conversely, an equation of form (7-13) may be expressed in form (7-11) or (7-12) by completing the squares, provided A and C have opposite signs. The only exception is the case in which the left-hand side has been expressed as the difference of two squares and the right-hand side is zero. In this case the left member can be factored into real linear factors, and the equation represents two intersecting lines. Otherwise it represents a hyperbola.

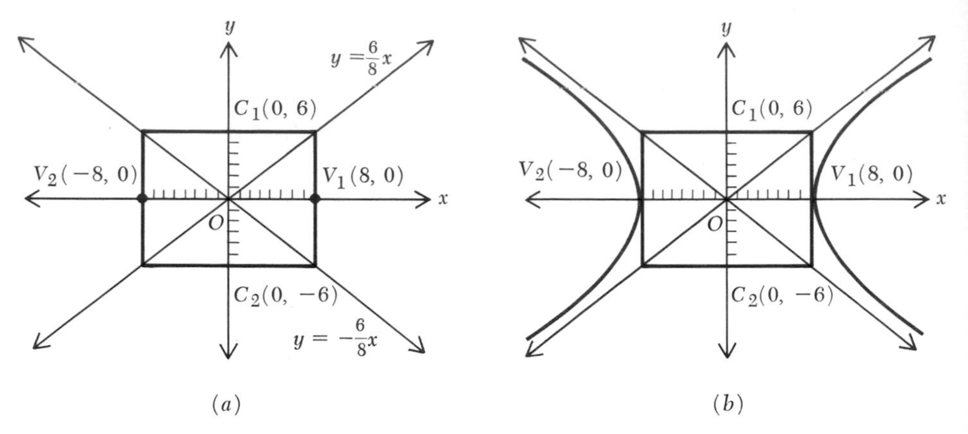

Fig. 7-14 (*a*) For $x^2/64 - y^2/36 = 1$ vertices are $V_1(8, 0)$ and $V_2(-8, 0)$ and equations of asymptotes are $y = \frac{6}{8}x$ and $y = -\frac{6}{8}x$. (*b*) Graph of the hyperbola $x^2/64 - y^2/36 = 1$.

Example

Sketch the curve $x^2/64 - y^2/36 = 1$.

Solution The graph is that of a hyperbola with $a = 8$, $b = 6$, $c = \sqrt{a^2 + b^2} = 10$, and center at the origin. The end points of the transverse axis are at $V_1(8, 0)$ and $V_2(-8, 0)$; those of the conjugate axis are at $C_1(0, 6)$ and $C_2(0, -6)$.

To find the asymptotes we draw a rectangle with center at the center of the hyperbola $O(0, 0)$ and containing the ends of the transverse and conjugate axes (Fig. 7-14*a*). The sides of the rectangle are of lengths $2a$ and $2b$, or 16 and 12, respectively, in this case. The lines containing the diagonals of the rectangle are the asymptotes. Their equations are

$$y = \tfrac{6}{8}x \qquad \text{and} \qquad y = -\tfrac{6}{8}x$$

A rough sketch may be made from this information (Fig. 7-14*b*).

7-6 EQUILATERAL HYPERBOLA

If $b = a$, the equation

$$\frac{x^2}{a^2} - \frac{y^2}{b^2} = 1$$

takes the form

$$x^2 - y^2 = a^2 \tag{7-14}$$

The asymptotes in this special case are perpendicular lines $y = x$ and $y = -x$. This hyperbola is called an *equilateral* or *rectangular hyperbola.* If the axes are rotated through $-45°$, Eq. (7-14) becomes

$$2xy = a^2 \tag{7-15}$$

The equation of the equilateral hyperbola is often encountered in the form

$$xy = k \tag{7-16}$$

Thus the law connecting the pressure and volume of a perfect gas under constant temperature is $pv = k$. The graph of Fig. 7-15 depicts the way in which the pressure of a given quantity of gas decreases as its volume increases.

A more general form of the equation of the equilateral hyperbola is

$$axy + by + cx + d = 0 \tag{7-17}$$

When solved for y, the equation becomes

$$y = -\frac{cx + d}{ax + b}$$

The line $x = -b/a$ is the vertical asymptote and the line $y = -c/a$ is the horizontal asymptote; this will be discussed in Chap. 8 (Fig. 8-18 is an example).

The equation

$$axy + by + cx + d = 0$$

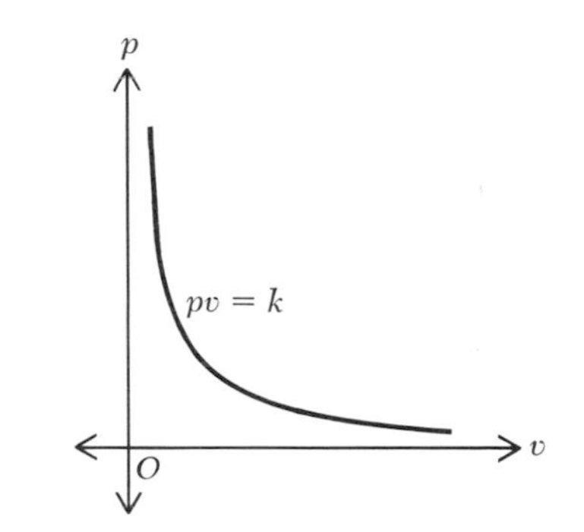

Fig. 7-15 Graph of $pv = k$ for $p > 0$ and $v > 0$.

can be changed into the form

$$x'y' = k$$

by making a translation of axes so as to move the origin to the center of the hyperbola. The required substitution is

$$x = x' - \frac{b}{a}$$

$$y = y' - \frac{c}{a}$$

7-7 APPLICATIONS

One of the important applications of the hyperbola is in range finding. If the precise times at which the sound of a gun reaches two listening posts F and F' are recorded and the difference in time is multiplied by the velocity of sound, the result is the difference of the distances of the gun from F and F'. In our notation this difference is $2a$, and the gun is somewhere on the hyperbola having foci at F and F' and transverse axis $2a$. The "branch" of the hyperbola on which it is located is known, since this depends only on which of the two stations received the sound first. By substituting a third station F'' for either F or F', the gun can be similarly located on a known branch of another hyperbola. Finally, a point of intersection of these curves gives the position of the gun.

PROBLEMS

1. Sketch the hyperbola represented by each equation, draw its asymptotes, and give the coordinates of its vertices and foci:

   (*a*) $x^2 - 4y^2 = 100$ (*b*) $3x^2 - 4y^2 = -144$
   (*c*) $16x^2 - 9y^2 = 144$ (*d*) $9y^2 - 4x^2 + 36 = 0$
   (*e*) $x^2 - 4y^2 = 36$ (*f*) $0.4x^2 - 0.6y^2 = 14.4$

2. What is the eccentricity of an equilateral hyperbola?
3. The two hyperbolas $x^2/a^2 - y^2/b^2 = 1$ and $y^2/b^2 - x^2/a^2 = 1$ are called *conjugate hyperbolas*. Show that the conjugate axis of either is the transverse of the other, and vice versa, and that they have the same asymptotes. Draw both on the same axes, for $a = 5$ and $b = 3$.
4. (*a*) Sketch the graph of the equation $x^2 - y^2 = 4$. Rotate the axes through 45° and find the new equation for the curve relative to these axes. (*b*) Show that when the axes are rotated through an angle of −45° the new equation for the hyperbola $x^2 - y^2 = a^2$ is $2x'y' = a^2$.

5. Find the equation of the hyperbola satisfying each of the following sets of conditions:

   (*a*) Vertices at $(\pm 6, 0)$ and $e = \frac{4}{3}$
   (*b*) Vertices at $(0, \pm 4)$ and foci at $(0, \pm 5)$
   (*c*) Foci at $(0, \pm 6)$ and $e = \frac{3}{2}$
   (*d*) Vertices at $(\pm 8, 0)$ and $e = \sqrt{2}$

6. Find the equation of the hyperbola satisfying each of the following sets of conditions:

   (*a*) Center at $(4, 2)$, one focus at $(4, 7)$, and $e = \frac{5}{3}$
   (*b*) Vertices at $(-2, 3)$ and $(6, 3)$ and one focus at $(-4, 3)$
   (*c*) Center at $(-1, -2)$, one vertex at $(-1, 1)$, and $e = 2$

7. A point moves so that the difference between its undirected distances from the origin and from the point $A(10, 0)$ is 6 units. Find the equation of its path and sketch the graph.
8. A line is drawn through a focus of the hyperbola $x^2/a^2 - y^2/b^2 = 1$ perpendicular to the transverse axis. It intersects the hyperbola in two points, P and Q. $\overline{PQ}$ is called the *latus rectum* of the hyperbola. Show that $|PQ| = 2b^2/a$.
9. Let d_1 and d_2 be the undirected distances of a point $P(x, y)$ from $A(-10, 0)$ and from $B(10, 0)$, respectively. Show that if $d_1 - d_2 = 12$, then the coordinates of P must satisfy the relation $x^2/36 - y^2/64 = 1$. Is it true that every point whose coordinates satisfy this equation has $d_1 - d_2 = 12$? What is the locus of all points for which $d_1 - d_2 = 12$?
10. Let d_1 and d_2 be the undirected distances of a point $P(x, y)$ from the origin and from $B(8, 0)$, respectively. Show that if $d_1 - d_2 = 4$, then the coordinates of P must satisfy the equation $3x^2 - y^2 = 12(2x - 3)$. Is it true that every point whose coordinates satisfy this equation has $d_1 - d_2 = 4$? What is the locus of all points for which $d_1 - d_2 = 4$?
11. Let d_1 and d_2 be the undirected distances of a point $P(x, y)$ from the points $A(0, -2)$ and $B(0, 8)$, respectively. Show that if $d_1 - d_2 = 6$, then the coordinates of P must satisfy the equation $16y^2 - 96y - 9x^2 = 0$. Is it true that every point whose coordinates satisfy this equation has $d_1 - d_2 = 6$? What is the locus of all points for which $d_1 - d_2 = 6$?
12. Express each of the following equations in standard form and sketch each graph:

   (*a*) $4x^2 - y^2 - 4y - 8x = 4$
   (*b*) $x^2 - y^2 - 2x - 4y - 7 = 0$
   (*c*) $3x^2 - y^2 + 18x - 2y + 14 = 0$
   (*d*) $5x^2 - 4y^2 = 40x$
   (*e*) $4x^2 - 9y^2 - 32x + 54y = 17$

In Probs. 13 to 18 translate the axes, choosing the new origin so that the resulting equation will not contain any terms of the first degree. Sketch the graph and show both sets of axes.

13. $xy + 2y - x + 1 = 0$
14. $xy + 4y - 2x - 3 = 0$
15. $2xy + 3y = 3x - 5$
16. $2y - 5x = 3xy + 2$
17. $y + 3x = 2 - 2xy$
18. $5xy + 3x = 6 - 2y$

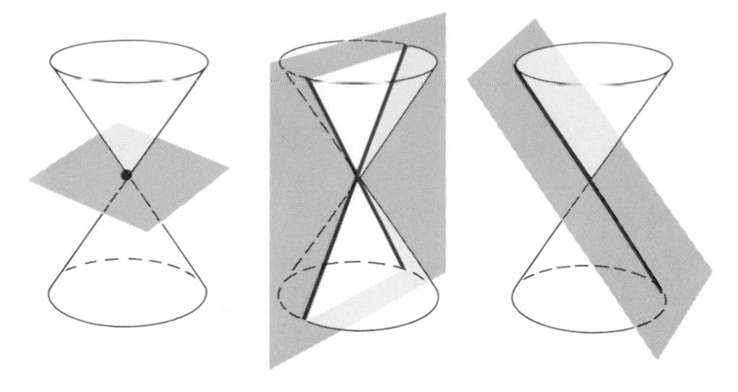

Fig. 7-16 Planes intersecting cones in one point, two lines, and one line.

7-8 SECTIONS OF A CONE

Now that we have the properties of a circle, an ellipse, a parabola, and a hyperbola, a more general discussion of all of these curves is possible; we can consider them as conic sections and as special cases of one particular form of an algebraic equation. It should be interesting and thought-provoking to see how the various isolated items of information are unified in the general theory of conics.

It can be proved that when a right circular cone (including both its upper and lower nappes) is cut by a plane, the section formed is a parabola, an ellipse (or circle), or a hyperbola if the plane does not pass through the vertex of the cone (Fig. 7-1). If the plane does pass through the vertex, the section may be a single point or two intersecting or coincident lines (see Fig. 7-16).

All these loci are called *conic sections,* or simply *conics.* The term *regular conic* may be used to designate sections cut by planes that do not pass through the vertex and the term *degenerate conic* to name sections cut by planes through the vertex. This last designation includes a pair of parallel lines. They could not be cut from a cone but could be cut from a cylinder, which is the limiting case of the cone.

7-9 THE GENERAL EQUATION OF THE SECOND DEGREE

The most general equation of the second degree in x and y is

$$Ax^2 + Bxy + Cy^2 + Dx + Ey + F = 0 \tag{7-18}$$

where A, B, and C are not all zero. The special case in which $B = 0$ is

$$Ax^2 + Cy^2 + Dx + Ey + F = 0 \tag{7-19}$$

When this equation has a locus, it is always a conic section. As has been shown, if the conic is not degenerate, then the conic is:

A parabola if A or $C = 0$
An ellipse (or circle) if A and C have the same sign
A hyperbola if A and C have opposite signs

Furthermore, in each case the axis or axes of the conic are parallel to the coordinate axes.

If the xy term is present (that is, if $B \neq 0$), it can be removed by rotating the axes through some particular angle. This means that the most general equation of second degree, Eq. (7-18), also represents a conic section (or no locus). When the xy term is present, the axis or axes of the conic are inclined to the coordinate axes.

Example 1
The equation $5x^2 + 4xy + 8y^2 - 36 = 0$ represents what type of conic?

Solution If the x and y axes are rotated through an angle $\theta = \arctan 2$ (Fig. 7-17), the xy term is eliminated, and the equation becomes, with respect to the new axes,

$$\frac{x'^2}{4} + \frac{y'^2}{9} = 1$$

which is the equation of an ellipse. The original equation, then, represents an ellipse whose minor axis makes an angle $\theta = \arctan 2$ with the x axis.

In order to establish a method for determining the correct angle of rotation to remove the xy term, consider the equation

$$Ax^2 + Bxy + Cy^2 + Dx + Ey + F = 0$$

The equations for rotation are, from Chap. 5,

$$x = x' \cos\theta - y' \sin\theta \qquad \text{and} \qquad y = x' \sin\theta + y' \cos\theta$$

The result of this transformation is a new equation of the form

$$A'x'^2 + B'x'y' + C'y'^2 + D'x' + E'y' + F' = 0 \tag{7-20}$$

where

$$\begin{aligned} A' &= A\cos^2\theta + B\sin\theta\cos\theta + C\sin^2\theta \\ B' &= B\cos 2\theta - (A - C)\sin 2\theta \\ C' &= A\sin^2\theta - B\sin\theta\cos\theta + C\cos^2\theta \\ D' &= D\cos\theta + E\sin\theta \\ E' &= E\cos\theta - D\sin\theta \\ F' &= F \end{aligned} \tag{7-21}$$

To eliminate the $x'y'$ term, let $B' = 0$; that is, choose θ such that

$$B \cos 2\theta - (A - C) \sin 2\theta = 0 \tag{7-22}$$

The solution of Eq. (7-22) is

$$\tan 2\theta = \frac{B}{A - C} \qquad \text{if } A \neq C \tag{7-23}$$

In Eq. (7-22), if $A = C$ (and $B \neq 0$), the required value of θ is given by $\cos 2\theta = 0$; that is, $\theta = \pm 45°$.

Example 2

Find the angle through which the axes must be rotated to remove the xy term from the equation

$$5x^2 + 4xy + 8y^2 - 36 = 0$$

Solution By Eq. (7-23),

$$\tan 2\theta = \frac{4}{5 - 8} = -\frac{4}{3}$$

Since

$$\tan 2\theta = \frac{2 \tan \theta}{1 - \tan^2 \theta}$$

we have

$$-\frac{4}{3} = \frac{2 \tan \theta}{1 - \tan^2 \theta}$$

and

$$2 \tan^2 \theta - 3 \tan \theta - 2 = 0$$

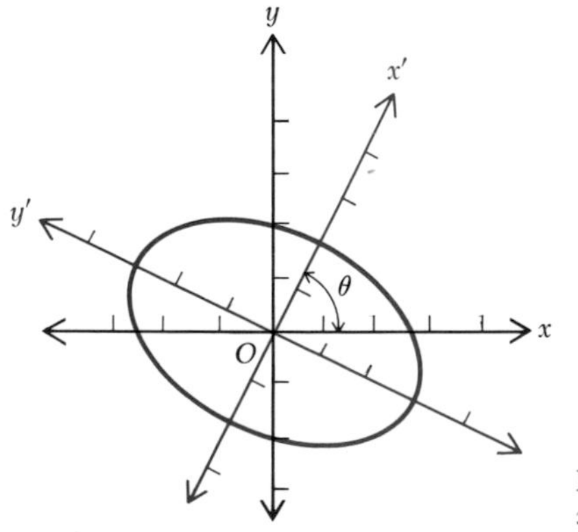

Fig. 7-17 Ellipse: $5x^2 + 4xy + 8y^2 - 36 = 0$ or $x'^2/4 + y'^2/9 = 1$.

Solving this equation gives

$$\tan\theta = 2 \qquad \text{or} \qquad \tan\theta = -\tfrac{1}{2}$$

The xy term will be eliminated if the axes are rotated through either of the angles $\theta = \arctan 2$ or $\theta = \arctan -\frac{1}{2}$ (Fig. 7-17).

7-10 IDENTIFICATION OF A CONIC

After the xy term has been removed from a general equation of second degree by rotation of the axes, the conic can be easily identified. If a graph exists, then the following conditions apply:

If A' or $C' = 0$, the conic is a parabola (in the degenerate case, a pair of parallel or coincident lines).
If A' and C' have the same sign, the conic is an ellipse (in the degenerate case, a single point).
If A' and C' have opposite signs, the conic is a hyperbola (in the degenerate case, a pair of intersecting lines).

It is possible to give rules by which the conic can be identified without the necessity of removing the xy term.

Theorem 7-1
The graph (if a graph exists) of the equation

$$Ax^2 + Bxy + Cy^2 + Dx + Ey + F = 0$$

is

(*a*) *A parabola (or its degenerate case) if* $B^2 - 4AC = 0$
(*b*) *An ellipse (or its degenerate case) if* $B^2 - 4AC < 0$
(*c*) *A hyperbola (or its degenerate case) if* $B^2 - 4AC > 0$

Proof The proof of this theorem depends upon the fact (see Prob. 5, page 201) that the expression $B^2 = 4AC$ is unchanged when the axes are rotated through any angle; that is, no matter what the angle of rotation θ may be,

$$B^2 - 4AC = B'^2 - 4A'C'$$

Now, if θ is chosen so that $B' = 0$, then

$$B^2 - 4AC = -4A'C'$$

This means that

(*a*) $B^2 - 4AC = 0$ if and only if $A' = 0$ or $C' = 0$.
(*b*) $B^2 - 4AC < 0$ if and only if A' and C' have the same sign.
(*c*) $B^2 - 4AC > 0$ if and only if A' and C' have opposite signs.

The theorem follows immediately from these three statements.

Example
Identify the conic represented by the equation

$$5x^2 + 4xy + 8y^2 - 36 = 0$$

Solution

$$B^2 - 4AC = 16 - 4(5)(8) = 16 - 160 = -144$$

Since $B^2 - 4AC$ is negative, the equation represents an ellipse.

7-11 SECOND DEFINITION OF A CONIC

There is an alternative definition of a regular conic. A regular conic is the locus of a point which moves so that the ratio of its undirected distance from a fixed point F to its undirected distance from a fixed line l is a positive constant e, the eccentricity of the conic.

If $e < 1$ the conic is an ellipse (a circle if $e = 0$).
If $e = 1$ the conic is a parabola.
If $e > 1$ the conic is a hyperbola.

In order to derive an equation for the conic from this definition, choose the coordinate axes so that the fixed line coincides with the y axis and

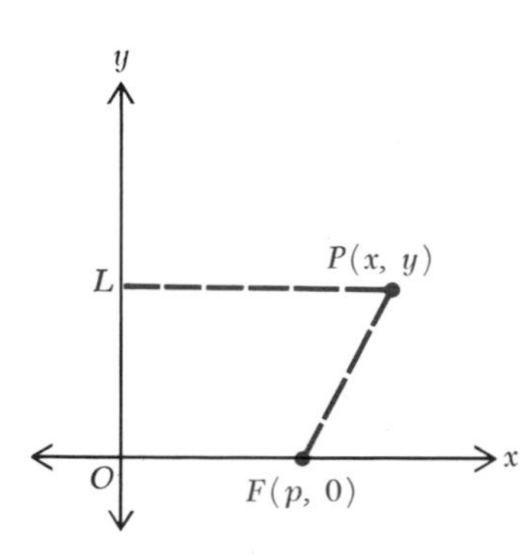

Fig. 7-18 Definition of a conic: $|PF|/|PL| = e$.

the fixed point F has coordinates $(p, 0)$ as indicated in Fig. 7-18. Let $P(x, y)$ be a point of the plane, and let d_1 and d_2 be its undirected distances from the point F and from the line (in this case the y axis), respectively. Then

$$d_1 = |FP| = \sqrt{(x-p)^2 + y^2} \qquad \text{and} \qquad d_2 = |LP| = |x| = \sqrt{x^2}$$

The coordinates of every point for which $d_1/d_2 = e$ (or $d_1 = ed_2$) must satisfy the equation

$$\sqrt{(x-p)^2 + y^2} = e\sqrt{x^2} \tag{7-24}$$

Square both sides and simplify to obtain the following results:

$$x^2 - 2px + p^2 + y^2 = e^2x^2 \tag{7-25}$$

or

$$(1 - e^2)x^2 + y^2 - 2px + p^2 = 0 \tag{7-26}$$

Thus the coordinates of every point for which $d_1 = ed_2$ must satisfy Eq. (7-26). It is easy to show, by reversing the above steps, that $d_1 = ed_2$ for every point whose coordinates satisfy (7-26). Thus if Eq. (7-26) is true for a point $P(x, y)$, then (7-25) must be true. If (7-25) is true, then (7-24) must be true, because the positive square roots of two equal positive numbers are equal.

It has been proved that Eq. (7-26) is the equation of the locus of a point which moves so that the ratio of its undirected distance from the point $F(p, 0)$ to its undirected distance from the y axis is always equal to a constant e. The following conclusions are warranted:

If $e < 1$ the coefficients of x^2 and y^2 have the same sign, and the locus is an ellipse. If $e = 0$ the coefficients of x^2 and y^2 are equal, and the locus is a circle.

If $e = 1$ the coefficient of x^2 is zero, and the locus is a parabola.

If $e > 1$ the coefficients of x^2 and y^2 have opposite signs, and the locus is a hyperbola.

The fixed point F is called a focus of the conic, and the fixed line l is called a directrix. In the case of the parabola ($e = 1$) these are precisely the focus and directrix of the previous definition in Sec. 7-2. In the case of the ellipse and hyperbola the "focus" as used in this definition coincides with one of the two foci of the previous definitions.

7-12 INTERIORS AND EXTERIORS OF CONICS

The *interiors* of a hyperbola and a parabola are shown shaded in Fig. 7-19. What is meant by the interior of an ellipse is immediately clear.

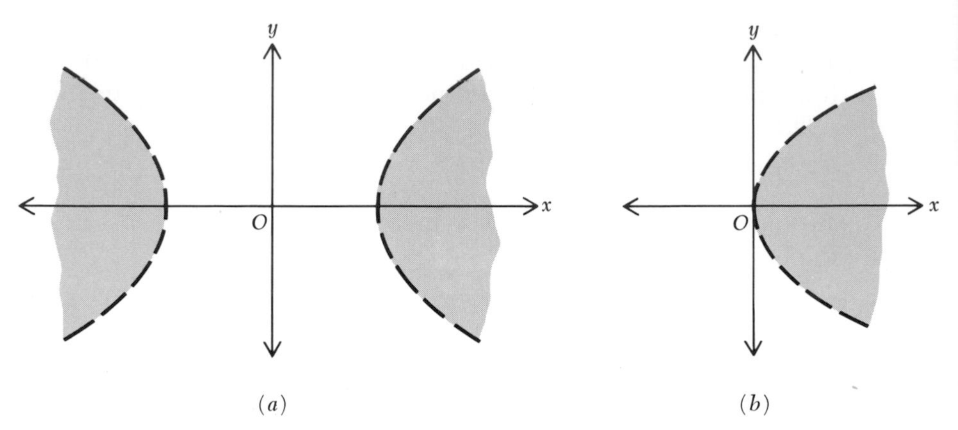

Fig. 7-19 Interiors of hyperbola and parabola (shaded).

In general, any line through a point in the interior of a conic cuts a conic in two distinct points. The equation of a conic is of the form

$$Ax^2 + Bxy + Cy^2 + Dx + Ey + F = 0$$

It can be proved that the interior of the conic can be represented by one of the inequalities

$$Ax^2 + Bxy + Cy^2 + Dx + Ey + F > 0 \tag{7-27}$$

or

$$Ax^2 + Bxy + Cy^2 + Dx + Ey + F < 0 \tag{7-28}$$

and the exterior by the other. To find which inequality represents the interior and which represents the exterior it is necessary only to test for any point not on the curve.

Example

Write the inequalities representing the sets of points in the exterior and interior of the conic $4x^2 - 9y^2 - 36 = 0$.

Solution The origin is known to be in the exterior (Fig. 7-19*a*), and substituting its coordinates gives -36 on the left side. This is less than zero. Hence $4x^2 - 9y^2 - 36 < 0$ is the inequality representing the exterior and $4x^2 - 9y^2 - 36 > 0$ is the inequality for the interior.

PROBLEMS

1. Name each conic, assuming its graph exists and that the conic is not degenerate.

   (*a*) $x^2 + y^2 = x + y$ (*b*) $x^2 - 3x + 2y^2 = 4$
   (*c*) $3x^2 = y^2 + y + 1$ (*d*) $4x - y = 4 + 2x^2$
   (*e*) $x^2 + 2 = 3x - 2y^2$ (*f*) $x^2 + 6x = y^2 + 2y - 3$

2. Show that each given equation has no graph or that it represents a degenerate conic:

   (*a*) $x^2 + y^2 - 4x + 8y + 26 = 0$ (*b*) $(x + y)^2 = 0$
   (*c*) $x^2 - 6x = y^2 - 6y$ (*d*) $(x + 2y)^2 + 4 = 0$

3. Identify the conic represented by each equation and find the positive acute angle through which the axes should be rotated so that the new equation does not contain an $x'y'$ term:

   (*a*) $4x^2 + 4xy + 7y^2 = 10$ (*b*) $5x^2 + 6xy - 3y^2 = 6$
   (*c*) $5x^2 + 24xy - 2y^2 = 30$ (*d*) $3x^2 + 4xy + 3y^2 = 8$
   (*e*) $8x^2 + 8xy - 7y^2 - 25$ (*f*) $34x^2 - 24xy + 41y^2 = 50$

4. For each equation rotate the axes through a positive acute angle, choosing the angle so that the resulting equation will not have an $x'y'$ term; draw the curve and show both sets of axes:

   (*a*) $3x^2 + 4xy - 4 = 0$ (*b*) $5x^2 + 4xy + 8y^2 - 36 = 0$
   (*c*) $16x^2 - 24xy + 9y^2 - 90x - 120y = 0$ (*d*) $3x^2 - 10xy + 3y^2 + 32 = 0$
   (*e*) $5x^2 + 8xy + 5y^2 = 25$

5. Show that $B'^2 - 4A'C' = B^2 = 4AC$ for any angle θ through which the axes may be rotated.
6. Show that $A' + C' = A + C$ for any angle θ through which the axes may be rotated.
7. For each equation evaluate $B^2 - 4AC$ and use this evaluation to identify the locus, if there is one:

   (*a*) $x^2 - xy - 6y^2 + x - 3y = 0$
   (*b*) $x^2 - 2xy + y^2 + x + 2y + 1 = 0$
   (*c*) $x^2 - xy - 2y^2 - x + 2y = 0$
   (*d*) $x^2 + xy + y^2 + 4y = 0$
   (*e*) $x^2 + 2xy + y^2 - 1 = 0$

8. A point moves so that its undirected distance from $A(5, 0)$ is always equal to one-half its undirected distance from the line $x + 4 = 0$. Find the equation of its path and sketch the graph.
9. A point moves so that its undirected distance from $A(3, 0)$ is always equal to one-half its undirected distance from the y axis. Find the equation of its path and sketch the graph.

10. A point moves so that the ratio of its undirected distances from the point $A(0, 5)$ and from the x axis is equal to $\frac{3}{2}$. Find the equation of its path and sketch the graph.
11. Find the equation of the locus of all points each of which is twice as far from the point $A(0, 6)$ as from the x axis and sketch the graph.
12. A point moves so that its undirected distance from $A(7, 0)$ is always equal to two-thirds its undirected distance from the line $x - 2 = 0$. Show that the sum of its undirected distances from $A(7, 0)$ and $B(15, 0)$ is equal to 12.
13. A point moves so that its undirected distance from the origin always equals one-third its undirected distance from the line $x + 2 = 0$. Find the coordinates of the point or points on its path that are farthest from the x axis.
14. A point moves so that its undirected distance from the point $A(4, 0)$ is always equal to four-thirds its undirected distance from the y axis. Find the equation of its path and sketch it.
15. Write the inequalities for the interior and exterior of each nondegenerate conic in Prob. 1.

7-13 POLAR EQUATIONS REPRESENTING CONIC SECTIONS

Polar equations for the parabola, ellipse, and hyperbola could be obtained by transforming the standard equations for these curves from rectangular coordinates, with $x = \rho \cos \theta$ and $y = \rho \sin \theta$. However, simpler polar equations are obtained by taking a focus as the origin. Choose the coordinate system as indicated in Fig. 7-20 and use the second definition of the conic section from Sec. 7-11. A parabola, an ellipse, or a hyperbola is the locus of a point P in a plane whose undirected distance from a fixed point of the plane is equal to the eccentricity e times its undirected distance from a fixed line of the plane. The curve is a parabola if $e = 1$, an ellipse if $0 < e < 1$, and a hyperbola if $e > 1$.

Let p be a positive number representing the distance from the directrix $\overleftrightarrow{DD'}$ in Fig. 7-20 to the focus at the pole O. Then the point $P(\rho, \theta)$ lies on the parabola, ellipse, or hyperbola if and only if

$$|OP| = e|MP| \qquad \text{for } e > 0$$

But $|OP| = \rho$ and $|MP| = p + \rho \cos \theta$. The condition is then

$$\rho = e(p + \rho \cos \theta)$$

Solving this equation for ρ results in the standard equation in polar form:

$$\rho = \frac{ep}{1 - e \cos \theta} \tag{7-29}$$

This equation represents an ellipse if $0 < e < 1$, a parabola if $e = 1$, and a hyperbola if $e > 1$.

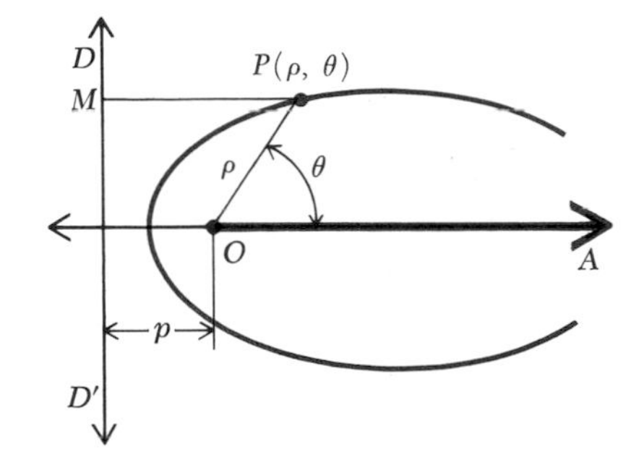

Fig. 7-20 Conic with a focus at the pole in polar coordinates: $\rho = ep/(1 - e \cos \theta)$.

Note that if the directrix is to the right of the pole instead of to the left as shown in Fig. 7-20, the corresponding equation is

(7-30) $$\rho = \frac{ep}{1 + e \cos \theta}$$

Finally, if the directrix is taken parallel to the polar axis, the equation is

(7-31) $$\rho = \frac{ep}{1 \pm e \sin \theta}$$

The positive sign applies if the directrix is above the pole, and the negative sign if it is below.

An equation of the form

(7-32) $$e = \frac{k}{a \pm b \cos \theta \text{ (or } \sin \theta)}$$

can, in general, be expressed in one of the above forms and, accordingly, represents a conic section (or a branch of one in the case of the hyperbola) with the focus at the origin.

In making a rough sketch of a conic from an equation in polar form it is often sufficient, after identifying the curve, to plot only the points corresponding to θ equal to 0, $\frac{1}{2}\pi$, π, and $\frac{3}{2}\pi$. The value of θ (if there is one) for which ρ "becomes infinite" can be found by equating the denominator of the right-hand member to zero and solving for θ. This gives the direction of the axis in the case of a parabola and the direction of the asymptotes in the case of a hyperbola. Observe that the asymptotes to the hyperbola pass through the center of the hyperbola, not through the origin. It is only their directions that are determined by equating the denominator to zero and solving for θ.

Example 1

Identify and sketch the curve whose equation is $\rho = 8/(2 - \cos \theta)$.

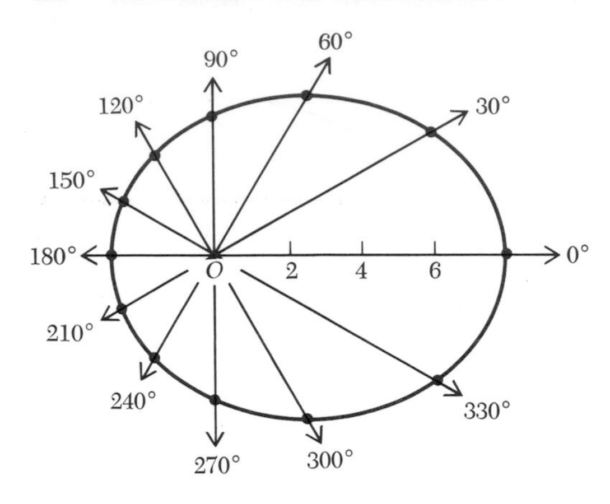

Fig. 7-21 Graph of the ellipse $\rho = 8/(2 - \cos\theta)$.

Solution First divide the numerator and the denominator by 2 in order to obtain the standard polar form:

$$\rho = \frac{4}{1 - \frac{1}{2}\cos\theta}$$

Since $e = \frac{1}{2}$, the equation represents an ellipse for which $ep = 4$, or $p = 8$. The graph is shown in Fig. 7-21.

Example 2

Identify and sketch the curve whose equation is $\rho = 8/(5 + 5\sin\theta)$.

Solution Divide the numerator and the denominator by 5 to obtain

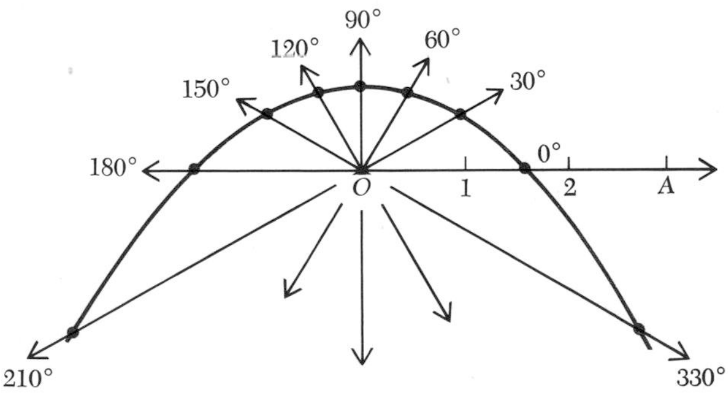

Fig. 7-22 Graph of the parabola $\rho = 8/(5 + 5\sin\theta)$.

the equation in standard polar form:

$$\rho = \frac{1.6}{1 + \sin \theta}$$

In this case $e = 1$, and the equation represents a parabola. The graph is shown in Fig. 7-22.

PROBLEMS

Sketch the graph of the equations in Probs. 1 to 14. Find the corresponding equation for each curve in rectangular coordinates.

1. $\rho(\cos \theta + \sin \theta) = 3$
2. $\rho(2 \cos \theta + 3 \sin \theta) = 8$
3. $\rho = \dfrac{12}{2 \cos \theta - 3 \sin \theta}$
4. $\rho = \dfrac{5}{\sin \theta - 2 \cos \theta}$
5. $\rho \cos \theta + 4 = 0$
6. $2 + \rho \sin \theta = 0$
7. $6 + \rho(3 \sin \theta - \cos \theta) = 0$
8. $\rho(4 \cos \theta + 5 \sin \theta) + 20 = 0$
9. $\rho + 3 \sin \theta = 0$
10. $\rho + 2 \cos \theta = 0$
11. $\rho = 8 \cos \theta + 6 \sin \theta$
12. $\rho + 2 \cos \theta = 3 \sin \theta$
13. $\rho^2 + 11 = 8\rho \cos \theta - 4\rho \sin \theta$
14. $\rho + 2.8 \sin \left(\theta + \dfrac{\pi}{6}\right) = 0$

Express the equations in Probs. 15 to 30 in standard form, and identify and sketch the curve. Find the corresponding equation in rectangular coordinates.

15. $\rho = \dfrac{8}{2 + \sin \theta}$
16. $\rho = \dfrac{12}{3 + 2 \cos \theta}$
17. $\rho = \dfrac{4.8}{1.6 + 1.2 \cos \theta}$
18. $\rho = \dfrac{6}{2 - \sin \theta}$
19. $\rho \; \dfrac{-6}{3 - 2 \sin \theta}$
20. $\rho = \dfrac{7}{3 \cos \theta - 4}$
21. $6\rho - 4\rho \sin \theta = 9$
22. $\rho + \rho \sin \theta = 4$
23. $\rho = \dfrac{8}{3 + 3 \cos \theta}$
24. $\rho = \dfrac{9}{2 - 2 \sin \theta}$
25. $3\rho - \dfrac{8}{1 + \sin \theta}$
26. $2\rho = \dfrac{9}{\cos \theta - 1}$
27. $\rho = \dfrac{6}{2 - 3 \cos \theta}$
28. $\rho = \dfrac{8}{4 + 3 \sin \theta}$
29. $3\rho + 6\rho \sin \theta = 16$
30. $\rho = \dfrac{13}{1 - 2 \sin \theta}$
31. Sketch the locus of all points (ρ, θ) which satisfy the equation $\rho \sin^2 \theta = 6 \cos \theta$. What is the corresponding equation in rectangular coordinates?
32. Sketch the locus of all points (ρ, θ) which satisfy the equation $\rho \sin^2 \theta = 4 \sin \theta$. What is the corresponding equation in rectangular coordinates?

33. Sketch the graph of the equation

$$\rho^2 = \frac{144}{9 \cos^2 \theta + 16 \sin^2 \theta}$$

What is the corresponding equation in rectangular coordinates?

34. Sketch the graph of the equation

$$\rho^2 = \frac{36}{4 - 13 \sin^2 \theta}$$

What is the corresponding equation in rectangular coordinates?

7-14 PATH OF A PROJECTILE

Assume that a projectile is fired with initial velocity v_0 ft/sec at an angle α with the horizontal. With the point from which the projectile is fired as the origin, choose coordinate axes as indicated in Fig. 7-23, with the x axis as the horizontal. If the projectile is acted on only by the gravitational force while it is in flight, its position at the end of t sec is given by the equations

$$x = (v_0 \cos \alpha)t \qquad (7\text{-}33)$$
$$y = (v_0 \sin \alpha)t - \tfrac{1}{2}gt^2$$

where g is a constant whose value is approximately 32 ft/sec². These equations, which give the coordinates of the projectile at any time t while it is in flight, are the parametric equations of its path.

In order to obtain the relation that holds between x and y for any point of the path, eliminate t by solving the first equation for t in terms of x and substituting this into the second. The result is

$$y = (\tan \alpha)x - \frac{g}{2v_0^2 \cos^2 \alpha} x^2 \qquad (7\text{-}34)$$

This is an equation of the form $y = Ax - Bx^2$, and it represents a parabola.

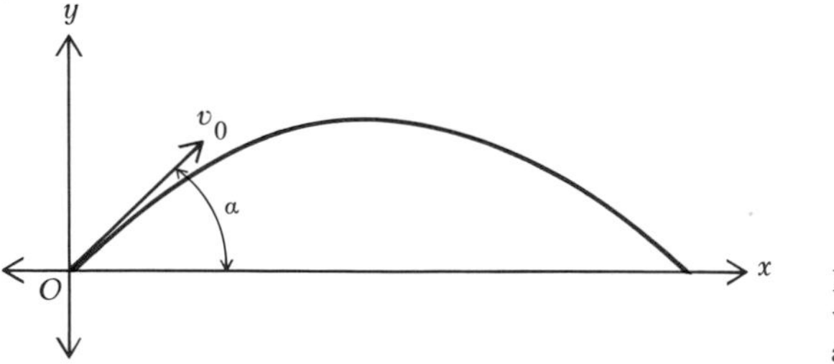

Fig. 7-23 Path of a projectile with equations $x = (v_0 \cos \alpha)t$ and $y = (v_0 \sin \alpha)t - \frac{1}{2}gt^2$.

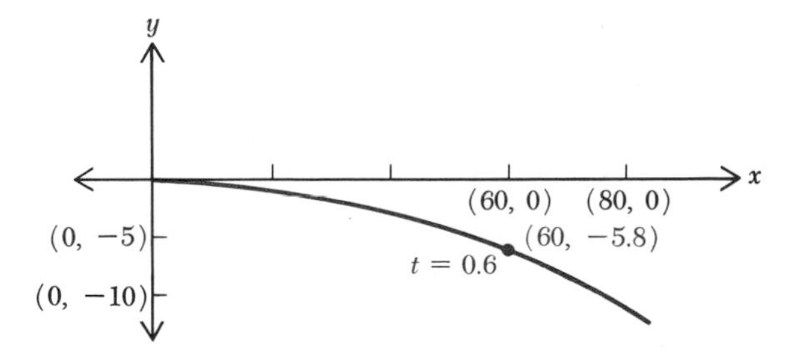

Fig. 7-24 Path of a pitched ball.

The equations for the case in which the projectile is fired horizontally can be obtained from the general case of Eqs. (7-33) by letting $\alpha = 0$. The parametric equations for this case are then

$$x = v_0 t \qquad (7\text{-}35)$$
$$y = -\tfrac{1}{2}gt^2$$

These equations hold in the case of a ball or stone that is thrown horizontally and is subject only to the force of gravity during its flight.

Example

A ball is pitched horizontally at a speed of 100 ft/sec. How far will it drop in a horizontal distance of 60 ft?

Solution With $v_0 = 100$ and $g = 32$, the parametric equations are, by Eqs. (7-35),

$$x = 100t$$
$$y = -16t^2$$

If $x = 60$ in $x = 100t$ then $t = 0.6$; that is, the ball travels 60 ft horizontally in the first 0.6 sec. If $t = 0.6$ in $y = -16t^2$, then $y = -5.76$ ft. This means that the ball would have approximately the coordinates $(60, -5.8)$ when $t = 0.6$ sec (Fig. 7-24).

PROBLEMS

Plot the graph from each set of parametric equations; then eliminate the parameter and, if possible, identify the curve:

1. $x = 4 - t$
   $y = 4t - t^2$

2. $x = 2(t - 3)$
   $y = 2t(t - 3)$

3. $x = \sin \frac{\pi}{2} t$

   $y = \cos^2 \frac{\pi}{2} t$

4. $x = 2 + \sin \frac{\pi}{4} t$

   $y = 1 - \cos^2 \frac{\pi}{4} t$

5. $x = 2 + 2 \cos t$

   $y = 1 - 3 \sin^2 t$

6. A baseball is thrown horizontally from a point 8 ft above the ground with a speed of 140 ft/sec. Write parametric equations for its path. Assuming the ground to be level, determine how far the ball will travel horizontally before striking it.
7. Assume that a baseball leaves the pitcher's hand at a point 5 ft above the ground, and that it is thrown horizontally so that $\alpha = 0$. What is the minimum initial speed of the ball if it must be at least 6 in. above the ground when it passes over home plate, which is 60 ft away?
8. From a point O on the side of a hill, a ball is thrown as indicated in Fig. 7-25. Find the coordinates, to the nearest foot, of the point at which it will strike the ground.

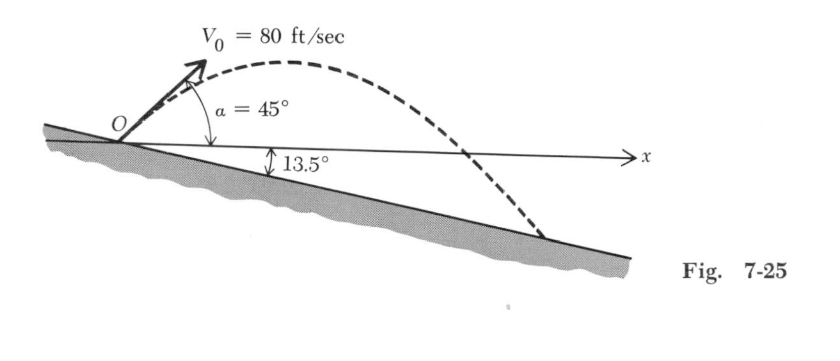

Fig. 7-25

SUMMARY

The standard equations for circles, ellipses, parabolas, and hyperbolas are all special cases of the general second-degree equation

$$Ax^2 + Bxy + Cy^2 + Dx + Ey + F = 0$$

These loci are the various sections that are intersections of a right circular cone and a plane and hence are called conic sections.

A parabola is the locus of all points that are equidistant from a fixed point and a fixed line; the fixed point is the focus, and the fixed line is the directrix of the parabola.

The equation $y^2 = 2px$ represents a parabola with focus $(\frac{1}{2}p, 0)$ and directrix $x = -\frac{1}{2}p$. The vertex is at the origin. The axis of the parabola, which is a line of symmetry, is the x axis. If p is positive the parabola opens to the right, and if p is negative it opens to the left.

The equation $(y - k)^2 = 2p(x - h)$ is the standard form of the equation of a parabola with vertex at $V(h, k)$ and axis parallel to the x axis. The parabola opens to the right if p is positive and to the left if p is negative.

A hyperbola is the locus of points such that the difference of the undirected distances from two fixed points is a constant; the two fixed points are the foci, and the point midway between them is the center.

The equation $x^2/a^2 - y^2/b^2 = 1$ represents a hyperbola with center at the origin (0, 0). The foci are at $(c, 0)$ and $(-c, 0)$, where $c^2 - a^2 = b^2$. The x intercepts (the vertices) are at $(a, 0)$ and $(-a, 0)$. For sufficiently large absolute values of x the graph of the hyperbola nearly coincides with that of the lines $y = (b/a)x$ and $y = (-b/a)x$, called the asymptotes of the curve.

The standard form of the equation of a hyperbola with center at $C(h, k)$ and transverse axis parallel to the x axis is $(x - h)^2/a^2 - (y - k)^2/b^2 = 1$.

Each conic may be represented by a general equation of second degree in x and y in the form

$$Ax^2 + Bxy + Cy^2 + Dx + Ey + F = 0$$

where A, B, and C are not all zero. Two methods may be used to easily identify the conic for an equation of this type. One uses the expression $B^2 - 4AC$, and the other uses the eccentricity c/a. The results are summarized in a table.

	$B^2 - 4AC$	$e = \frac{c}{a}$
Parabola	Equal to 0	Equal to 1
Ellipse	Less than 0	Less than 1
Hyperbola	Greater than 0	Greater than 1

In polar coordinates, the equation $\rho = ep/(1 - e \cos \theta)$ represents a conic with a focus as the pole. This equation represents an ellipse if $0 < e < 1$, a parabola if $e = 1$, and one branch of a hyperbola if $e > 1$.

REVIEW PROBLEMS

1. In each case find the coordinates of the vertex and focus and the equation of the directrix of the parabola whose equation is given; then sketch the graph:

   (a) $y^2 = 10x$ (b) $(x-3)^2 = -8(y+4)$ (c) $3y^2 + 12y - 5x + 17 = 0$

2. Find the equation of the parabola satisfying the given conditions:

   (*a*) Vertex at $V(3, -1)$ and focus at $F(3, -2)$
   (*b*) Vertex on the y axis and axis parallel to the x axis and passing through $R(-4, 1)$ and $S(-1, -1)$

3. Find the equation of a parabola with axis parallel to the y axis and passing through $K(1, 1)$, $L(3, 0)$, and $M(4, -4)$.
4. Find the points of intersection of $3y^2 + 4x + 12y - 12 = 0$ and $2x + 3y + 6 = 0$. Sketch the curves.
5. A ball is thrown downward from a height of 10 ft with a velocity of 12 ft/sec. When will it strike the ground?
6. Sketch each hyperbola whose equation is given, draw its asymptotes, and give the coordinates of its vertices and foci:

   (*a*) $9x^2 - 16y^2 - 144 = 0$ (*b*) $9x^2 - 16y^2 = 36x - 96y - 36$

7. Find the equation of the hyperbola satisfying the following conditions:

   (*a*) Asymptote $x - 3y = 0$ and one end of conjugate axis at $C(0, 3)$
   (*b*) Foci at $F(4, -2)$ and $F'(4, 10)$ and $e = 3$

8. Sketch the graphs of the following equations and determine their points of intersection algebraically:

   (*a*) $x^2 - y^2 + 3x - y + 8 = 0$ (*b*) $x^2 - y^2 + 4x - 4y + 16 = 0$

9. Identify the conic represented by the equation $7x^2 + 12xy - 2y^2 = 10$. Rotate the axes through a positive acute angle so the resulting equation will not have an $x'y'$ term.
10. Find the equation of the parabola with focus at the origin and having as directrix the directrix (which is graphed in quadrants I and IV) of the hyperbola $3x^2 - y^2 = 108$.

FOR EXTENDED STUDY

1. When a ball is thrown with initial velocity v_0 ft/ sec at an angle of 45° with the horizontal, it travels a path whose equation is approximately

$$y = x - \frac{32x^2}{v_0^2}$$

   If $v_0 = 96$ ft/sec, find the horizontal distance traveled and the maximum height reached by the ball.

2. If the equation $axy + by + cx + d = 0$ is solved for y in terms of x, the result is

$$y = -\frac{cx + d}{ax + b}$$

Thus the lines $x = -b/a$ and $y = -c/a$ are vertical and horizontal asymptotes, respectively, to the hyperbola represented by the equation. The center is then the point $(-b/a, -c/a)$. Show that if the axes are translated so that this point becomes the new origin, then the new equation has the form $x'y' = k$, where $k = (cb - ad)/a^2$.

3. Prove that the product of the undirected distances from any point on a hyperbola to the asymptotes is a constant.
4. Prove that in a hyperbola an asymptote, a directrix, and a line from the corresponding focus perpendicular to the asymptote are concurrent.
5. Prove that the equation $x^{\frac{1}{2}} + y^{\frac{1}{2}} = a^{\frac{1}{2}}$ represents a parabola.
6. Prove that in the general equation of a conic if $B \neq 0$ and either $A = 0$ or $C = 0$ (and not all other constants are zero), then the equation represents a hyperbola.
7. Prove that if a line is parallel to an asymptote of a hyperbola, it intersects the hyperbola in exactly one point.
8. Find the equations of the asymptotes to the hyperbola $(x-3)^2/9 - (y-2)^2/4 = 1$.
9. Let F be the focus of the parabola $y^2 = 2px$ for $p > 0$. Draw the line $x = -b$ for $b > 0$. If d_1 is the undirected distance from any point $P(x, y)$ on the parabola to the line $x = -b$ and d_2 is the undirected distance from P to F, prove $d_1 - d_2 = b - p/2$.
10. Prove that the perpendicular distance r (see Fig. 7-12) between a point $P_1(x_1, (b/a)x_1)$ on the asymptote $y = (b/a)x$ to a point on the hyperbola $x^2/a^2 - y^2/b^2 = 1$ is less than the difference d between the ordinate of a point $P_1(x_1, (b/a)x_1)$ on the asymptote and the ordinate $P_2(x_1, (b/a)\sqrt{x_1^2 - a^2})$ of the hyperbola.

POLYNOMIALS AND RATIONAL FRACTIONAL FUNCTIONS 8

8-1 DEFINITION

A *polynomial in x* is a function that can be written in the form

$$a_0x^n + a_1x^{n-1} + a_2x^{n-2} + \cdots + a_{n-1}x + a_n \tag{8-1}$$

where the a's are real-number constants and the exponents are positive integers. The polynomial is said to be of *degree n* if $a \neq 0$ (a constant is usually classed as a polynomial of degree zero). Thus

$$2x + 5 \qquad (x^2 - 2)^3 \qquad x^5 - \sqrt{2}x^3 + 4$$

are polynomials in x of degree 1, 6, and 5, respectively. The functions

$$\sqrt{x} \qquad \frac{x^2 - 2}{x^2 + 2} \qquad x^3 + 4\sqrt{x} + 5$$

are *not* polynomials, because not all terms of each function are expressed with exponents that are positive integers.

The equations discussed in this chapter are of the form

$$y = a_0x^n + a_1x^{n-1} + a_2x^{n-2} + \cdots + a_{n-1}x + a_n \tag{8-2}$$

The equation $y = mx + b$ is a special case that has already been studied in Chap. 4; the graph is a straight line. A second special case is the equation $y = ax^2 + bx + c$, in which the right-hand member is a polynomial of second degree. Its graph, a parabola, was studied in Chap. 7.

8-2 POLYNOMIALS OF DEGREE GREATER THAN TWO

Figures 8-1 and 8-2 show three graphs of polynomials of fourth degree. If a polynomial is expressed as a product of linear factors, or if such factors can be found easily, they can be used in sketching the graph. For example, the right-hand member of the equation

$$y = x^4 - 11x^3 + 37x^2 - 45x + 18 \tag{8-3}$$

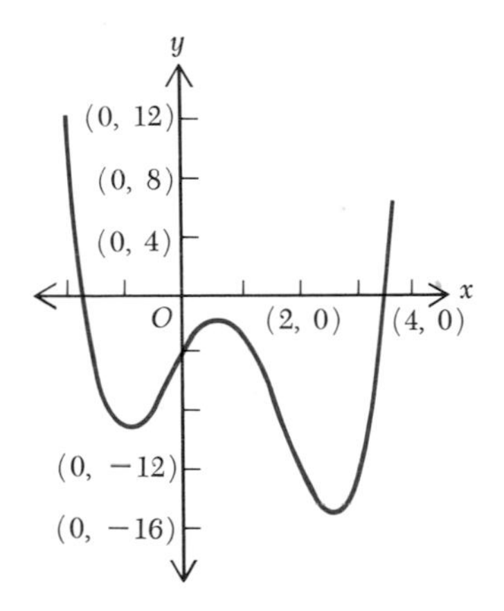

Fig. 8-1 Graph of $y = x^4 - 3x^3 - 3x^2 + 6x - 4$.

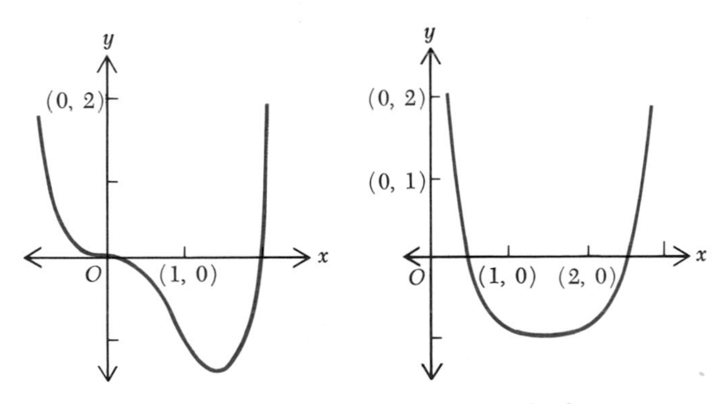

Fig. 8-2 (*a*) Graph of $y = x^4 - 2x^3$. (*b*) Graph of $y + 1 = (x - \frac{3}{2})^4$.

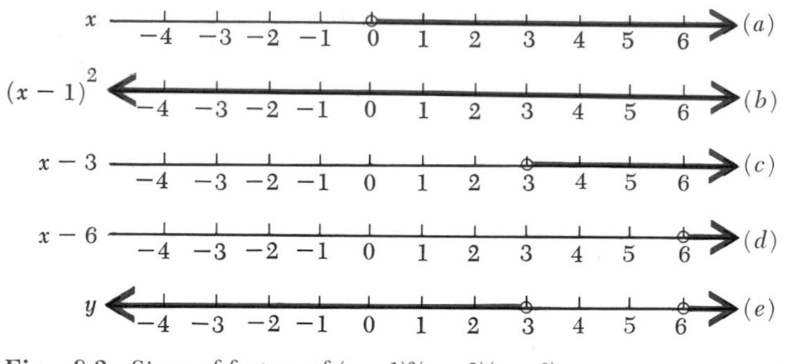

Fig. 8-3 Signs of factors of $(x-1)^2(x-3)(x-6)$.

can be factored as

$$y = (x-1)^2(x-3)(x-6) \tag{8-4}$$

It is much easier to sketch the graph from Eq. (8-4) than from Eq. (8-3) because the intercepts on the x axis can be readily obtained from (8-4) and the general shape of the graph can be visualized. Thus it is seen that

$$y = 0 \qquad \text{when } x = 1,\ x = 3, \text{ or } x = 6$$

These are the x intercepts. Further information on the graph is provided by the signs of the factors in the product $(x-1)^2(x-3)(x-6)$ for the intervals $x > 6$, $3 < x < 6$, $1 < x < 3$, and $x < 1$, and from their graphs on a number line (Fig. 8-3).

The number line in Fig. 8-3*a* is shaded for x positive; it is shaded in Figs. 8-3*b* to *d* for values of x that make the factor positive. A small circle indicates an x intercept or that the y coordinate of a point is neither positive nor negative. The sign of y in Fig. 8-3*e*, for Eq. (8-3), is evident from the graph of each of its factors. Note that

For $x > 6$ all factors are positive, so y is positive.

For $3 < x < 6$ the factor $x - 6$ alone is negative, so y is negative.

For $1 < x < 3$ the factors $x - 3$ and $x - 6$ are negative, while $(x-1)^2$ is positive, so y is positive.

For $x < 1$ the factors have the same signs as for $1 < x < 3$, so y is positive.

Now, *excluded regions,* or *empty bands,* may be shown for the graph of $y = (x-1)^2(x-3)(x-6)$. These regions are those where the graph cannot be drawn. For $x > 6$, y is positive; hence it cannot be negative. For $3 < x < 6$, y is negative; hence it cannot be positive. For $x < 3$, y is positive; hence it cannot be negative. These excluded regions are shaded in Fig. 8-4.

Observe that the graph of $y = (x - 1)^2(x - 3)(x - 6)$ (Fig. 8-5) *crosses* the x axis at $x = 3$ and $x = 6$. These numbers are *single* roots of the equation $(x - 1)^2(x - 3)(x - 6) = 0$. However, the graph touches the x axis without crossing it at $x = 1$, which is a *double* root. The cause of this behavior is readily explained. The factor $x - 6$, for example, is equal to zero for $x = 6$, negative for $x < 6$, and positive for $x > 6$; the sign of this factor $x - 6$, and consequently the sign of y, therefore changes when x increases from less than 6 to greater than 6. On the other hand, the factor $(x - 1)^2$ does not change sign when x increases from less than 1 to greater than 1. It is positive for $x > 1$, zero for $x = 1$, and positive again for $x < 1$. Thus the graph must touch the x axis at $x = 1$, but it does not *cross* the axis there.

If the factor had been $(x - 1)^3$, the graph would have crossed the x axis at $x = 1$. In this case the curve would have had roughly the shape shown in Fig. 8-6 in the neighborhood of (1, 0); it would have been tangent to the x axis at the crossing point.

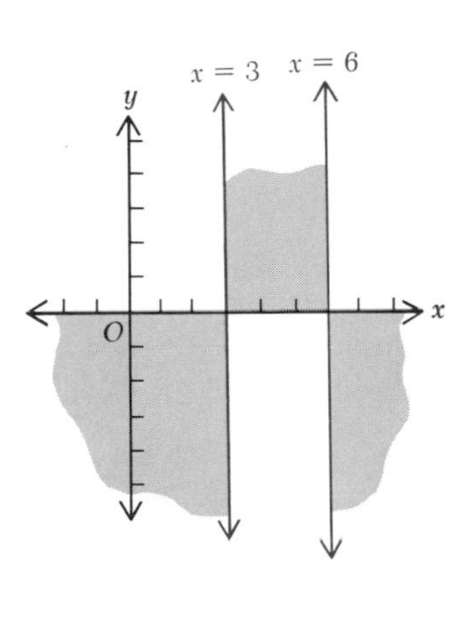

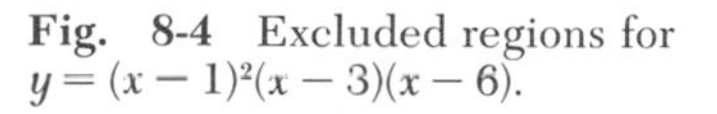
Fig. 8-4 Excluded regions for $y = (x - 1)^2(x - 3)(x - 6)$.

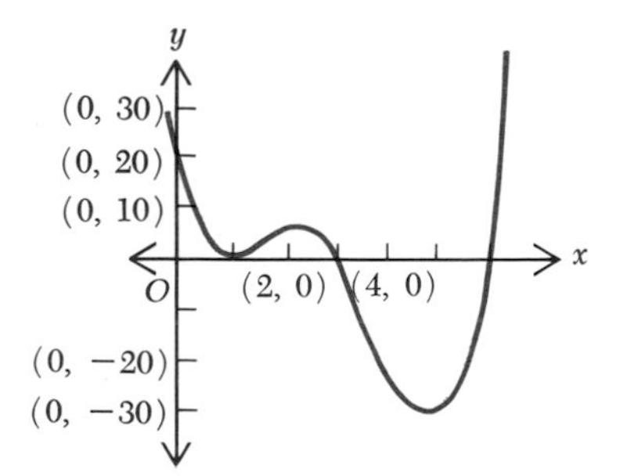

Fig. 8-5 Graph of $y = (x - 1)^2(x - 3)(x - 6)$.

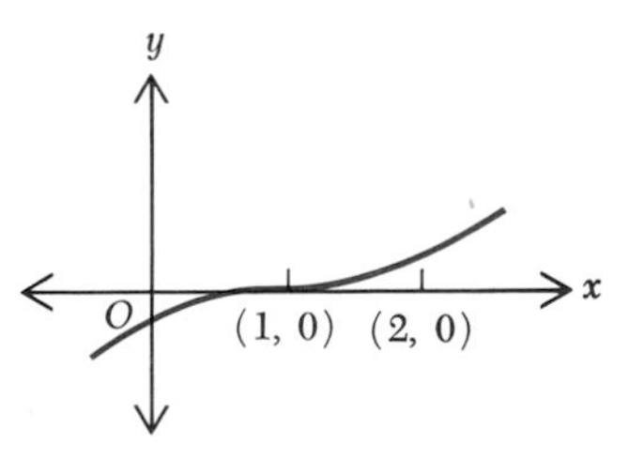

Fig. 8-6 Curve tangent to and crossing x axis at (1, 0).

Example

A right circular cylinder of radius x in. is inscribed in a right circular cone of radius 12 in. and height 9 in. Express the volume V of the cylinder as a function of x. Find (approximately) the value of x for which V is greatest.

Solution The cylinder inscribed in a cone is shown in Fig. 8-7. Let its radius be x in. and its height h in. The volume V of the cylinder is $\pi x^2 h$, where h is to be replaced by a function of x. A cross section of the cylinder inscribed in the cone is pictured in Fig. 8-8. From similar triangles,

$$\frac{h}{12 - x} = \frac{9}{12}$$

or

$$h = \tfrac{3}{4}(12 - x)$$

Hence for $0 \leqslant x < 12$

$$V = \tfrac{3}{4}\pi x^2 (12 - x) \qquad \text{cu in.}$$

This expression for V in terms of x is a polynomial of third degree. It may be graphed as follows.

Intercepts If $V = 0$, then $0 = \frac{3}{4}\pi x^2(12 - x)$. Hence $x = 12$ or $x = 0$, with a double root at $x = 0$. The intercepts are (0, 0) and (12, 0).

Empty bands In the interval $0 \leqslant x \leqslant 12$, V is never negative because each factor x^2 and $12 - x$ is zero or positive for every value of x in the domain. The required graph will be in the first quadrant.

A table of values will be helpful in locating a few points in the graph:

x	0	2	4	6	8	10	12
V	0	30π	96π	162π	192π	150π	0

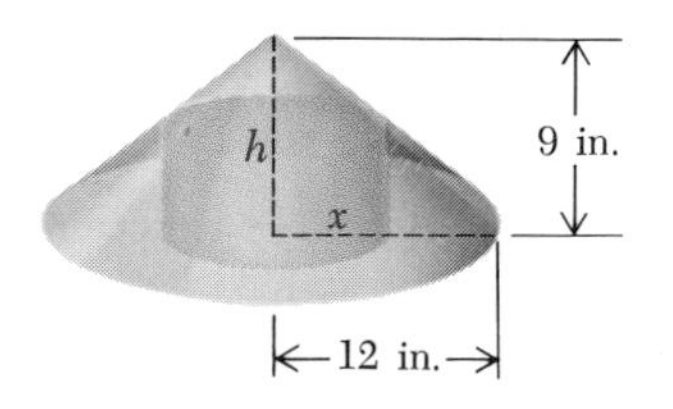

Fig. 8-7 Right circular cylinder inscribed in a right circular cone.

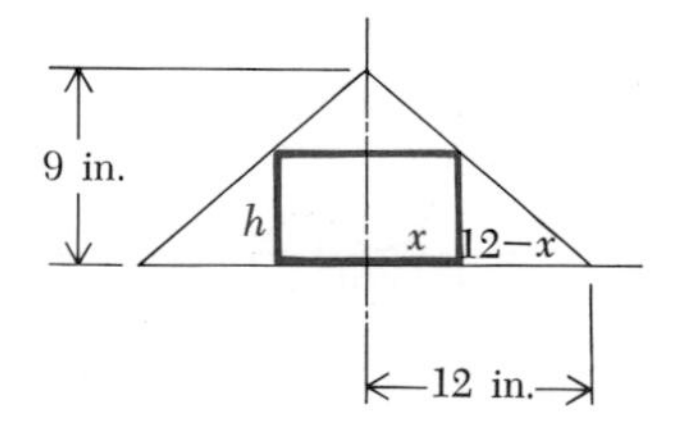

Fig. 8-8 Cross section of right circular cylinder inscribed in a right circular cone.

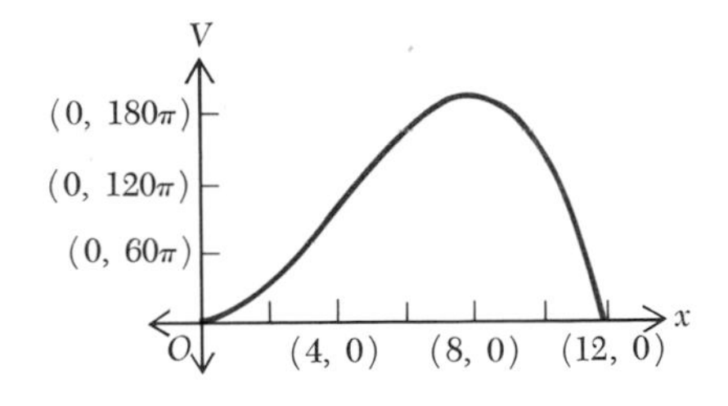

Fig. 8-9 Graph of $V = \frac{3}{4}\pi x^2(12 - x)$.

A scale of multiples of 60π is chosen for the V axis and multiples of 2 for the x axis because of values in the table. The graph is shown in Fig. 8-9. Even though different scales were used on the axes, it is possible to estimate from the graph that V would be a maximum for x in the neighborhood of 8. Hence the greatest inscribed cylinder is one with a radius of approximately 8 in.

PROBLEMS

1. Which of the following are polynomials in x? Give the degree of each polynomial:

   (*a*) $x^2(x^2 - 1)$ (*b*) $x^3/(3x + 1)$
   (*c*) $x^3 + 5x^2 + 2\sqrt{x}$ (*d*) $x^3 + 5x^2 + \sqrt{2}x$

Graph the equations in Probs. 2 to 7, and for each draw a picture similar to Fig. 8-3 from which excluded regions may be obtained.

2. $y = x(x - 2)(x + 4)$
3. $y = (x + 1)(x + 5)(x - 3)$
4. $y = \frac{1}{2}x(x^2 - 7x + 10)$
5. $y = x^2(x^2 - 9)$
6. $f(x) = \frac{1}{4}x(16 - x^2)$
7. $f(x) = 8x^3 - 3x^4$

8. In Probs. 2 to 4 replace the sign = by > and graph the resulting inequality.
9. In Probs. 5 to 7 replace the sign = by ≤ and graph the resulting inequality.
10. Graph each of the following sets of points:

    (*a*) $P = \{(x, y) \mid y = x(x - 2)(x + 3)\}$
    (*b*) $Q = \{(x, y) \mid y = -x(3 - x)(5 + 2x)\}$
    (*c*) $R = \{(x, y) \mid y - [x^2(x - 2)(x + 3)^2] = 0\}$
    (*d*) $S = \{(x, y) \mid 4y + x^2(x + 1)(x - 4) = 0\}$
    (*e*) $T = \{(x, f(x)) \mid f(x) = \frac{1}{2}x(x - 3)^2(x - 5)\}$
    (*f*) $U = \{(x, f(x)) \mid 4f(x) = (x + 3)^2(x^2 - 4)\}$

11. A farmer has 900 ft of fence and wishes to enclose a rectangular plot and then partition it into two congruent regions with a fence joining the midpoints of two sides. Express the area A enclosed as a function of the width

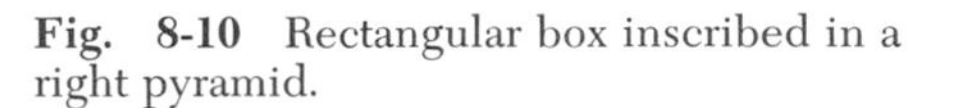

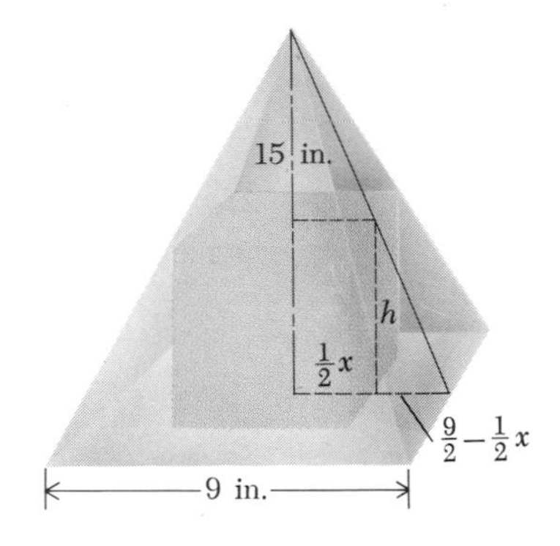

Fig. 8-10 Rectangular box inscribed in a right pyramid.

x. Draw the graph of the function. What dimensions give the greatest enclosed area?

12. It is desired to enclose a rectangular field along a stream. No fence is needed along the stream, and 180 ft of fencing is available. Express the area A enclosed as a function of the width x of the field. Draw the graph of the function and find the dimensions that give the greatest enclosed area.
13. A projectile is fired with velocity v_0 ft/sec at an angle α with the horizontal. If air resistance is neglected, the equation of its path can be shown to be

$$y = (\tan \alpha)x \left(- \frac{g}{2v_0^2 \cos^2 \alpha} x^2 \right)$$

where g is the gravitational constant (32 ft/sec², approximately). Sketch the path for the case in which $v_0 = 80$ ft/sec and $\alpha = 45°$. Find the approximate maximum height reached by the projectile.

14. If air resistance is neglected, the equation of the path traveled by a baseball that is thrown horizontally can be obtained by putting $\alpha = 0$ in the equation of Prob. 13. What is the minimum initial speed if the ball is to drop not more than $4\frac{1}{2}$ ft in a horizontal distance of 60 ft?
15. The base of a right pyramid 15 in. high is a square region with sides 9 in. long. A rectangular box whose base is a square region with sides x in. long is inscribed in the pyramid as shown in Fig. 8-10. Express the volume V of the box as a function of x. Draw the graph of the function. Estimate the value of x for which V is greatest.
16. A right circular cylinder of height y is inscribed in a sphere of diameter 12 in. Express the volume V of the cylinder as a function of y. Draw the graph. Estimate the value of y for which V is greatest.

8-3 RATIONAL FRACTIONAL FUNCTIONS

A function of the form

$$\frac{N(x)}{D(x)} \tag{8-5}$$

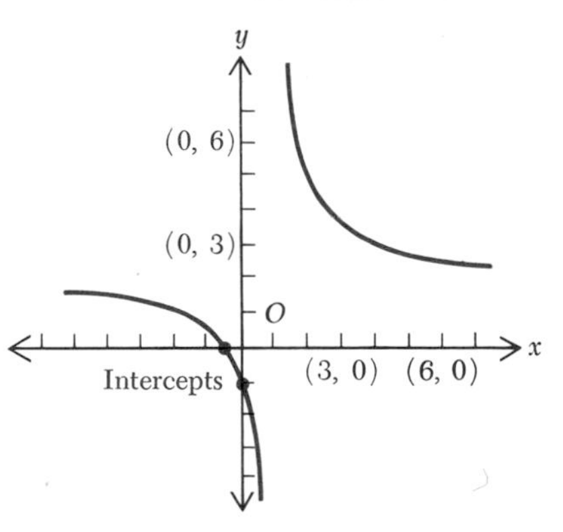

Fig. 8-11 Graph of $y = (2x + 1)/(x - 1)$.

where $N(x)$ and $D(x)$ are polynomials in x, is called a *rational fractional function of x*. Examples are

$$\frac{x+1}{x^2-4} \qquad \frac{6}{x} \qquad \frac{x^2+2x-7}{x^3+8}$$

In many applications an accurately plotted graph of $y = N(x)/D(x)$ is not needed; all that is necessary is a reasonably accurate sketch. The equation

$$y = \frac{2x+1}{x-1}$$

whose graph is shown in Fig. 8-11, will be used as an illustrative example in the next four sections.

8-4 INTERCEPTS

The *intercepts* of the graph on the x axis are the abscissas of the points where the graph touches or crosses the x axis. As described previously, these can be found by setting $y = 0$ in the equation for the graph and solving for x. Similarly, the intercepts on the y axis are the ordinates of the points where the graph touches or crosses the y axis. These can be found by setting $x = 0$ and solving for y. In this manner it is found that the graph of the equation

$$y = \frac{2x+1}{x-1}$$

crosses the x axis at $x = -\frac{1}{2}$ and the y axis at $y = -1$.

In connection with the problem of finding the intercepts, it is important to remember that a fraction $N(x)/D(x)$ is equal to zero for those values of x, and only those values, for which its numerator is zero and its denominator is not zero. Thus to find the x intercept of the equation

$$y = \frac{2x+1}{x-1}$$

first set the numerator equal to zero and solve for x; $2x + 1 = 0$ if and only if $x = -\frac{1}{2}$. Then make certain that the denominator is *not* zero for this value of x before concluding that $x = -\frac{1}{2}$ is the solution.

8-5 SYMMETRY

A curve is said to be *symmetrical with respect to a line* l if the line drawn from each point P on the curve perpendicular to l cuts the curve at another point P' such that l is the perpendicular bisector of $\overline{PP'}$. The curve in Fig. 8-12*a* is symmetrical about line l. The line l is sometimes called the *axis of symmetry*, or the *axis of reflection*. The point P' is often referred to as the reflection of P in l.

A curve is said to be *symmetrical with respect to a point* O if the line joining any point P of the curve to O intersects the curve at another point P' such that $|PO| = |OP'|$. The curve in Fig. 8-12*b* is symmetrical with respect to the point O.

In general, it is difficult to determine from a given equation whether or not there exists some line or point of symmetry. It is easy, however, to determine whether or not the graph is symmetrical with respect to one of the coordinate axes or the origin. The tests are as follows.

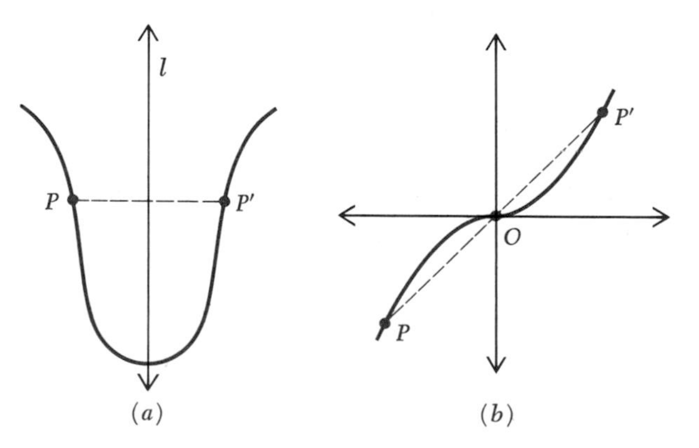

Fig. 8-12 (*a*) Curve symmetrical to line l. (*b*) Curve symmetrical to point O.

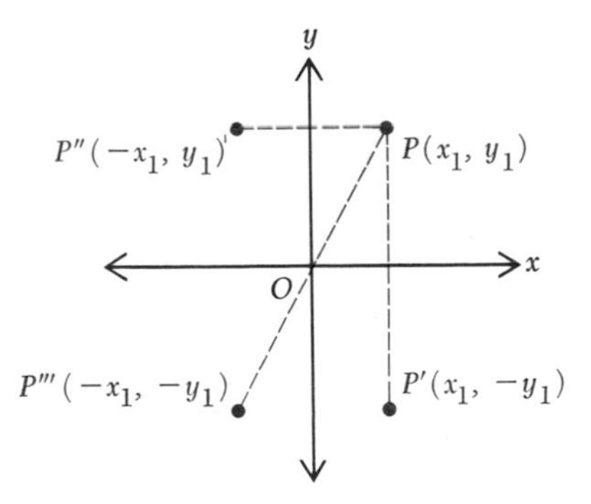

Fig. 8-13 Symmetry with respect to x axis, y axis, and origin.

Symmetry with respect to the x axis The graph of an equation is symmetrical with respect to the x axis if and only if there is for each point $P(x_1, y_1)$ on the graph another point $P'(x_1, -y_1)$ also on the graph (Fig. 8-13). That is, for $f(x, y) = 0$ to be symmetrical with respect to the x axis $f(x, y)$ must equal $f(x, -y)$. See Fig. 8-14*a* for curves that are symmetrical to the x axis.

Example 1
Show that the graph of the equation $y^2 - 4x = 0$ is symmetrical with respect to the x axis.

Solution The equation that results when y is replaced by $-y$ is $(-y)^2 - 4x = 0$ or $y^2 - 4x = 0$. Hence for every point (x, y) that satisfies the equation there exists a point $(x, -y)$ that also satisfies the equation, and the graph is symmetrical with respect to the x axis.

Symmetry with respect to the y axis The graph of an equation is symmetrical with respect to the y axis if and only if there is for each point $P(x_1, y_1)$ on the graph another point $P''(-x_1, y_1)$ also on the graph (Fig. 8-13). That is, for $f(x, y) = 0$ to be symmetrical with respect to the y axis $f(x, y)$ must equal $f(-x, y)$. See Fig. 8-14*b* for curves that are symmetrical to the y axis.

Symmetry with respect to the origin The graph of an equation is symmetrical with respect to the origin if and only if the equation that results when x is replaced by $-x$ and y is replaced by $-y$ is equivalent to the original equation. That is, $f(x, y)$ must equal $f(-x, -y)$. In Fig. 8-13, $P(x_1, y_1)$ and $P'''(-x_1, -y_1)$ are symmetrical with respect to the origin. See Fig. 8-14*c* for curves that are symmetrical to the origin. Note that

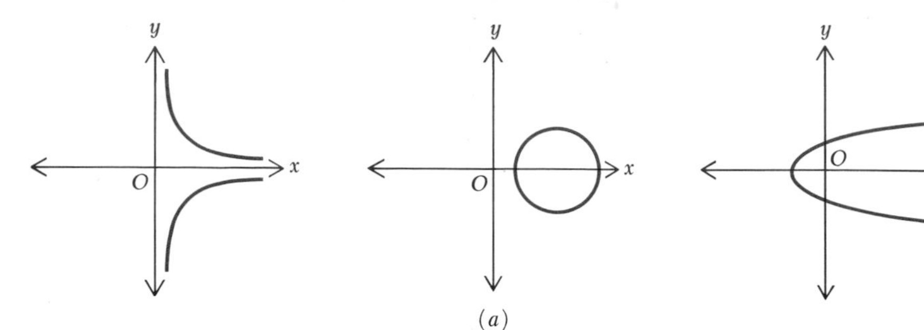

(a)

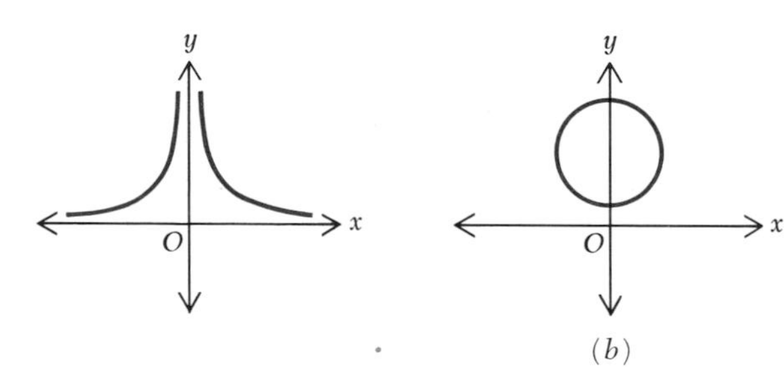

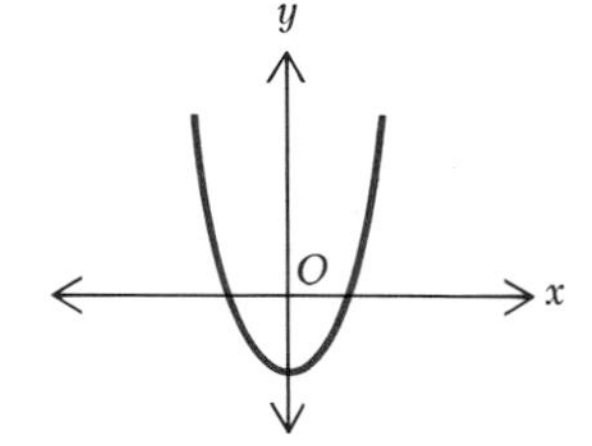

(b)

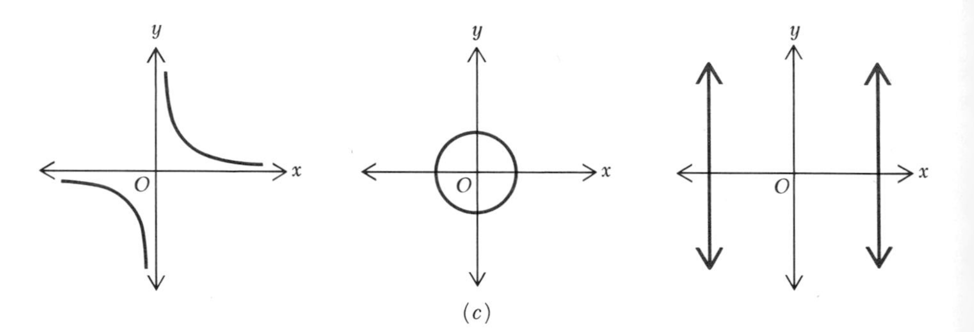

(c)

Fig. 8-14 (*a*) Curves symmetrical to x axis. (*b*) symmetrical to y axis. (*c*) Curves symmetrical to origin.

a graph does not have to be symmetrical with respect to the axes to be symmetrical with respect to the origin.

Example 2

Show that the graph of the equation $y^3 + 2y = x^3$ is symmetrical with respect to the origin.

Solution If x is replaced by $-x$ and y is replaced by $-y$, the equation becomes

$$(-y)^3 + 2(-y) = (-x)^3$$

This is equivalent to $y^3 + 2y = x^3$, which is the original equation.

8-6 VERTICAL ASYMPTOTES

Consider again the equation

$$y = \frac{2x+1}{x-1}$$

and let x be replaced by each number in the sequence

$$1\tfrac{1}{10},\ 1\tfrac{1}{100},\ 1\tfrac{1}{1,000},\ 1\tfrac{1}{10,000},\ 1\tfrac{1}{100,000},\ 1\tfrac{1}{1,000,000},\ \ldots$$

The corresponding values of y are as shown:

x	$1\frac{1}{10}$	$1\frac{1}{100}$	$1\frac{1}{1,000}$	$1\frac{1}{10,000}$	$1\frac{1}{100,000}$	$1\frac{1}{1,000,000}$
y	32	302	3,002	30,002	300,002	3,000,002

Observe that as x approaches closer and closer to 1, the corresponding values of y become greater and greater beyond bound. This situation results from the fact that the denominator of the fraction is approaching zero while the numerator is approaching a number different from zero, namely, 3. In a similar manner, as x approaches 1 through a sequence of values less than 1, values of y again become greater and greater in absolute value but in this case are always negative. This fact is verified by the following table:

x	$\frac{9}{10}$	$\frac{99}{100}$	$\frac{999}{1,000}$	$\frac{9,999}{10,000}$	$\frac{99,999}{100,000}$	$\frac{999,999}{1,000,000}$
y	−28	−298	−2,998	−29,998	−299,998	−2,999,998

The behavior of the graph in the neighborhood of $x = 1$ is that indicated in Fig. 8-15. Note that there is no value of y, and consequently no point on the graph, corresponding to $x = 1$. With the exclusion of this one point, however, there are values of y for x as arbitrarily near 1 as we care to make them. Furthermore, for x sufficiently near 1, the value of y is arbitrarily great in absolute value. The situation is described by saying that the curve approaches the line $x = 1$ *asymp-*

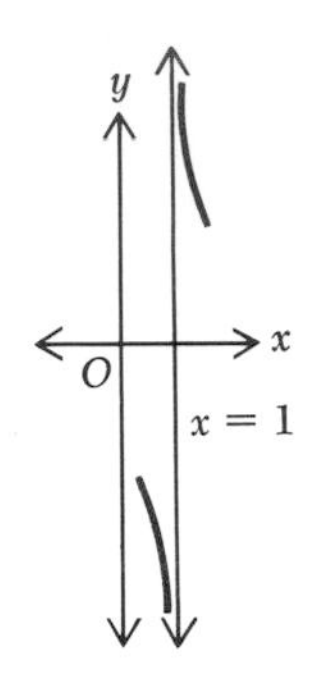

Fig. 8-15 Graph of $y = (2x + 1)/(x - 1)$ in the neighborhood of $x = 1$.

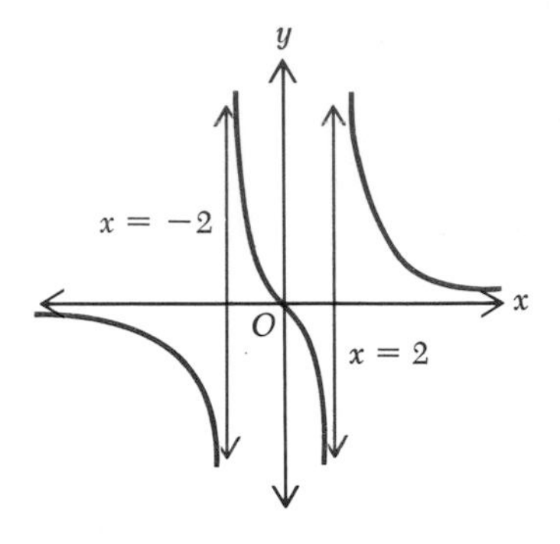

Fig. 8-16 Graph of $y = 2x/(x^2 - 4)$.

totically. The line $x = 1$ is called an *asymptote* to the curve. It can be argued that the graph of the equation

$$y = \frac{N(x)}{D(x)} \tag{8-6}$$

will have one of these vertical asymptotes at each value of x for which the denominator is zero and the numerator is other than zero. If the fraction in Eq. (8-6) is in simplest form (numerator and denominator have no common factor), then the numerator cannot be zero for any value of x for which the denominator is zero. This case may be summarized as follows: To find the vertical asymptotes to the graph of Eq. (8-6), set $D(x) = 0$ and solve for x.

Example

Find the vertical asymptotes of the equation $y = 2x/(x^2 - 4)$.

Solution The graph of the equation $y = 2x/(x^2 - 4)$ will have vertical asymptotes at $x = 2$ and $x = -2$, the roots for $x^2 - 4 = 0$. The behavior of the curve near the asymptotes is seen in the graph in Fig. 8-16.

8-7 HORIZONTAL ASYMPTOTES

Consider again the equation

$$y = \frac{2x + 1}{x - 1}$$

and let x take on the values 10, 100, 1,000, . . . , that is, become greater and greater. The corresponding values of y are as shown:

x	10	100	1,000	10,000	100,000	1,000,000
y	$2\frac{1}{3}$	$2\frac{1}{33}$	$2\frac{1}{333}$	$2\frac{1}{3,333}$	$2\frac{1}{33,333}$	$2\frac{1}{333,333}$

It seems that as x gets greater and greater, the value of y approaches 2. Actually, the difference between the value of y and 2 approaches zero. Similarly, if x is negative and becomes greater and greater in absolute value, the value of y again approaches 2. The line $y = 2$ is a horizontal asymptote to the curve. The behavior of the curve for y coordinates close to 2 is that indicated in Fig. 8-17.

Using the ideas concerning intercepts, symmetry, and asymptotes, the graph of $y = (2x + 1)/(x - 1)$ may be sketched as in Fig. 8-18.

As stated above, the graph of Eq. (8-6) may have several vertical asymptotes if $D(x) = 0$ has several real roots. However, it has either one horizontal asymptote or none; to find the horizontal asymptote, if there is one, we need only determine what happens to the value of y as x becomes arbitrarily great in absolute value. For $N(x)/D(x)$ there are three possibilities.

Case 1 Degree of N lower than that of D In this case the value of $N(x)/D(x)$ approaches zero as x becomes greater and greater in absolute value, and the graph is asymptotic to the x axis.

Example 1
The graph of the equation

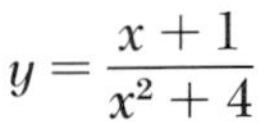

$$y = \frac{x + 1}{x^2 + 4}$$

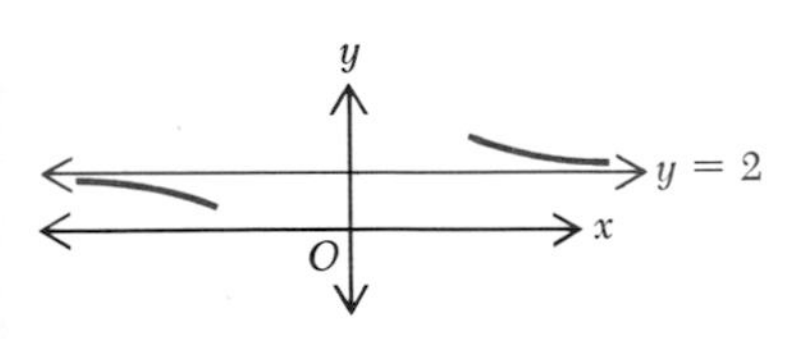

Fig. 8-17 Graph of $y = (2x + 1)/(x - 1)$ in the neighborhood of $y = 2$.

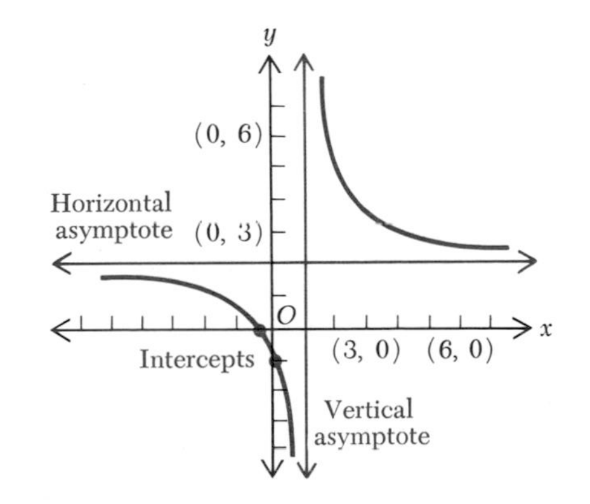

Fig. 8-18 Graph of $y = (2x + 1)/(x - 1)$.

has the x axis as a horizontal asymptote; that is, the value of y approaches zero as x becomes greater in absolute value. The reader may compute y for $x = 10$, $x = 100$, and $x = 1{,}000$ (see also Prob. 21, page 229).

Case 2 Degree of N the same as that of D In this case, as x becomes greater and greater in absolute value, the value of the fraction $N(x)/D(x)$ approaches a number which is equal to the quotient of the coefficients of the terms of highest degree in $N(x)$ and $D(x)$. If

$$N(x) = a_0x^n + a_1x^{n-1} + \cdots + a_n$$

and

$$D(x) = b_0x^n + b_1x^{n-1} + \cdots + b_n$$

the horizontal asymptote is the line $y = a_0/b_0$.

Example 2

The graph of the equation

$$y = \frac{6x^2 + x + 4}{2x^2 - 1}$$

is asymptotic to the horizontal line $y = \frac{6}{2} = 3$. Note that for very great values of x the fraction is nearly the same as $6x^2/2x^2$, because the other terms are small in value in comparison with these two. Observe also that, by long division,

$$\frac{6x^2 + x + 4}{2x^2 - 1} = 3 + \frac{x + 7}{2x^2 - 1}$$

As x becomes greater and greater in absolute value, the second term on the right approaches zero (by Case 1), and the right-hand side must therefore approach 3. This indicates how a proof for Case 2 can be carried out after Case 1 has been proved (see also Prob. 22, page 229).

Case 3 Degree of N higher than that of D In this case, as x becomes greater and greater in absolute value, the absolute value of y also becomes arbitrarily great. There is no horizontal asymptote.

Example 3

The graph of the equation

$$y = \frac{x^2 + 1}{x - 1}$$

has no horizontal asymptote. By division,

$$\frac{x^2 + 1}{x - 1} = x + 1 + \frac{2}{x - 1}$$

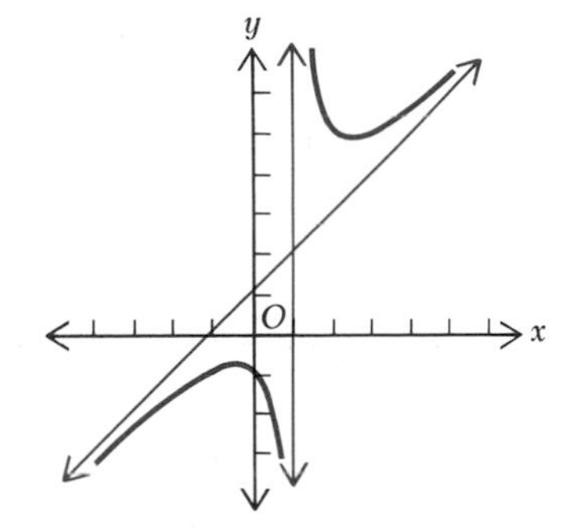

Fig. 8-19 Graph of $y = (x^2 + 1)/(x - 1)$.

As x becomes greater and greater in absolute value, the fraction $2/(x - 1)$ on the right approaches zero. This means that for values of x the value of the fraction $(x^2 + 1)/(x - 1)$ is nearly the same as that of the linear function $x + 1$. The line $y = x + 1$ is, in fact, an inclined asymptote to the curve (Fig. 8-19). This discussion should make intuitively clear the fact that there is, in general, an inclined straight line to which the curve is asymptotic if the degree of the numerator exceeds that of the denominator by exactly 1.

8-8 CURVE DISCUSSION

Discussing the graph of an equation of the form $y = N(x)/D(x)$ means finding the intercepts, testing for symmetry, locating all vertical and horizontal asymptotes, and determining excluded regions. Any other information that can be obtained easily from the equation, and that will be useful in sketching the graph, should be included in the discussion. In the problems of the next set first write out the discussion in the brief form given in the following example. Then use the discussion, in conjunction with a few plotted points, to sketch the graph.

Example

Discuss and sketch

$$y = \frac{x}{(1 - x)(x - 3)}$$

Solution

Intercepts On x axis if $y = 0$, $x = 0$. On y axis if $x = 0$, $y = 0$.

Symmetry About x axis, none. About y axis, none. About origin, none.

Asymptotes Vertical, $x = 1$ and $x = 3$, from $(1 - x)(x - 3) = 0$. Hori-

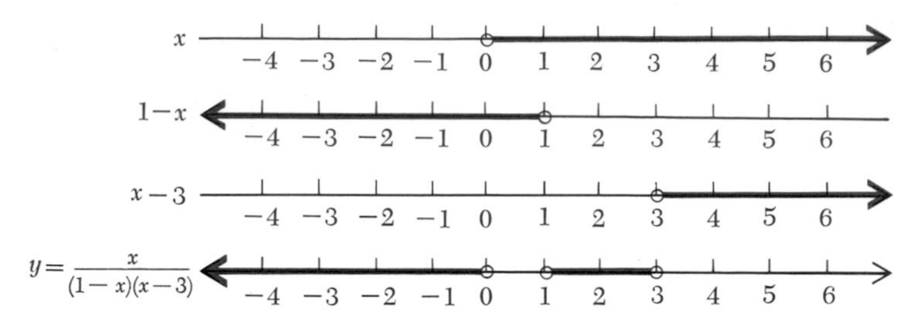

Fig. 8-20 Sign of y for values of x for $y = x/[(1-x)(x-3)]$.

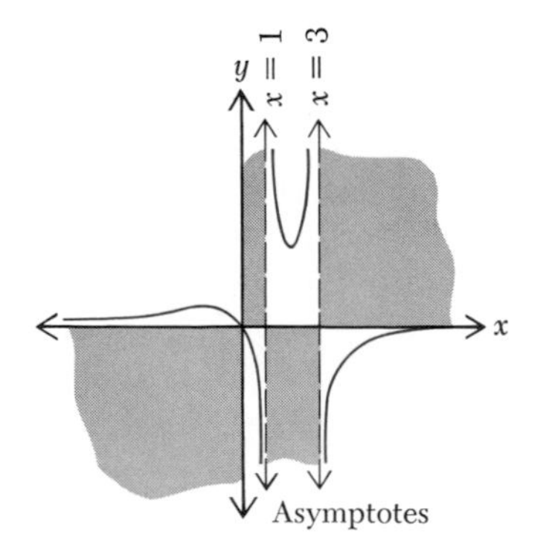

Fig. 8-21 Graph of $y = x/[(1-x)(x-3)]$.

zontal, $y = 0$, because y approaches 0 as x increases in absolute value. Other information Excluded regions (see Fig. 8-20). First the intercept is plotted, the excluded regions are shaded, and the asymptotes are labeled. Then the graph is sketched as in Fig. 8-21. If necessary, a few points for x less than 0, between 1 and 3, and greater than 3 may be plotted to determine the shape of the curve in these three regions.

PROBLEMS

Discuss and sketch the graphs of the equations in Probs. 1 to 18.

1. $y = \dfrac{2}{3x}$
2. $y = \dfrac{8}{3x^2}$
3. $y = \dfrac{9}{x-3}$
4. $y = \dfrac{6}{2x+3}$
5. $y = \dfrac{5-2x}{3x}$
6. $y = \dfrac{2x}{(x+2)^2}$
7. $y = \dfrac{5}{x^2+2}$
8. $y = \dfrac{4x}{x^2+4}$
9. $y = \dfrac{6}{x^2-4}$
10. $y = \dfrac{6x}{x^2-4}$
11. $y = \dfrac{3x}{2x^2-9}$
12. $y = \dfrac{9-x^2}{x^2-5}$

13. $y = \dfrac{x^2 + 4}{x^2 - 4}$ 14. $y = \dfrac{x^2 - 4}{x^2 + 4}$ 15. $y = \dfrac{x^2}{x^2 - 12}$

16. $y = \dfrac{x^3}{3x + 6}$ 17. $y = \dfrac{x(x-1)}{(x^2 - 4)(x - 6)}$ 18. $y = \dfrac{x^2}{(x-2)(x-4)}$

19. A right circular cone of height y is circumscribed about a sphere whose radius is 6 in. (Fig. 8-22). Express the volume V of the cone as a function of y and sketch the graph showing how V varies with y. *Hint:* $V = \frac{1}{3}\pi x^2 y$, and from similar triangles (Fig. 8-23), $x/y = 6/\sqrt{(y-6)^2 - 6^2}$.
20. A rectangular box (prism) with square base is to enclose 108 cu ft. The costs per square foot of material for bottom, top, and sides are 2, 4, and 6 cents, respectively. Express the cost C of material as a function of the length x of a side of the base. Draw the graph of the function and estimate the value of x for which C is least.
21. Consider the fraction

$$\frac{3x^2 + x + 4}{5x^3 + 7x^2 - 6}$$

Dividing both numerator and denominator by x^3 (which is the highest power of x that occurs in the fraction) results in the identity

$$\frac{3x^2 + x + 4}{5x^3 + 7x^2 - 6} = \frac{3/x + 1/x^2 + 4/x^3}{5 + 7/x - 6/x^3} \qquad \text{for } x \neq 0$$

Now, as x becomes greater and greater in absolute value, each term in the numerator on the right approaches zero. The second and third terms in the denominator also approach zero. The numerator therefore approaches zero, while the denominator approaches 5, and the fraction consequently approaches zero. Use this idea in the general case to write the proof for Case 1 of Sec. 8-7.
22. Using the method suggested in Prob. 21, write a proof for Case 2 of Sec. 8-7.

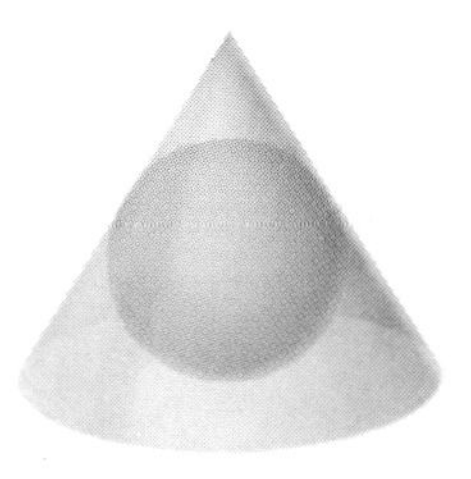

Fig. 8-22 Sphere inscribed in a right circular cone.

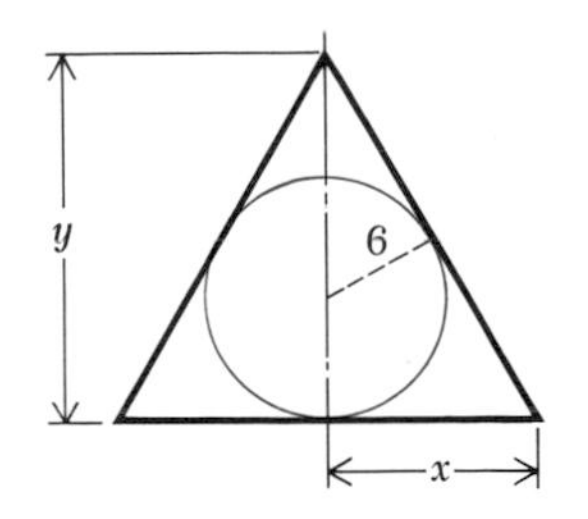

Fig. 8-23 Cross section of a sphere inscribed in a right circular cone.

Discuss and sketch the graphs of the equations in Probs. 23 to 36 after solving for y in terms of x:

23. $2xy - 4y + 3x - 8 = 0$
24. $xy + 2y = 4x$
25. $4y = x^2y + 4x$
26. $2xy + 4y = x^2 + 4$
27. $y(x^2 + 1) = (x + 1)^2$
28. $xy = x^2 - 4$
29. $x^2y + 3y = 2x^2 + 7$
30. $x^2y = 4x - y + 3$
31. $2xy - x^2 = 4y + 2x + 6$
32. $x^2y - 9y = x^2 + 4x$
33. $(x^2 - 1)^2y = 2x$
34. $x^2y + xy = 3x - y + 6$
35. $(x + 1)^2y = 8x$
36. $(x^2 - 9)^2y = 4x(x^2 - 4)$

SUMMARY

A polynomial in x is a function of x that can be written in the form

$$a_0x^n + a_1x^{n-1} + a_2x^{n-2} + \cdots + a_{n-1}x + a_n$$

where the a's are real numbers and the exponents are positive integers.

A function of the form $N(x)/D(x)$, where $N(x)$ and $D(x)$ are polynomials in x, is called a rational fractional function of x. Graphing these functions by determining number pairs (x, y) is often laborious. Hence the following important properties of the graph are obtained and a sketch is made for $y = f(x)$ or $y = h(x)/g(x)$. Intercepts on the y axis are found by replacing x by 0 and solving the resulting equation for y. Intercepts on the x axis are found by replacing y by 0 and solving the resulting equation for x.

Excluded regions appear in the xy plane where there is no graph for a given equation. If $f(x)$ is factorable, the excluded regions are determined by examining the signs of those factors for given values of x.

A curve is symmetrical with respect to:

1. The x axis if for every point $P_1(x_1, y_1)$ on the graph another point $P_2(x_1, -y_1)$ is also on the graph; that is, $f(x, y) = f(x, -y)$
2. The y axis if for every point $P_1(x_1, y_1)$ on the graph another point $P_2(-x_1, y_1)$ is also on the graph; that is, $f(x, y) = f(-x, y)$
3. The origin if for every point $P_1(x_1, y_1)$ on the graph another point $P_2(-x_1, -y_1)$ is also on the graph; that is, $f(x, y) = f(-x, -y)$

If $y = N(x)/D(x)$, a rational fractional function in x, vertical asymptotes are found for each value of x for which $D(x) = 0$ and $N(x) \neq 0$. For

$y = N(x)/D(x)$ there will be one or no horizontal asymptotes. The asymptote is found for the value that y approaches when x becomes greater and greater in absolute value.

REVIEW PROBLEMS

In Probs. 1 to 3 draw a number-line picture to show excluded regions and then sketch the graph of the equation.

1. $y = x(x-1)(x+5)$
2. $y = x^3 - 9x$
3. $f(x) = (x-1)^2(x^2+4)$

For the equations in Probs. 4 to 9 list the intercepts and asymptotes, indicate symmetry with respect to the coordinate axes and the origin, and show excluded regions on a number line. Then sketch the graph of the equation.

4. $y = \dfrac{3}{x-2}$
5. $y = \dfrac{-2x}{9-x^2}$
6. $y = \dfrac{x(x+3)}{(x-1)(x-4)(x-7)}$
7. $y = \dfrac{x^3}{x^2-x-6}$
8. $y = \dfrac{x^2+4}{x^2-x-12}$
9. $x^2 - xy - 2x + 2y = 0$
10. A right circular cone that has altitude h is inscribed in a sphere of radius 1. (*a*) Express the volume V of the cone as a function of h. (*b*) Draw the graph. (*c*) Estimate the value of h for which V is greatest.

FOR EXTENDED STUDY

1. Develop rules for symmetry with respect to the lines $y = x$ and $y = -x$.
2. May a set of points have symmetry with respect to exactly two points? Prove your answer.
3. Given the line $ax + by + c = 0$ and the point $P_1(x_1, y_1)$, find the coordinates of $P_2(x_2, y_2)$ such that P_1 and P_2 are symmetric with respect to the line.
4. Find the meanings of the terms "even function" and "odd function." Discuss the symmetry of each type of function with respect to the x axis, the y axis, and the origin.
5. The graph of $y = x^3$ is symmetrical with respect to $P(0, 0)$. Discuss the point symmetry of $y = (x-1)^3$, $y = (x+4)^3$, $y = (x-a)^3$, $y = -(x-1)^3$, $y = -(x+4)^3$, and $y = -(x-a)^3$.

9 ALGEBRAIC CURVES OF HIGHER DEGREE

9-1 DEFINITIONS

The algebraic curves considered in this chapter are those that are the loci of equations $f(x, y) = 0$, where $f(x, y)$ is a polynomial in x and y. The general first-degree equation of this type is

$$Ax + By + C = 0$$

Its graph is a straight line. The general equation of second degree is

$$Ax^2 + Bxy + Cy^2 + Dx + Ey + F = 0$$

Such an equation may have no locus at all, but if it has one, it is a conic section.

It would be natural as a next step to consider polynomials of higher degree. Some special cases were discussed in Chap. 8. For example, the equation

$$y = \frac{x + 2}{x(x^2 + 1)}$$

is equivalent to

$$x^3y + xy - x - 2 = 0$$

The left-hand member of this last equation is a fourth-degree polynomial in x and y. It is, however, of only the first degree in y and thus is characteristic of the equations studied in Chap. 8. In this chapter we consider cases of a more general nature.

9-2 DISCUSSION OF ALGEBRAIC CURVES

The sketching of an algebraic curve is often facilitated by a discussion similar to that employed in Chap. 8. We need more general methods for finding horizontal and vertical asymptotes, and certain other properties of the curve are extracted.

Intercepts Intercepts are found as before by substituting zero for y and solving the resulting equation $f(x, 0) = 0$ for x, and then similarly substituting zero for x and solving $f(0, y) = 0$ for y.

Symmetry The tests previously given for symmetry with respect to the coordinate axes and to the origin apply to any equation in x and y. In addition, a graph is symmetrical with respect to the line $y = x$ if the equation that results when x is replaced by y and y is replaced by x is identical with the original equation. For instance, the equations $3xy = 5$ and $x^2 - 7xy + y^2 = 10$ are symmetrical about the line $y = x$ because in each case replacing x by y and y by x results in an identical equation.

Asymptotes If an equation can be solved for y in terms of x and thus put into the form $y = F(x)/G(x)$, then vertical asymptotes can be found by the method of Chap. 8; that is, if there are any linear factors $ax + b$ for which the denominator $G(x)$ is zero and the numerator $F(x)$ is not zero, then $ax + b = 0$ is a vertical asymptote. Similarly, if the equation can be solved for x in terms of y and thus put into the form $x = H(y)/J(y)$, then horizontal asymptotes can be found by determining the values of y for which the denominator $J(y)$ is zero and the numerator $H(y)$ is not zero.

Example 1

If the equation $x^2y^2 - 4y^2 - x^2 = 0$ is solved for y in terms of x and for x in terms of y, the resulting equations are

$$y = \pm\sqrt{\frac{x^2}{x^2 - 4}}$$

$$x = \pm\sqrt{\frac{4y^2}{y^2 - 1}}$$

From the first of these it is seen that the lines $x = 2$ and $x = -2$ are vertical asymptotes, and from the second it is seen that the lines $y = 1$ and $y = -1$ are horizontal asymptotes.

Since it is sometimes difficult, or even impossible, to solve an equation for y and for x in this manner, a more general method is needed for finding the horizontal and vertical asymptotes. The following reasoning yields such a method. The root of the equation $ax + b = 0$ becomes arbitrarily great if and only if a approaches zero. Similarly, one root of the quadratic equation $ax^2 + bx + c = 0$ is arbitrarily great if and only if a approaches zero; one root of the cubic equation $ax^3 + bx^2 + cx + d = 0$ becomes arbitrarily great if and only if a (the coefficient of the term of highest degree in x) approaches zero, and so on. Now consider the equation

$$(y - 2)x^3 + 3yx^2 + y^3x + 6 = 0$$

The left member is a polynomial in x and y, arranged in descending powers of x. The coefficient of the x^3 term approaches zero (and so one root of the equation becomes arbitrarily great) only when the value of y approaches 2. Thus the line $y = 2$ is a horizontal asymptote and there are no others.

Similarly, if the polynomial is arranged in descending powers of y, the result is

$$xy^3 + (3x^2 + x^3)y - 2x^3 + 6 = 0$$

The coefficient of the highest power of y is x, and it is when and only when this coefficient approaches zero that one of the corresponding values of y becomes indefinitely great. This means that the line $x = 0$ is a vertical asymptote and that there is no other vertical asymptote.

Thus we may formulate the following general rule. To find the equations of the horizontal asymptotes of an algebraic curve, equate to zero the coefficient of the highest power of x that occurs in the original equation and solve the resulting equation for y. To find the equations of the vertical asymptotes, equate to zero the coefficient of the highest power of y that occurs in the equation and solve the resulting equation for x. In each case the real roots of the equation being solved yield the corresponding asymptotes. If the equation has no real roots, then the algebraic curve has no horizontal or no vertical asymptotes, as the case may be. In particular, there is no horizontal asymptote if the coefficient of the highest power of x is a constant, and there is no vertical asymptote if the coefficient of the highest power of y is constant.

Example 2
The highest power of x in the equation $x^2y^2 - 2y^2 + 3x^2 + 4x = 0$ is x^2. Its coefficient is $y^2 + 3$. The equation $y^2 + 3 = 0$ has no real roots, so the curve has no horizontal asymptote.

The highest power of y is y^2. Its coefficient is $x^2 - 2$. The equation $x^2 - 2 = 0$ has $x = \pm\sqrt{2}$ as real roots, so the curve has two vertical asymptotes, the lines $x = \sqrt{2}$ and $x = -\sqrt{2}$.

Example 3
In the equation $x^3y + 2y^3 + 3xy + y^2 = 6$ the coefficient of the highest power of x is y. The curve therefore has one horizontal asymptote, the line $y = 0$. The coefficient of the highest power of y is 2. This means that the curve has no vertical asymptote.

Tangent lines at the origin An algebraic curve passes through the origin if and only if the constant term in its equation is zero. For example, the curve whose equation is

$$x^3y + x^2 - 4y^2 = 0$$

passes through the origin, but the curve whose equation is

$$x^3y + x^2 - 4y^2 + 5 = 0$$

does not, for the coordinates (0, 0) satisfy the first equation but do not satisfy the second.

If the curve passes through the origin, its behavior near that point may be determined by striking out from the equation all terms except the one or ones of lowest degree and studying the resulting equation. Thus if the equation has one or more terms of first degree, strike out all terms except those of first degree. If the equation has no terms of first degree, but has one or more terms of second degree, strike out all terms except those of second degree, and so on. It can be shown in particular that if the resulting equation represents one or more straight lines, then these lines are the tangent lines to the curve at the origin.

Example 4
Consider the curve whose equation is

$$x^2y + y^2 + 3x - y = 0$$

The curve passes through the origin and has the line $3x - y = 0$ as its tangent line at that point.

Example 5

The curve whose equation is

$$x^3y + x^2 - 4y^2 = 0$$

passes through the origin. Striking out the fourth-degree term results in

$$x^2 - 4y^2 = 0$$

This can be written as

$$(x - 2y)(x + 2y) = 0$$

and it represents the two lines $y = \frac{1}{2}x$ and $y = -\frac{1}{2}x$. These are the tangent lines to the curve at the origin. Incidentally, the curve has the line $y = 0$ as its only horizontal asymptote, and it has no vertical asymptote (since the coefficient of the highest power of y is a constant). The graph is shown in Fig. 9-1.

We can strike out all terms except those of lowest degree in studying the behavior of a graph near the origin, because for values of x and y close to zero the values of terms of higher degree are small in absolute value compared with those of lowest degree. Thus, for example, the equation $x^2 - 4y^2 = 0$ is satisfied by the coordinates (0.02, 0.01). These coordinates almost satisfy the equation $x^3y + x^2 - 4y^2 = 0$, because at the point (0.02, 0.01) the value of the x^3y term is $(0.02)^3(0.01)$, or 0.00000008.

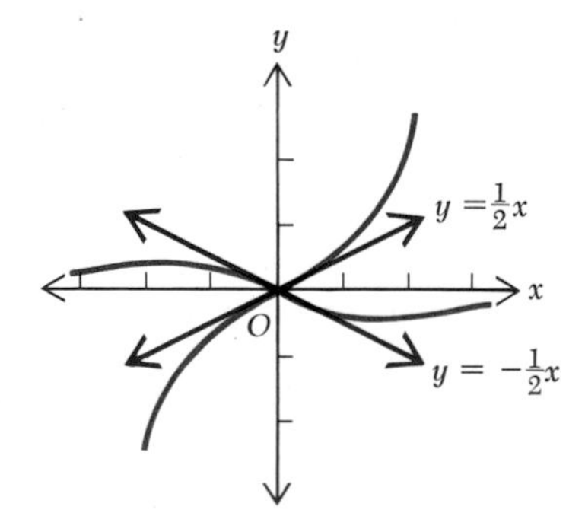

Fig. 9-1 $y = \frac{1}{2}x$ and $y = -\frac{1}{2}x$ are tangent lines at O to $x^3y + x^2 - 4y^2 = 0$.

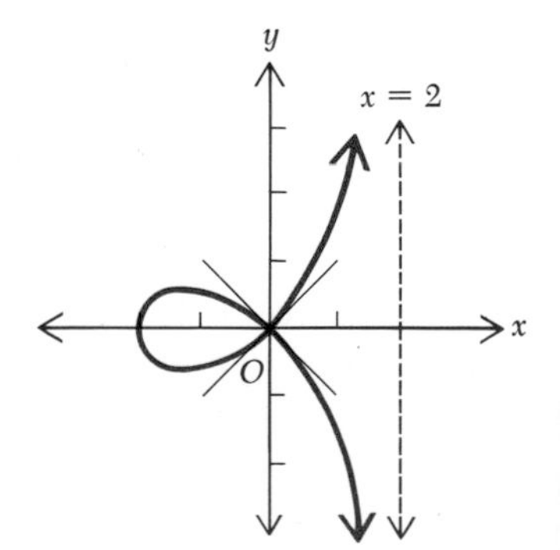

Fig. 9-2 Graph of strophoid: $x^3 + xy^2 + 2x^2 - 2y^2 = 0$.

Excluded intervals It may be possible to determine from the given equation that there will be no real value of y corresponding to values of x in a certain interval. For example, in the case of the equation $y^2 = 4 - x$ there are no real values of y corresponding to any value $x > 4$. For the equation $x^2 + y^2 = 25$ there are no real values of y corresponding to $|x| > 5$; thus the entire graph lies in the interval $-5 \leqslant x \leqslant 5$.

If an equation can be solved for y in terms of x, the result may be such that terms involving x appear under square-root signs. If the expression under such a radical sign is negative for a certain domain of x, then there will be no real value of y corresponding to values of x in this interval. Such a domain of x is called an *excluded interval* on the x axis. Similarly, it may be that there are excluded intervals for certain values of y.

Example 6
If the equation $x^3 + xy^2 + 2x^2 - 2y^2 = 0$ is solved for y in terms of x, then

$$y = \pm x \sqrt{\frac{x+2}{2-x}}$$

If $x > 2$, the fraction under the radical is negative because the denominator is negative and the numerator is positive. If $x < -2$, the fraction under the radical is again negative because the numerator is negative and the denominator is positive. There are then no real values of y corresponding to values of x that are greater than 2 or less than -2.

It would be difficult to solve the equation $x^3 + xy^2 + 2x^2 - 2y^2 = 0$ for x in terms of y in order to investigate the possibility of excluded intervals along the y axis. However, from the following consideration, it can be concluded that there is no such interval. If any value y_1 is substituted for y in the given equation, the resulting equation is a cubic in x with real numbers as coefficients. Such a cubic must have at least one real root, so there must be at least one real value of x corresponding to every real value of y. The graph shown in Fig. 9-2 is that of a curve called the *strophoid*. Its general equation is $x^3 + xy^2 + Kx^2 - Ky^2 = 0$, where K is a constant.

Solutions to the problems of the next set should be arranged in an orderly fashion as indicated in the following examples.

Example 7
Discuss and sketch the graph of the equation

$$x^3 + xy^2 = 2(3x^2 - y^2)$$

Solution

Intercepts If $y = 0$, then $x^3 = 6x^2$, and $x = 6$ or $x = 0$ (double root). If $x = 0$, then $0 = -2y^2$ and $y = 0$ (double root). Hence the intercepts are (0, 0) and (6, 0).

Symmetry If y is replaced by $-y$, then $x^3 + x(-y)^2 = 2[3x^2 - (-y)^2]$, or $x^3 + xy^2 = 2(3x^2 - y^2)$. Hence there is symmetry with respect to the x axis. There is no other symmetry.

Asymptotes The coefficient of x^3 (the highest power of x) is 1; hence there are no horizontal asymptotes. The coefficient of y^2 (the highest power of y) is $x + 2$; hence $x + 2 = 0$ is a vertical asymptote.

Excluded intervals If $x^3 + xy^2 = 2(3x^2 - y^2)$ is solved for y, the result is

$$y = \pm\sqrt{\frac{x^2(6 - x)}{x + 2}}$$

For $x > 6$ the radical is negative and for $x < -2$ the radical is negative. Hence there are no real values of y for $x > 6$ or $x < -2$. For $-2 < x < 6$ there are two values of y, one the additive inverse of the other, for each value of x.

Tangents at the origin Retaining only terms of lowest degree in x and y and setting them equal to 0, the result is $2(3x^2 - y^2) = 0$. Thus the equations of tangent lines at the origin are

$$y = \sqrt{3}x \qquad \text{or} \qquad y = -\sqrt{3}x$$

The curve of Fig. 9-3 may now be sketched. Coordinates of a few points may be determined if more careful plotting is desired. The curve is called a *trisectrix* (see Prob. 10, page 240).

Example 8

Discuss and sketch the graph of the equation

$$x^2y + y - 4x = 0$$

Solution

Intercepts If x is replaced by 0, then $y = 0$. If y is replaced by 0, then $x = 0$. Hence the only intercept on the axes is the origin.

Symmetry If x is replaced by $-x$ and y by $-y$, the resulting equation is

$$(-x)^2(-y) + (-y) - 4(-x) = 0 \qquad \text{or} \qquad -x^2y - y + 4x = 0$$

This is equivalent to the original equation and indicates symmetry with respect to the origin.

Asymptotes The coefficient of x^2 (the highest power of x) is y. There-

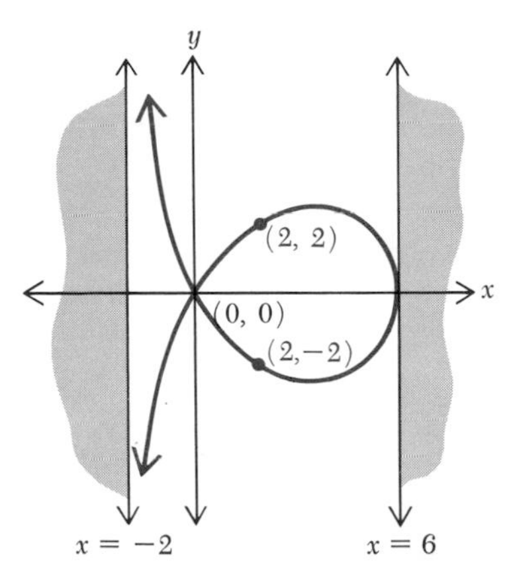

Fig. 9-3 Graph of trisectrix: $x^3 + xy^2 = 2(3x^2 - y^2)$.

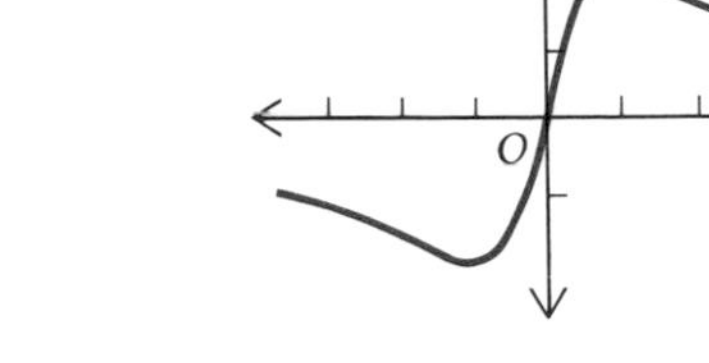

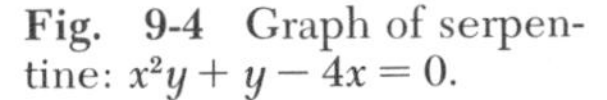

Fig. 9-4 Graph of serpentine: $x^2y + y - 4x = 0$.

fore $y = 0$ is a horizontal asymptote. The coefficient of y is $x^2 + 1$. Since $x^2 + 1 = 0$ has no real roots, there is no vertical asymptote.

Tangents at the origin Striking all terms in $x^2y + y - 4x = 0$ except those of first degree results in the equation $y - 4x = 0$. This is the equation of the tangent at the origin.

Excluded intervals Upon solving the original equation for y we have

$$y = \frac{4x}{x^2 + 1}$$

Thus for every value of x there is one value for y. Upon solving for x we have

$$x = \frac{2 \pm \sqrt{4 - y^2}}{y}$$

The radical is negative if $|y| > 2$. Hence there are no real values for x for $|y| > 2$.

The curve called a *serpentine* is sketched in Fig. 9-4; its general equation is $x^2y + b^2y - a^2x = 0$, where a and b are constants.

Example 9

Discuss and sketch the graph of the equation

$$x^3 + y^3 = 3xy$$

Solution

Intercepts On the x axis, (0, 0). On the y axis, (0, 0).

Symmetry Symmetrical with respect to the line $y = x$ (since interchanging x and y leaves the equation unaltered).

Asymptotes Vertical, none. Horizontal, none.

Tangents at origin $xy = 0$, so $x = 0$ or $y = 0$ (both axes).

Excluded intervals None (since substituting a value for x or y results in a cubic equation in the other variable).

This example illustrates the fact that in some cases the analysis does not yield enough information to sketch the curve without plotting a large number of points. Observe also that when a value is substituted for x, a cubic equation must be solved for corresponding values of y. This is, of course, a laborious process, and it is partly for this reason that algebraic curves of higher degree are often studied by means of parametric equations. A further discussion of this curve, called the *folium of Descartes,* and its graph are given in Sec. 9-3.

PROBLEMS

In Probs. 1 to 6 the equation of a parabola, a circle, an ellipse, or a hyperbola is given. Identify the curve and sketch it after discussing its intercepts, symmetry, asymptotes, tangent lines at the origin, and excluded intervals.

1. $x^2 + y^2 + 3x + 3y = 0$
2. $x^2 + 4y^2 + 2x - 8y = 0$
3. $9x^2 + 16y^2 = 72x$
4. $2x^2 + xy + 2y^2 + 6x + 6y = 0$
5. $xy + 2x + 2y = 0$
6. $x^2 + 4xy + 4y^2 + 4x + 8y = 0$
7. Describe and sketch the locus of all points $P(x, y)$ for which the sum of the two coordinates is equal to the sum of their cubes, $x + y = x^3 + y^3$.
8. Describe and sketch the locus of all points $P(x, y)$ for which the sum of the squares of the two coordinates is equal to the sum of their cubes, $x^2 + y^2 = x^3 + y^3$.
9. Describe and sketch the locus of all points $P(x, y)$ for which the product of the squares and the sum of the squares of the two coordinates are equal, $x^2y^2 = x^2 + y^2$.
10. Draw a line through the origin with inclination θ and draw a line through the point $P(2a, 0)$ with inclination 3θ. Show that the locus of the point of intersection of these lines consists of the x axis and a curve having the equation $x^3 + xy^2 = a(3x^2 - y^2)$. Discuss and sketch this curve. It is called a *trisectrix.* In what way could it be used to trisect a given angle?

In Probs. 11 to 20 discuss and sketch the graph of the given equation:

11. $x^2y + 4y = 8$
12. $x^2y + 6y = 12x$
13. $x^2y - 4y - 4x = 0$
14. $x^2y = x^2 + 4y$
15. $y^2 = x(x - 2)(x - 4)$
16. $x(x^2 + y^2) = 4y^2$

17. $x^4 + y^4 = 16$
18. $x^2y^2 - 9y^2 = 4x^2$
19. $y^2 = x^2(x - 4)$
20. $x^2 = 2y^2(y + 4)$
21. Discuss and sketch the curve whose equation is $x^2y + 4y - 4x = 0$. By translating the axes so that the point $P(2, 1)$ on the curve becomes the new origin, investigate the direction of the tangent line to the curve at this point.
22. A curve consists of all points $P(x, y)$ for which $x(x^2 + y^2) = 5(y^2 - x^2)$. Show that no point of the curve lies outside the interval $-5 \leqslant x < 5$.
23. Discuss and sketch the curve whose equation is $xy^2 - 4y = 9x$.
24. Discuss and sketch the curve whose equation is $y^4 - 4y^3 + 4x^2 = 0$.
25. Discuss and sketch the curve whose equation is $x^2 + y^2 = 16x^2y$.

9-3 PARAMETRIC EQUATIONS OF ALGEBRAIC CURVES

As indicated in Sec. 9-2, it is sometimes easier to graph an algebraic curve of higher degree from its parametric equations than from the direct relationship between x and y. This is illustrated by the *folium of Descartes*, which has as its equation $x^3 + y^3 = 3xy$. The only useful information garnered from the analysis in Sec. 9-2 is that the only intercept is (0, 0), the curve is symmetrical with respect to the line $y = x$, the axes are tangents at the origin, and there are no horizontal or vertical asymptotes and no excluded intervals.

The parametric equations of the folium may be derived by finding, in terms of m, the coordinates of its points of intersection with a line of slope m through the origin. The following sequence of steps shows the simultaneous solution of $x^3 + y^3 = 3xy$ and $y = mx$. If y is replaced by mx in $x^3 + y^3 = 3xy$, the result is

$$x^3 + m^3x^3 = 3x(mx)$$

This may be factored as

$$x^2[x(m^3 + 1) - 3m] = 0$$

Hence $$x = 0 \qquad \text{or} \qquad x = \frac{3m}{m^3 + 1}$$

The solution $x = 0$ is not considered, as it does not contain terms in m. If $x = 3m/(m^3 + 1)$ is substituted in $y = mx$, then

$$y = \frac{3m^2}{m^3 + 1}$$

and the parametric equations corresponding to the rectangular equation

HISTORICAL NOTE

Some of the algebraic curves studied in this chapter are of significant interest. The Greeks as early as 450 B.C., in their study of geometry, were attempting to solve three famous problems:

Squaring the circle, constructing a square region whose area is that of a given circular region
Trisecting an angle, constructing for any angle an angle whose measure is one-third that of the measure of the original angle
Duplicating the cube, constructing the edge of a cube with volume twice that of a given cube

They failed to find solutions to these problems with only a compass and straightedge. Actually, at a much later date it was proved that the constructions are impossible with just these instruments. However, in seeking the solutions the Greeks studied carefully certain algebraic curves, and by discovering properties of these curves they did succeed in solving the problems with the restrictions removed.

One of the curves, the *cissoid*, invented by Diocles about 200 B.C., gives a simple solution to the problem of duplicating the cube. It is drawn by first constructing a circle with center O and diameter $\overline{AB}$ (Fig. 9-5). Points T and T' are located such that $\overline{OT} \cong \overline{OT'}$. Points D and D' on the circle are such that $\overline{DT}$ and $\overline{D'T'}$ are perpendicular to $\overline{AB}$. If $\overline{AD'}$ is drawn, the intersection of $\overline{AD'}$ and $\overline{DT}$ is a point on the cissoid. The curve is the colored one in Fig. 9-5.

Trisection of any angle is accomplished by means of the *conchoid* of Nicomedes, another Greek who lived about 200 B.C. This curve is constructed by drawing $\overleftrightarrow{AB}$ (Fig. 9-6) and a fixed line m perpendicular to it. Rays with end point A are drawn intersecting line m at Q. The distance on the ray from Q to point P is a constant. The conchoid is the colored curve in Fig. 9-6.

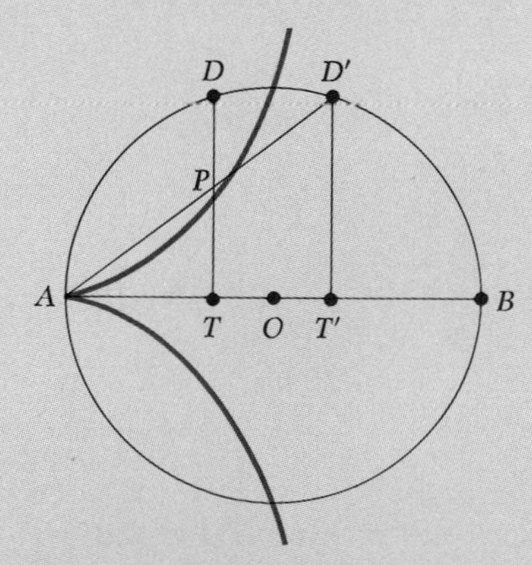

Fig. 9-5 Construction of cissoid.

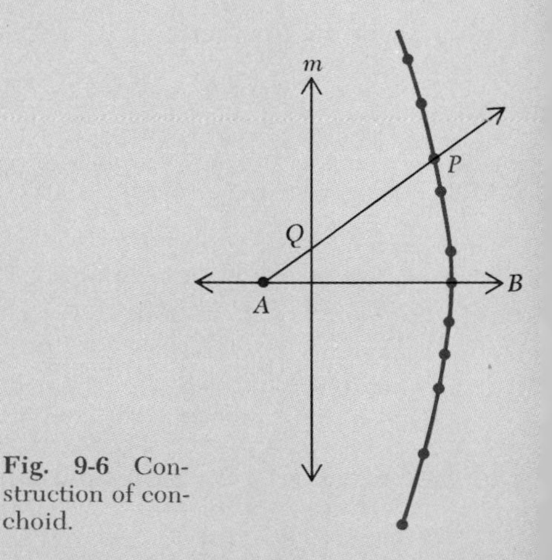

Fig. 9-6 Construction of conchoid.

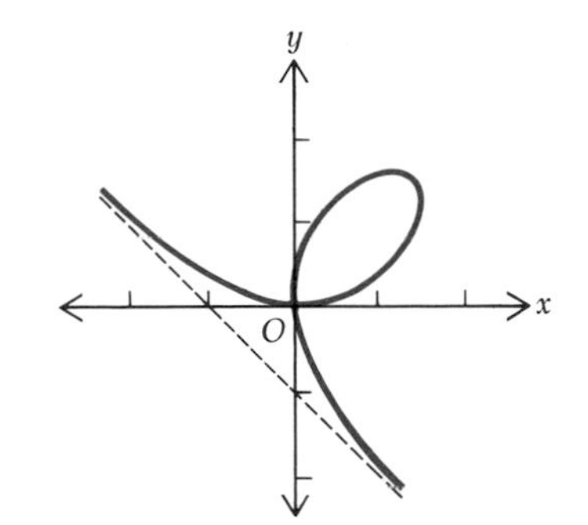

Fig. 9-7 Graph of folium of Descartes: $x^3 + y^3 = 3xy$.

$x^3 + y^3 = 3xy$ are

$$x = \frac{3m}{m^3 + 1}$$

$$y = \frac{3m^2}{m^3 + 1}$$

Now, when a number is substituted for m, the equations yield number pairs (x, y) that correspond to a point on the curve. Thus for $m = 2$, $x = \frac{2}{3}$ and $y = \frac{4}{3}$, and $P(\frac{2}{3}, \frac{4}{3})$ is a point on the graph. A number of such ordered pairs, along with the analysis (page 240), permit drawing of the folium (Fig. 9-7).

A method for determining the parametric form from the rectangular form may not always be obvious. The simple substitution above, by which the parametric equations of the folium were obtained, could actually be the result of much trial and error. However, changing the parametric form to rectangular form, while it requires insight, is often less difficult.

Example

Express the parametric equations

$$x = t^2(3 - 2t)$$
$$y = t^3(3 - 2t)$$

in rectangular form.

Solution Observe that in the second equation the expression for y is simply t times the expression for x. Hence $y = tx$ or $t = y/x$. Substituting this in $x = t^2(3 - 2t)$ or $y = t^3(3 - 2t)$ results in the rectangular form $x^4 - 3xy^2 + 2y^3 = 0$.

PROBLEMS

In Probs. 1 to 4 plot the graph from the parametric equations and then eliminate the parameter.

1. $x^2 = 16 \sin t$, $y^2 = 16 \cos t$
2. $x = 2\sqrt{1/t}\,\sqrt{1-t}$, $y = 2\sqrt{t}\,\sqrt{1-t}$
3. $x^2 = \sqrt{1/t}\,\sqrt{6-t}$, $y^2 = \sqrt{t}\,\sqrt{6-t}$
4. $x = 4 \cot t$, $y = 4 \sin^2 t$

In Probs. 5 to 7 let $y = tx$, and in Prob. 8 let $x = ty$, and express the given equation in parametric representation. Draw the graph of each from parametric equations.

5. $x^4 + y^3 = 4x^2y$
6. $x^4 - 3xy^2 + 2y^3 = 0$
7. $x^3 + xy^2 + 2y^2 - 6x^2 = 0$
8. $x^2 - 2y^3 - 8y^2 = 0$

In Probs. 9 to 16 draw the graph from the parametric equations.

9. $x = t(4 - t^2)$, $y = t^2(4 - t^2)$
10. $x = 1 + t^2$, $y = 4t - t^3$
11. $x = \frac{1}{8}t^3 + t$, $y = \frac{1}{8}t^3 - t$
12. $x = 4 - 0.25t^2$, $y = 0.2t^3 - 1.8t$
13. $x = 4t^3(2 - t)$, $y = 4t^2(2 - t)$
14. $x = \dfrac{16t}{(1+t^2)^2}$, $y = \dfrac{16t^2}{(1+t^2)^2}$
15. $x = \dfrac{4t^2 - 1}{1 - t^3}$, $y = \dfrac{4t^3 - t}{1 - t^3}$
16. $x = 1 - t$, $y = \dfrac{t^2 - 1}{t^2}$

17. Express the equations of Probs. 9 to 16 in rectangular coordinates.

9-4 THE CYCLOID

If a circle of radius a rolls along a fixed line without slipping, the curve traced by a point P on a circle is called a *cycloid*. The curve is shown in Fig. 9-8. Expressions for the coordinates of P can be derived in terms of the number of radians θ through which the circle has turned from its starting position when the point P was at the origin.

If the radius of the rolling circle is a, then

$$OA = \text{arc } PA = a\theta$$

The coordinates of P are then

$$\begin{aligned} x &= OD = OA - DA = OA - PB = a\theta - a \sin\theta \\ y &= DP = AB = AC - BC = a - a\cos\theta \end{aligned}$$

The parametric equations of a cycloid are therefore

$$\begin{aligned} x &= a(\theta - \sin\theta) \\ y &= a(1 - \cos\theta) \end{aligned} \tag{9-1}$$

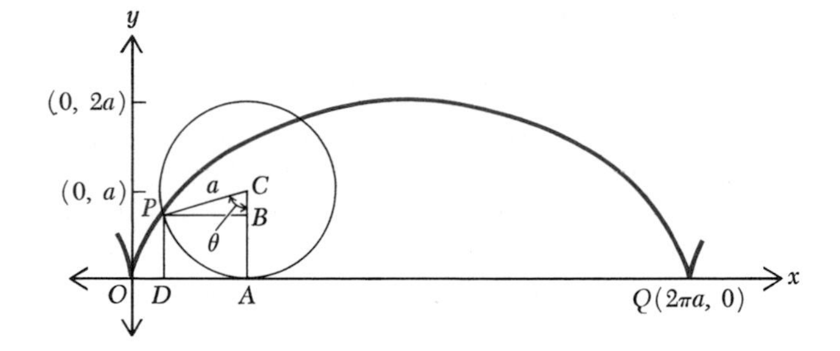

Fig. 9-8 Construction of cycloid: parametric equations are $x = a(\theta - \sin\theta)$ and $y = a(1 - \cos\theta)$.

It is possible to eliminate the parameter θ in Eqs. (9-1) by solving the second equation for $\cos\theta$ in terms of y and substituting this result into the first. However, the resulting rectangular equation is quite complicated, and the parametric equations (9-1) of the cycloid are almost always used.

9-5 THE INVOLUTE

A string is wound about a circle, one end of the string being initially at A in Fig. 9-9. The string is then unwound in the plane of the circle while it is held taut. The curve traced by the end of the string as it unwinds is called the *involute* of the circle. The parametric equations for this curve may be derived as follows. When the string that originally lay along the arc AB has been unwound, the end of the string is

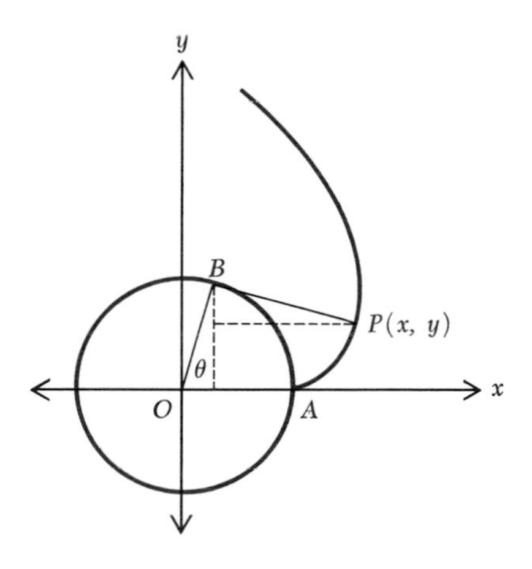

Fig. 9-9 Construction of involute: parametric equations are $x = a(\cos\theta + \theta\sin\theta)$ and $y = a(\sin\theta - \theta\cos\theta)$.

HISTORICAL NOTE

Many great mathematicians have studied the properties of the cycloid. Galileo, for example, used the cycloid to construct an area exactly equal to the area of a circle. He determined that the length of one arch O to P to Q of the cycloid in Fig. 9-10 is equal to four times the length of the diameter of the generating circle. He found further that the area between one arch and the line on which the circle rolls is three times the area of the generating circle. It follows that the shaded region on either side of the circular region in Fig. 9-10 is exactly equal to the area of the circular region.

The Bernoulli brothers studied the problem of quickest descent in the seventeenth century. This problem may be stated as follows: "If a small body moves under the influence of gravity along a curve from A to B (Fig. 9-11), what curve will provide a path such that the time to travel from A to B is a minimum?" Reasonable guesses might be a straight line, a parabola, or even the arc of a circle. The Bernoullis determined that the curve of quickest descent was the cycloid.

In the seventeenth century the French mathematician Pascal carried the study of the cycloid far beyond the point previously reached and in so doing solved problems equivalent to those now solved by integral calculus. Pascal said that some of the ideas occurred to him while he was suffering from a toothache and that after the discovery the toothache stopped.

Various forms of cycloids are used in gear and cam construction, and the involute of the circle is sometimes used as the shape of gear teeth.

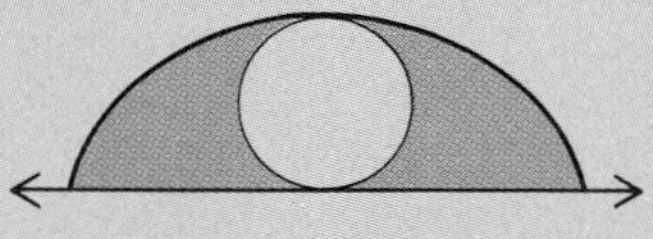

Fig. 9-10 Region between cycloid and line is separated into three regions with the same area.

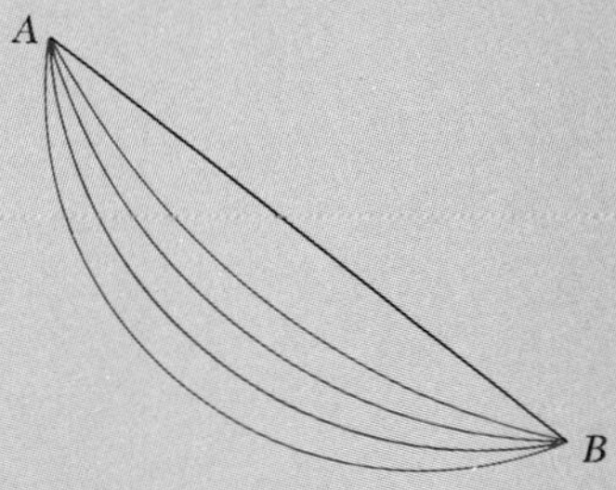

A

B

Fig. 9-11 Of all curves between A and B, the cycloid is the one of quickest descent.

at P, where $\overline{PB}$ is perpendicular to $\overline{OB}$; also, if θ is the angle (radians) subtended at O by arc AB, then

$$BP = \text{arc } AB = a\theta$$

where $a = |OB|$ is the radius of the circle. If the coordinates of P are x and y, then

$$x = OB \cos\theta + BP \sin\theta = a\cos\theta + a\theta\sin\theta$$
$$y = OB \sin\theta - BP \cos\theta = a\sin\theta - a\theta\cos\theta$$

These equations give the coordinates of any point P on the curve in terms of the angle θ; the curve is accordingly defined by the parametric equations

$$x = a(\cos\theta + \theta\sin\theta) \qquad \text{(9-2)}$$
$$y = a(\sin\theta - \theta\cos\theta)$$

PROBLEMS

In Probs. 1 to 7 sketch the curve represented by the given parametric equation. Eliminate the parameter and thus find the rectangular equation of the curve.

1. $x = 1 + \cos\theta$, $y = 2 - \sin\theta$
2. $x = 3 + 5\sin\theta$, $y = 4 + 5\cos\theta$
3. $x = 2\sin\theta$, $y = 2(1 + \cos\theta)$
4. $x = 3 - 3\cos\theta$, $y = 3\sin\theta$
5. $x = 3\cos\theta - 2$, $y = 5\sin\theta + 2$
6. $x = 4(2 + \cos\theta)$, $y = 2\sin\theta$
7. $x = 3\sin\theta$, $y = 1 + \cos\theta$
8. A rod AB of length r rotates in the xy plane about end A, which is fixed at the origin. Show that if B is initially on the x axis, and if the rod rotates at ω rad/sec, then the coordinates of B at the end of t sec are

$$x = r\cos\omega t$$
$$y = r\sin\omega t$$

9. Show that if the rod in Prob. 8 initially makes an angle of α rad with the x axis, then the coordinates of B at the end of t sec are

$$x = r\cos(\omega t + \alpha)$$
$$y = r\sin(\omega t + \alpha)$$

10. Find a parametric representation for the cycloid using the highest point of an arch as the origin.
11. If the tracing point on the rolling circle generating a cycloid is at a distance b from the center, with $b \neq a$, show that the equations of its path are

$$x = a\theta - b\sin\theta$$
$$y = a - b\cos\theta$$

This curve is called a *prolate cycloid* if $b > a$ and a *curtate cycloid* if $b < a$.

12. Draw the curtate cycloid (see Prob. 11) for which $a = 8$ and $b = 4$.
13. A circle of radius $\frac{1}{4}a$ rolls inside a circle of radius a as shown in Fig. 9-12. Show that the path traversed by a point P on the rolling circle is defined by the equations

$$x = a \cos^3 \theta$$
$$y = a \sin^3 \theta$$

where θ is the angle shown. This curve is called the *hypocycloid of four cusps*. *Hint:* Arc AP = arc AB; hence angle $ACP = 4\theta$. Then

$$x = OD + LP$$
$$y = DC - LC$$

but $OD = OC \cos \theta = \frac{3}{4}a \cos \theta$. Observe that angle $LCP = \frac{1}{2}\pi - 3\theta$.
14. The hypocycloid is the curve traced by a point fixed on the circumference of a circle of radius b when this circle rolls along the inside of a circle of radius a, with $a > b$. With a figure similar to Fig. 9-12, show that the equations of this curve are

$$x = (a - b) \cos \theta + b \cos \frac{a - b}{b} \theta$$

$$y = (a - b) \sin \theta - b \sin \frac{a - b}{b} \theta$$

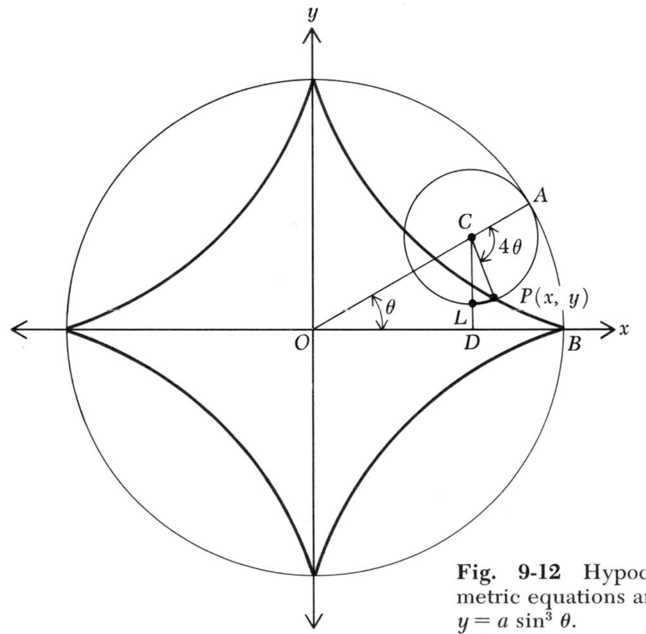

Fig. 9-12 Hypocycloid of four cusps: parametric equations are $x = a \cos^3 \theta$ and $y = a \sin^3 \theta$.

15. The *epicycloid* is the curve traced by a point fixed on a circle of radius b when this circle rolls along the outside of a circle of radius a. Show that the parametric equations of this curve are

$$x = (a + b)\cos\theta - b\cos\frac{a+b}{b}\theta$$

$$y = (a + b)\sin\theta - b\sin\frac{a+b}{b}\theta$$

Use a figure similar to Fig. 9-12 with the rolling circle on the outside of the fixed circle.

16. A circle of radius a is drawn tangent to the x axis with one diameter coinciding with the y axis (Fig. 9-13); the tangent to the circle at A is then drawn. Through the origin a secant line is drawn, meeting the circle at B and the tangent at C. The horizontal line through B and the vertical line through C meet at $P(x, y)$. Show that the coordinates of P are

$$x = 2a\tan\phi$$
$$y = 2a\cos^2\phi$$

where ϕ is the parameter shown in the figure. The curve that is the locus of P is called the *witch of Agnesi*, named for Maria Agnesi (1718–1799), one of the first women to be considered a distinguished mathematician. By eliminating the parameter, show that its rectangular equation is

$$x = \frac{8a^3}{x^2 + 4a^2}$$

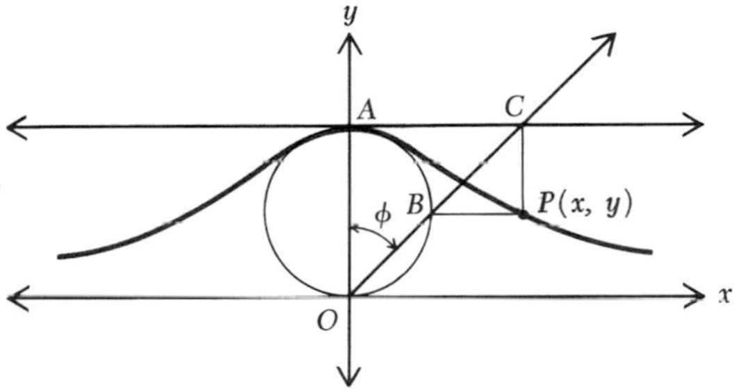

Fig. 9-13 Witch of Agnesi: parametric equations are $x = 2a\tan\phi$ and $y = 2a\cos^2\phi$.

9-6 ANALYSIS OF POLAR EQUATIONS

The problem of sketching the graph of a given equation in polar coordinates is often facilitated by an appropriate analysis of the equation. This analysis may include the determination of intercepts, tests for symmetry, and extraction of other properties that were included in the discussion of equations in rectangular coordinates.

Example 1
Sketch the graph of $\rho = 3 + 3 \cos \theta$.

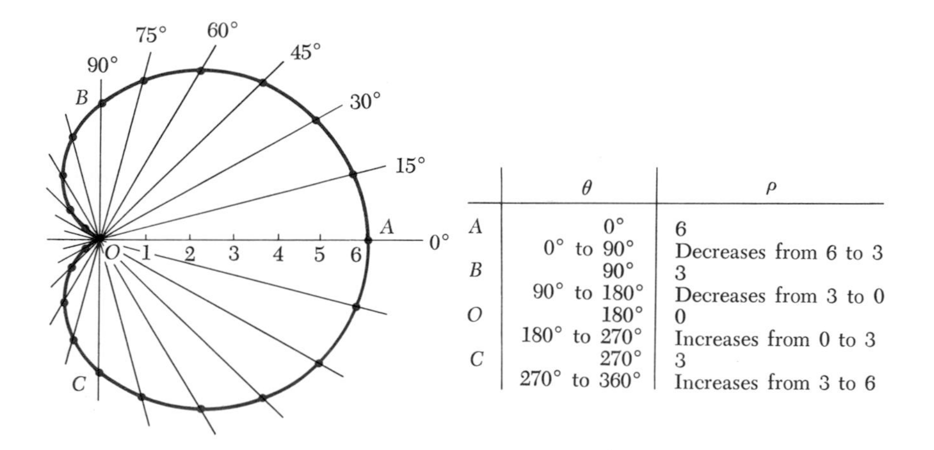

	θ	ρ
A	0°	6
	0° to 90°	Decreases from 6 to 3
B	90°	3
	90° to 180°	Decreases from 3 to 0
O	180°	0
	180° to 270°	Increases from 0 to 3
C	270°	3
	270° to 360°	Increases from 3 to 6

Fig. 9-14 Cardioid: $\rho = 3 + 3 \cos \theta$.

Solution Points on the curve for θ equal to 0°, 90°, 180°, and 270°, which often provide a general idea of the shape of a curve, are shown in the table accompanying Fig. 9-14 and are graphed as points A, B, O, and C. Valuable information can frequently be obtained by noting how ρ varies with θ as θ increases from 0 to 90°, 90 to 180°, 180 to 270°, and 270 to 360°. In the case at hand, when $\theta = 0$, $\cos \theta = 1$ and $\rho = 6$. Since 1 is the greatest value for $\cos \theta$, 6 is the greatest value for ρ. Now, as θ increases from 0 to 90°, $\cos \theta$ decreases from 1 to 0, and consequently, ρ decreases from 6 to 3. As θ increases from 90 to 180°, $\cos \theta$ decreases from 0 to -1, and consequently, ρ decreases from 3 to 0. As θ continues to increase from 180 to 360°, $\cos \theta$ increases from -1 to 1, and consequently, ρ increases from 0 to 6. This information is also summarized in the table in Fig. 9-14. With this analysis, a sketch of the curve may be made (Fig. 9-14).

Example 2
Sketch $\rho = 4 + 4 \sin \frac{1}{2}\theta$.

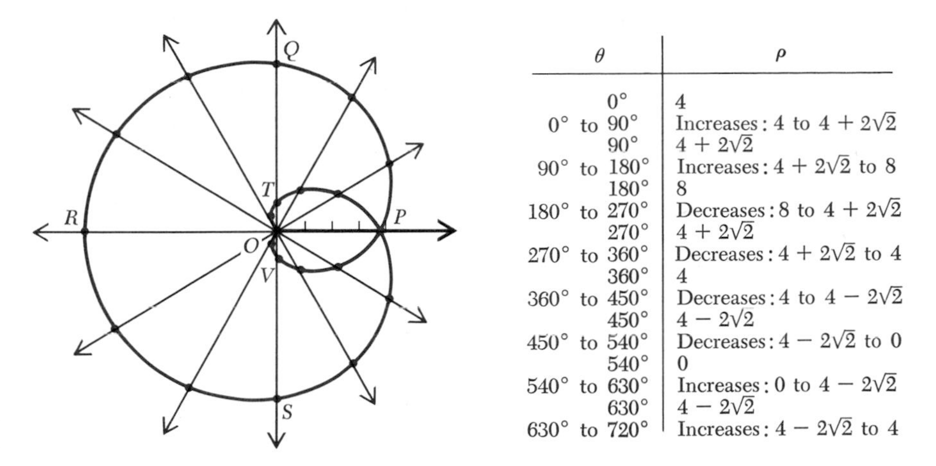

θ	ρ
0°	4
0° to 90°	Increases: 4 to $4 + 2\sqrt{2}$
90°	$4 + 2\sqrt{2}$
90° to 180°	Increases: $4 + 2\sqrt{2}$ to 8
180°	8
180° to 270°	Decreases: 8 to $4 + 2\sqrt{2}$
270°	$4 + 2\sqrt{2}$
270° to 360°	Decreases: $4 + 2\sqrt{2}$ to 4
360°	4
360° to 450°	Decreases: 4 to $4 - 2\sqrt{2}$
450°	$4 - 2\sqrt{2}$
450° to 540°	Decreases: $4 - 2\sqrt{2}$ to 0
540°	0
540° to 630°	Increases: 0 to $4 - 2\sqrt{2}$
630°	$4 - 2\sqrt{2}$
630° to 720°	Increases: $4 - 2\sqrt{2}$ to 4

Fig. 9-15 Analysis and graph of $\rho = 4 + 4 \sin \frac{1}{2}\theta$.

Solution Values of ρ for θ equal to 0°, 90°, 180°, and 270° are tabulated in Fig. 9-15 and plotted as points *P*, *Q*, *R*, and *S*. Values of θ equal to 450°, 540°, and 630° must be used to locate the points *T*, *O*, and *V* as shown in the table and the graph, because in the equation, ρ is a function of $\frac{1}{2}\theta$ rather than θ; that is, $\sin \frac{1}{2}\theta$ is periodic with period 720°, or 4π rad.

Study of the variation of ρ as θ increases reveals that when $\theta = 0$, $\sin \frac{1}{2}\theta = 0$ and $\rho = 4$. As θ increases from 0 to 180°, $\sin \frac{1}{2}\theta$ increases from 0 to 1, and consequently, ρ increases from 4 to 8. As θ increases from 180 to 360°, $\sin \frac{1}{2}\theta$ decreases from 1 to 0, and therefore ρ decreases from 8 to 4. From this analysis the "outside" part of the graph (Fig. 9-15) is sketched. Now as θ increases from 360 to 540°, $\sin \frac{1}{2}\theta$ decreases from 0 to -1, and therefore ρ decreases from 4 to 0; finally, as θ increases from 540 to 720°, $\sin \frac{1}{2}\theta$ increases from -1 to 0 and ρ increases from 0 to 4. This information is summarized in Fig. 9-15. From this analysis, the "inner" part of the graph is sketched.

9-7 THE CARDIOID AND LIMAÇON

In this and the next few sections we shall consider briefly several equations whose graphs are of special interest. The first of these is the polar coordinate equation

$$\rho = a + a \cos \theta \qquad (9\text{-}3)$$

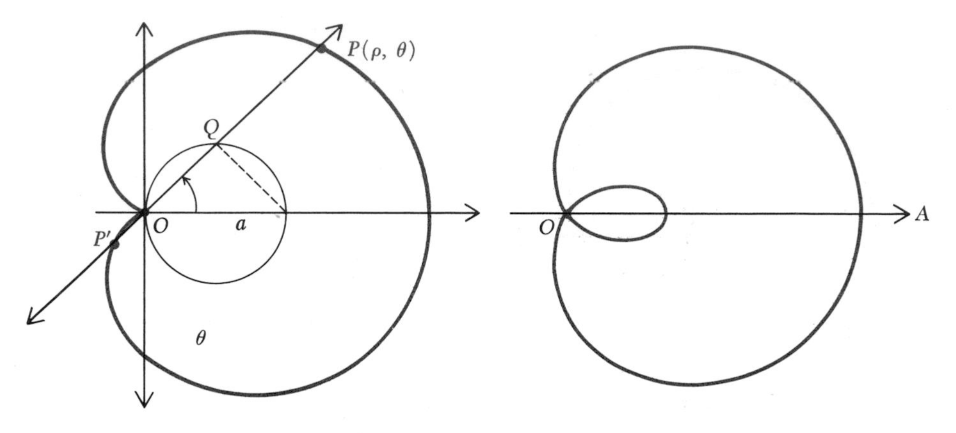

Fig. 9-16 Construction of cardioid $|QP| = |QP'| = a$. $\rho = a + a \cos \theta$.

Fig. 9-17 Graph of limaçon with equation $\rho = 2 + 4 \cos \theta$.

whose graph is illustrated by Fig. 9-16. Because of its heartlike shape, this curve is called a *cardioid*. It is defined geometrically as follows. Draw a circle of diameter a (Fig. 9-16). From one end point O of a diameter draw any line intersecting the circle at Q; on this line mark off two points, P and P', such that

$$|QP| = |QP'| = a$$

The locus of P and P' is the cardioid. The reader may show that the points on the curve so constructed satisfy the above equation and that other equations representing cardioids are

$$\rho = a(1 - \cos \theta) \tag{9-4}$$

and

$$\rho = a(1 \pm \sin \theta) \tag{9-5}$$

The construction may be generalized by making $|QP| = |QP'| = b$, where b is a positive number that is in general not equal to a. This more general curve, of which the cardioid is a special case, is called a *limaçon*. If $b > a$, the curve has the equation

$$\rho = b + a \cos \theta \tag{9-6}$$

In this case ρ is positive for all values of θ. If $b < a$, the curve has an inner loop. The graph of the limaçon whose equation is $\rho = 2 + 4 \cos \theta$ is shown in Fig. 9-17.

9-8 THE ROSE CURVES

The graphs of the equations

$$\rho = a \cos n\theta \qquad (9\text{-}7)$$
$$\rho = a \sin n\theta$$

where n is an integer, are called *rose curves*. They have the general shape illustrated by Fig. 9-18. With negative values of ρ allowed, the graph has n leaves if n is odd and $2n$ leaves if n is even. Thus the graph of the equation $\rho = a \cos 2\theta$ has four leaves; the graph of $\rho = a \sin 3\theta$ (Fig. 9-18) has three leaves.

In graphing these curves it is often helpful to find the greatest and least values for ρ.

Example
Find the greatest and least values of ρ if $\rho = a \sin 3\theta$.

Solution Two required values occur when $3\theta = 90° + n(360°)$ or $3\theta = 270° + n(360°)$, where n is an integer, since at those values $\sin 3\theta$ takes on its maximum and minimum, respectively. All greatest and least values of ρ are obtained as follows:

$$\begin{array}{lll} \sin 3\theta = 1 & & \sin 3\theta = -1 \\ 3\theta = 90° \quad \text{or} & 3\theta = 450° & 3\theta = 270° \\ \theta = 30° & \theta = 150° & \theta = 90° \\ \rho = a & \rho = a & \rho = -a \end{array}$$

It is convenient in sketching the rose curve to draw a circle of radius a

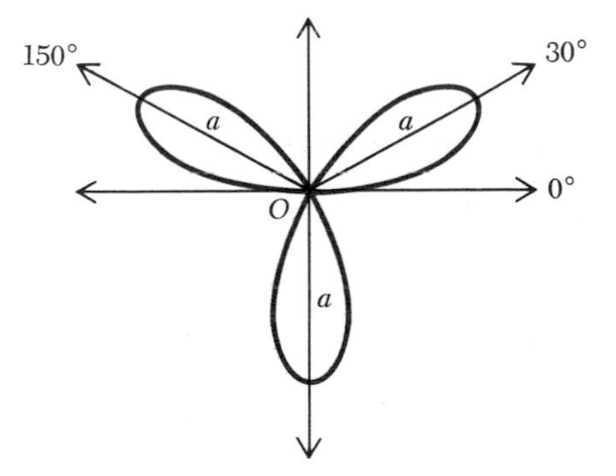

Fig. 9-18 Graph of three-leaved rose: $\rho = a \sin 3\theta$.

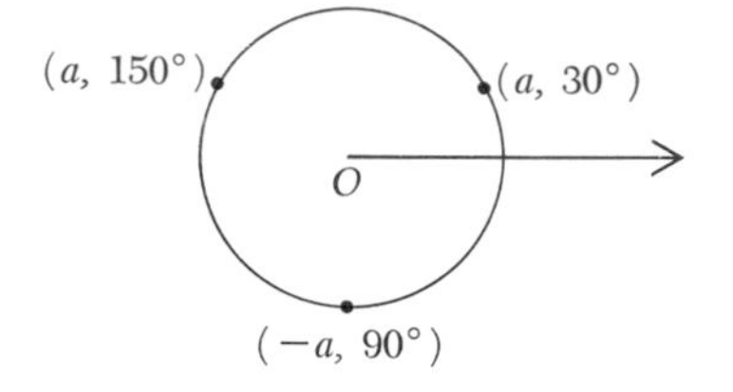

Fig. 9-19 Greatest values of ρ for $\rho = a \sin 3\theta$.

and graph the points $(a, 30°)$, $(a, 150°)$, and $(-a, 90°)$ as shown in Fig. 9-19 (note that additional values of θ provide no new points with greatest or least ρ). These points are of considerable help in completing the graph of $\rho = a \sin 3\theta$ (Fig. 9-18).

9-9 THE LEMNISCATE

The locus of the equation

$$\rho^2 = a^2 \cos 2\theta \tag{9-8}$$

is called a *lemniscate*. The graph is shown in Fig. 9-20. The equation $\rho^2 = a^2 \sin 2\theta$ represents a curve with the same general shape but having intercepts on the line $\theta = 90°$ instead of on the line $\theta = 0°$.

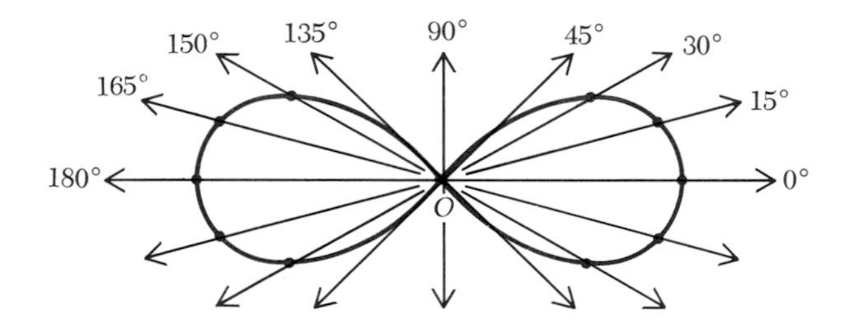

Fig. 9-20 Lemniscate: $\rho^2 = a^2 \cos 2\theta$.

9-10 THE SPIRALS

The polar equation that has the same form as the equation $y = ax$ in rectangular coordinates is

$$\rho = a\theta \tag{9-9}$$

Its graph is called a *spiral of Archimedes*. The curve is shown in Fig. 9-21 for the case in which $a > 0$.

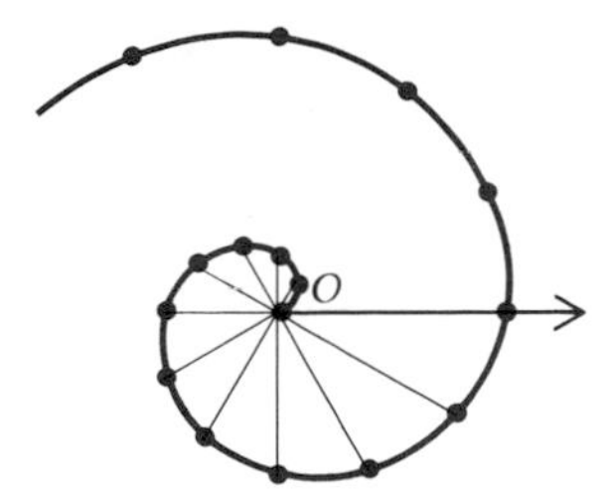

Fig. 9-21 Spiral of Archimedes: $\rho = a\theta$ for $a > 0$.

Other polar equations whose graphs are spirals are

$$\rho = e^{a\theta} \tag{9-10}$$

or

$$\log \rho = a\theta \tag{9-11}$$

the graph of which is called a *logarithmic spiral,* and

$$e^{\theta} = a$$

whose graph is called a *hyperbolic,* or *reciprocal,* spiral. The construction of these curves is left for the exercises.

9-11 INTERSECTIONS OF POLAR CURVES

If a pair of equations in polar coordinates are solved simultaneously, the number pairs (ρ, θ) that satisfy both equations are found. The corresponding points are, of course, points of intersection of the two graphs. However, this procedure may not yield all points of intersection for the reason that a given point has infinitely many pairs of polar coordinates, and a point is on the graph of a given equation if any one of its pairs of coordinates satisfies the equation. It is quite possible that a point of intersection of two graphs may have a pair of coordinates that satisfies one of the two equations and a different pair that satisfies the other, but no pair which satisfies both.

Example
Find the intersections of the graphs of the system

$$\rho = 2 \cos \theta$$
$$\rho = 2 \sin \theta$$

Solution First eliminate ρ and solve for θ. Then

$$2 \sin \theta = 2 \cos \theta$$
$$\tan \theta = 1 \qquad \text{provided } \cos \theta \neq 0$$

So $\theta = \pi/4$ or $\theta = 5\pi/4$, or in general

$$\theta = \pi/4 + n\pi \qquad \text{for } n \text{ an integer}$$

If $\theta = \pi/4$ is substituted in either equation, then $\rho = \sqrt{2}$. The curves then intersect at $(\sqrt{2}, \pi/4)$. Furthermore, every pair of coordinates representing this point of intersection satisfies both equations. The graphs are shown in Fig. 9-22. They obviously also intersect at the origin. The number pair (0, 0) satisfies the second equation and the pair $(0, \pi/2)$ satisfies the first, but neither pair satisfies both equations.

A given curve can have more than one equation in polar coordinates. For example, the equations

$$\rho = 4 + 4 \sin \tfrac{1}{2}\theta \qquad \text{or} \qquad \rho = 4 - 4 \sin \tfrac{1}{2}\theta$$

represent the same curve (the second equation was obtained from the first by replacing θ by $\theta + 2\pi$). The points of intersection of this curve with the curve whose equation is $\rho = 3 - 2 \cos \theta$ are found by solving either of the following systems (Fig. 9-23):

$$\rho = 4 + 4 \sin \tfrac{1}{2}\theta \qquad \text{and} \qquad \rho = 3 - 2 \cos \theta \tag{9-12}$$

or

$$\rho = 4 - 4 \sin \tfrac{1}{2}\theta \qquad \text{and} \qquad \rho = 3 - 2 \cos \theta \tag{9-13}$$

In Eqs. (9-12), with the restriction $\rho > 0$, the value of θ at any point of intersection other than the origin must satisfy the relation

$$4 + 4 \sin \tfrac{1}{2}\theta = 3 - 2 \cos \theta \qquad \text{or} \qquad 4 \sin \tfrac{1}{2}\theta = -1 - 2 \cos \theta$$

The result after squaring both sides and replacing $\sin^2 \frac{1}{2}\theta$ by $\frac{1}{2}(1 - \cos \theta)$ is

$$8(1 - \cos \theta) = 1 + 4 \cos \theta + 4 \cos^2 \theta$$

This equation may be expressed as

$$4 \cos^2 \theta + 12 \cos \theta - 7 = 0$$

or

$$(2 \cos \theta + 7)(2 \cos \theta - 1) = 0 \tag{9-14}$$

This means that θ must be such that $\cos \theta = \frac{1}{2}$. If $\theta = \frac{1}{3}\pi$, then $\rho = 6$ in one equation and $\rho = 2$ in the other. If $\theta = \frac{1}{3}\pi + 2\pi$ or $\frac{7}{3}\pi$, then $\rho = 2$

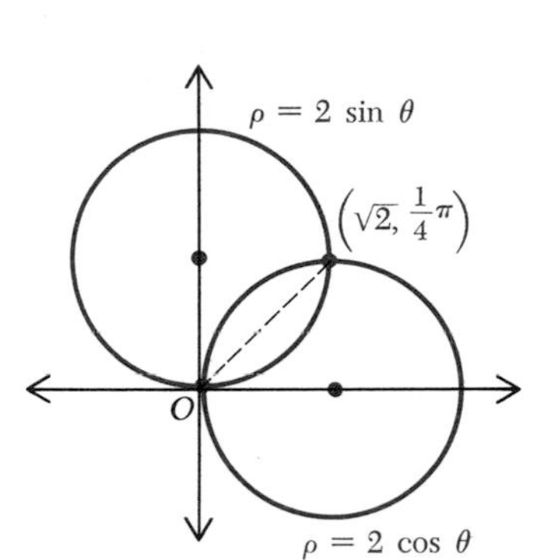

Fig. 9-22 Intersections of $\rho = 2 \sin \theta$ and $\rho = 2 \cos \theta$.

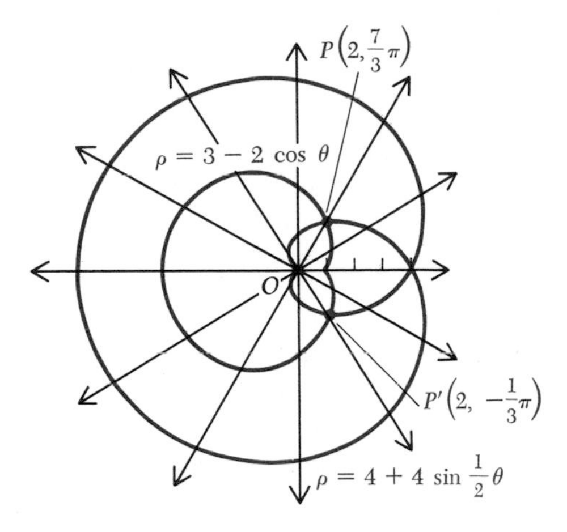

Fig. 9-23 Intersections of $\rho = 3 - 2 \cos \theta$ and $\rho = 4 + 4 \sin \frac{1}{2}\theta$.

in each equation. Thus the number pair $(2, \frac{7}{3}\pi)$ satisfies both equations. Similarly, the pair $(2, -\frac{1}{3}\pi)$ satisfies both equations. These are coordinates of the points P and P' of intersection shown in Fig. 9-23.

Equation (9-14) is the same for the system (9-13) as well. In this case $\theta = \frac{1}{3}\pi$ gives $\rho = 2$ in each equation, while $\theta = \frac{7}{3}\pi$ gives $\rho = 6$ in one equation and $\rho = 2$ in the other.

PROBLEMS

1. Show that the graph of an equation $\rho = f(\theta)$ is symmetrical with respect to the line $\theta = 0°$ if the equation that results when θ is replaced by $-\theta$ has the same graph as the given equation. What is the corresponding statement regarding symmetry with respect to the line $\theta = 90°$?

In Probs. 2 to 9 draw the graph of the given equation.

2. $\rho = 1 + \cos \theta$
3. $\rho = 2(1 - \sin \theta)$
4. $\rho = 3 - 3 \cos \theta$
5. $\rho = 2 + \sin \theta$
6. $\rho = 4 + 3 \cos \theta$
7. $\rho = 5 - 3 \cos \theta$
8. $\rho = 2 - 4 \sin \theta$
9. $\rho = 5 \cos \theta - 3$

10. Draw the graphs of $\rho = 4 \sin \theta + 2$ and $\rho = 4 \sin \theta - 2$.
11. Describe the graph of the equation $\rho = 3 \sin \theta - 3$ for the coordinate system in which $\rho \geqslant 0$. Describe the graph if negative values of ρ are allowed.

In Probs. 12 to 21 sketch the graph of the given equation for $\rho \geqslant 0$.

12. $\rho = 3 \cos 2\theta$
13. $\rho = 5 \sin 2\theta$
14. $\rho = 3 \sin 5\theta$

15. $\rho = \cos 5\theta$
16. $\rho = \frac{1}{2}\theta$
17. $\rho = -2\theta$
18. $\rho\theta = \pi$
19. $\rho = e^{\theta/4}$
20. $\rho^2 = 9 \cos 2\theta$
21. $\rho^2 = 8 \sin 2\theta$

22. Show that the transformation equations for rotation of axes in polar coordinates are $\rho = \rho'$ and $\theta = \theta' + \alpha$, where α is the angle through which the polar axes are to be rotated.

In Probs. 23 to 30 sketch the two curves whose equations are given and find the coordinates of all points of intersection:

23. $\rho = 4 \sin \theta;\quad \rho = 3 \cos \theta$
24. $\rho \sin \theta = 3;\quad \rho = 6$
25. $\rho = 4 \sin \theta;\quad \rho \sin \theta = 2$
26. $\rho = 2 \sin \theta;\quad \rho = 2 + 4 \cos \theta$
27. $\rho = \sin \theta;\quad \rho = \sin 2\theta$
28. $\rho = \cos 2\theta;\quad \rho = 1 + \sin \theta$
29. $\rho = 3 \cos \theta;\quad \rho = 1 + \cos \theta$
30. $\rho^2 = \sin 2\theta;\quad \rho = \sin \theta$

SUMMARY

Special algebraic curves graphed in this chapter are those that are the locus of $f(x, y) = 0$, where $f(x, y)$ is a polynomial in x and y. For graphing, an analysis is recommended that is similar to that suggested in Chap. 8 for studying intercepts, excluded regions, symmetry, and asymptotes. Because of the increased "complexity" of the equations, some new procedures are needed for rapid sketching. Vertical asymptotes are found, for example, by setting the coefficient of the highest power of y in $f(x, y) = 0$ equal to zero and solving for x. If this coefficient is $g(x)$, each linear factor of $g(x)$ is set equal to zero. Such equations are equations of vertical asymptotes. Horizontal asymptotes are found by a similar procedure. Furthermore, if a curve passes through the origin, tangents to the curve at that point are readily determined and provide the slope of the curve there.

Parametric equations of algebraic curves are especially useful for graphing some curves. This is evident in graphing an equation such as $x^3 + y^3 = 3xy$. Finding number pairs is rather difficult, and making the analysis does not materially aid the rapid sketching. Additional points on the graph are readily determined from the parametric form

$$x = \frac{3m}{1 + m^3}$$

$$y = \frac{3m^2}{1 + m^3}$$

and the sketch may be completed. Because of numerous applications in physical science and engineering, cycloids and involutes are studied carefully. Here again the parametric equations offer the easiest form for graphing and determining characteristics of the curve.

Sketching curves when their equations are given in polar form is also facilitated by special procedures:

1. Determining the ρ coordinate for θ equal to 0° and multiples of 90°
2. Finding how the ρ coordinate varies as θ changes from 0 to 90°, 90 to 180°, 180 to 270°, and so on
3. Finding the values of θ at which the ρ coordinate is least and greatest

In solving pairs of polar equations simultaneously, all points of intersection may not be named by a single pair (ρ, θ), because there is not a one-to-one correspondence between points and number pairs (ρ, θ). For example, the point of intersection on one curve may be (ρ, θ), but this same point may have coordinates $(-\rho, \theta + \pi)$ on the other curve.

REVIEW PROBLEMS

For each of Probs. 1 to 4 discuss the graph of the equation by listing intercepts, symmetry, asymptotes, tangents at the origin, and excluded intervals; then sketch the graph.

1. $y = \dfrac{4 - x^2}{1 - x^2}$ 2. $y^2 = \dfrac{2x}{x + 1}$ 3. $9y^2 = x^3$ 4. $x^2y^2 - 9x^2 - 8y = 0$

For Probs. 5 to 7 draw the graph from the given parametric equations, and then express the relationship between x and y by rectangular coordinates.

5. $x = t + 2$
$y = \dfrac{2}{t(t + 4)}$

6. $x = \dfrac{1}{t} - t$
$y - \dfrac{1}{t} + t$

7. $x = \sin \theta$
$y = \cos 2\theta$

8. For the equation $x^2y + 4a^2y - 8a^3 = 0$ make the substitution $x = 2a \tan \phi$ and solve for y. Write the parametric equations for $x^2y + 4a^2y - 8a^3 = 0$.
9. Sketch the graph of the following equations expressed in polar form after determining the way in which ρ varies as θ increases:

(*a*) $\rho = 5 \sin \theta - 3$ (*b*) $\rho = 2 \cos 4\theta$
(*c*) $\rho = 1 - 2 \cos \theta$ (*d*) $\rho = \frac{3}{2}\theta$

10. In each of the following cases sketch the two curves whose equations are

given and find the coordinates of their points of intersection:

(*a*) $\rho = \sin\theta$; $\rho = \cos\theta$ (*b*) $\rho = 4\sin\theta$; $\rho = 4\cos\theta$

11. The midpoints of the segments through the focus of the parabola $x^2 = 8y$ with end points on the parabola lie on a curve. Write the equation of the curve in parametric form. *Hint:* Find the coordinates of the midpoint in terms of the slope m of a line through the focus.

FOR EXTENDED STUDY

1. A point moves so that the product of its undirected distances from the points $(-a, 0)$ and $(a, 0)$ is equal to a constant b^2. Show that its locus has the equation

$$(x^2 + y^2 + a^2)^2 - 4a^2x^2 = b^4$$

Sketch the curve for the case in which $a = 2$ and $b = 2$.

2. Eliminate the parameter from the equations of a cycloid to obtain

$$x = a \arccos \frac{a - y}{a} \mp \sqrt{2ay - y^2}$$

where the minus sign is used for first- and second-quadrant values of the arccosine and the plus sign for third- and fourth-quadrant values.

3. Two tests for symmetry in polar coordinates are as follows: (*a*) If an equivalent equation is obtained by replacing (ρ, θ) with $(\rho, -\theta)$, the graph is symmetrical with respect to the polar axis; (*b*) if an equivalent equation is obtained by replacing (ρ, θ) with $(-\rho, \theta)$, the graph is symmetrical with respect to the pole. The first of these tests holds for $\rho = 3 + 3\cos\theta$ but fails for $\rho = 4 + 4\sin\frac{1}{2}\theta$. Both are symmetrical with respect to the polar axis. The second tests holds for $\rho^2 = a^2\cos 2\theta$ and fails for $\rho = a\cos 2\theta$. Both are symmetrical with respect to the pole. Explain these results.
4. Graph $y = x - 3$ and $y = (x^2 - 9)/(x + 3)$. What is the difference between the graphs?
5. Show that the parametric equations

$$x = \frac{at(3 - t^2)}{(1 + t^2)^2}$$

$$y = \frac{at^2(3 - t^2)}{(1 + t^2)^2}$$

define the rose curve whose polar equation is $\rho = a\sin 3\theta$.

6. A curve has the equation $x^2 - 2y^3 - 8y^2 = 0$ in rectangular coordinates. Find its equation in polar coordinates. Draw the graph from either equation.
7. A curve has the equation $x^4 - 3xy^2 + 2y^3 = 0$ in rectangular coordinates. Find its equation in polar coordinates and draw the graph from this polar equation.

THE TRIGONOMETRIC CURVES 10

10-1 TRIGONOMETRIC FUNCTIONS

The trigonometric ratios were introduced in Chap. 2. As indicated there, given any real number x, we may construct an angle of x rad with vertex at the origin, choose a point P on the terminal side, and divide the ordinate of P by the undirected distance from the origin to P. The real number $\sin x$ is the quotient so constructed. Thus the function $y = \sin x$ associates a real number y for $-1 \leq y \leq 1$ with every real number x. Cos x can be similarly defined.

Consider, for example, the function

$$W = 5 \sin \frac{2\pi t}{3}$$

If t is any real number, then $2\pi t/3$ is a real number. The function thus yields a real number W for every real number t. It is a simple matter in this example to find W for certain special values of t. Thus, if $t = 2$,

$$W = 5 \sin \frac{2\pi}{3}(2) = 5 \sin \tfrac{4}{3}\pi = 5\left(-\frac{\sqrt{3}}{2}\right) = -\frac{5}{2}\sqrt{3}$$

HISTORICAL NOTE

Both the ancient Egyptians and the Babylonians are given some credit for beginning the study of trigonometry. The Greek Hipparchus, who lived about 140 B.C., is often called the father of trigonometry. At that time the chief importance of trigonometry was in astronomy. During the Dark Ages and Middle Ages the Hindus and Arabians continued to develop trigonometry as a tool for astronomy.

It was not until 1533 that Regiomontanus produced the first European text that developed trigonometry (both plane and spherical) independently of astronomy. By the end of the sixteenth century trigonometry was well systematized, and excellent tables of trigonometric functions were available. Francois Vieta (1540–1603) contributed much to the development of analytic trigonometry, and it is the analytic approach to the subject that has become predominant in modern texts (the formulas of trigonometry are reviewed in the Appendix).

Trigonometric curves have many applications. Both an alternating current and its corresponding electromotive force, for example, can be represented by sine or cosine curves. A sine curve can be used to represent a simple musical sound such as that produced by a tuning fork.

If $t = 1.8$,

$$\frac{2\pi t}{3} \approx \frac{2(3.1416)(1.8)}{3} \approx 3.7699$$

$$\begin{aligned} W &\approx 5 \sin 3.7699 \approx -5 \sin 0.6283 \\ &\approx -5(0.5878) \\ &\approx -2.939 \end{aligned}$$

The formula $\sin(\theta + \pi) = -\sin \theta$ was used to replace sin 3.7699 with $-\sin 0.6283$. The value of sin 0.6283 can be obtained from a table (0.63 rad is approximately equal to 36°).

In working with radian and degree measurements of angles the relationship between the systems should be kept in mind. Since $1° = \pi/180$, or approximately 0.017453 rad, $n° = n\pi/180$, approximately $n(0.017453)$ rad:

$$30° = 30\frac{\pi}{180} = \frac{\pi}{6} \text{ rad}$$

$$270° = 270\frac{\pi}{180} = \frac{3\pi}{2} \text{ rad}$$

10-2 THE GRAPHS OF THE SINE AND COSINE FUNCTIONS

The graphs of sin x and cos x can be sketched by substituting numbers for x, finding the corresponding values of y from tables, and plotting the points. Thus in the case of sin x, tables may be used to find

$$\sin 0 = 0,\ \sin 0.5 \approx 0.479,\ \sin 1 \approx 0.841,\ \sin 1.5 \approx 0.977,\ \ldots$$

The graph of $y = \sin x$ for $0 \leq x \leq 2.5$ is shown in Fig. 10-1.

In constructing a number scale along the x axis it is more convenient to mark points such as

$$\frac{\pi}{6}, \frac{\pi}{4}, \frac{\pi}{3}, \frac{\pi}{2}, \ldots$$

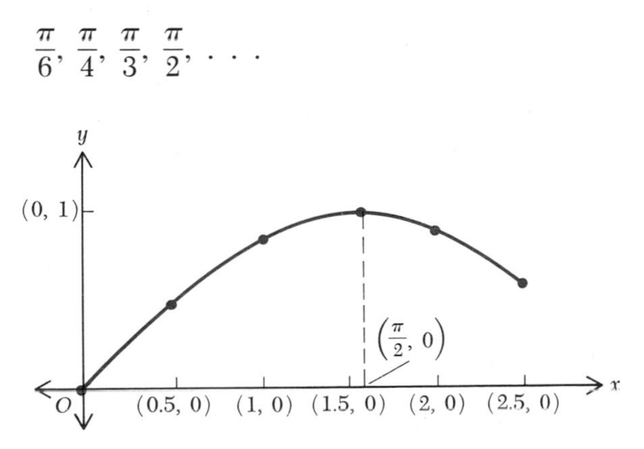

Fig. 10-1 Graph of function $y = \sin x$ for $0 \leq x \leq 2.5$.

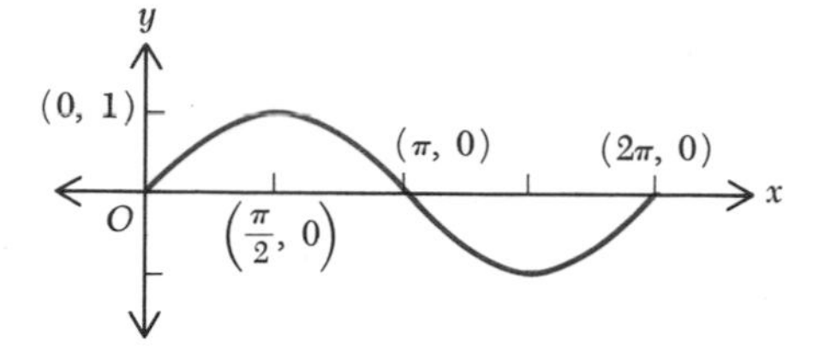

Fig. 10-2 Graph of function $y = \sin x$ for $0 \leqslant x \leqslant 2\pi$.

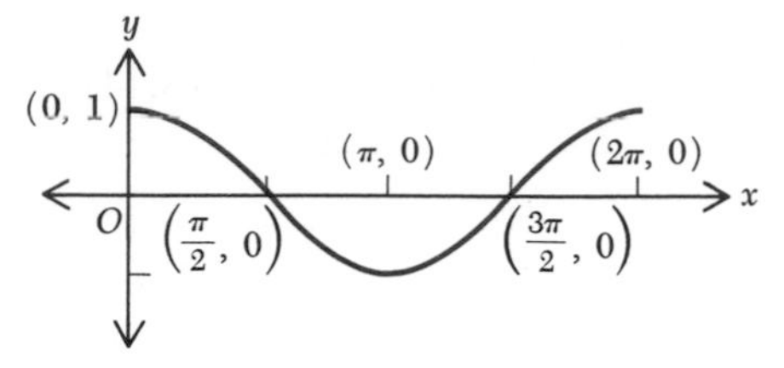

Fig. 10-3 Graph of function $y = \cos x$ for $0 \leqslant x \leqslant 2\pi$.

when graphing trigonometric functions. The graphs of $y = \sin x$ and $y = \cos x$ are shown with such a scale in Figs. 10-2 and 10-3. It should be observed that since $\sin (x + \frac{1}{2}\pi) = \cos x$, the graph of $\cos x$ is identical to that of $\sin (x + \frac{1}{2}\pi)$. This graph, in turn, is simply the graph of $\sin x$ with the axes translated so that the point $(\frac{1}{2}\pi, 0)$ becomes the new origin. The sine and cosine curves might be said to be $\frac{1}{2}\pi$, or 90°, "out of phase."

10-3 PERIODIC FUNCTIONS

A function $f(x)$ is said to be *periodic*, with period k, if

$$f(x + k) = f(x)$$

holds for all values of x for which $f(x)$ is defined. Since $\sin (x + 2\pi) = \sin x$ for all values of x, the function $\sin x$ is periodic with period 2π. The graph duplicates itself over an interval of 2π, and every such interval along the x axis is called a *complete cycle* of the curve. This means, for example, that for $2\pi \leqslant x \leqslant 4\pi$ and $-2\pi \leqslant x \leqslant 0$ the curves shown in Figs. 10-2 and 10-3 would be duplicated.

It follows from the definition of period that for $\sin x$ any number $2\pi n$, where n is an integer other than zero, could be considered as the period. Ordinarily the period of a periodic function $f(x)$ is defined as the least positive number k for which $f(x + k) = f(x)$.

10-4 THE GRAPHS OF THE FUNCTIONS $a \sin nx$ AND $a \cos nx$

The maximum and minimum values of both $\sin \theta$ and $\cos \theta$ are 1 and -1 for θ a real number. The corresponding maximum and minimum values of $a \sin nx$ and $a \cos nx$, for $a > 0$, are $\pm a$. The constant a is called the *amplitude* of the function.

Example 1
The maximum value of $5 \sin \frac{2}{3}x$ is 5, and its minimum value is -5. If x takes on values in the interval $0 \leq x \leq 3\pi$, the maximum value of the function occurs when x is such that $\frac{2}{3}x = \frac{1}{2}\pi$; that is, when $x = \frac{3}{4}\pi$. The minimum value occurs when $\frac{2}{3}x = 3\pi/2$, or $x = 9\pi/4$.

Theorem 10-1
$2\pi/n$ is the period of $a \sin nx$ or $a \cos nx$.

Proof

$$\sin n\left(x + \frac{2\pi}{n}\right) = \sin (nx + 2\pi)$$
$$= \sin nx$$

Example 2
Find the amplitude and period of the function $\sin 3x$.

Solution The amplitude is 1. By Theorem 10-1, the period is $\frac{2}{3}\pi$. Hence in any interval of 2π units along the x axis there are three complete cycles of the curve.

Example 3
Find the amplitude and period of the function $\frac{3}{2} \cos \frac{1}{2}x$.

Solution The amplitude is $\frac{3}{2}$, and by Theorem 10-1, its period is $2\pi/\frac{1}{2} = 4\pi$. Hence there is only one-half a complete cycle for $0 \leq x \leq 2\pi$.

Example 4
Graph the equation $y = 5 \sin \frac{2}{3}x$.

Solution Some information can be secured by the methods used for graphing algebraic curves.
Intercepts To find the x intercepts set $y = 0$ and solve for x:

$$0 = 5 \sin \tfrac{2}{3}x$$

Hence

$$0 = \sin \tfrac{2}{3}x$$

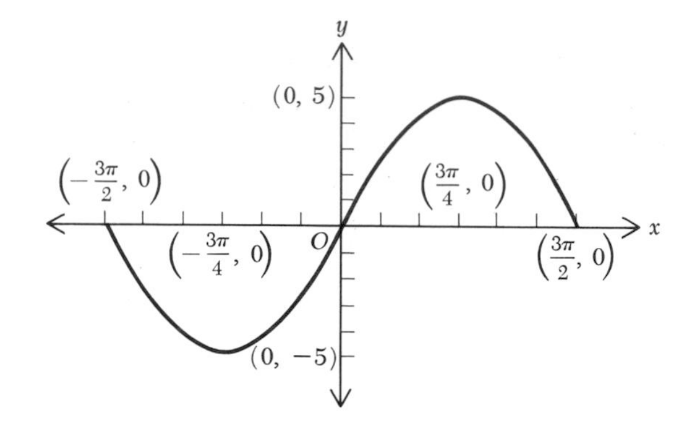

Fig. 10-4 Graph of function $5 \sin \frac{2}{3}x$ for $-3\pi/2 \leq x \leq 3\pi/2$.

Then, $\frac{2}{3}x$ may be any member of the set $\{. . . -2\pi, -\pi, 0, \pi, 2\pi, . . .\}$ and x may be any member of the set $\{-6\pi/2, -3\pi/2, 0, 3\pi/2, 6\pi/2, . . .\}$.

Tangents There are no horizontal or vertical tangents, because y is defined by all real values of x and x is defined for all real values of y such that $-5 \leq y \leq 5$. The test for tangents at the origin is not applicable.

Empty bands There are empty bands for $y > 5$ and $y < -5$. This is concluded from the fact that the amplitude of the curve is 5.

The graph of $y = 5 \sin \frac{2}{3}x$ in the interval $-3\pi/2 \leq x \leq 3\pi/2$ is shown in Fig. 10-4.

Example 5

Graph the equation $y = 1.4 \sin \frac{2}{3}\pi t$.

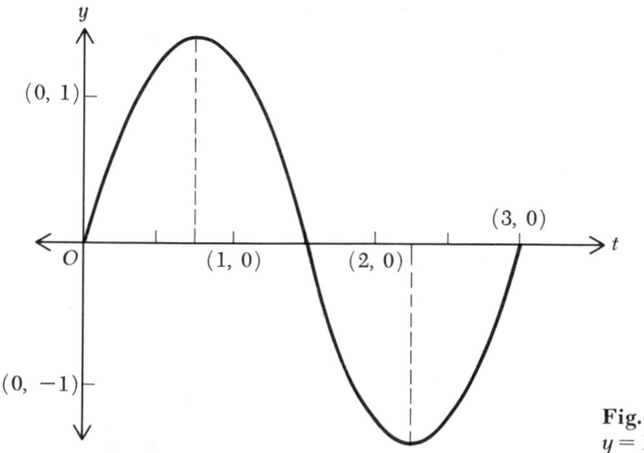

Fig. 10-5 Graph of function $y = 1.4 \sin \frac{2}{3}\pi t$ for $0 \leq t \leq 3$.

Solution The function $1.4 \sin \frac{2}{3}\pi t$ has an amplitude of 1.4 and a period of $2\pi/\frac{2}{3}\pi = 3$. Hence in any interval of 3 units along the t axis there is one complete cycle of the curve. With this information a sketch of the graph of $y = 1.4 \sin \frac{2}{3}\pi t$ may be made. It is shown in Fig. 10-5 for $0 \leq t \leq 3$.

In summary, the graphs of the equations $y = a \sin nx$ and $y = a \cos nx$ are sine and cosine curves, respectively, with amplitude a and period $2\pi/n$.

Example 6

The graph of the equation $y = 2 \cos \frac{2}{3}x$ is like that of $y = \cos x$, but the amplitude is 2 rather than 1, and the period is $2\pi/\frac{2}{3} = 3\pi$ rather than 2π. The graphs of $y = \cos x$ and $y = 2 \cos \frac{2}{3}x$ are both shown in Fig. 10-6 for $0 \leq x \leq 3\pi$.

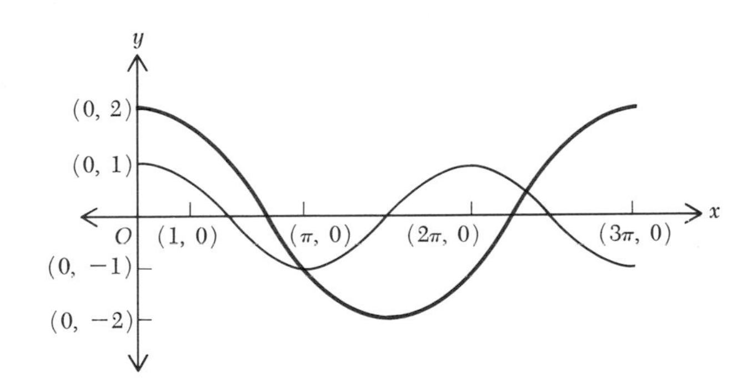

Fig. 10-6 Graphs of functions $y = 2 \cos \frac{2}{3}x$ and $y = \cos x$ for $0 \leq x \leq 3\pi$.

PROBLEMS

1. A function $f(x)$ is said to be an *even function* if the relation $f(-x) = f(x)$ holds for all values of x for which $f(x)$ is defined. Similarly, $f(x)$ is said to be an *odd function* if $f(-x) = -f(x)$. Which of the following functions are even, which are odd, and which are neither?

   (*a*) $\sin x$ (*b*) $\cos 3x$ (*c*) 2^x (*d*) x^3 (*e*) $x^2 - 3x$

2. (*a*) Find the value of the function $5 \sin \frac{1}{3}\pi t$ for each of the following values of t: -0.5, 0, 0.5, 0.75, 1, and 2. (*b*) What is the maximum value that the function can have, and what is the least positive value of t for which it has this maximum value?
3. (*a*) Find the value of the function $4 \sin \frac{2}{3}t$ for each of the following values

of t: 0, 0.5, 1, $\frac{1}{2}\pi$, 2, 3, and π. (*b*) What is the maximum value of the function, and what is the least positive value of t for which it has this maximum value?

4. What are the maximum and minimum values of the function $4 + 5 \cos \frac{1}{3}\pi t$? Find a value of t for which the function is a maximum and a value of t for which it is a minimum.
5. If $W = 4 \sin \frac{1}{3}\pi t + 4 \cos \frac{2}{3}\pi t$, find the value of W: (*a*) when $t = 2$; (*b*) when $t = -1$.

In Probs. 6 to 15 find the period and amplitude of the given function and sketch its graph. Mark the units on both axes.

6. $3 \cos \frac{2}{3}x$
7. $5 \sin \frac{1}{2}\pi x$
8. $\frac{1}{2} \sin \frac{1}{2}\pi t$
9. $2 \cos \frac{3}{2}t$
10. $5 \cos 2t$
11. $2.6 \sin 0.25\pi x$
12. $2.5 \sin \frac{1}{2}x$
13. $3.4 \cos \pi t$
14. $\{(x, y) \mid y = 6 \cos \frac{3}{4}x\}$
15. $\{(x, y) \mid y = 10 \sin \frac{5}{4}\pi x\}$

16. A certain quantity E varies with time in accordance with the formula $E = 2 + 2 \sin \frac{1}{5}\pi t$, where t is in seconds. Show that for all values of t $0 \leqslant E \leqslant 4$. Show that E goes through a complete cycle of values every 10 sec. Sketch the graph showing how E varies with t.
17. A certain quantity E varies with time in accordance with the formula $E = 40 \cos wt$, where t is in seconds. For what value of w will E go through a complete cycle of values in $\frac{1}{60}$ sec.?
18. Sketch the graph of the function $\sin x$, and on the same axes sketch the graph of $\sin^2 x$. Show that $\sin^2 x$ is periodic with period π.
19. A certain quantity Q varies with time in accordance with the formula $Q = 2 \sin \frac{1}{2}\pi t + 3 \cos \frac{2}{3}\pi t$, where t is in seconds. Is Q a periodic function of t, and if so, what is the period?
20. A certain quantity E varies with time in accordance with the formula $E = 1.8 \sin 0.4\pi t + 3.6 \cos \frac{2}{3}\pi t$, where t is in seconds. Show that E will go through a complete cycle of values every 15 sec. What would be the period if the number π were deleted from this formula?

10-5 THE FUNCTIONS $a \sin (nx + \alpha)$ AND $a \cos (nx + \alpha)$

The graphs of the equations $y = \sin x$ and $y = \sin (x + \frac{1}{4}\pi)$ are shown in Fig. 10-7. They are identical, of course, except that the curve $y = \sin (x + \frac{1}{4}\pi)$ *leads* $y = \sin x$ by $\frac{1}{4}\pi$. If x represents time in seconds, for example, the function $\sin (x + \frac{1}{4}\pi)$ reaches its maximum value $\frac{1}{4}\pi$ sec earlier. Similarly, the graph of the equation $y = \sin (x - \frac{1}{4}\pi)$ would be

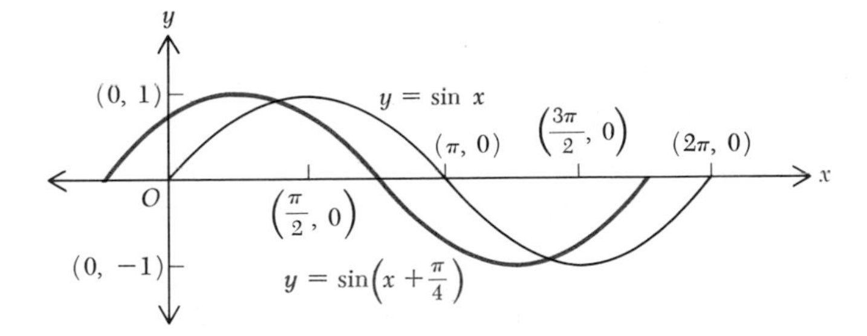

Fig. 10-7 Graphs of functions $y = \sin x$ and $y = \sin(x + \frac{1}{4}\pi)$; $y = \sin(x + \frac{1}{4}\pi)$ leads $y = \sin x$ by $\frac{1}{4}\pi$.

identical with that of $y = \sin x$ but displaced to the right by $\frac{1}{4}\pi$. Displacement to the right is usually called a *lag*.

Theorem 10-2
The graph of the function sin $(nx + \alpha)$ *is identical with that of sin* nx *but has a phase displacement (lag or lead) of* α/n.

Proof In the equation

$$y = \sin(nx + \alpha)$$

let $x = x' - \alpha/n$. This translates the origin to the left or right by α/n, according to whether α is positive or negative, and the equation with respect to x' is

$$y = \sin\left[n\left(x' - \frac{\alpha}{n}\right) + \alpha\right] = \sin nx'$$

Similar considerations apply to the function $\cos(nx + \alpha)$.

The conclusion is that the graphs of the functions $a \sin(nx + \alpha)$ and $a \cos(nx + \alpha)$ are sine and cosine curves, respectively, having amplitude a, a period $2\pi/n$, and a lead (if $\alpha > 0$) or lag (if $\alpha < 0$) of α/n.

Example
Draw the graph of the equation $y = \frac{3}{2}\sin(2x + \frac{1}{2}\pi)$.

Solution This is a sine curve with amplitude $\frac{3}{2}$, period $2\pi/2$ or π,

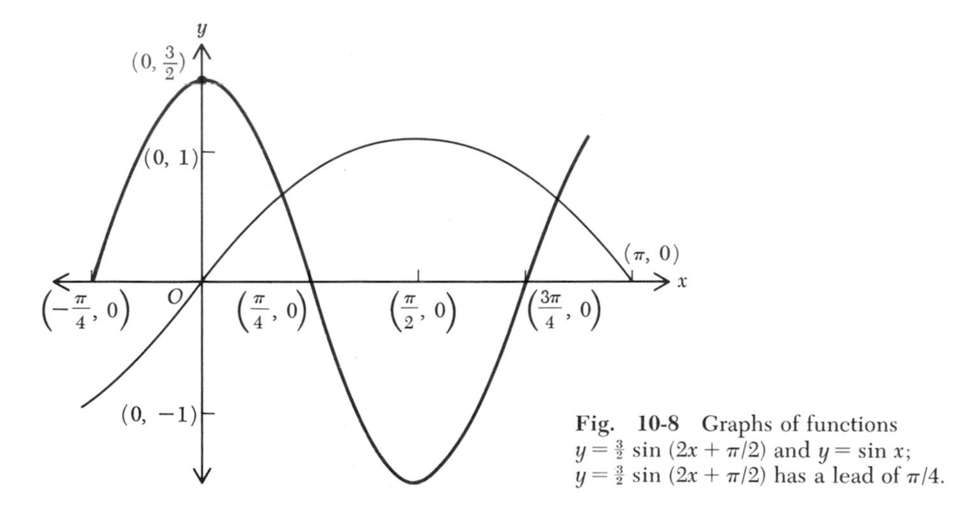

Fig. 10-8 Graphs of functions $y = \frac{3}{2}\sin(2x + \pi/2)$ and $y = \sin x$; $y = \frac{3}{2}\sin(2x + \pi/2)$ has a lead of $\pi/4$.

and a lead of $\frac{1}{2}\pi/2$, or $\pi/4$. One complete cycle is shown in Fig. 10-8, along with a partial cycle of $y = \sin x$ for comparison.

10-6 THE FUNCTION $a \sin nx + b \cos nx$

The graphs of the functions $\sin x$ and $\cos x$ are drawn in black in Fig. 10-9. The graph of the equation

$$y = \sin x + \cos x$$

can be constructed graphically by adding the ordinates for each value of x (Fig. 10-9). Thus, if for some value x_1 we have $y_1 = \sin x_1$ and $y_2 = \cos x_1$, then

$$y_1 + y_2 = \sin x_1 + \cos x_1$$

The curve resulting from the addition of the ordinates for a sine wave and a cosine wave with equal periods appears to be another sine or cosine wave with a greater amplitude and with a lag or lead. In order to show that this actually is the case, observe that for all values of x

$$\sin x + \cos x = \sqrt{2}\left(\frac{1}{\sqrt{2}}\sin x + \frac{1}{\sqrt{2}}\cos x\right)$$

Now, $$\sin\frac{\pi}{4} = \cos\frac{\pi}{4} = \frac{1}{\sqrt{2}}$$

so $$\sin x + \cos x = \sqrt{2}\left(\sin x \cos\frac{\pi}{4} + \cos x \sin\frac{\pi}{4}\right)$$

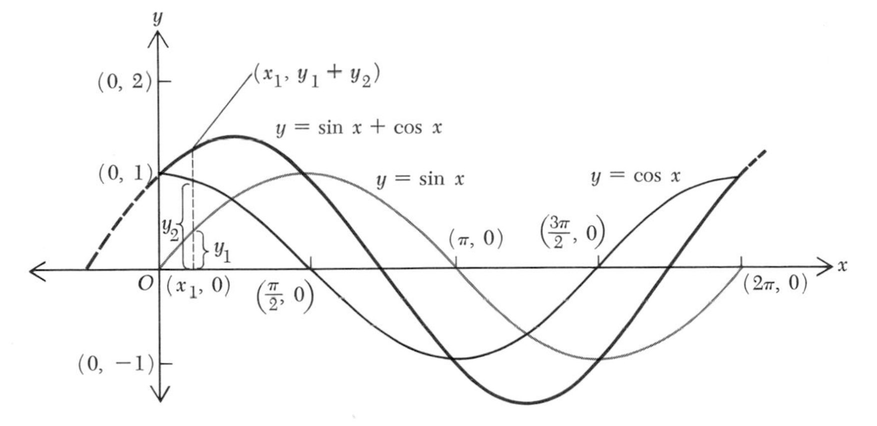

Fig. 10-9 Graph of function $y = \sin x + \cos x$ by method of addition of ordinates.

$$= \sqrt{2} \sin \left(x + \frac{\pi}{4} \right)$$

Thus the graph of the function $\sin x + \cos x$ is identical with that of the function $\sqrt{2} \sin (x + \pi/4)$. It is a sine curve with amplitude $\sqrt{2}$, period 2π, and lead $\pi/4$.

The more general case is that of addition of the ordinates of a sine wave of amplitude a and a cosine wave of amplitude b, both having the same period. The equation

$$y = a \sin nx + b \cos nx$$

is equivalent to the equation

$$y = \sqrt{a^2 + b^2} \left(\frac{a}{\sqrt{a^2 + b^2}} \sin nx + \frac{b}{\sqrt{a^2 + b^2}} \cos nx \right) \tag{10-1}$$

Let α denote a number such that the sine and cosine of α rad are $b/\sqrt{a^2 + b^2}$ and $a/\sqrt{a^2 + b^2}$, respectively, as indicated in Fig. 10-10; then

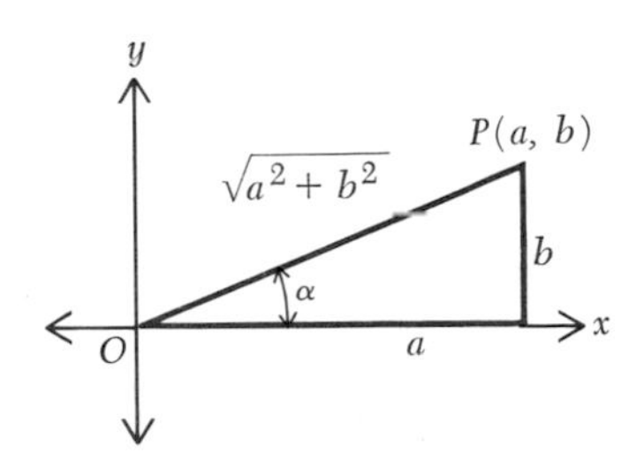

Fig. 10-10 $\sin \alpha = b/\sqrt{a^2 + b^2}$ and $\cos \alpha = a/\sqrt{a^2 + b^2}$.

Eq. (10-1) becomes

$$(10\text{-}2)\qquad \begin{aligned} y &= \sqrt{a^2+b^2}\,(\sin nx \cos\alpha + \cos nx \sin\alpha) \\ &= \sqrt{a^2+b^2}\,\sin(nx+\alpha) \end{aligned}$$

This means that the graph of the function $a \sin nx + b \cos nx$ is a sine curve with amplitude $\sqrt{a^2+b^2}$, period $2\pi/n$, and a lag or lead of α/n, where the value of α depends upon a and b as indicated above.

The addition of ordinates is applicable even when the functions have unequal periods. One special case is considered in the next section.

10-7 ADDITION OF ORDINATES OF SINE AND COSINE CURVES

In Fig. 10-11 the graphs of the functions $\sin x$ and $\frac{1}{3}\sin 3x$ are drawn over the interval $0 \leqslant x \leqslant \pi$. By adding the two ordinates for each value of x, the graph of the equation

$$y = \sin x + \tfrac{1}{3}\sin 3x$$

may be obtained. The line $y = \pi/4$ is also drawn in Fig. 10-11. Observe that the composite curve lies rather close to this line throughout most of the interval $0 \leqslant x \leqslant \pi$.

In Fig. 10-12 are the graphs of the functions $\sin x$, $\frac{1}{3}\sin 3x$, and $\frac{1}{5}\sin 5x$. The graph of the equation

$$y = \sin x + \tfrac{1}{3}\sin 3x + \tfrac{1}{5}\sin 5x$$

can be obtained by adding the three ordinates for each value of x. It

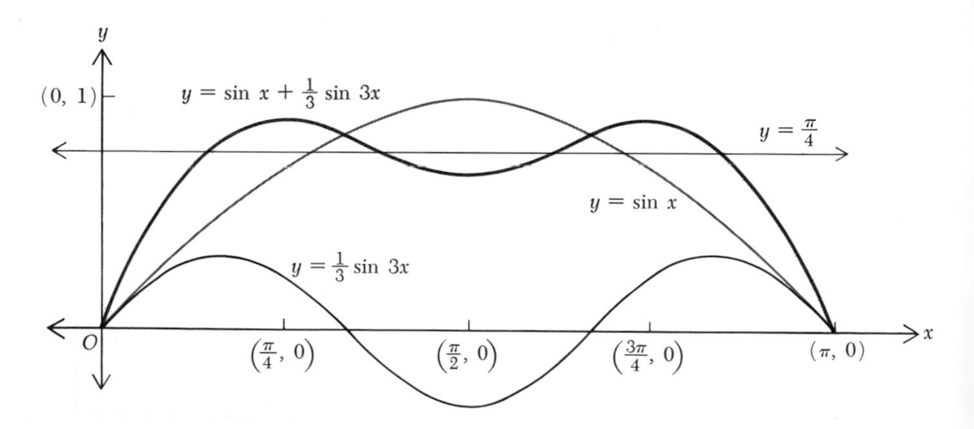

Fig. 10-11 Graph of function $y = \sin x + \frac{1}{3}\sin 3x$ drawn by adding ordinates of $y = \sin x$ and $y = \frac{1}{3}\sin 3x$.

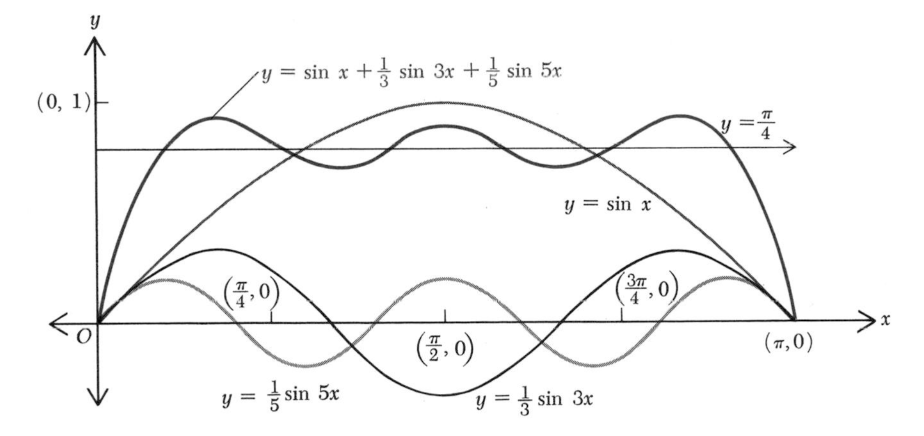

Fig. 10-12 Graph of function $y = \sin x + \frac{1}{3} \sin 3x + \frac{1}{5} \sin 5x$ drawn by method of addition of ordinates.

may again be observed that the curve lies near the line $y = \pi/4$ throughout most of the interval. We can prove, by advanced methods, that if the graphs of the equations

$$y = \sin x + \tfrac{1}{3} \sin 3x + \tfrac{1}{5} \sin 5x + \tfrac{1}{7} \sin 7x$$
$$y = \sin x + \tfrac{1}{3} \sin 3x + \tfrac{1}{5} \sin 5x + \tfrac{1}{7} \sin 7x + \tfrac{1}{9} \sin 9x$$
$$\cdots\cdots\cdots\cdots\cdots$$

are constructed successively, they come progressively nearer to coinciding with the line $y = \pi/4$ throughout any interval that lies within the interval $0 \leq x \leq \pi$. In the interval $\pi \leq x \leq 2\pi$ the curve would similarly approximate the line $y = -\pi/4$.

In the study of alternating current and voltage, in the analysis of vibrations, and, in fact, in the problems of many branches of applied mathematics, use is made of the fact that a function $f(x)$ defined over an interval $a \leq x \leq b$ and satisfying certain conditions can be approximated by a series of sine and cosine terms. One example is called a *Fourier series,* which is an infinite series similar in form to the functions of x which were graphed in Figs. 10-11 and 10-12.

10-8 THE TANGENT, COTANGENT, SECANT, AND COSECANT FUNCTIONS

The graph of the function tan x is shown in Fig. 10-13. The lines

$$x = \frac{\pi}{2},\ x = \frac{3\pi}{2},\ \ldots,\ x = \frac{(2n+1)\pi}{2}$$

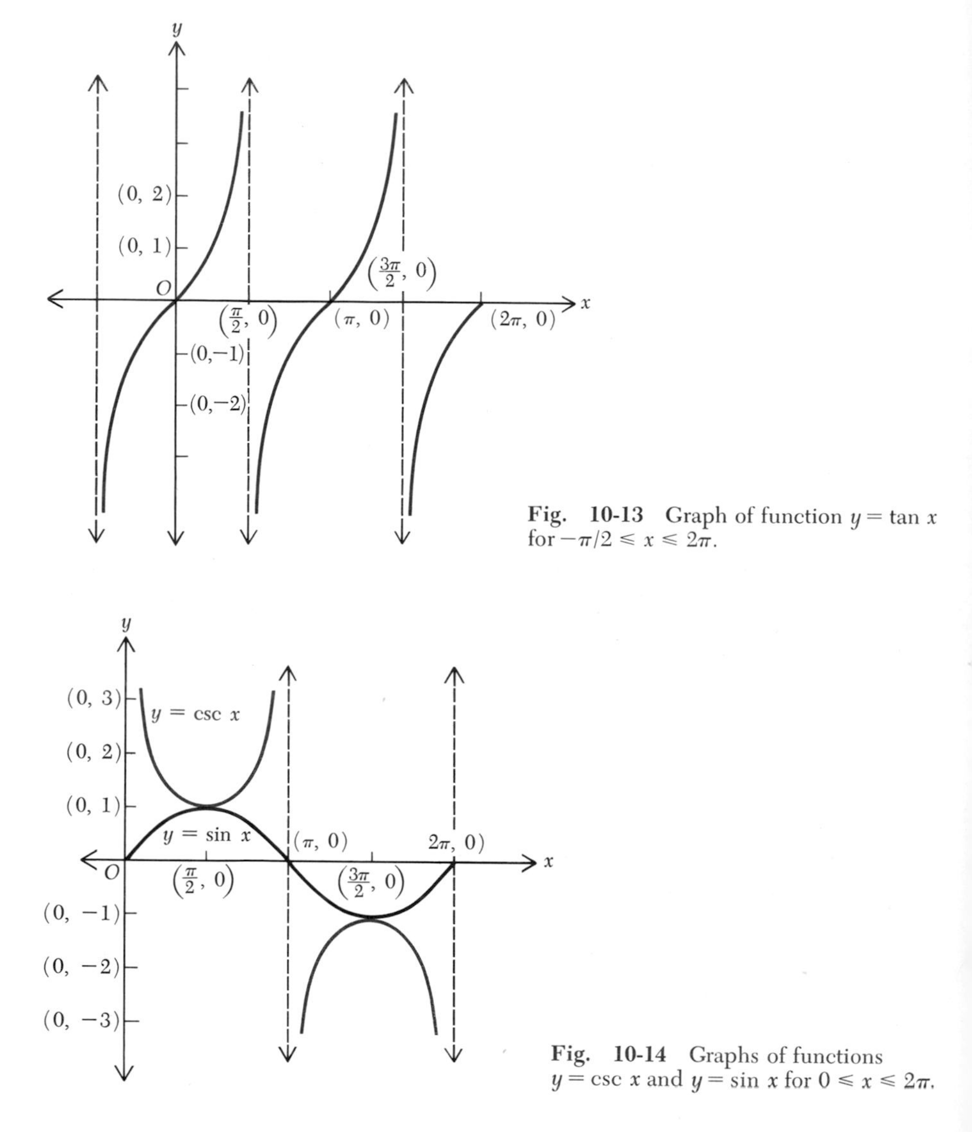

Fig. 10-13 Graph of function $y = \tan x$ for $-\pi/2 \leqslant x \leqslant 2\pi$.

Fig. 10-14 Graphs of functions $y = \csc x$ and $y = \sin x$ for $0 \leqslant x \leqslant 2\pi$.

where n is an integer, are vertical asymptotes. The function is periodic with period π.

For the graph of the equation $y = \frac{1}{2} \tan (\pi t/4)$ the first vertical asymptote to the right of the y axis would be at $t = 2$, because if

$$\frac{\pi t}{4} = \frac{\pi}{2}$$

then $t = 2$. The period for this function is 4. There is then a complete cycle of the graph in an interval of 4 units along the t axis.

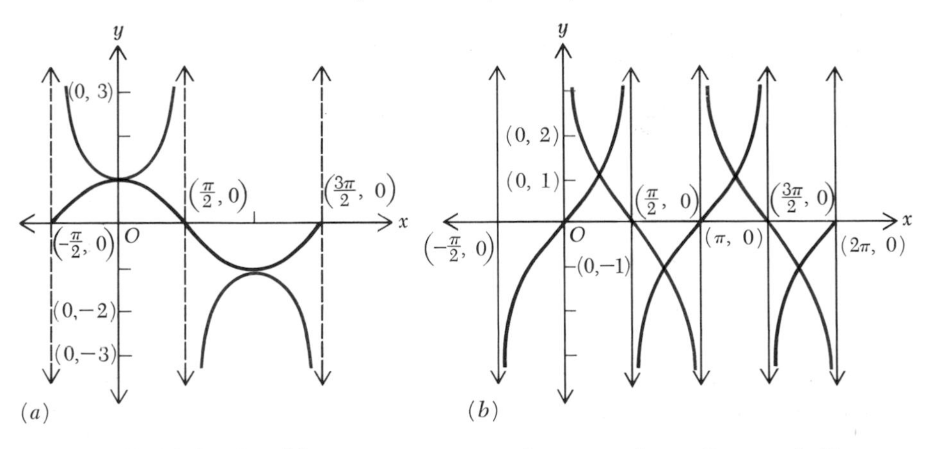

Fig. 10-15 (*a*) Graphs of functions $y = \cos x$ and $y = \sec x$ for $-\pi/2 \leq x \leq 3\pi/2$. (*b*) Graphs of functions $y = \tan x$ and $y = \cot x$ for $-\pi/2 \leq x \leq 2\pi$.

Similar considerations apply in the case of the function cot x; the details are left to the reader.

In Fig. 10-14 the graph of the function sin x is plotted on the same axes as that of the function csc x. The latter graph is readily sketched and easily remembered because of the relation

$$\csc x = \frac{1}{\sin x}$$

Thus where $\sin x = \frac{1}{2}$, $\csc x = 2$; where $\sin x = 1$, $\csc x = 1$; where sin x approaches zero, csc x becomes arbitrarily large. This last statement means that the lines

$$x = 0, \quad x = \pi, \quad \ldots, \quad x = n\pi \qquad \text{for } n \text{ an integer}$$

are vertical asymptotes to the graph of the function csc x.

The same reciprocal relation exists between the functions sec x and cos x. Hence the graph of sec x bears the same relation to that of cos x as the graph of csc x bears to that of sin x (see Fig. 10-15*a*). Similarly, tan x and cot x are reciprocals of each other. So the graphs of tan x and cot x (see Fig. 10-15*b*) are also related in the same manner as sin x and csc x.

PROBLEMS

In Probs. 1 to 8 graph the given equation.

1. $y = \sin\left(x - \frac{\pi}{6}\right)$ 2. $y = 2\cos\left(x + \frac{\pi}{4}\right)$

3. $y = 1.2 \sin\left(x + \frac{2\pi}{3}\right)$
4. $y = 1.6 \sin\left(2t - \frac{\pi}{2}\right)$
5. $y = 2.8 \cos\left(\frac{1}{2}t + \frac{\pi}{2}\right)$
6. $y = 3 \sin(0.6t + 3.6\pi)$
7. $y = \frac{3}{2} \sin \pi(x + 1)$
8. $y = 2.4 \cos \frac{\pi}{2}(4x - 1)$

Each of the equations in Probs. 9 to 16 is of the form

$$y = a \cos mx + b \sin nx$$

In each case sketch the graphs of the separate functions $a \cos mx$ and $b \sin nx$ on the same axes, and then obtain the graph of the given equation by adding ordinates.

9. $y = 3 \sin x + \cos x$
10. $y = 3 \cos x + 4 \sin x$
11. $y = 2 \cos x + \sin 2x$
12. $y = 2 \cos \frac{1}{2}x + 3 \sin x$
13. $y = 2 \sin \frac{\pi}{2} x + \cos \pi x$
14. $y = 3 \cos \frac{\pi}{3} x - 2 \sin \frac{2\pi}{3} x$
15. $y = 3 \cos \frac{\pi}{2} x - 4 \sin \frac{\pi}{2} x$
16. $y = \sin \frac{\pi}{2} x - 2 \cos \frac{\pi}{3} x$
17. Using the method of Sec. 10-6, show that the function $\sin 2x + \cos 2x$ is equivalent to $\sqrt{2} \sin(2x + \pi/4)$. Hence infer that its graph is a sine curve with amplitude $\sqrt{2}$, period π, and lead $\pi/8$. Sketch the graph.
18. Using the method of Sec. 10-6, show that the function $4 \sin x + 3 \cos x$ is equivalent to $5 \sin(x + \alpha)$, where α is a number such that $\sin \alpha = \frac{3}{5}$ and $\cos \alpha = \frac{4}{5}$. Sketch the graph.

In Probs. 19 to 24 sketch the graph.

19. $y = \frac{1}{2} \tan \frac{1}{2}x$
20. $y = \cot x$
21. $y = \sec x$
22. $y = \sec \frac{\pi}{2} t$
23. $\left\{(x, y) \mid y = 2 \csc \frac{\pi}{4} x\right\}$
24. $\left\{(x, y) \mid y = \tan\left(x + \frac{\pi}{4}\right)\right\}$

In Probs. 25 to 28 solve the given pair of equations to find the coordinates of the points of intersection of their graphs in the given interval. Illustrate by drawing the graphs.

25. $y = \sin x$ and $y = \frac{1}{2} \tan x$ for $0 \leq x < 2\pi$
26. $y = \sin 2x$ and $y = \cos x$ for $0 \leq x < 2\pi$
27. $y = 3 \tan \frac{1}{4}x$ and $y = 2 \sin \frac{1}{2}x$ for $0 \leq x \leq 4\pi$
28. $y = \sin \frac{\pi}{2} t$ and $y = \cos \frac{\pi}{4} t$ for $0 \leq t < 8$

10-9 THE INVERSE TRIGONOMETRIC RELATIONS

The notation

$$y = \arcsin x \qquad \text{or} \qquad y = \sin^{-1} x \tag{10-3}$$

means that y is the measure of an angle whose sine is x. It is understood, unless otherwise indicated, that the measure of the angle is given in radians. The domain of the relation is $-1 \leq x \leq 1$ for x a real number; to each x in the domain there corresponds an indefinite number of values of y. For example, if $x = \frac{1}{2}$, then $y = \arcsin \frac{1}{2}$ is the number of radians in an angle whose sine is $\frac{1}{2}$; hence

$$y = \begin{cases} \dfrac{\pi}{6} + 2\pi n \\ \dfrac{5\pi}{6} + 2\pi n \end{cases} \qquad \text{for } n \text{ an integer}$$

If $y = \arcsin x$ is solved for x in terms of y, the result is

$$x = \sin y$$

The graph of $y = \arcsin x$ (Fig. 10-16) is then identical with that of $x = \sin y$, which in turn is simply the graph of $y = \sin x$ with the x and y axes interchanged.

Corresponding to a certain number x_1 in the interval $-1 \leq x \leq 1$ there is exactly one value of $\arcsin x_1$ which lies in the interval $-\pi/2 \leq y \leq \pi/2$. If the interval of the variables is thus restricted, then the graph of $y = \arcsin x$ is the curve AB in Fig. 10-16. In this case

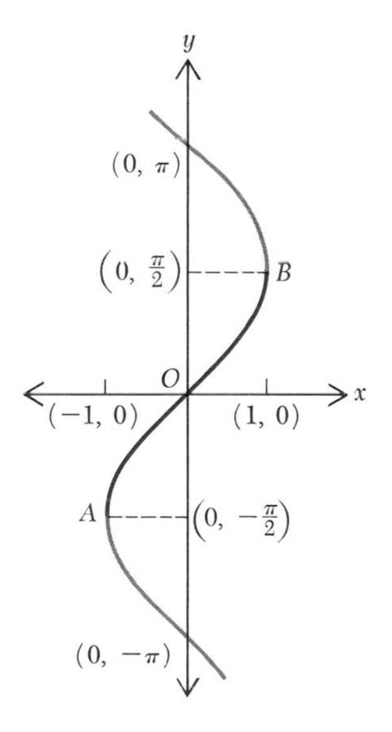

Fig. 10-16 Graph of relation $y = \arcsin x$; curve AB is the graph of function $y = \text{Arcsin } x$.

$$\arcsin \frac{1}{2} = \frac{\pi}{6} \qquad \arcsin 1 = \frac{\pi}{2} \qquad \arcsin -1 = -\frac{\pi}{2} \quad \left(\text{not } \frac{3\pi}{2}\right)$$

The curve AB is called the *principal branch* of the graph, and the y coordinate for any point on curve AB is called the *principal value* of $\arcsin x$.

For most practical purposes it is desirable to restrict the symbol $\arcsin x$ to represent its principal value. If this convention is applied, there will be exactly one solution for y in an equation such as $y = \arcsin (\sqrt{3}/2)$ rather than an infinite number of solutions. The principal value of the arcsin relation is ordinarily expressed as Arcsin (using a capital A). Then $\text{Arcsin } x$ is a function and $y = \text{Arcsin } x$ has a domain of $-1 \leqslant x \leqslant 1$ and a range of $-\pi/2 \leqslant y \leqslant \pi/2$.

Example 1

Find the radian measure for y if $y = \text{Arcsin } (\sqrt{3}/2)$ and if $y = \arcsin (\sqrt{3}/2)$.

Solution For $y = \text{Arcsin } (\sqrt{3}/2)$ there is one solution. Since $-\pi/2 \leqslant y \leqslant \pi/2$, the angle y whose sine is $\sqrt{3}/2$ equals $\pi/3$.

For $y = \arcsin (\sqrt{3}/2)$ the solution is

$$y = \frac{\pi}{3} + 2\pi n \quad \text{or} \quad y = \frac{2\pi}{3} + 2\pi n \qquad \text{for } n \text{ an integer}$$

The situation regarding the equation $y = \arccos x$ is entirely analogous.

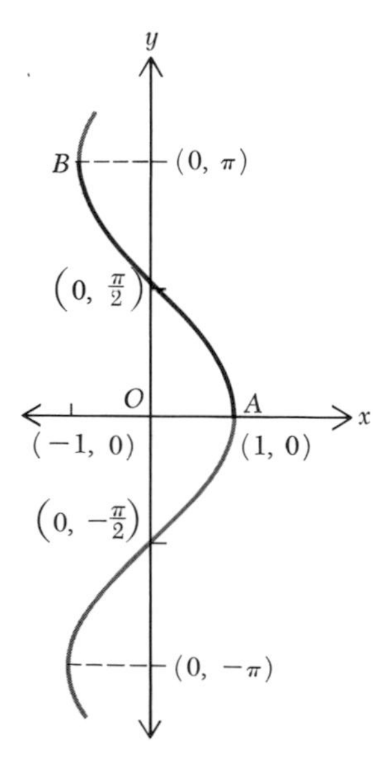

Fig. 10-17 Graph of relation $y = \arccos x$; curve AB is the graph of function $y = \text{Arccos } x$.

The equation is equivalent to the equation $x = \cos y$, so its graph is that of $y = \cos x$ with the axes interchanged (Fig. 10-17). Curve AB is the principal branch, and the values of arccos x that correspond to points on this curve are the principal values. They are the values of arccos x that lie in the interval $0 \leqslant y \leqslant \pi$. Thus for this branch

$$\arccos \tfrac{1}{2} = \tfrac{1}{3}\pi, \ \arccos -\tfrac{1}{2} = \tfrac{2}{3}\pi, \ \arccos 0 = \tfrac{1}{2}\pi, \ \ldots$$

Thus when arccos x is considered a function, the symbol Arccos x will be used, and it will be understood that solutions to the equation $y =$ Arccos x are restricted to $0 \leqslant y \leqslant \pi$.

The corresponding situation regarding the functions $y = \arctan x$ and $y = \operatorname{arccot} x$ are indicated in Figs. 10-18 and 10-19. In each case the principal branch is the curve AB. The functions are defined for all values of x. The principal value of arctan x, expressed as Arctan x,

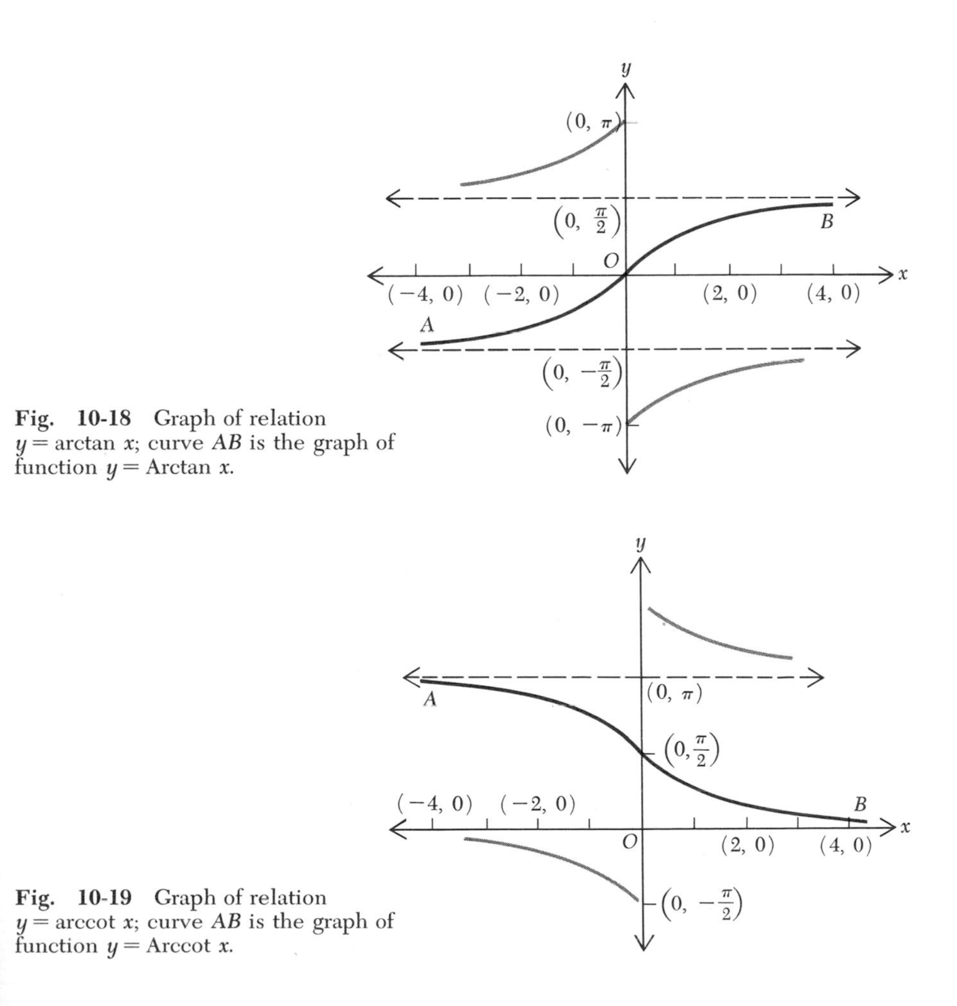

Fig. 10-18 Graph of relation $y = \arctan x$; curve AB is the graph of function $y = \operatorname{Arctan} x$.

Fig. 10-19 Graph of relation $y = \operatorname{arccot} x$; curve AB is the graph of function $y = \operatorname{Arccot} x$.

lies in the interval $-\pi/2 \leq y \leq \pi/2$, and that of arccot x, expressed as Arccot x, lies in the interval $0 \leq y \leq \pi$.

The range of the function $y = \text{Arcsec } x$ is $0 \leq y \leq \pi$, and the range of the function $y = \text{Arccsc } x$ is $-\pi/2 \leq y \leq \pi/2$. These definitions follow from the reciprocal relationships that exist between cos x and sec x and between sin x and csc x. These functions can usually be avoided by substituting Arccos $(1/x)$ for Arcsec x and Arcsin $(1/x)$ for Arccsc x.

The following examples and the problems of the next set illustrate various ways in which the inverse trigonometric functions are used.

Example 2
Find the value of $\sin(2 \text{ Arctan } \frac{2}{3})$.

Solution The problem is to find $\sin 2\theta$, where

$$\theta = \text{Arctan } \tfrac{2}{3}$$

That is, first solve the equation $\theta = \text{Arctan } \frac{2}{3}$, and then find $\sin 2\theta$.

Since $\tan \theta = \frac{2}{3}$, and $\tan \theta$ is in quadrant I (Fig. 10-20),

$$\sin \theta = \frac{2}{\sqrt{13}} \qquad \text{and} \qquad \cos \theta = \frac{3}{\sqrt{13}}$$

Then, since $\sin 2\theta = 2 \sin \theta \cos \theta$,

$$\sin 2\left(\text{Arctan } \frac{2}{3}\right) = 2\left(\frac{2}{\sqrt{13}}\right)\left(\frac{3}{\sqrt{13}}\right) = \frac{12}{13}$$

Example 3
Prove that $\text{Arctan } \frac{1}{4} + \text{Arctan } \frac{3}{5} = \pi/4$.

Solution The problem is to prove that the number of radians in the angle whose tangent is $\frac{1}{4}$ plus the number of radians in the angle whose

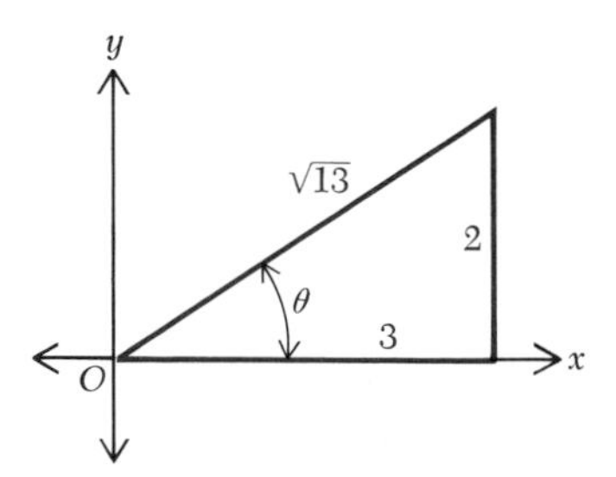

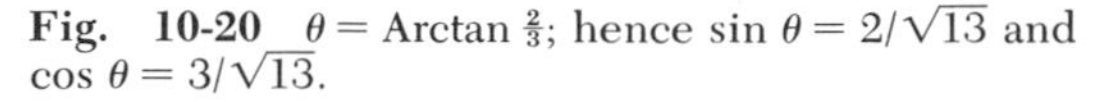
Fig. 10-20 $\theta = \text{Arctan } \frac{2}{3}$; hence $\sin \theta = 2/\sqrt{13}$ and $\cos \theta = 3/\sqrt{13}$.

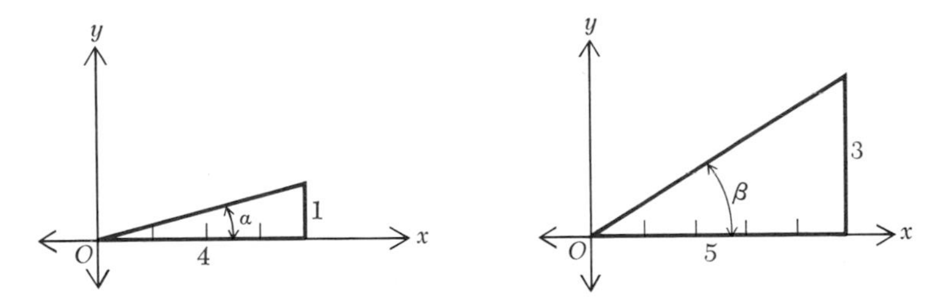

Fig. 10-21 $\alpha = \text{Arctan } \frac{1}{4}$ and $\beta = \text{Arctan } \frac{3}{5}$.

tangent is $\frac{3}{5}$ (principal values, of course) is equal to $\frac{1}{4}\pi$. Let $\alpha = \text{Arctan } \frac{1}{4}$ and $\beta = \text{Arctan } \frac{3}{5}$, as indicated in Fig. 10-21. A first thought might be to find α and β from tables and add them. This would be unsatisfactory, because interpolation would probably be necessary and the result would be only an approximation that is near $\frac{1}{4}\pi$. It is necessary instead to use the following plan of attack. If α and β are substituted for Arctan $\frac{1}{4}$ and Arctan $\frac{3}{5}$, then

$$\tan(\alpha + \beta) = \frac{\tan\alpha + \tan\beta}{1 - \tan\alpha\tan\beta}$$
$$= \frac{\frac{1}{4} + \frac{3}{5}}{1 - \frac{1}{4}(\frac{3}{5})}$$
$$= \frac{\frac{17}{20}}{\frac{17}{20}} = 1$$

Therefore $\alpha + \beta = \pi/4$.

PROBLEMS

Find the value of the expression in each of Probs. 1 to 10.

1. sin (Arccos $\frac{4}{5}$)
2. tan (Arcsin 0.3)
3. sin (Arccos $-\frac{1}{2}$)
4. cos (Arctan $-$ 1)
5. sec (Arctan $-\frac{3}{4}$)
6. tan (Arccot 0.2)
7. cos (Arcsin 1)
8. sin (Arccot 0)
9. 2 sin (Arctan 2) $-$ 6 cos (Arccot $-\frac{1}{2}$)
10. 5 cos² (Arcsin 1) + 3 sin² (Arccos $\frac{1}{2}$)

In Probs. 11 to 19 evaluate the given expression.

11. tan (2 Arcsin $\frac{3}{5}$)
12. sin (2 Arctan $\frac{1}{2}$)
13. cos (2 Arcsin $-\frac{2}{3}$)
14. tan (2 Arctan $-\frac{5}{2}$)
15. 2 tan ($\frac{1}{2}$ Arctan 1)
16. 4 sin ($\frac{1}{2}$ Arccos $-\frac{1}{2}$)
17. sin (Arcsin 1 + Arctan $\frac{3}{4}$)
18. cos (Arctan $\frac{12}{5}$ + Arccos $-\frac{4}{5}$)
19. sin (2 Arcsin $-$ 1 + Arccos $-\frac{4}{5}$)

20. Prove that Arctan $\frac{1}{3}$ + Arctan $\frac{1}{5}$ = Arctan $\frac{4}{7}$.
21. Prove that Arctan $\frac{1}{2}$ + Arctan $\frac{1}{3} = \pi/4$.
22. Solve Arcsin $2x$ − Arcsin $x = \frac{1}{3}\pi$ for x.
23. Solve Arctan $2x$ + Arctan $3x = \frac{1}{4}\pi$ for x.
24. The radius of the base of a right circular cone is 4 ft. The height is 3 ft at time $t = 0$, and it increases at 0.2 ft/min. Express the vertex angle θ as a function of t.

In Probs. 25 to 30 draw the graph.

25. $y = 3 \arcsin 2x$
26. $y = 2 \arccos 2x$
27. $y = \text{Arctan}\,(x - 1)$
28. $y = \text{Arcsin}\,(x - 2)$
29. $y = 2\,\text{Arccos}\,(x + 1)$
30. $\{(x, y) \mid y = 3\,\text{Arcsin}\,\frac{1}{6}x\}$

SUMMARY

Curves whose equations are stated in terms of the trigonometric functions are extremely useful because of their many applications.

The graph of $y = \sin x$ is made by studying the function $\sin x$. Its greatest and least values are 1 and −1, respectively. Therefore the y coordinate varies between 1 and −1. For $y = a \sin x$ the greatest value of y is a and the least is $-a$. The curve $y = a \sin x$ is said to have an amplitude of a. Similar generalizations can be made about $y = \cos x$.

The graph of $y = \sin x$ passes through $(0, 0)$, $(2\pi, 0)$, $(4\pi, 0)$, . . . , $(2n\pi, 0)$ for n an integer. The graphs in the intervals $0 \leqslant x \leqslant 2\pi$, $2\pi \leqslant x \leqslant 4\pi$, . . . are duplications of each other. It is thus said that the curve is periodic, with a period of 2π. The function $\sin nx$ is also periodic with a period of $2\pi/n$. This is because $\sin n\,(x + 2\pi/n) = \sin\,(nx + 2\pi) = \sin nx$. Similar statements can be made concerning the cosine function.

In general, the period of a periodic function (this includes tangent, cotangent, secant, and cosecant functions) is defined as the least positive number k for which $f(x + k) = f(x)$.

The graph of $y = A \sin\,(nx + \alpha/n)$ is a sine curve. Its graph differs from that of $y = A \sin nx$ in that it reaches its maximum α/n units later (a lead) if $\alpha > 0$ and α/n units earlier (a lag) if $\alpha < 0$. The graph of $y = \cos x$ has a lead of $\pi/2$ over the graph of $y = \sin x$.

The graph of the function $a \sin nx + b \cos nx$ is a sine curve because it can be expressed in the equivalent form $\sqrt{a^2 + b^2} \sin\,(nx + \alpha)$. The amplitude of this curve is $\sqrt{a^2 + b^2}$, its period is $2\pi/n$, and it has a lag or lead of α/n, where α is defined by the relation $\sin \alpha = b/\sqrt{a^2 + b^2}$.

Graphing the curve for an equation such as $y = \sin x + \frac{1}{3} \sin x$ is accomplished most readily by graphing

$$y_1 = \sin x \qquad \text{and} \qquad y_2 = \tfrac{1}{3} \sin 3x$$

on the same axes. Then the y coordinate for any x on $y = \sin x + \frac{1}{3} \sin 3x$ is found by adding the y coordinate y_1 of $y_1 = \sin x$ and the y coordinate y_2 of $y_2 = \frac{1}{3} \sin 3x$. This is called "graphing by adding ordinates."

The functions $\tan x$, $\cot x$, $\sec x$, and $\csc x$ are graphed by considering their periodic nature, values of x for which they are undefined and hence their vertical asymptotes, and reciprocal relations such as $\csc x = 1/\sin x$.

The inverse relations $y = \arcsin x$ and $y = \arccos x$ are equivalent to $x = \sin y$ and $x = \cos y$. Their graphs are identical to $y = \sin x$ and $y = \cos x$, respectively, with the x and y axes interchanged. Their domain is $-1 \leq x \leq 1$, and to each x there corresponds an indefinite number of values for y. So that these inverse relationships may be functions, the range for y is restricted by the inequality $-\pi/2 \leq y \leq \pi/2$ for $y = \text{Arcsin}\, x$ and by the inequality $0 \leq y \leq \pi$ for $y = \text{Arccos}\, x$. These values are called principal values of the relation.

REVIEW PROBLEMS

1. On the same set of axes sketch the graphs of the equations:

   (*a*) $y = \sin 2x$ (*b*) $y = \cos 2x$ (*c*) $y = -\sin 2x$ and $y = -\cos 2x$

2. Find the amplitude and the period of each function and sketch its graph:

   (*a*) $4 \sin \frac{3}{4}\pi x$ (*b*) $5 \cos \frac{5}{4}x$

3. Use the method of addition of ordinates to graph each of the following:

   (*a*) $y = \sin x + 2 \cos x$ (*b*) $y = 3 \sin (\pi x/6) + 2 \sin (\pi x/3)$
   (*c*) $y = \sin (\pi x/3) - \cos (\pi x/3)$

4. Express the function $5 \sin x + 12 \cos x$ in the form $r \sin (x + \alpha)$.
5. Draw the graph of each equation:

   (*a*) $y = 2 \text{ Arcsin } \frac{1}{2}x$ (*b*) $y = 3 \text{ Arccos } \frac{1}{2}x$

6. Evaluate each of the following expressions:

   (*a*) $\text{Sin} [\text{Arcsin} (-\sqrt{3}/2)]$ (*b*) $\cos (\text{Arctan} -1)$
   (*c*) $\text{Arccos } \frac{1}{2} + \text{Arcsin } \frac{1}{2}$ (*d*) $\text{Arcsin} [\cos (\pi/4)]$
   (*e*) $\cos [\text{Arcsin} -\frac{1}{2} - \text{Arccos} (-\sqrt{3}/2)]$

7. Graph $y = x + \sin x$ by the method of addition of ordinates (x is a real number).
8. Solve the pair of equations $y = \sin x$ and $y = \cos 2x$ simultaneously to find the

coordinates of their points of intersection in the interval $0 \leq x < 2\pi$. Verify your results by graphing the equations on the same axes.

9. Graph each of the following equations:

$$(a)\ y = 3 \sin (x - \pi/6) \qquad (b)\ y = 4 \cos (x + 2\pi/3)$$

FOR EXTENDED STUDY

1. Between what maximum and minimum values does the function $y = 7 - 5 \sin x \cos x$ oscillate? Find a value of x for which the value of the function is greatest and a value of x for which it is least.
2. E_1 and E_2 vary with time t in accordance with the formulas

$$E_1 = 3 \sin \frac{\pi}{4} t \qquad \text{and} \qquad E_2 = -6 \cos \frac{\pi}{8} t$$

where t is in seconds. For what value or values of t between 0 and 16 is E_1 equal to E_2?

3. E varies with time t in accordance with the formula

$$E = 3.5 + 1.2 \cos \frac{\pi}{6} t - 1.6 \sin \frac{\pi}{6} t$$

where t is in seconds. Show that E is never greater than 5.5 or less than 1.5, and that E goes through a complete cycle of values every 12 sec. Sketch the graph.

4. Sketch the graph of the equation

$$y = 2 \sin x - \sin 2x + \tfrac{2}{3} \sin 3x$$

by sketching the graphs of three equations and then adding ordinates as illustrated in Fig. 10-12.

5. A picture 2 ft high hangs on a wall with its lower edge 3 ft above the observer's eye (Fig. 10-22). Express the number of radians in the angle θ subtended by the picture at the eye as a function of the distance x (feet) of the observer from the wall. Sketch a graph showing approximately how θ varies with x.

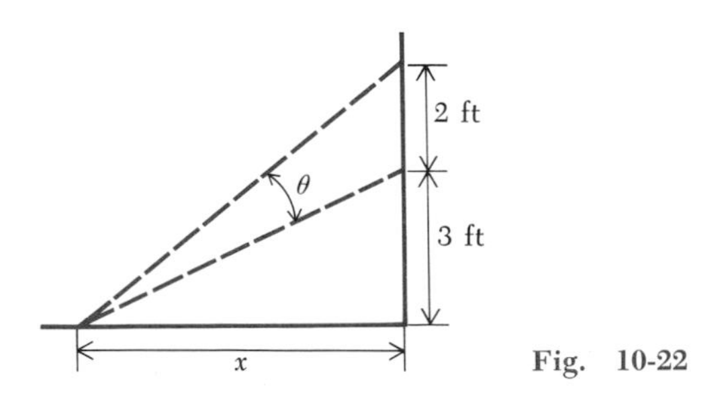

Fig. 10-22

6. Test each of the following for symmetry with respect to the origin:

(a) $y = \sin x$ (b) $y = \sin nx$
(c) $y = a \cos nx$ (d) $y = a \cos (nx + \pi/2)$

7. Graph each of the following:

(a) $y = x \sin x$ (b) $y = |\sin x|$
(c) $y < \sin^2 x$ (d) $y = |\sin x| + |\cos x|$
(e) $y < \sin x$ (f) $y \geqslant \cos x$

EXPONENTIAL AND LOGARITHMIC CURVES 11

11-1 THE EXPONENTIAL FUNCTION

Given any positive number a, the symbol a^x represents a definite real number for every real number x. Thus a^x is a function, with the real numbers and the positive real numbers as its domain and range; an approximation to its graph can be drawn by the usual procedure of preparing a table of values, plotting the points, and drawing a smooth curve through them. Thus for the equation

$$y = 1.5^x$$

the following table can be constructed (all values of y are shown correct to two decimal places):

x	0	1	2	3	4	5	−1	−2	−3	−4
y	1	1.50	2.25	3.38	5.06	7.59	0.67	0.44	0.30	0.20

The corresponding graph is shown in Fig. 11-1. Points between those actually plotted can be found by using nonintegral values of x.

HISTORICAL NOTE

While the discovery of the logarithmic and exponential functions cannot be credited to any one man, because they were developed and refined over a period of time, some mathematicians did make useful contributions.

Nicole Oresme, who lived during the fourteenth century, is given credit for first using fractional exponents. Francois Vieta helped develop the modern notation, and John Wallis, who lived during the seventeenth century, explained the significance of zero and negative exponents.

Logarithms were the invention of John Napier (1550–1617). The common logarithms in use today were the work of Henry Briggs. Oddly enough, the discovery of logarithms came before exponents were in general use.

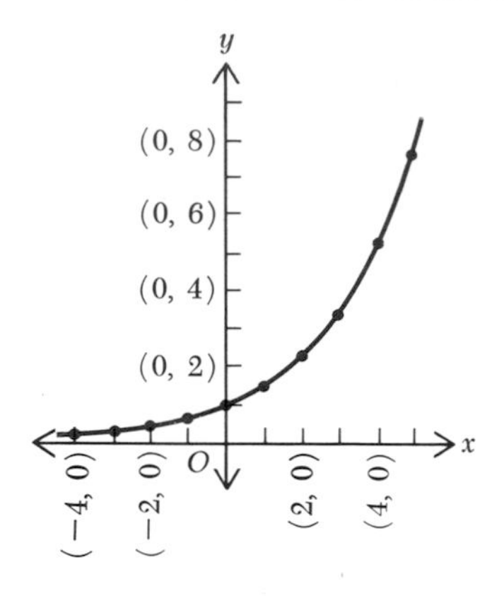

Fig. 11-1 Graph of $y = 1.5^x$.

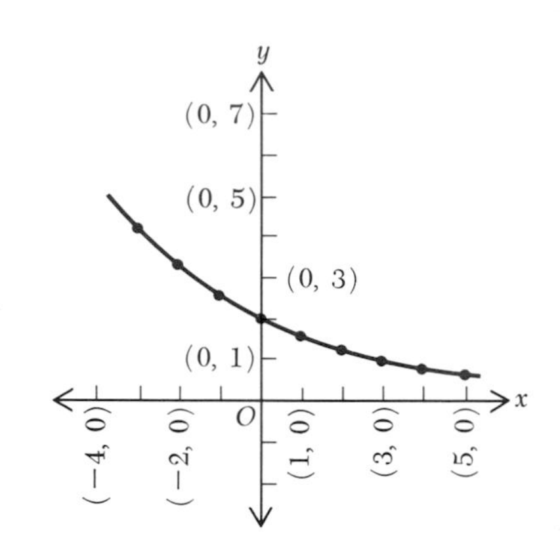

Fig. 11-2 Graph of $y = 2(0.6)^{\frac{1}{2}x}$.

For example, corresponding to $x = 2.5$,

$$y = 1.5^{2.5} = 1.5^2\sqrt{1.5} \approx 2.76$$

The function a^x, where a is positive and not equal to 1, is called the *exponential function*. The value of the function is positive for all real values of x, and it increases as x increases if $a > 1$, with the graph having the general shape shown in Fig. 11-1. For $0 < a < 1$, a^x decreases as x increases, and the graph has the general form shown in Fig. 11-2.

11-2 THE LOGARITHMIC FUNCTION

It is possible to express any given positive real number N in the form 10^x by properly choosing the exponent x. For example, $437 = 10^x$ for one and only one value of x. Since $10^2 = 100$ and $10^3 = 1{,}000$, this value of x must be between 2 and 3; it is, in fact, 2.64, correct to three significant figures. The number 2.64 is called the *common logarithm* of 437 or the *logarithm* 437 *to the base* 10, and we may write, in the standard notation,

$$\log_{10} 437 \approx 2.64$$

since

$$10^{2.64} \approx 437$$

More generally, it can be proved that if a is any positive real number different from 1, and N is any given positive real number, there exists one and only one real number x such that $a^x = N$. The number x is called the *logarithm of N to the base a*. For a given base $a > 1$, x increases as N increases.

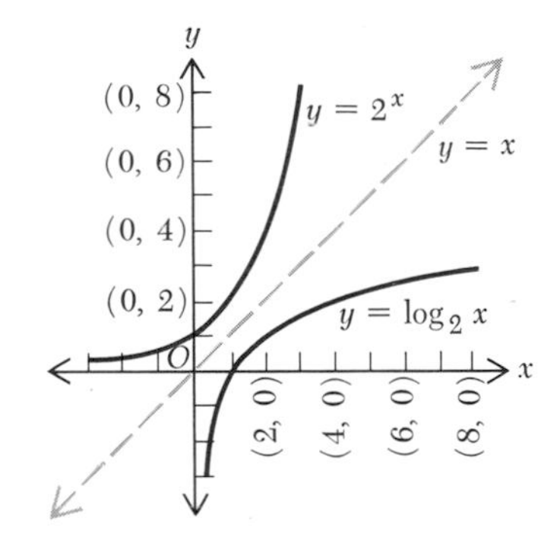

Fig. 11-3 Graphs of $y = 2^x$ (color) and $y = \log_2 x$ (black); the curves are symmetrical with respect to $y = x$.

The equation $y = \log_a x$ implies that $x = a^y$. These equations therefore have the same graph, and this graph is simply the graph of the equation $y = a^x$ with the axes interchanged. The graphs of the equations $y = 2^x$ and $y = \log_2 x$ are shown in Fig. 11-3. Note that the curves are symmetrical with respect to the line $y = x$. Each is a monotone-increasing function; that is, $f(x)$ continues to increase, without ever decreasing, as x increases.

11-3 THE FUNCTIONS e^x AND $\log_e x$

In the expression

$$(1 + v)^{1/v}$$

let v approach closer and closer to zero; that is, let $v = 1$, then 0.1, then 0.01, and so on. It can be proved that the corresponding values of this expression approach a certain irrational number which, to five significant figures, is 2.7183.

If $v = 1$, then $(1 + v)^{1/v} = 2^1 = 2$.
If $v = 0.1$, then $(1 + v)^{1/v} = (1.1)^{10} \approx 2.5937$.
If $v = 0.01$, then $(1 + v)^{1/v} = (1.01)^{100} \approx 2.7048$.
If $v = 0.001$, then $(1 + v)^{1/v} = (1.001)^{1,000} \approx 2.7169$.

The proof of the fact that the sequence of numbers on the right approaches arbitrarily near to a definite number as v approaches zero is not elementary and is not given here. The number is symbolized by the letter e, which, as previously stated, is approximately equal to 2.7183.

The logarithmic function $\log_e x$, usually written $\ln x$, is called the *natural logarithm* of x. Figure 11-4 compares the graphs of $y = \ln x$ and $y = \log_2 x$.

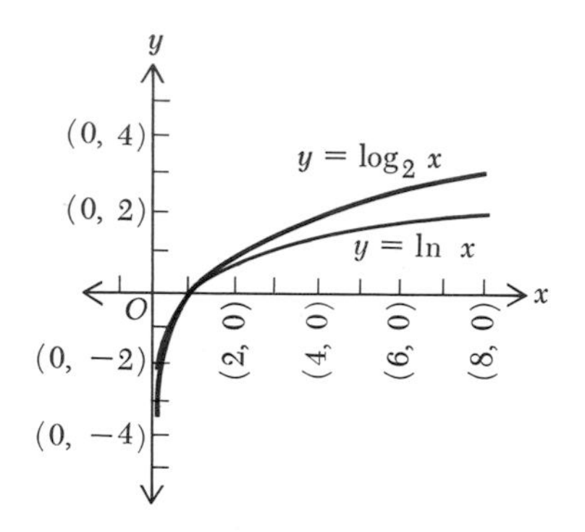

Fig. 11-4 Graphs of $y = \log_2 x$ (color) and $y = \ln x$ (black).

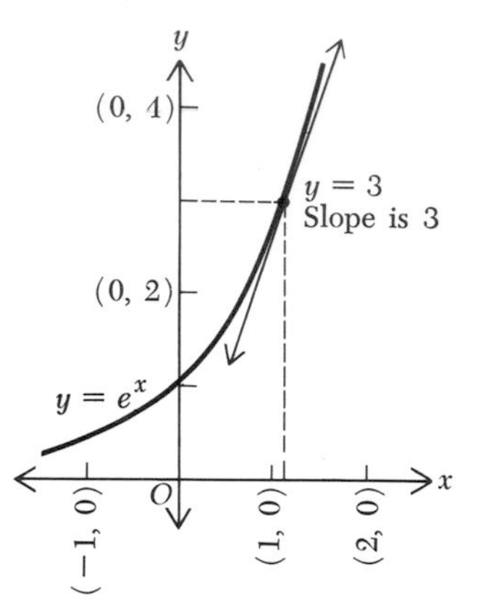

Fig. 11-5 Graph of $y = e^x$; slope of curve at any point is the y coordinate of that point.

Since any positive number a can be expressed in the form e^k, the logarithmic function a^{rx} is equivalent to $(e^k)^{rx}$, or e^{krx}. Thus it is always possible to use e as the base for the exponential function.

Note: It is shown in calculus that the function e^x is simpler than 2^x or 3^x or 7^x in the following respect. The line drawn tangent to the graph of the equation $y = e^x$ at any point P on the curve has a slope which is equal to the ordinate at that point. Thus at the point where $y = 3$ the slope of the tangent line is 3, and at the point where $y = 5$ the slope is 5 (Fig. 11-5). In the more general case of $y = a^x$ the slope of the tangent is proportional to y, that is, equal to ky; but it is only when $a = e$ that $k = 1$.

It is shown in calculus that the graph of $y = \ln x$ is simpler in the following respect. The line drawn tangent to the graph of the equation $y = \ln x$ at any point P on the curve has a slope which is equal to the reciprocal of the abscissa of that point. Thus where $x = 4$, the slope is $\frac{1}{4}$. In the general case of the function $\log_a x$ this slope is proportional to $1/x$, that is, equal to $k(1/x)$, but it is only when $a = e$ that $k = 1$.

Two problems are sometimes encountered in studying logarithmic functions:

1. If $r = \ln N$, what is $\log N$ ($\log N$ means $\log_{10} N$)?
2. If $s = \log N$, what is $\ln N$?

Both questions are answered by the formula

$$\ln N = (\ln 10)(\log N)$$

The proof is as follows. Let $x = \log N$. Then $10^x = N$. If the logarithms of 10^x and N are taken to the base e, the result is

$$\ln 10^x = \ln N$$

or

$$x \ln 10 = \ln N$$

But $x = \log N$, so

$$\ln N = (\ln 10)(\log N) \tag{11-1}$$

If Eq. (11-1) is solved for $\log N$,

$$\log N = \frac{1}{\ln 10} \ln N \tag{11-2}$$

Correct to five decimal places,

$$\ln 10 = 2.30259 \qquad \text{and} \qquad \frac{1}{\ln 10} = 0.43429$$

These approximations may be substituted in formulas (11-1) and (11-2) when computations are performed.

Example
Express the function $5^{2.64x}$ in the form e^{kx}.

Solution From tables $\ln 5 \approx 1.609$, and it follows that $5 \approx e^{1.609}$. Then

$$5^{2.64x} \approx (e^{1.609})^{2.64x} \approx e^{1.609(2.64)x} \approx e^{4.25x}$$

PROBLEMS

In Probs. 1 to 8 sketch the graph of the given equation.

1. $y = 2^{-1/2x}$
2. $y = 1.4^x$
3. $y = 4(1.2)^{2x}$
4. $y = 8(0.25)^{-x}$
5. $y = 3(\frac{3}{2})^{-2x}$
6. $y = 1.5^{1-x}$
7. $8y = 2^{x+2}$
8. $\frac{1}{4}y = 2^{1-2x}$

In Probs. 9 to 20 sketch the graph of the given equation. Use Table IV, pages 412–413, and Table V, page 414, as needed.

9. $y = e^{0.5x}$
10. $y = \frac{1}{2}e^{-0.5x}$
11. $y = 0.5e^{3x/2}$
12. $y = 0.6e^{-0.2x}$
13. $y = e^x + e^{-x}$
14. $y = 3e^{-x/4}$
15. $y = \ln(1 + x)$
16. $y = \log_2(1 - x)$
17. $y = 1 - x + 1.5^x$
18. $y = 2^x - x^2$
19. $x^3 + y = 3^x$
20. $y = \frac{1}{2}x + 2^{-x}$

21. Show that the graph of $y = \log_a bx$ is the graph of $y = \log_a x$ shifted vertically a distance of $\log_a b$.
22. Prove that $\log_b N = \log_a N/\log_a b$.
23. Graph $y = 2^x$. Now change the scale on the y axis so the graph drawn is that of $y = 3(2)^x$.
24. Prove each of the following statements:

    (a) $b^{\log_b x} = x$ (b) $\log_a b = 1/\log_b a$ (c) $x^x = b^{x \log_b x}$ for $x > 0$

25. Determine the values of a and c for which the curve $y = ce^{ax}$ passes through the points $(3, e)$ and $(4, 2e)$.
26. Show that the function e^{ax+b} can be written in the form ce^{ax}, where $c = e^b$.
27. Write the function $4.6^{2.8x}$ in the form e^{kx}.
28. Express the following in powers of e:

    (a) 2^x (b) 1.06^{4n} (c) $3^{0.1t}$ (d) $10^{-0.3x}$

29. For what values of k is the function 10^{ax} equivalent to the function e^{kax}?
30. If \$700 is invested at a rate of 3% per annum compounded continuously, the compound amount at the end of t years is given by $A = 700e^{0.03t}$. Make a graph showing how A increases with t. Find the amount, to the nearest dollar, at the end of 10 years.
31. Find the compound amount if \$300 is invested for 24 years at a rate of 2.5% per annum compounded continuously.
32. The number of bacteria in a culture increases at a rate which is always 6% of the number present. If the original number of bacteria was 1,000, the number N after t hr is $N = 1{,}000e^{0.06t}$. (a) Find the number of bacteria after 10 hr. (b) Find the time t to the nearest hour when the number of bacteria has doubled.
33. Radium decomposes at a rate (per century) which equals 3% of the quantity Q remaining. If t represents a number of centuries, and the original amount of radium (at $t = 0$) was 100 mg, then $Q = 100e^{-0.03t}$. (a) How much radium will remain after 2,000 years? (b) When will half the radium remain (answer to the nearest 100 years)?
34. An approximation for the pressure p (millimeters) of mercury at a height h (kilometers) above sea level is given by $p = 760e^{-0.144h}$. (a) Find h when $p = 7.6$. (b) Find the height for which the pressure is one-half that at sea level.

11-4 BOUNDARY CURVES: DAMPED VIBRATIONS

In Fig. 11-6 the graphs of $y = 3e^{-x/4}$ and $y = \sin(\pi x/2)$ are drawn in black. The latter curve has period of 4 and therefore has two complete cycles in the interval $0 \leqslant x \leqslant 8$. By multiplying the ordinates for each value of x, the graph (in color) of

$$y = 3e^{-x/4} \sin \frac{\pi x}{2} \tag{11-3}$$

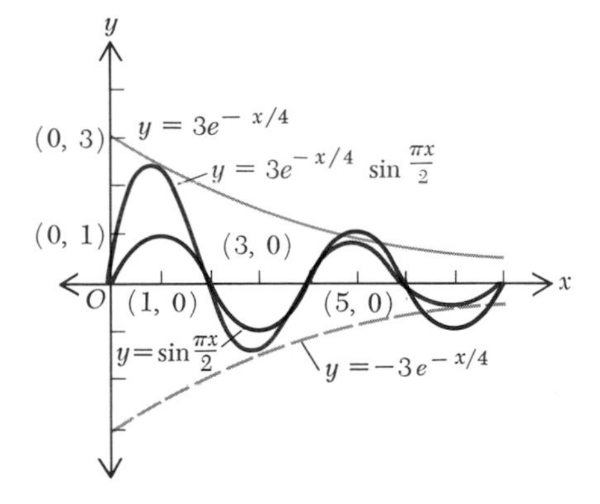

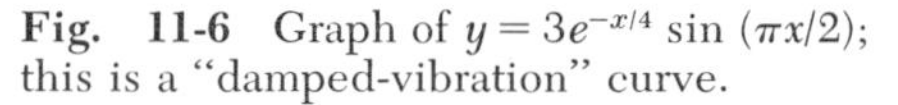
Fig. 11-6 Graph of $y = 3e^{-x/4} \sin (\pi x/2)$; this is a "damped-vibration" curve.

Fig. 11-7 Weight supported by a spring.

is obtained. In connection with this multiplication of ordinates in the interval $0 \leqslant x \leqslant 8$, the following facts should be observed:

1. At every point where the factor $\sin (\pi x/2)$ is equal to zero the product is zero; hence the graph of Eq. (11-3) crosses the x axis wherever x is a member of the set $\{. . . , -4, -2, 0, 2, 4, 6, . . . , 2n, . . .\}$ for n an integer.

2. Since $\sin (\pi x/2)$ is never greater than 1, the product curve is never above the graph of $y = 3e^{-x/4}$; at each point where $\sin (\pi x/2)$ is equal to 1, the product is equal to $3e^{-x/4}$. In fact, the graph of Eq. (11-3) is tangent to the curve $y = 3e^{-x/4}$ wherever x is a member of the set $\{. . . , 1, 5, 9, 13, . . . , 4n + 1, . . .\}$ for n an integer.

3. At points where $\sin (\pi x/2)$ is equal to -1 the product is the negative of $3e^{-x/4}$. Thus wherever x is a member of the set $\{. . . , 3, 7, 11, . . . , 4n - 1, . . .\}$ for n an integer the graph of Eq. (11-3) is tangent to the dashed curve whose equation is $y = -3e^{-x/4}$. The two curves $y = \pm 3e^{-x/4}$ therefore form *boundaries* between which the graph of (11-3) oscillates, its "amplitude" decreasing as x increases.

Figure 11-7 shows a weight W which is supported by a spring. If the weight is pulled down some initial distance and then released, it will oscillate up and down. If the resistance of the air or other medium to the motion is proportional to the velocity, the equation governing the oscillations has the general form

$$y = Ae^{-kt} \sin (nt + \alpha) \qquad (11\text{-}4)$$

where y is the vertical displacement of the weight (from its position of equilibrium) at time t, and the other symbols represent constants whose values depend upon the stiffness of the spring, the viscosity of the

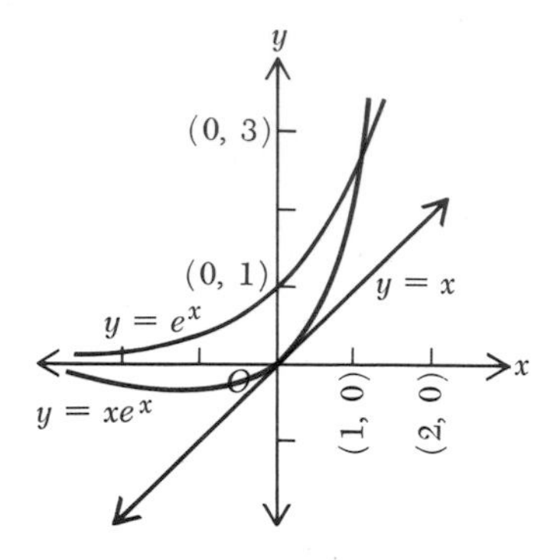

Fig. 11-8 $y = xe^x$ graphed by a method of multiplication of ordinates.

medium, and the initial displacement and velocity of the weight. The curve shown in Fig. 11-6 is a *damped-vibration curve* of the form (11-4) which might describe the motion of the weight. Such curves are important not only in the study of mechanical vibrations in machines and structures, but also in certain electric-circuit phenomena.

The method of multiplication of ordinates is often valuable. In the case of equations such as $y = x \cos x$, $y = (1/x) \sin x$, or $y = xe^x$ the general character of the graph can usually be visualized by sketching the graphs of the separate factors and mentally multiplying the ordinates. The graphs of $y = x$ and $y = e^x$ are drawn in Fig. 11-8. Multiplying the ordinates results in the graph of the equation

$$y = xe^x$$

In connection with the use of multiplication of ordinates, observe the following facts:

1. Where both factors are positive or negative, the product is positive; that is, where both curves lie above or below the x axis, the curve for the product is above the x axis. Where one is above and the other below the x axis, the product curve is below.

2. Where either factor is zero, the product is zero if the other factor is finite.

3. Where either factor has the value 1, the product is equal to the other factor. Thus in Fig. 11-8 the curve for the product crosses the graph of $y = e^x$ at $x = 1$.

4. Where both factors are greater than 1 in absolute value, the absolute value of the product is greater than that of either factor; where the absolute value of one factor is greater than 1 and that of the other is less than 1, that of the product is between the two. Thus in Fig. 11-8 the graph of $y = xe^x$ lies above that of either factor for $x > 1$ but lies between the two for $0 < x < 1$. Finally, where both factors are less than 1 in

absolute value, the absolute value of the product is less than that of either factor. Thus in Fig. 11-8 the ordinates of the curve $y = xe^x$ in the interval $-1 < x < 0$ are less in absolute value than those of either the line $y = x$ or the curve $y = e^x$.

11-5 THE HYPERBOLIC FUNCTIONS

Certain combinations of the exponential functions e^x and e^{-x} occur sufficiently often in scientific applications to make it desirable to assign names to them and tabulate their values. These functions are called *hyperbolic functions*. It can be shown that they are related to the equilateral hyperbola in somewhat the same way that the trigonometric functions are related to the circle, and they are accordingly given similar names. The definitions are as follows:

Hyperbolic sine of x: $$\sinh x = \frac{e^x - e^{-x}}{2}$$

Hyperbolic cosine of x: $$\cosh x = \frac{e^x + e^{-x}}{2}$$

Hyperbolic tangent of x: $$\tanh x = \frac{e^x - e^{-x}}{e^x + e^{-x}}$$

The hyperbolic cosecant (csch), secant (sech), and cotangent (coth) are defined, respectively, as the reciprocals of the hyperbolic sine, cosine, and tangent. It should be emphasized that the hyperbolic functions are not new functions; they are simply names given to certain frequently occurring combinations of the exponential functions.

The graphs of $y = \sinh x$ and $y = \cosh x$ are shown in Fig. 11-9. They could be constructed by drawing the separate graphs of $y = \frac{1}{2}e^x$ and $y = \frac{1}{2}e^{-x}$, and then adding or subtracting the ordinates. Table V, page 414, gives values of sinh x, cosh x, and tanh x.

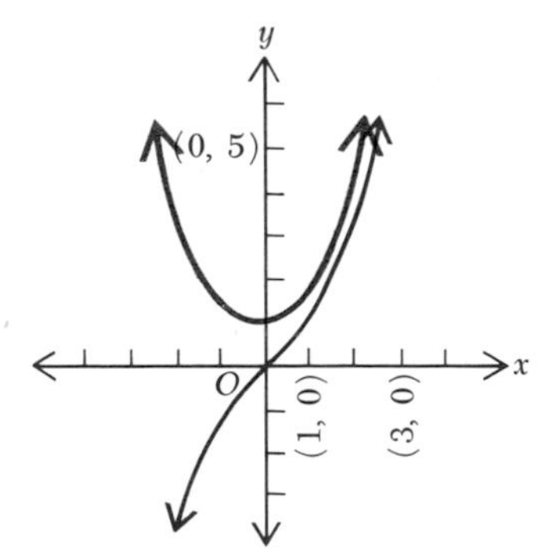

Fig. 11-9 Graphs of $y = \sinh x$ (black) and $y = \cosh x$ (color).

It can be shown in terms of principles of mechanics that a completely flexible uniform cable supported at its ends hangs in a curve whose equation is of the form

$$y = \frac{a}{2}\left(e^{x/a} + e^{-x/a}\right) \tag{11-5}$$

where a is a constant whose value depends upon the amount of sag allowed. This curve is called a *catenary;* its equation can be written in the form

$$y = a \cosh \frac{x}{a} \tag{11-6}$$

The graph of $y = \cosh x$ (Fig. 11-9) is then the catenary in which $a = 1$.

PROBLEMS

In Probs. 1 to 8 sketch the graph of the given equation, using the method of multiplication of ordinates.

1. $y = 3e^{-x/4} \cos \frac{1}{2}\pi x$
2. $y = 2e^{0.2t} \cos \frac{1}{2}\pi t$
3. $y = 5e^{-0.1t} \sin \frac{1}{4}\pi t$
4. $y = 5e^{-x/2} \cos \pi x$
5. $y = 3e^{x-1} \cos \frac{1}{2}\pi x$
6. $y = 2e^{0.1x} \sin \frac{1}{4}\pi x$
7. $y = 2e^{-0.2x} \cos \frac{1}{2}x$
8. $y = 3e^{0.2x} \sin \frac{1}{2}x$

In Probs. 9 to 15 sketch the graph of the given equation.

9. $y = \dfrac{2}{\sqrt{x}} \sin \dfrac{1}{2} \pi x$ for $0 < x \leqslant 8$
10. $y = \dfrac{2}{x} \cos \pi x$ for $0 < x \leqslant 4$
11. $y = \frac{1}{4}x \sin \frac{1}{2}\pi x$ for $-4 \leqslant x \leqslant 8$
12. $y = 0.4x \cos \frac{1}{2}\pi x$ for $-4 \leqslant x \leqslant 8$
13. $y = x(2)^{-(x/2)^2}$ for $-4 \leqslant x \leqslant 4$
14. $y = 2^{1-(x/2)^2} \sin \frac{1}{2}\pi x$ for $-4 \leqslant x \leqslant 4$
15. $y = 2e^{-(0.2t)^2} \cos \frac{1}{2}\pi t$ for $-4 \leqslant t \leqslant 4$
16. Draw the catenary whose equation is $y = 5(e^{0.1x} + e^{-0.1x})$ for the interval $-10 \leqslant x \leqslant 10$.
17. Draw the catenary whose equation is $y = 2.5(e^{0.2x} + e^{-0.2x})$ for the interval $-8 \leqslant x \leqslant 8$.
18. Prove (*a*) that the curve $y = \sinh x$ is symmetrical with respect to the origin; (*b*) that the curve $y = \cosh x$ is symmetrical with respect to the y axis.
19. Prove that $\sinh 2x = 2 \sinh x \cosh x$.
20. The height y ft of a certain cable above the ground at a horizontal distance x ft from the middle is given by the equation $y = 100 \cosh 0.01x - 70$. (*a*) How high is it at one end where $x = 50$ ft? (*b*) How high is it in the middle?

SUMMARY

The function a^x, where $a > 0$ and $a \neq 1$, is called the exponential function. There is a definite real number for a^x for each value of x. The graph of $y = a^x$ is approximated by preparing a table of values. Some characteristics of the graph of $y = ba^x$, where $b > 0$ and $a > 1$, are as follows:

1. The y coordinate is always positive because b and a^x are always positive.
2. If $|x|$ is great and x is negative, y approaches zero and the x axis is an asymptote.
3. If $|x|$ is great and x is positive, y is great.

A commonly used exponential function is e^x, where e is the limiting value of $(1 + v)^{1/v}$ as v approaches zero. To five significant figures, e is 2.7183. One important property of the graph of $y = e^x$ is that the slope of the tangent line to the curve at any point is equal to the ordinate of the point.

The exponential and logarithmic functions are related in that $y = a^x$ and $x = \log_a y$ are equivalent equations and the graphs $y = a^x$ and $y = \log_a x$ are symmetrical with respect to the line $y = x$.

Certain equations such as $y = be^{-ax} \sin (nx + \alpha)$, where $a > 0$, are found in special applications involving damped vibrations. The graph crosses the x axis at regular intervals, but the factor e^{-ax} becomes smaller with increasing values of x, and hence the amplitude decreases as x increases. Because of this property it is called a damped-vibration curve. The method of graphing

$$y = be^{-ax} \sin (nx + \alpha)$$

is to graph on the same axes

$$y_1 = be^{-ax} \qquad \text{and} \qquad y_2 = \sin (nx + \alpha)$$

Then the ordinate y is equal to $y_1 y_2$.

Sums, differences, and quotients of terms involving e^{ax} occur so frequently in science that special functions, called hyperbolic functions, have been defined. A completely flexible uniform cable supported at the ends hangs on a curve whose equation is

$$y = \frac{a}{2} (e^{x/a} + e^{-x/a})$$

This is often written

$$y = a \cosh \frac{x}{a}$$

where $\cosh (x/a)$ is the hyperbolic cosine of x/a.

REVIEW PROBLEMS

In Probs. 1 to 8, sketch the graph of the given equation.

1. $y = 3^x$
2. $y = 3(1.2)^{2x}$
3. $y = e^{2x}$
4. $y = 0.2e^{-x}$
5. $y = \log (2 + x)$
6. $y = 2 \ln 2x$
7. $y = x + e^x$
8. $y = e^x \sin \dfrac{\pi x}{2}$
9. Write the function $3^{2.4x}$ in the form e^{kx}.
10. Show that the graph of $y = \ln bx$ for $b > 0$ is the graph of $y = \ln x$ shifted vertically a distance of $\ln b$.
11. If \$100 is invested at 2% per annum compounded continuously, the accumulated amount y at the end of x years is given by the equation $y = 100e^{0.02x}$. Sketch the graph of this equation and from the graph estimate the accumulated amount, to the nearest dollar, at the end of (*a*) 2 years; (*b*) 5 years; (*c*) 8 years. (*d*) Estimate the time required for the original amount to double.

FOR EXTENDED STUDY

1. (*a*) If money can be invested at 1% per interest period, how many interest periods will it take for the money to double? (*b*) Write the function and sketch the graph.
2. Prove that $\cosh^2 x - \sinh^2 x = 1$.
3. In the equation $y = (a/2)(e^{x/a} + e^{-x/a})$, replace e by $10^{\log e}$ (which is, of course, equal to e by the definition of $\log e$). From this show that if 10 is used as the base for the exponential function, the equation of the catenary has the form

$$y = \frac{k}{2 \ln 10} (10^{x/k} + 10^{-x/k})$$

4. Solve the equation $y = \frac{1}{2}(e^x - e^{-x})$ for x in terms of y. The resulting function of y is called the *inverse hyperbolic sine* of y.
5. Graph $y = \ln |x|$.
6. Graph $y = |\ln x|$.
7. Compare the graphs of $y = (\ln x)^2$, $y = \ln x^2$, and $y = 2 \ln x$.

8. Solve the equation $y = \frac{1}{2}(e^x + e^{-x})$ for x in terms of y. The resulting function of y is called the ***inverse hyperbolic cosine*** of y.
9. Solve the equation $y = (e^x - x^{-x})/(e^x + e^{-x})$ for x in terms of y. The resulting function of y is called the ***inverse hyperbolic tangent*** of y.
10. Prove that $x = a \cosh \phi$ and $y = a \sinh \phi$, where ϕ is a parameter, are parametric equations for a hyperbola with center at the origin and transverse axis of length $2a$ along the x axis.

CURVE FITTING 12

12-1 INTRODUCTION

In various branches of science a set of corresponding values for two variables is obtained by observation or experiment. It may then be important to find an equation that "fits" this table of values in a satisfactory way. Thus we might measure the atmospheric pressure at various altitudes in a given locality at a given time and then try to devise an equation that would fit the observed data. Such an equation would express, at least in an approximate way, the atmospheric pressure as a function of the altitude.

A set of corresponding values of two variables in the form of ordered pairs can, of course, be plotted as points. The problem then becomes that of finding the equation of a curve that passes through or near these points in such a way as to indicate their general trend. An equation determined in such a manner is called an *empirical equation.* The process of finding it is called *curve fitting.*

There is no unique solution to the problem of fitting a curve to every set of points. Both the straight line and the curve shown in Fig. 12-1 fit the given set of points approximately, and it is obvious that other lines and curves could be drawn that might serve equally well.

The first problem in fitting a curve to a given set of data is to decide

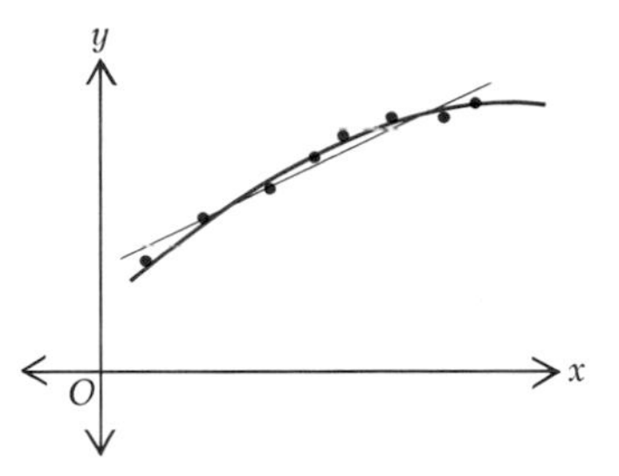

Fig. 12-1 Straight line and curve fit the set of points in an approximate way.

upon the type of curve to be employed. In some cases this is determined from theoretical considerations. When this is not the case, the selection may be based on considerations discussed later in this chapter. It is generally desirable to use the simplest type of equation with which a reasonably good fit can be obtained. The types to be considered in this chapter are

(12-1)	$y = ax + b$	linear
(12-2)	$y = ax^2 + bx + c$	parabolic
(12-3)	$y = ab^x$	exponential
(12-4)	$y = ax^n$	power

After the type of equation has been chosen, the remaining problem is to determine the constants such that the equation fits the given data satisfactorily. Thus if the given points lie approximately along a straight line, we would ordinarily choose the linear form, Eq. (12-1), and proceed to determine a and b so as to secure a satisfactory fit.

Two methods of evaluating the constants are commonly used. The first is the *method of average points*, and the second is the *method of least squares*. The first is simpler and easier to apply, but the second ordinarily provides a result which describes the data more accurately.

12-2 DEFINITION OF AVERAGE POINT

The *average point* of a given set of points is the point whose abscissa is the average of the abscissas of the set and whose ordinate is the average of the ordinates. Thus if the coordinates of the given points are

$$(x_1, y_1), (x_2, y_2), (x_3, y_3), \ldots, (x_n, y_n)$$

then

$$x_a = \frac{x_1 + x_2 + x_3 + \cdots + x_n}{n} \tag{12-5}$$

and

(12-6)
$$y_a = \frac{y_1 + y_2 + y_3 + \cdots + y_n}{n}$$

are the coordinates (x_a, y_a) of the average point for the set.

12-3 LINEAR TYPE: METHOD OF AVERAGE POINTS

When the given points lie approximately along a line, as in Fig. 12-1, a linear equation that will fit the data reasonably well can be obtained by the method of average points. First the given set of points is partitioned into two sets with approximately the same number of points in each set. Next the coordinates of the average point for each set are found, and the line determined by these two points is the required line.

There are no fixed rules for partitioning the set of points into two sets. The examples in this chapter use the criterion of partitioning the table of data so that the two sets of points appear on the left and right of the graphs. While this procedure is customary, it is only a convenience.

Example
Find the equation of a line that fits the points shown in Fig. 12-2, with coordinates as follows:

x	1	2	3	4	5	6	7	8	9	10
y	1.48	1.76	2.78	3.32	3.86	4.15	4.75	5.66	6.18	6.86

Solution Take the first five points as the first set; by Eqs. (12-5) and (12-6), the average point for this set has the coordinates

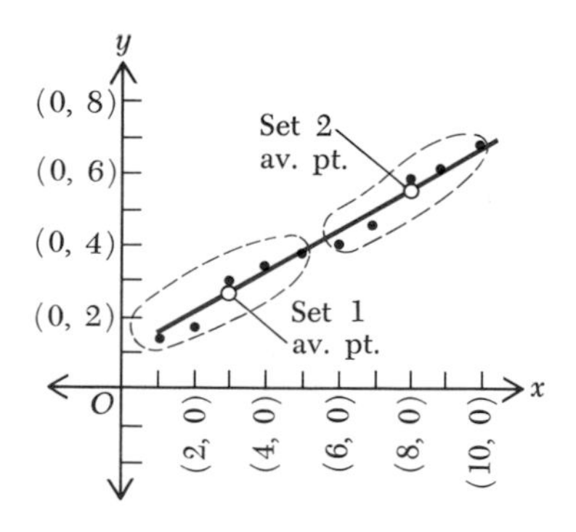

Fig. 12-2 Method of average points for fitting a line to a set of points.

$$x_a = \frac{1+2+3+4+5}{5} = 3$$

$$y_a = \frac{1.48 + 1.76 + 2.78 + 3.32 + 3.86}{5} = 2.64$$

Then take the remaining five points as the second set; the coordinates of the average point for this set are

$$x'_a = \frac{6+7+8+9+10}{5} = 8$$

$$y'_a = \frac{4.15 + 4.75 + 5.66 + 6.18 + 6.86}{5} = 5.52$$

The slope of the line through these two average points is

$$\frac{5.52 - 2.64}{8 - 3} = 0.576$$

and its equation is

$$y - 2.64 = 0.576(x - 3)$$

or

$$y = 0.576x + 0.912$$

The result obtained by this procedure depends upon the way in which the points are grouped. It is often necessary to choose the sets with care in order to obtain a good fit. It can be shown that the various choices of the sets result in lines all of which pass through the average point of the entire set.

12-4 RESIDUALS: CRITERIA FOR GOODNESS OF FIT

Assume that in Fig. 12-3 the curve $y = f(x)$ has been fitted to the points $P_1(x_1, y_1)$, $P_2(x_2, y_2)$, and so on. The *residual* of any one of the plotted

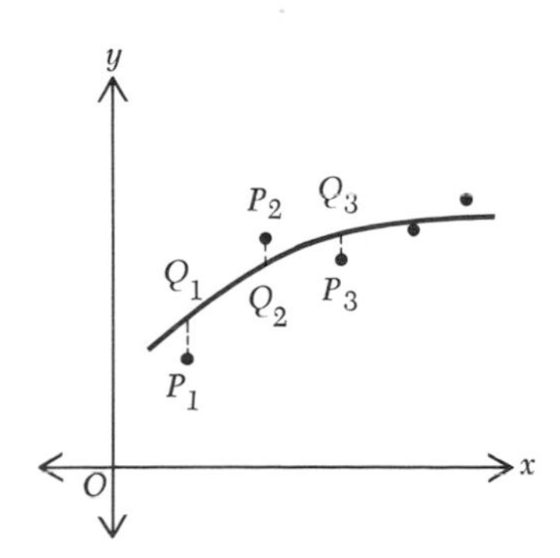

Fig. 12-3 Residual of point P_1 is Q_1P_1, that of point P_2 is Q_2P_2, and that of point P_3 is Q_3P_3.

points with respect to the fitted curve is defined to be the directed vertical distance from the curve to the point. Thus the residuals of the points P_1, P_2, and P_3 are, respectively,

$$Q_1P_1 = y_1 - f(x_1)$$
$$Q_2P_2 = y_2 - f(x_2)$$
$$Q_3P_3 = y_3 - f(x_3)$$

The residual of a point is positive if the point is above the curve and negative if below.

The expression "a good fit" has been used intuitively in connection with the fitting of a curve to a given set of points. Formally, a good-fitting curve is defined as a curve such that the absolute values of the residuals are small.

It can be shown that when a straight line is fitted to a given set of points by the method of average points, the result, whatever the choice of subsets, is a line for which the algebraic sum of the residuals is zero. This does not mean, however, that we necessarily have a good fit; the residuals could all be great in absolute value, with the positive residuals balancing the negative ones. Thus the line which best fits two points is the one passing through them. The perpendicular bisector of the segment joining them is also a line for which the algebraic sum of the residuals is zero, and this line would, of course, be a poor fit. It is evident, then, that the algebraic sum of the residuals could not be used as a criterion of the goodness of fit.

The squares of the residuals, however, are always nonnegative. The sum of the squares of the residuals could not be zero (unless the curve actually passed through all the points), and there would obviously be a good fit if this sum were small. So in fitting a curve of a specified type to a given set of points the attempt is usually made to determine the constants such that the sum of the squares of the residuals is as small as possible.

Definition

The best-fitting curve of a given type is the one in which the constants are determined such that the sum of the squares of the residuals is a minimum.

The justification for calling this the best-fitting curve of the type lies in the theory of probability and is not discussed here.

The best-fitting curve cannot, in general, be found by the method of average points but requires the method of *least squares* given at the end of this chapter. The calculations involved in this latter method are sometimes rather tedious, and for this reason the method of average points is often used where a good fit, but not necessarily the best fit, is required.

PROBLEMS

In Probs. 1 to 3 fit a straight line to the given data by the method of average points, using the partition indicated by the double bar.

1.

x	2	4	6	8 ‖	10	12	14	16
y	6.4	7.1	8.0	9.1 ‖	9.8	10.8	11.6	12.2

2.

x	1.4	2.6	3.0	3.8	4.4 ‖	4.8	5.8	7	8.2
y	9.3	8.6	7.7	8.0	7.6 ‖	7.6	7.1	6.6	5.9

3.

x	−2	−1	0	1	2 ‖	3	4	5	6
y	−3.2	−2.5	−1.8	−1.3	−0.6 ‖	0.1	0.4	1.0	1.5

4. For the data in Prob. 1 fit a straight line by the method of average points, using the first, third, fifth, and seventh columns of the table for one set and the second, fourth, sixth, and eighth columns for the other.
5. (*a*) Find the sum of the squares of the residuals for the straight line in Prob. 1. (*b*) Find the sum of the squares of the residuals in Prob. 4. (*c*) Is the line in Prob. 1 or Prob. 4 a better fit for the data in Prob. 1?
6. For the data in Prob. 2 fit a straight line by the method of average points, using the first four columns of the table for the first set.
7. Is the line in Prob. 2 or Prob. 6 a better fit for the data in Prob. 2?
8. The table below gives the weight W (grams) of a salt that dissolved in 100 g of water at various temperatures T (degrees) during an experiment. Find the equation of a straight line to fit these data, using the partition shown:

T	10	20	25	30 ‖	40	50	55	60
W	59	64	66	68.5 ‖	74	80	81	84

9. The resistances R (ohms) of a certain coil of copper wire at several temperatures T (degrees centigrade) were found by measurement to be as follows:

T	10.2°	19.8°	28.6°	39.8°	49.5° ‖	58.4°	71.2°	80.5°	88.8°
R	13.12	13.54	14.08	14.72	15.06 ‖	15.58	16.24	16.66	17.24

(*a*) Find the equation of a line to fit these data, using the partition shown. (*b*) Use the result to estimate the resistance at 45°.

10. The length L (inches) of a certain coiled steel spring under the action of a tensile pull was found to increase with the pull P (pounds) as shown below:

P	0	5	10	15	20	25	30	35	40	45
L	10.25	11.45	12.80	14.35	15.75	17.30	18.55	19.55	21.20	23.00

(*a*) Find the equation of a straight line to fit these data. (*b*) Use the result to estimate the length under a pull of 22 lb.

11. The cost C (dollars) of printing a pamphlet varies with the number N that are printed, as shown in the following table:

N	1,000	2,000	3,000	4,000	5,000	6,000
C	$167.50	$201.50	$234.00	$265.50	$295.00	$323.00

(*a*) Find the equation of a straight line to fit these data. (*b*) Use the result to estimate the cost of printing 10,000 copies.

12. The planetary data on gravity (gravity for earth is 1) and escape velocity in miles per second are given in the following table for eight planets:

	Gravity	Escape Velocity
Mercury	0.37	2.6
Venus	0.89	6.4
Earth	1.00	7.0
Mars	0.38	3.1
Jupiter	2.65	37.9
Saturn	1.14	23.0
Uranus	0.96	13.7
Neptune	1.53	15.5

(*a*) Find the equation of a line to fit these data. (*b*) Does the nature of the data make a straight line a good choice for a type of curve to fit to it?

12-5 PARABOLIC TYPE

Average points can be used similarly to fit a parabola

$$y = ax^2 + bx + c$$

to a given set of points. In this case there are three constants to determine. The given set of points is accordingly partitioned into three subsets, the average point of each subset is located, and the parabola passing through these three points is determined.

Example
Fit a parabola of the form $y = ax^2 + bx + c$ to the following data:

x	−2	−1	0	1	2	3	4	5	6	7	8
y	−3.5	0.4	2.5	4.2	5.8	6.6	7.8	8.0	8.6	7.6	6.2

Solution Take the first three points as set 1, the next four as set 2, and the last four as set 3. The coordinates of the corresponding average points are

Set 1: $(-1, -0.2)$
Set 2: $(2.5, 6.1)$
Set 3: $(6.5, 7.6)$

In order to pass the parabola through these three points, substitute their coordinates for x and y in the equation $y = ax^2 + bx + c$ and obtain the following equations in a, b, and c:

$$a - b + c = -0.2$$
$$6.25a + 2.5b + c = 6.1$$
$$42.25a + 6.5b + c = 7.6$$

Then $a = -0.190$, $b = 2.085$, and $c = 2.075$. The desired equation is

$$y = -0.190x^2 + 2.085x + 2.075 \tag{12-7}$$

The curve of Eq. (12-7) and the given set of points are shown in Fig. 12-4. It is evident that a different partition of the experimental points may result in a parabola that differs appreciably from this one. A study of the figure suggests that a better fit would probably be obtained by taking the first four points as set 1, the next four as set 2, and the last three as set 3.

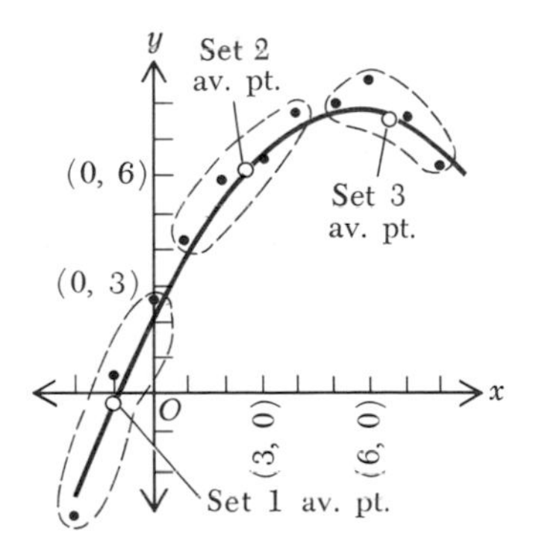

Fig. 12-4 The method of average points for fitting a parabola to a set of points.

PROBLEMS

1. Fit a parabola of the form $y = ax^2 + bx + c$ to the following data, using the partitions shown:

x	0	1	2	3	4	5	6	7
y	−5.4	−5.2	−4.4	−2.8	−1.2	1.4	3.6	6.6

2. Fit a parabola of the form $y = ax^2 + bx + c$ to the following data, using the partitions shown:

x	−1	0	1	2	3	4	5	6
y	3.2	4.2	4.1	3.8	2.8	1.4	−1.1	−3.8

3. (*a*) For Prob. 1 fit another parabola to the data, partitioning the table after the third and sixth columns. (*b*) Is the parabola in Prob. 1 or Prob. 3 a better fit?
4. (*a*) For Prob. 2 fit another parabola to the data, partitioning the table after the second and fourth columns. (*b*) Is the parabola in Prob. 2 or 4 a better fit?
5. An experimental rocket was designed to be dropped from a plane, with the rocket engine to fire slightly later. Tracking stations fed a computer the following data about the horizontal and vertical distance (in feet) of the rocket from its initial position during the first phase of its flight. Fit a parabola of the form $y = ax^2 + bx + c$ to the data, using the partition shown:

x	0	3	6	9	12	15	18	21
y	0	−1.2	−2.9	−6.0	−11.1	−18.6	−27.0	−37.1

6. If all forces acting on a projectile except gravity are neglected, the distance y (feet) of the projectile above the earth at any time t (seconds) is given by the formula

$$y = at + bt^2$$

where a and b depend on such factors as initial velocity. Fit the equation to the following data; the values for y indicate thousands of feet:

t	10	20	30	40	50	60	70	80
y	14	26	34	38	40	38	34	26

7. The following table shows the rate R (pounds per hour) at which a piece of

ice was melting t (hours) after being placed in a room. Fit the equation $R = at^2 + bt + c$ to these data:

t	0	1	2	3	4	5	6	7
R	0	9	17	21	23	25	24	20

8. The table shows, correct to the nearest million, the labor force in the United States in 9 consecutive recent years:

Year	1	2	3	4	5	6	7	8	9
Labor force (millions)	68	69	69	71	72	72	73	74	75

(*a*) Fit a parabola to these data. (*b*) Estimate the labor force 3 years after the last year in the table.

12-6 THE POWER TYPE

Consider now the problem of fitting an equation of the type

$$y = ax^n \tag{12-8}$$

to a given table of number pairs (x, y). Take the common logarithms of both members of Eq. (12-8) to obtain

$$\log y = \log a + n \log x \tag{12-9}$$

This is obviously a linear relation between $\log x$ and $\log y$. Let

$$\log x = u$$

and

$$\log y = v$$

so that

$$v = nu + \log a$$

where $\log a$ is a constant. It follows that if the relation between x and y is of the form $y = ax^n$, the relation between $\log x$ and $\log y$ is linear. We may then proceed as follows. First make a new table of data in which the entries are the logarithms of the numbers in the original table; denote these new entries by u and v, where $u = \log x$ and $v = \log y$. Then plot the points (u, v) on rectangular-coordinate paper. If they tend to lie along a straight line, the equation $y = ax^n$ is applicable. Finally, fit a straight line $v = nu + L$, where $L = \log a$, to this set of data, using the method of average points (or the method of least squares, to be discussed later). The value of n in the required equation $y = ax^n$ is the coefficient of u in the linear equation, and the value of a can be found from the fact that $L = \log a$.

Example

Show that an equation of the form $y = ax^n$ is suitable for the data given below, and find values for a and n:

x	1	2	3	4	5	6	7	8
y	0.36	0.90	1.92	3.44	4.78	6.74	9.40	11.8

Solution The following table shows the common logarithms of the given numbers:

$u = \log x$	0.0000	0.3010	0.4771	0.6021	0.6990	0.7782	0.8451	0.9031
$v = \log y$	−0.4437	−0.0458	0.2833	0.5238	0.6794	0.8287	0.9731	1.0719

These points, when plotted on rectangular-coordinate paper, tend to lie along a line as shown in Fig. 12-5. This means that an equation of the form $y = ax^n$ is applicable.

To fit a line to these points choose the first four as set A and the last

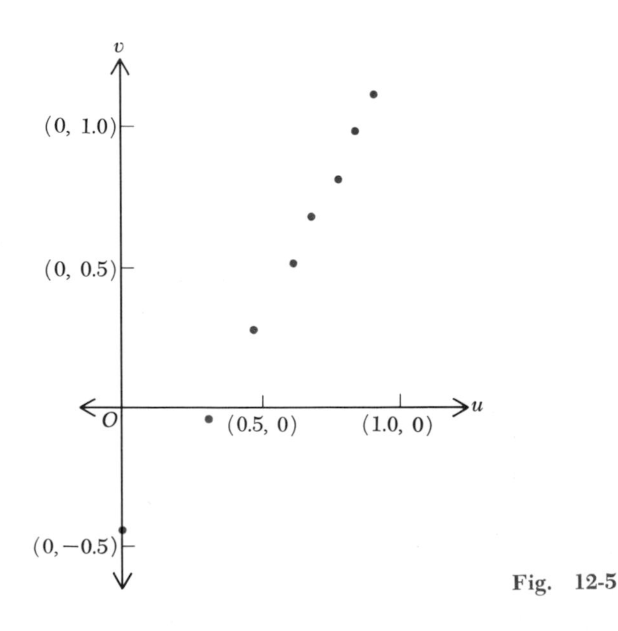

Fig. 12-5

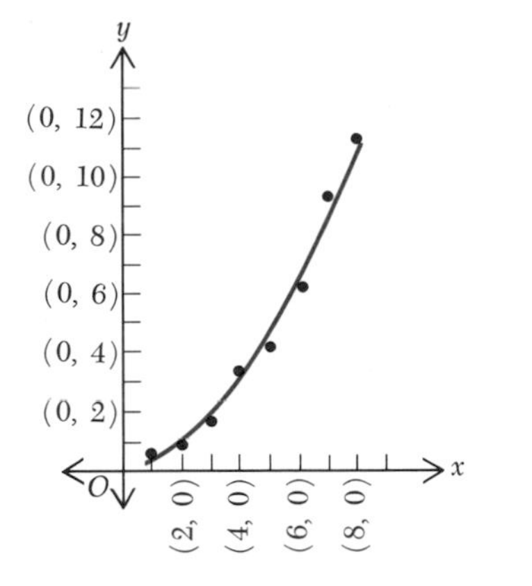

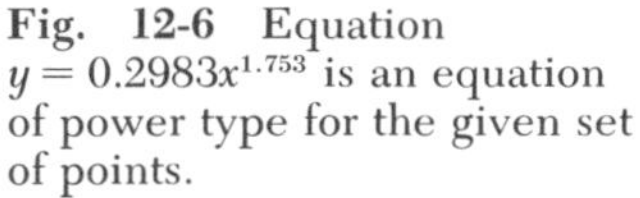
Fig. 12-6 Equation $y = 0.2983x^{1.753}$ is an equation of power type for the given set of points.

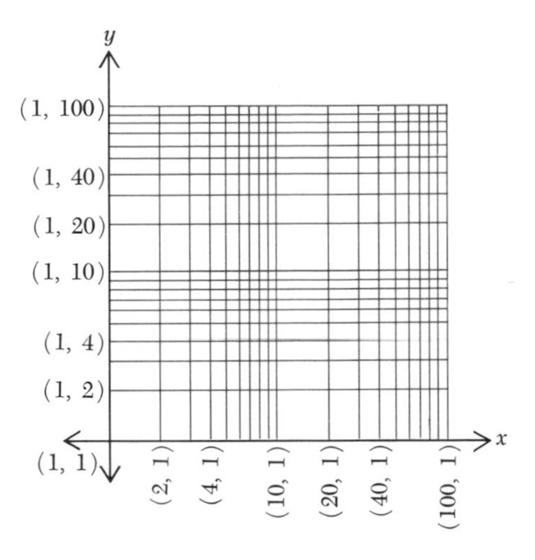

Fig. 12-7 Logarithmic-coordinate paper.

four as set B. The coordinates of the corresponding average points are (0.3450, 0.0794) and (0.8064, 0.8883); the resulting line has the equation

$$v = 1.753u - 0.5254$$

This is of the form $v = nu + L$, where $n = 1.753$ and

$$L = \log a = -0.5254 = 9.4746 - 10$$

From tables, $a = 0.2983$, correct to four significant figures. The required empirical relation between x and y is then

$$y = 0.2983x^{1.753} \tag{12-10}$$

The graph of Eq. (12-10) and the given points are shown in Fig. 12-6.

The test for applicability of the formula $y = ax^n$ can be made more quickly by plotting the given data on logarithmic-coordinate paper rather than determining logarithms and using rectangular-coordinate paper.

Logarithmic-coordinate paper, as shown in Fig. 12-7, has its horizontal and vertical rulings not at distances of 1, 2, 3, . . . from the origin, but at distances of log 1, log 2, log 3, ... The sample shows two "cycles," the first from 1 to 10, and the second from 10 to 100. The next cycle to the right along the x axis would be from 100 to 1,000; the next one to the left of those shown would be from 0.1 to 1.

Plotting the given data directly on this paper is, of course, equivalent to plotting the logarithms on rectangular-coordinate paper. If the points tend to lie on a line, an equation of the form $y = ax^n$ is applicable.

12-7 THE EXPONENTIAL TYPE

The problem of fitting an exponential equation of the form

$$y = a(10)^{kx} \tag{12-11}$$

to a given table of values of x and y is quite similar to that just discussed. Again, take the common logarithms of both sides of (12-11), obtaining

$$\log y = \log a + kx \tag{12-12}$$

This is a linear relation between x and $\log y$. Let $\log y = v$, so that

$$v = kx + \log a$$

where again $\log a$ is a constant. It follows that if the relation between x and y is of the form $y = a(10)^{kx}$, that between x and $\log y$ is linear.

Then make a new table in which the entries are x and $\log y$, denoting $\log y$ by v, and plot the points (x, v) on rectangular-coordinate paper. If they tend to lie in a straight line, the equation $y = a(10)^{kx}$ is applicable. Finally, fit a straight line $v = kx + L$ (where $L = \log a$) to this set of data and write the required equation $y = a(10)^{kx}$.

One method of determining values for a and k is described in the following example.

Example

Show that an equation of the form $y = a(10)^{kx}$ is suitable for the data given below and find correct values for a and k:

x	1	2	3	4	5	6	7	8
y	0.72	1.08	1.68	3.24	5.28	8.64	13.8	22.6

Solution The table below shows the logarithms of the given values of y:

x	1	2	3	4	5	6	7	8
$v = \log y$	−0.1427	0.0334	0.2253	0.5106	0.7226	0.9365	1.1399	1.3541

Verify that these points, when plotted, tend to lie along a line. We may fit a line to them, using the method of average points, to obtain

$$v = 0.2204x - 0.3944 \tag{12-13}$$

This equation is of the form $v = kx + L$, where $k = 0.2204$ and

$$L = \log a = -0.3944 = 9.6056 - 10$$

From tables, $a = 0.4033$, correct to four significant figures. The required empirical relation between x and y is then

$$y = 0.4033(10)^{0.2204x} \tag{12-14}$$

The graph of Eq. (12-14) and the given points are shown in Fig. 12-8.

It is often convenient to use the exponential equation in the form $y = ae^{kx}$, where e is the base of the natural logarithms. With 10 replaced by its approximate value $e^{2.303}$, Eq. (12-14) becomes

$$y = 0.4033(e^{2.303})^{0.2204x} \tag{12-15}$$

Then, since $(e^m)^n = e^{mn}$,

$$y = 0.4033e^{0.5076x} \tag{12-16}$$

Of course, the result can also be easily expressed in the general exponential form $y = ab^x$. Thus, since

$$10^{0.2204x} = (10^{0.2204})^x = 1.661^x$$

Eq. (12-16) can be written in the form

$$y = 0.4033(1.661)^x \tag{12-17}$$

In practice it is usually desirable to use either 10 or e as the base.

The test for applicability of the formula $y = a(10)^{kx}$ can be made more quickly by plotting the given data directly on semilogarithmic-coordinate paper, as is done in Fig. 12-9, instead of determining the logarithms of the given values of y and using ordinary coordinate paper. As shown

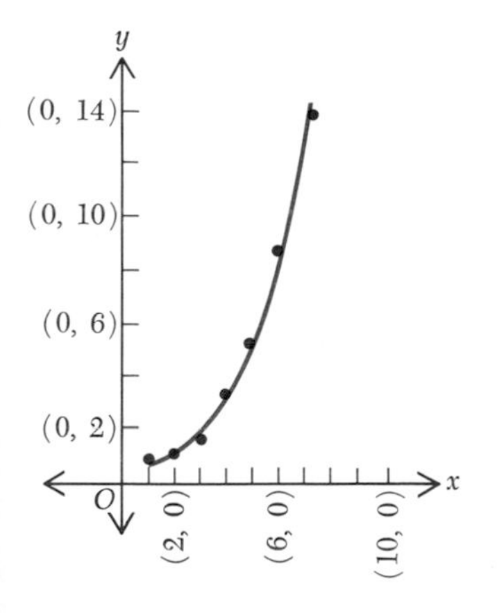

Fig. 12-8 Equation $y = 0.4033(10)^{0.2204x}$ is an equation of exponential type for the given set of points.

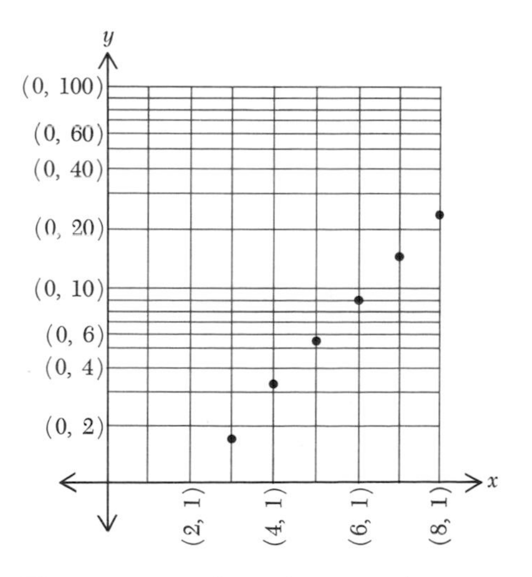

Fig. 12-9 Semilogarithmic-coordinate paper.

in the figure, this paper has uniformly spaced vertical rulings and logarithmically spaced horizontal rulings. Plotting the given points (x, y) on such paper is equivalent to plotting the points $(x, \log y)$ on ordinary coordinate paper. If the points tend to lie on a line, the equation $y = a(10)^{kx}$ (or ae^{kx}) is applicable.

PROBLEMS

In Probs. 1 to 4 show that an equation of the form $y = ax^n$ fits the data and find values for a and n.

1.

x	1	2	3	4	5	6
y	1	3	8	14	22	31

2.

x	1	2	3	4	5	6	7	8
y	0.54	1.42	2.68	3.92	5.66	7.34	9.28	11.4

3.

x	2	4	6	8	10	12	14	16
y	3.6	5.5	7.1	8.3	9.4	10.5	11.5	12.4

4.

x	1	2	3	4	5	6
y	26.4	9.3	5.2	3.6	2.5	1.9

In Probs. 5 to 8 show that an equation of the form $y = a(10)^{kx}$ fits the data and find values for a and k. Express the final result also in the form $y = ae^{k'x}$.

5.

x	1	2	4	6	8	10
y	3.3	5.4	14.8	40.2	109.2	296.8

6.

x	1	2	3	4	5	6
y	1.3	2.8	6.2	13.8	33.6	64.2

7.

x	0	1	2	3	4	5	6
y	42	26	16	10	7	4	3

8.

x	0	2	5	8	12	16
y	80	59	35	21	10	5

9. The time t (seconds) required for an object to fall a distance s (feet) was observed, and the data were recorded as shown:

s	5	10	15	20	25	30	35	40
t	0.58	0.84	1.03	1.19	1.34	1.47	1.60	1.71

Show that an equation of the form $t = as^n$ fits the data and find a and n.

10. In a certain chemical reaction it was found that of an initial 60 g of a certain substance, the amount M (grams) remaining after t (minutes) was given:

t	0	3	6	9	12	15	18	21
M	60	38.2	24.4	15.6	9.9	6.4	4.1	2.6

Show that an equation of the form $M = a(10)^{kt}$ is applicable and fit such an equation to the data. Express the result also in the form $M = ae^{k't}$.

11. The air resistance R (pounds) acting against a moving automobile was found to vary with the speed V (miles per hour) as shown:

V	10	20	30	40	50	60
R	8	29	72	128	197	274

Show that an equation of the form $R = aV^n$ is applicable and find values for a and n.

12. The number N of bacteria per unit volume in a culture at the end of t (hours) was found to be as follows:

t	1	2	3	4	5	6	7	8
N	52	64	76	95	114	130	158	187

Show that the equation $N = a(10)^{kt}$ is applicable and find values for a and k. Use the result to estimate the value of N when $t = 5.4$.

13. The distances d (in units such that the earth is 1 unit) of the planets from the sun and their periods of revolution t (years) are given in the table below:

	Mercury	Venus	Earth	Mars	Jupiter	Saturn	Uranus	Neptune
t	0.240	0.625	1.00	1.88	11.9	29.5	84.0	165.0
d	0.387	0.723	1.00	1.52	5.20	9.54	19.2	30.1

Fit a curve of the form $d = at^n$ to these data.

14. The pressure p exerted by an expanding volume v of gas is:

v	8	10	20	30	40	50	60	70	80
p	125	92.8	36.8	21.5	14.6	10.9	8.5	6.9	5.8

Fit a curve of the form $p = av^n$ to these data.

15. A cold plate was taken into a warm tunnel. The difference d between the temperature of the air and the plate is shown for various times t:

t	0	0.5	1.0	1.5	2.0
d	29.4	22.1	16.6	12.5	9.4

Fit a curve of the form $d = ae^{kt}$ to these data.

16. The pressure of a gas expanding adiabatically was found to vary with its volume as follows:

v	2.0	2.6	4.2	4.6	5.2	6.0
p	138	96	49	43	36	30

Show that a relation of the form $p = av^n$ is applicable, and determine a and n.

12-8 LINEAR TYPE: METHOD OF LEAST SQUARES

It has been pointed out in previous sections that when the method of average points is used to fit a straight line to a given set of data, the result is a line for which the algebraic sum of the residuals is zero. There are indefinitely many such lines; some of them are good fits and some are not. As discussed in Sec. 12-4, the sum of the squares of the residuals may be calculated and used as an indication of the goodness of fit.

The method of least squares gives the equation of the line for which the sum of the squares of the residuals is a minimum. There is only one such line for a given set of points, so the result does not depend upon any choice of subsets. In this section we shall discuss how the equation is found; the proof is given in Sec. 12-10.

In Sec. 12-3 the following data were given:

x	1	2	3	4	5	6	7	8	9	10
y	1.48	1.76	2.78	3.32	3.86	4.15	4.75	5.66	6.18	6.86

On the assumption that the equation $y = mx + b$ fits the data, we start by listing two sets of equations. The first set, shown on the left below, is obtained by substituting the given coordinates for x and y in $y = mx + b$; the equations of the second set, shown on the right, are formed by multiplying both members of each equation of the first set by the coefficient of m in it:

$$
\begin{array}{rl@{\qquad\qquad}rl}
1.48 = & m + b & 1.48 = & m + b \\
1.76 = & 2m + b & 3.52 = & 4m + 2b \\
2.78 = & 3m + b & 8.34 = & 9m + 3b \\
3.32 = & 4m + b & 13.28 = & 16m + 4b \\
3.86 = & 5m + b & 19.30 = & 25m + 5b \\
4.15 = & 6m + b & 24.90 = & 36m + 6b \\
4.75 = & 7m + b & 33.25 = & 49m + 7b \\
5.66 = & 8m + b & 45.28 = & 64m + 8b \\
6.18 = & 9m + b & 55.62 = & 81m + 9b \\
6.86 = & 10m + b & 68.60 = & 100m + 10b \\
\hline
40.80 = & 55m + 10b & 273.57 = & 385m + 55b
\end{array}
$$

Add in each column of equations as indicated.

$$
\begin{aligned}
40.80 &= 55m + 10b \\
273.57 &= 385m + 55b
\end{aligned}
$$

Hence $m = 0.596$ and $b = 0.802$.

The best-fitting line (according to the least-squares criterion) has the equation

$$y = 0.596x + 0.802$$

If both sides of the equation $40.80 = 55m + 10b$ (which was obtained by adding the columns on the left above) are divided by 10, the result is

$$4.08 = 5.5m + b$$

The average point of the given set of data is (5.5, 4.08), and the equation $4.08 = 5.5m + b$ states the condition that the line $y = mx + b$ pass through the point (5.5, 4.08). It follows that the line $y = 0.596x + 0.802$, which has been called the best-fitting line, is one of the many lines for which the algebraic sum of the residuals is zero. It is also the line for which the sum of the squares of the residuals is as small as possible. Figure 12-10 shows both the line $y = 0.596x + 0.802$ and the line shown in Fig. 12-2.

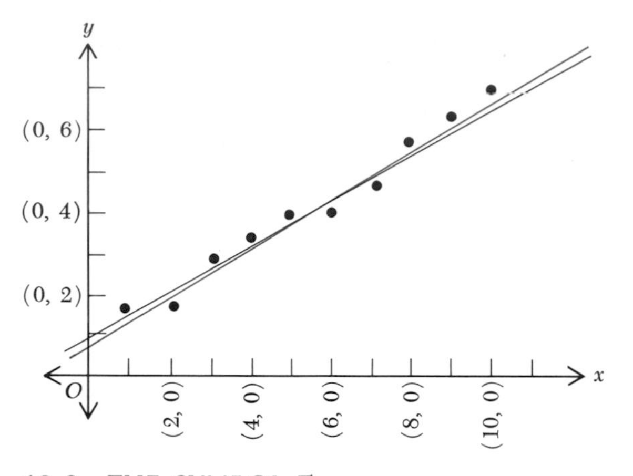

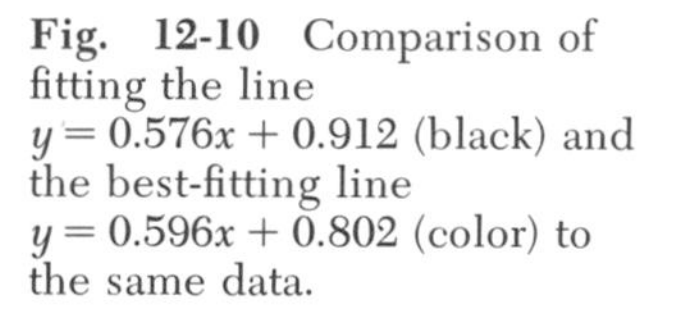
Fig. 12-10 Comparison of fitting the line $y = 0.576x + 0.912$ (black) and the best-fitting line $y = 0.596x + 0.802$ (color) to the same data.

12-9 THE SYMBOL Σ

The following notation will simplify the equations of the next section. Given a set of data $(x_1, y_1), (x_2, y_2), \ldots, (x_n, y_n)$, denote the sum of all the x coordinates by Σx and the sum of all the y coordinates by Σy; that is,

$$\Sigma x = x_1 + x_2 + \cdots + x_n$$
$$\Sigma y = y_1 + y_2 + \cdots + y_n$$

Then the coordinates of the average point of the set are

$$\left(\frac{\Sigma x}{n}, \frac{\Sigma y}{n}\right)$$

Similarly, the sum of the squares of the abscissas is Σx^2; the sum of the products of the abscissas and the corresponding ordinates is Σxy.

12-10 DERIVATION OF THE EQUATIONS FOR THE METHOD OF LEAST SQUARES

If the line $y = mx + b$ is fitted to the points $(x_1, y_1), (x_2, y_2), \ldots, (x_n, y_n)$, the residuals are

$$r_1 = y_1 - (mx_1 + b)$$
$$r_2 = y_2 - (mx_2 + b)$$
$$\cdots\cdots\cdots\cdots$$
$$r_n = y_n - (mx_n + b)$$

The squares of the residuals are

$$r_1^2 = y_1^2 - 2my_1x_1 - 2y_1b + m^2x_1^2 + 2mx_1b + b^2$$
$$r_2^2 = y_2^2 - 2my_2x_2 - 2y_2b + m^2x_2^2 + 2mx_2b + b^2$$
$$\cdots\cdots\cdots\cdots\cdots\cdots\cdots\cdots\cdots\cdots$$
$$r_n^2 = y_n^2 - 2my_nx_n - 2y_nb + m^2x_n^2 + 2mx_nb + b^2$$

The result of adding these and arranging the right-hand member in the form of a quadratic expression in b is

$$\Sigma r^2 = nb^2 + 2(m\Sigma x - \Sigma y)b + (m^2\Sigma x^2 - 2m\Sigma xy + \Sigma y^2)$$

If Σr^2 is plotted against b, the graph is a parabola lying entirely above the b axis (since Σr^2 is certainly positive); the parabola would open upward, since the coefficient of b^2 is positive, and its minimum point* would be at

$$b = -\frac{m\Sigma x - \Sigma y}{n}$$

This implies that b and m must satisfy the relation

$$\Sigma y = m\Sigma x + nb \tag{12-18}$$

in order that the sum of the squares of the residuals may be a minimum. Observe that Eq. (12-18) is precisely the equation obtained in Sec. 12-8 by adding the members of the equations in the left-hand column. Observe also that dividing both sides of the above equation by n results in

$$\frac{\Sigma y}{n} = m\frac{\Sigma x}{n} + b$$

which is the condition that the average point $(\Sigma x/n, \Sigma y/n)$ of the set be on the line.

If the expression for the sum of the squares of the residuals is arranged in the form of a quadratic in m, the result is

$$\Sigma r^2 = (\Sigma x^2)m^2 + 2(b\Sigma x - \Sigma xy)m + (\Sigma y^2 - 2b\Sigma y + nb^2)$$

For Σr^2 to be a minimum m must equal $-(b\Sigma x - \Sigma xy)/\Sigma x^2$, which implies that m and b must also satisfy the relation

$$\Sigma xy = m\Sigma x^2 + b\Sigma x \tag{12-19}$$

This is the equation obtained by adding the right-hand column in Sec. 12-8. The two equations (12-18) and (12-19) determine uniquely the values of m and b in the equation $y = mx + b$ in order that Σr^2 shall be a minimum by the least-squares criterion.

PROBLEMS

In Probs. 1 to 5 find the best-fitting straight line, using the method of least squares. Plot the points and draw the line.

* The graph of the equation $y = Ax^2 + Bx + C$ is a parabola with its axis parallel to the y axis. If $A > 0$, it opens upward and has a minimum point (the vertex) at $x = -B/2A$.

1.

x	1	2	3	4	5	6
y	−3.9	−2.5	−1.4	0.2	1.5	2.5

2.

x	1	2	3	4	5	6	7
y	15.8	15.2	14.5	13.8	12.9	12.3	11.5

3.

x	0	3	6	9	12	15
y	2.4	4.0	5.8	7.8	9.3	11.2

4.

x	1	2	3	4	5	6
y	0.13	0.28	0.43	0.55	0.69	0.85

5. The pressure P (centimeters of mercury) of a fixed quantity of gas was found to vary as follows with T (centigrade temperature) when its volume was held constant:

T	20	30	40	50	60	70	80
P	26.8	29.0	30.7	32.3	33.8	35.6	37.2

Express P in terms of T, using the method of least squares.

6. The rate of rotation r of a wheel decreased as shown in the table for various times t after the power was cut off:

t	1	10	20	30	40
r	3,000	1,348	606	273	122

Fit a curve of the form $r = ae^{kt}$ to these data, using the method of least squares.

In Probs. 7 to 10 the given data can be fitted by an equation of the form $y = ax^n$ or $y = a(10)^{kx}$. Determine which equation is applicable and use the method of least squares to fit the data.

7.

x	1	2	3	4	5	6
y	2.5	7.7	15.2	24.8	35.2	48.0

8.

x	2	5	7	8	10	12
y	34	42	46	47	50	52

9.

x	1	2	3	4	5	6
y	3.96	5.92	9.18	13.7	20.8	31.6

10.

x	5	10	15	20	25	30
y	32.1	21.3	14.6	9.2	6.4	4.5

SUMMARY

An equation of a curve that passes through or near a set of points in such a way as to indicate the general trend is called an empirical equation. The process of finding a suitable empirical equation for a set of points is called curve fitting.

Two decisions are basic in curve fitting applications: the type of curve to employ and how to evaluate the constants in the equation. A table is provided for various common methods of curve fitting studied.

Type of Curve	Method of Evaluating Constants	Procedure
$y = ax + b$	Average points	1. Partition given set into two sets. 2. Find coordinates of average point for each set. 3. Find equation of line through two average points.
$y = ax^2 + bx + c$	Average points	1. Partition given set into three sets. 2. Find coordinates of average point for each set. 3. Find equation of parabola through three average points.
$y = ax^n$	Average points	1. Take the logarithm of both members of the equation to obtain $\log y = \log a + n \log x$. 2. Make a new table in which the entries are the logarithms of the given coordinates. 3. Fit a straight line to this new set of data.

Continued on next page

$y = a(10)^{kx}$	Average points	1. Take the logarithm of both members of the equation to obtain $\log y = \log a + kx$. 2. Make a new table in which the entries are x and $\log y$. 3. Fit a straight line to these new data.
$y = mx + b$	Least squares	1. Write a set of equations of the form $y = mx + b$, obtained by substituting the coordinates given for x and y. 2. Write a second set of equations derived from the first by multiplying both sides of each equation by the coefficient m. 3. From each set of equations, obtain one equation in m and b by adding the left- and right-hand members. 4. Solve the resulting two equations for m and b.

A good-fitting curve can be defined as a curve for which the absolute values of the residuals are small. The residual of a point is the difference between its ordinate and that of the corresponding point on the curve. The best-fitting curve of a given type is the one in which the constants are determined so that the sum of the squares of the residuals is a minimum. In general, the best-fitting curve cannot be obtained by the method of average points, but requires the method of least squares.

The decision as to which type of curve to use can sometimes be made easier when special coordinate paper is used. If the points tend to lie on a line when plotted on logarithmic-coordinate paper, a power-type curve ($y = ax^n$) may be used; this paper has logarithmically spaced horizontal and vertical rulings. If the points tend to lie on a line when plotted on semilogarithmic-coordinate paper (with uniformly spaced vertical rulings and logarithmically spaced horizontal rulings) a curve of the exponential type [$y = a(10)^{kx}$] is appropriate.

REVIEW PROBLEMS

1. Fit a straight line to the data given below by the method of average points; use two sets of points with the first three pairs (x, y) in one set:

x	1	2	3	4	5	6
y	3.6	4.7	5.5	7.5	8.7	9.9

2. Fit a parabola of the form $y = ax^2 + bx + c$ to the following data; use three pairs (x, y) in each of three sets to find the average points:

x	−1	0	1	2	3	4	5	6	7
y	3.6	2.4	1.7	1.3	1.2	0.9	1.1	1.2	1.8

3. Show that an equation of the form $y = ax^n$ is applicable and find values for a and n by the method of averages:

x	2	4	8	16
y	5	6	8	10

4. Show that an equation of the form $y = a(10)^{kx}$ is applicable and find values for a and k by the method of averages; use two sets with the first two pairs (x, y) in one set and express the final result in the form $y = as^{k'x}$:

x	−3	−1	1	3	5
y	0.8	1.5	2.7	4.9	9.0

5. A business showed profits at the end of each year for 4 years as follows:

Year	1	2	3	4
Profit	\$100,000	\$120,000	\$130,000	\$150,000

(*a*) Determine the best-fitting straight line by the method of least squares. Plot the points and draw the line. (*b*) Estimate from the graph the profit at the end of the fifth year. (*c*) Calculate from the equation the profit at the end of the fifth year.

6. A body falls y ft in x sec. (*a*) For the following data find the best-fitting straight line by the least-squares method. (*b*) Show that an equation in the form $y = ax^n$ is applicable and fit such an equation to the data.

x	0.51	0.8	1.02	1.26	1.51
y	4	10	16	25	36

FOR EXTENDED STUDY

1. Given a set of data, show that every line for which the algebraic sum of the residuals is zero passes through the average point of the given set, and conversely. *Hint:* Let the given points be (x_1, y_1), (x_2, y_2), . . . , (x_n, y_n). Assuming the equation $y = ax + b$ and equating to zero the sum of the residuals, the result is

   $$(a) \quad [y_1 - (ax_1 + b)] + [y_2 - (ax_2 + b)] + \cdots + [y_n - (ax_n + b)] = 0$$

   Show that this may be simplified to

   $$(b) \qquad \frac{y_1 + y_2 + \cdots + y_n}{n} = a\,\frac{x_1 + x_2 + \cdots + x_n}{n} + b$$

   This means that the point

   $$\left(\frac{x_1 + x_2 + \cdots + x_n}{n}, \frac{y_1 + y_2 + \cdots + y_n}{n}\right)$$

   must be on the line if (a) is to be satisfied. Conversely, show that (a) follows from (b).
2. Perform the following experiments to determine some data as sets of ordered pairs. Choose an appropriate type of curve; then determine the constants by the method of average points. Draw the graph. (a) Read the same paragraph over eight times and record the number of seconds it takes you each time. (b) Hold a thermometer in ice water and record the changes in temperature until equilibrium, choosing some convenient unit of time. Use eight sets of coordinates. (c) Keep a record of the number of cars passing a certain point on a street or road during various 5-min periods. If possible, use the same 5-min period on several consecutive days.
3. Devise a method for fitting a polynomial curve of the form $y = ax^3 + bx^2 + cx + d$ to a set of data by the method of average points.
4. Fit a polynomial curve of the type in Prob. 3 to the following data:

x	1	2	3	4	5	6
y	−3	14	10	−1	−15	−40

5. Devise a general method for fitting a polynomial curve of the form $y = ax^n + bx^{n-1} + \cdots + k$ to a set of data by the method of average points.

ANALYTIC GEOMETRY OF THREE DIMENSIONS: DEFINITIONS AND FORMULAS 13

13-1 RECTANGULAR COORDINATES IN SPACE

The rectangular coordinate system is extended to three dimensions by drawing three mutually perpendicular lines through a point O, which becomes the *origin*. These lines are called the x, y, and z axes; positive directions on them are chosen as indicated by the arrows in Fig. 13-1.

The plane determined by the x and y axes is called the xy plane, the plane determined by the x and z axes is called the xz plane, and that determined by the y and z axes is called the yz plane. These three *coordinate planes,* which intersect at the origin, may be visualized as the floor and two adjacent walls of a room. The floor corresponds to the xy plane, and the two walls correspond to the xz and yz planes. The walls intersect along the vertical z axis.

A point P in space is located by specifying its directed distances from the three coordinate planes. Its directed distance from the yz plane, measured along or parallel to the x axis, is its x *coordinate.* The y and z coordinates of the point are defined similarly. The coordinates of a point P are written as an *ordered triple* (x, y, z). (See Fig. 13-1.)

The coordinate planes partition space into eight *octants* identified by the signs of the coordinates. The octant in which all three coordinates are positive is called the *first octant.* The others are usually not

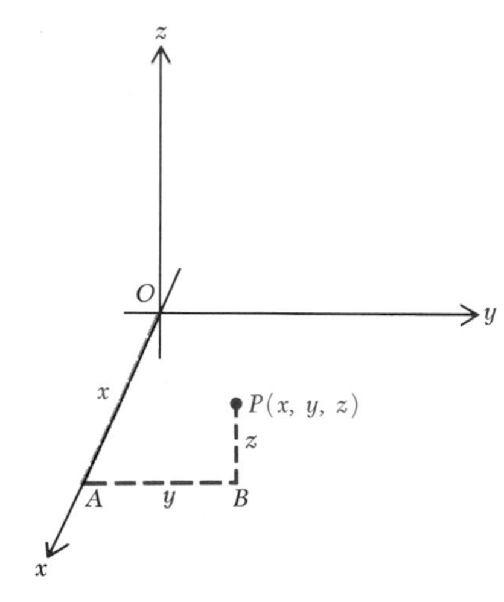

Fig. 13-1 Rectangular coordinate system in space: graph of $P(x, y, z)$.

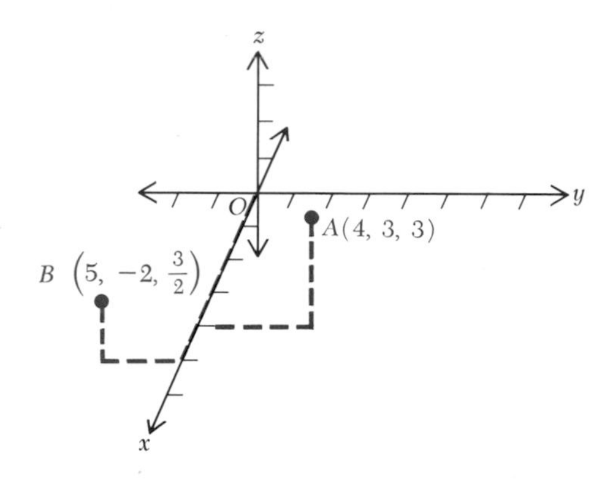

Fig. 13-2 Graph of $A(4, 3, 3)$ and $B(5, -2, \frac{3}{2})$.

numbered. In Fig. 13-2, $A(4, 3, 3)$ is in the first octant; $B(5, -2, \frac{3}{2})$ is in the octant in which the coordinates have, respectively, the signs +, −, and +.

13-2 CONSTRUCTION OF FIGURES

One of the tasks confronting the student of analytic geometry is that of drawing three-dimensional figures on two-dimensional paper. There are several ways of doing this; the method to be employed here is illustrated by Fig. 13-3. The y and z axes are drawn perpendicular to each other. The x axis is shown at an angle of about 120° with the y axis. Equal units are shown on all three axes. Sometimes the unit on the x axis is made about 0.7 as long. This *foreshortening* of the dimensions along the x axis is used to minimize the distortion in the figure. Here we have shown some lines with varying thickness to add dimensionality.

In Fig. 13-3 the point $P(6, 5, 4)$ has been plotted, and a rectangular box has been drawn in which $\overline{OP}$ is a diagonal. Observe that while $\overline{OP}$ is actually longer than $\overline{OA}$ in space, it appears shorter in the figure. In space the triangle OAP is a right triangle in which $\overline{OP}$ is the hypotenuse; the right angle is angle OAP.

You will have to learn to visualize the actual space relations by thinking of the figure in space. Any representation of a three-dimensional

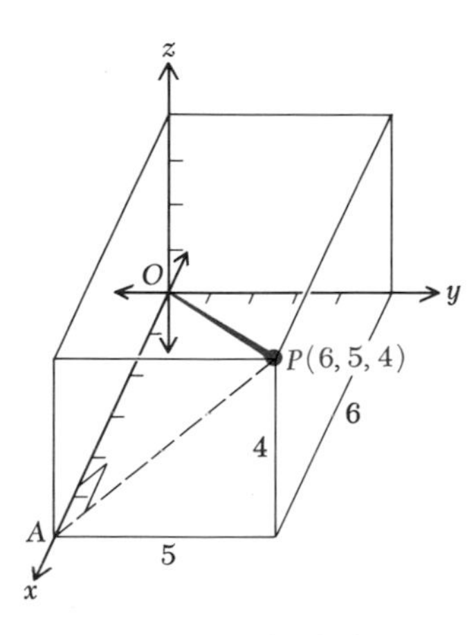

Fig. 13-3 Graph of $P(6, 5, 4)$.

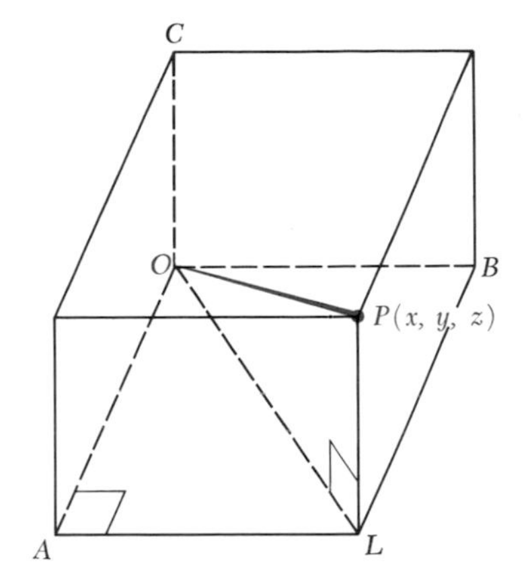

Fig. 13-4 Distance from origin to point P: $|OP| = \sqrt{x^2 + y^2 + z^2}$.

object on a flat sheet of paper is intended to serve only as an aid in visualizing the true relations in space.

13-3 THE DISTANCE FORMULA

Figure 13-4 represents a rectangular box whose edges are $\overline{OA}$, $\overline{OB}$, and $\overline{OC}$. To find the measure of a diagonal $\overline{OP}$ observe that in right triangle OAL

$$(OL)^2 = (OA)^2 + (AL)^2$$

Then, since $\overline{OP}$ is the hypotenuse of right triangle OLP,

$$\begin{aligned}(OP)^2 &= (OL)^2 + (LP)^2 \\ &= (OA)^2 + (AL)^2 + (LP)^2\end{aligned}$$

But $AL = OB$ and $LP = OC$. Hence

$$(OP)^2 = (OA)^2 + (OB)^2 + (OC)^2$$

The square of the measure of the diagonal of a rectangular box is equal to the sum of the squares of the measures of its three edges. If O is the origin and P has coordinates (x, y, z), then $OA = x$, $OB = y$, and $OC = z$. The undirected distance from O to P is then

$$|OP| = \sqrt{x^2 + y^2 + z^2} \tag{13-1}$$

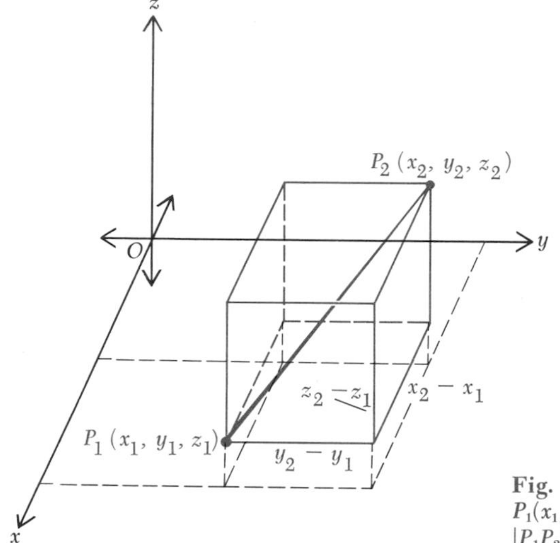

Fig. 13-5 Distance between $P_1(x_1, y_1, z_1)$ and $P_2(x_2, y_2, z_2)$: $|P_1P_2| = \sqrt{(x_2 - x_1)^2 + (y_2 - y_1)^2 + (z_2 - z_1)^2}$.

Now let $P_1(x_1, y_1, z_1)$ and $P_2(x_2, y_2, z_2)$ be any two points, and let it be required to find the undirected distance $|P_1P_2| = d$. As shown in Fig. 13-5, the segment $\overline{P_1P_2}$ may be regarded as the diagonal of a rectangular box whose edges have the measures $|x_2 - x_1|$, $|y_2 - y_1|$, and $|z_2 - z_1|$, respectively. The required undirected distance is then

$$d = \sqrt{(x_2 - x_1)^2 + (y_2 - y_1)^2 + (z_2 - z_1)^2} \tag{13-2}$$

13-4 DIRECTION ANGLES AND DIRECTION COSINES OF THE RADIUS VECTOR *OP*

The segment $\overline{OP}$ with end points at the origin and any point $P(x, y, z)$ is called the *radius vector* of P. If its measure is ρ, then, from Eq. (13-1),

$$\rho = |OP| = \sqrt{x^2 + y^2 + z^2}$$

The angles α, β, and γ which $\overrightarrow{OP}$, or the segment directed O to P, makes with the positive directions on the x, y, and z axes, respectively, are called its *direction angles*. In Fig. 13-6 they are

$$\begin{aligned} \alpha &= \text{angle } AOP \\ \beta &= \text{angle } BOP \\ \gamma &= \text{angle } COP \end{aligned} \tag{13-3}$$

Each of these is a nonnegative angle not greater than 180°.

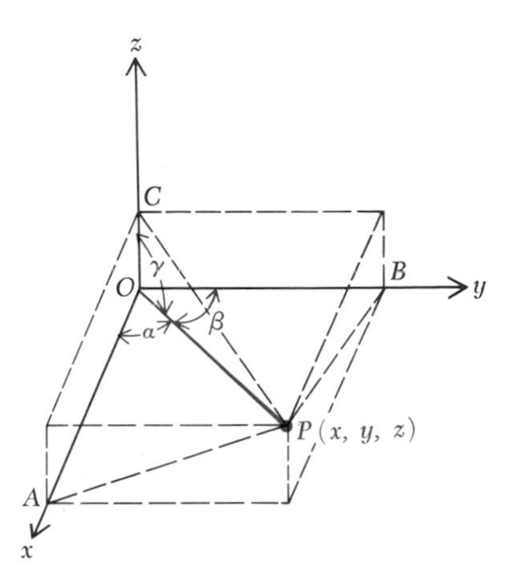

Fig. 13-6 Direction cosines of $\overrightarrow{OP}$ are $[\cos\alpha, \cos\beta, \cos\gamma]$: $\cos\alpha = OA/|OP|$, $\cos\beta = OB/|OP|$, $\cos\gamma = OC/|OP|$.

The cosines of the direction angles (13-3) are called the *direction cosines* of $\overrightarrow{OP}$. They are (from Fig. 13-6)

$$\cos\alpha = \frac{OA}{|OP|} = \frac{x}{\rho} \qquad \cos\beta = \frac{OB}{|OP|} = \frac{y}{\rho} \qquad \cos\gamma = \frac{OC}{|OP|} = \frac{z}{\rho} \tag{13-4}$$

These may be written as the ordered triple $[\cos\alpha, \cos\beta, \cos\gamma]$. Squaring the direction cosines (13-4) and adding gives

$$\cos^2\alpha + \cos^2\beta + \cos^2\gamma = \frac{x^2}{\rho^2} + \frac{y^2}{\rho^2} + \frac{z^2}{\rho^2} = \frac{x^2 + y^2 + z^2}{\rho^2} = \frac{\rho^2}{\rho^2} = 1 \tag{13-5}$$

Thus the sum of the squares of the direction cosines of $\overrightarrow{OP}$ is equal to 1.

If point P is in the first octant, its coordinates are all positive. In this case the direction cosines of $\overrightarrow{OP}$ are all positive, and its direction angles all measure less than 90°. If P has one or more negative coordinates, the corresponding direction cosines of $\overrightarrow{OP}$ are negative. The direction angles having negative cosines are between 90° and 180° (including the latter value). These definitions and observations are extensions to three dimensions of the concepts of direction angles and direction cosines for a line in the plane (Chaps. 2 and 4).

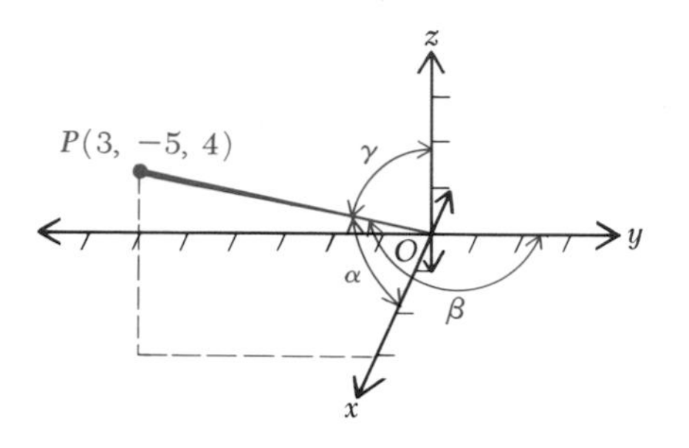

Fig. 13-7 $|OP| = \sqrt{3^2 + (-5)^2 + 4^2} = \sqrt{50}$: $\cos\alpha = 3/\sqrt{50}$, $\cos\beta = -5/\sqrt{50}$, $\cos\gamma = 4/\sqrt{50}$.

Example
Find the direction cosines and direction angles for $\overrightarrow{OP}$ if the point P has coordinates $(3, -5, 4)$.

Solution **If the undirected distance from O to P is designated by ρ (Fig. 13-7), then**

$$\rho = \sqrt{(-3)^2 + (-5)^2 + 4^2} = \sqrt{50} = 5\sqrt{2}$$

The direction cosines and direction angles of $\overrightarrow{OP}$ are, by Eqs. (13-3) and (13-4),

$$\cos\alpha = \frac{x}{\rho} = \frac{3}{5\sqrt{2}} \approx 0.42426 \qquad \alpha \approx 64°54'$$
$$\cos\beta = \frac{y}{\rho} = \frac{-5}{5\sqrt{2}} \approx -0.70711 \qquad \beta \approx 135°$$
$$\cos\gamma = \frac{z}{\rho} = \frac{4}{5\sqrt{2}} \approx 0.56568 \qquad \gamma \approx 55°33'$$

13-5 APPLICATIONS OF VECTORS

As was the case with the plane, it is convenient to define unit vectors in a three-dimensional system. The three unit vectors called **i**, **j**, and **k** have a common initial point at the origin, and terminal points, respectively, at $(1, 0, 0)$, $(0, 1, 0)$, and $(0, 0, 1)$ (Fig. 13-8a). Now each point $P(x, y, z)$ may be associated with a vector **OP** whose initial point is O and whose terminal point is P. Vector **OP** may be associated with the diagonal of a rectangular prism (Fig. 13-8b). Hence

$$\mathbf{OP} = x\mathbf{i} + y\mathbf{j} + z\mathbf{k} \tag{13-6}$$

As with unit vectors **i** and **j** in a plane, there are certain relationships among the unit vectors **i**, **j**, and **k** in space:

$$\mathbf{i} \cdot \mathbf{i} = \mathbf{j} \cdot \mathbf{j} = \mathbf{k} \cdot \mathbf{k} = 1$$

and
$$\mathbf{i} \cdot \mathbf{j} = \mathbf{i} \cdot \mathbf{k} = \mathbf{j} \cdot \mathbf{k} = \mathbf{j} \cdot \mathbf{i} = \mathbf{k} \cdot \mathbf{i} = \mathbf{k} \cdot \mathbf{j} = 0$$

Now it is possible to show for a nonzero vector **a** that if $\mathbf{a} = x\mathbf{i} + y\mathbf{j} + z\mathbf{k}$, then

$$\mathbf{a} \cdot \mathbf{i} = x$$

This is proved as follows:

$$\begin{aligned} \mathbf{a} \cdot \mathbf{i} &= (x\mathbf{i} + y\mathbf{j} + z\mathbf{k}) \cdot \mathbf{i} \\ &= x(\mathbf{i} \cdot \mathbf{i}) + y(\mathbf{j} \cdot \mathbf{i}) + z(\mathbf{k} \cdot \mathbf{i}) \\ &= x(1) + y(0) + z(0) \\ &= x \end{aligned}$$

With these ideas in mind, the discussion of direction cosines in Sec. 13-4 could have been phrased in the language of vectors. By definition

$$\mathbf{a} \cdot \mathbf{i} = |\mathbf{a}|\,|\mathbf{i}| \cos\,(\mathbf{a}, \mathbf{i})$$

Hence
$$\cos\,(\mathbf{a}, \mathbf{i}) = \frac{\mathbf{a} \cdot \mathbf{i}}{|\mathbf{a}|\,|\mathbf{i}|} = \frac{x}{|\mathbf{a}|}$$

Similarly, $\cos\,(\mathbf{a}, \mathbf{j}) = y/|\mathbf{a}|$ and $\cos\,(\mathbf{a}, \mathbf{k}) = z/|\mathbf{a}|$.

Note that these definitions are equivalent to the definitions of direc-

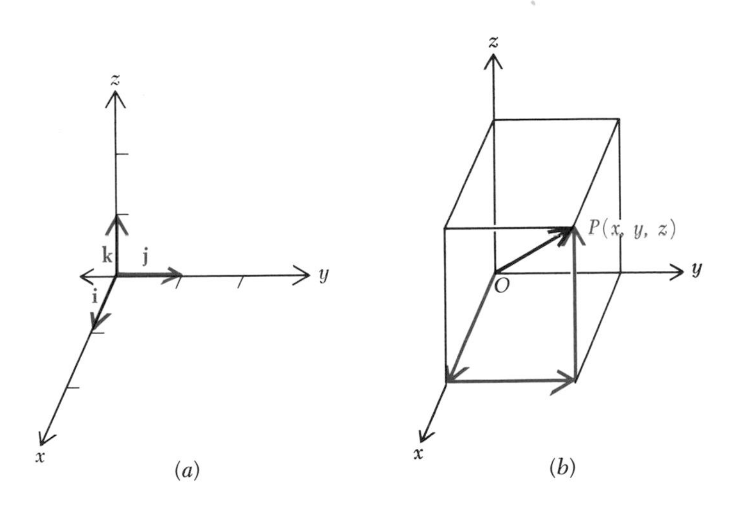

Fig. 13-8 (*a*) Unit vectors **i**, **j**, and **k**. (*b*) $\mathbf{OP} = x\mathbf{i} + y\mathbf{j} + z\mathbf{k}$.

tion cosines derived in Sec. 13-4. For example, the vector **OP**, with $P(3, 4, -2)$ is, by Eq. (13-6),

$$\mathbf{OP} = 3\mathbf{i} + 4\mathbf{j} - 2\mathbf{k}$$

The magnitude of **OP** is

$$|\mathbf{OP}| = \sqrt{3^2 + 4^2 + (-2)^2} = \sqrt{29}$$

Now

$$\frac{\mathbf{OP}}{|\mathbf{OP}|} = \frac{3}{\sqrt{29}}\mathbf{i} + \frac{4}{\sqrt{29}}\mathbf{j} - \frac{2}{\sqrt{29}}\mathbf{k} \tag{13-7}$$

Thus the vector form Eq. (13-7) shows that the coefficients of **i**, **j**, and **k** are the direction cosines of **OP**.

In problems of mechanics it is frequently necessary to find the magnitudes of the components of a given force in the directions of the coordinate axes.

Example

A pull of 50 lb is exerted in the direction determined by the ray $\overrightarrow{OP}$ in Fig. 13-9. Find the components of this pull in the directions of the coordinates axes.

Solution The force along **OP** is represented by the vector shown in the figure, with its length representing 50 lb. The required components are then represented by the lengths of the edges of the rectangular box as shown. If the magnitude of the given force is denoted by F, and if the components are denoted by F_x, F_y, and F_z, then

$$\begin{aligned} F_x &= F \cos \alpha \\ F_y &= F \cos \beta \\ F_z &= F \cos \gamma \end{aligned} \tag{13-8}$$

The direction cosines of $\overrightarrow{OP}$ are, from Eq. (13-1),

$$|OP| = \sqrt{6^2 + 4^2 + 3^2} = \sqrt{61}$$

and from Eq. (13-4),

$$\cos \alpha = \frac{6}{\sqrt{61}} \qquad \cos \beta = \frac{4}{\sqrt{61}} \qquad \cos \gamma = \frac{3}{\sqrt{61}}$$

[remember that the vector ends at the point $P(6, 4, 3)$]. The magnitudes

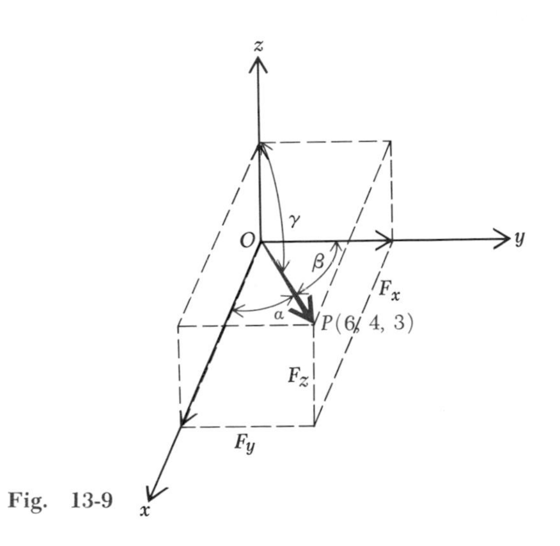

Fig. 13-9

of the components of the 50-lb force are then, from Eqs. (13-8),

$$F_x = 50\left(\frac{6}{\sqrt{61}}\right) \approx 38.4 \text{ lb}$$

$$F_y = 50\left(\frac{4}{\sqrt{61}}\right) \approx 25.6 \text{ lb}$$

$$F_z = 50\left(\frac{3}{\sqrt{61}}\right) \approx 19.2 \text{ lb}$$

PROBLEMS

1. Plot the point *P*, draw the radius vector $\overline{OP}$, find its length, and find the direction cosines and direction angles of $\overrightarrow{OP}$:

   (*a*) $P(2, 2, 4)$ (*b*) $P(-12, -3, 4)$ (*c*) $P(4, -3, 1)$
   (*d*) $P(5, 5, -5)$ (*e*) $P(-3, 2, -6)$ (*f*) $P(6, -3, -6)$

2. For a certain line $\overleftrightarrow{OP}$, $\cos\alpha = \frac{2}{3}$ and $\cos\beta = \frac{1}{3}$. Find the two possible values for $\cos\gamma$ and draw $\overleftrightarrow{OP}$.
3. A ray with end point at the origin is drawn so as to form congruent angles with the positive directions of the three coordinate axes. Find its direction cosines and draw the ray if the point (1, 1, 1) is on it.
4. Describe and sketch the locus of points $P(x, y, z)$ that satisfy each of the following sets of conditions:

   (*a*) $x = 4$ (*b*) $x = 0, z = 0$ (*c*) $x = 3, y = 4$
   (*d*) $z = x$ (*e*) $x + y = 4$ (*f*) $x + y = 4, z = 0$

5. Determine the coordinates of the point that is symmetrical to the point $P(x, y, z)$ with respect to (*a*) the origin; (*b*) the xy plane; (*c*) the x axis.
6. Plot the given points, draw the segment joining them, and compute its measure:

   (*a*) $A(2, 6, 1)$, $B(8, 3, 7)$ (*b*) $P(-4, 5, 2)$, $Q(2, 0, 5)$
   (*c*) $M(-4, -3, 2)$, $N(4, 1, -6)$ (*d*) $U(0, -5, 3)$, $V(-4, 3, 6)$

7. Draw the triangle and find the measures of its sides:

   (*a*) $A(0, 4, 1)$, $B(6, 5, 3)$, $C(10, 0, 6)$
   (*b*) $P(-3, -2, 3)$, $Q(0, 3, 0)$, $R(4, 0, 0)$

8. Draw the triangle and show that it is a right triangle:

   (*a*) $A(2, 6, 2)$, $B(5, 8, 5)$, $C(8, 2, 6)$
   (*b*) $P(-4, -2, 5)$, $Q(4, -1, 8)$, $R(6, 4, 1)$

9. Show that the coordinates of the midpoint of a segment with end points $P_1(x_1, y_1, z_1)$ and $P_2(x_2, y_2, z_2)$ are $(\frac{1}{2}(x_1 + x_2), \frac{1}{2}(y_1 + y_2), \frac{1}{2}(z_1 + z_2))$.
10. Find the coordinates of the points that partition the segment with end points $A(2, 7, 1)$ and $B(8, -2, 5)$ into three congruent segments.
11. Find the equation of the surface all points of which are at a distance of 5 units from the origin. What is the name of this surface?
12. Find the equation of the surface all points of which are at a distance of 5 units from the point $C(3, 2, 5)$. Describe the surface.
13. A point $P(x, y, z)$ moves so that its undirected distance from $F(4, 0, 0)$ is always equal to its distance from the yz plane. Find the equation of the surface on which it moves. Describe and sketch the surface.
14. A point $P(x, y, z)$ moves so that its distance from $A(2, 5, 1)$ is always equal to its distance from $B(8, 1, 6)$. Find the equation of the surface on which it moves and describe the surface.
15. Find the equation of the locus of a point the sum of whose undirected distances from $A(4, 0, 0)$ and $B(-4, 0, 0)$ is equal to 10.
16. Find the coordinates of the point on the x axis that is equidistant from $P(4, 3, 1)$ and $Q(-2, -6, 2)$.
17. A force of the given magnitude acts away from the origin along $\overrightarrow{OP}$; find its components along the axes:

   (*a*) 100 lb; $P(6, 6, 3)$ (*b*) 70 lb; $P(12, -4, -3)$
   (*c*) 60 lb; $P(3, 0, -4)$ (*d*) 200 lb; $P(-6, 3, 5)$

18. Pulls of 27 and 35 lb are applied at a point as shown in Fig. 13-10. Find the total vertical force and the forces in the x and y directions. What is the magnitude and what are the direction angles of a single force that would produce the same effect as these two forces?
19. Solve Prob. 18 for forces of 44 and 20 lb along $\overrightarrow{OA}$ and $\overrightarrow{OB}$, respectively, and with the coordinates $A(9, 6, 2)$ and $B(-8, 6, 0)$.

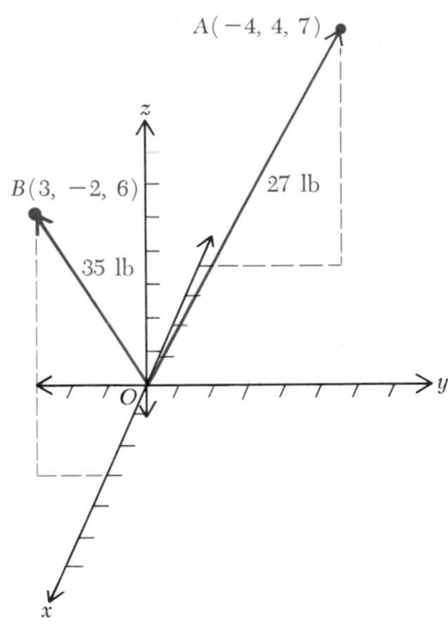

Fig. 13-10

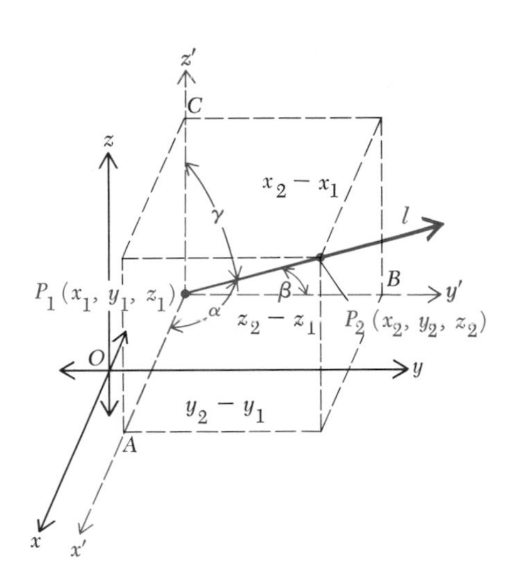

Fig. 13-11 Direction cosines of $\overrightarrow{P_1P_2}$ are $\left[\dfrac{x_2 - x_1}{|P_1P_2|}, \dfrac{y_2 - y_1}{|P_1P_2|}, \dfrac{z_2 - z_1}{|P_1P_2|}\right]$.

13-6 DIRECTION COSINES OF ANY DIRECTED LINE

Let l be any directed line in space (Fig. 13-11). Then the direction angles of l are defined as follows. Let $P_1(x_1, y_1, z_1)$ and $P_2(x_2, y_2, z_2)$ be any two points on l such that the direction from P_1 to P_2 is considered positive. Through P_1 draw rays $\overrightarrow{P_1A}$, $\overrightarrow{P_1B}$, and $\overrightarrow{P_1C}$ parallel to, and in the positive directions of, the x, y, and z axes, respectively. Then the direction angles of l are α, β, and γ, where

$$\alpha = \text{angle } AP_1P_2 \qquad \beta = \text{angle } BP_1P_2 \qquad \gamma = \text{angle } CP_1P_2 \tag{13-9}$$

Each of these is a nonnegative angle not greater than 180°.

The cosines of these direction angles are the direction cosines of the directed line l. To compute their values we first find the directed distance $|P_1P_2|$:

$$d = |P_1P_2| = \sqrt{(x_2 - x_1)^2 + (y_2 - y_1)^2 + (z_2 - z_1)^2} \tag{13-10}$$

Then, from Fig. 13-11,

$$\cos\alpha = \frac{P_1A}{|P_1P_2|} = \frac{x_2 - x_1}{d}$$

(13-11)
$$\cos\beta = \frac{P_1B}{|P_1P_2|} = \frac{y_2 - y_1}{d}$$

$$\cos\gamma = \frac{P_1C}{|P_1P_2|} = \frac{z_2 - z_1}{d}$$

Since d is positive, each of these direction cosines is positive or negative, depending upon the sign of the numerator. Observe that the effect of reversing the positive direction on the line is to change the signs of all the direction cosines and to replace the direction angles by their supplements.

Example

Find the direction cosines of the directed line determined by the points $P_1(-2, 1, 8)$ and $P_2(2, 8, 3)$ if the positive direction is from P_1 to P_2.

Solution The undirected distance $d = |P_1P_2|$ (Fig. 13-12) is, by Eq. (13-10),

$$d = \sqrt{[2-(-2)]^2 + (8-1)^2 + (3-8)^2}$$
$$= \sqrt{90} = 3\sqrt{10}$$

Then, from Eqs. (13-11),

$$\cos\alpha = \frac{2-(-2)}{3\sqrt{10}} = \frac{4\sqrt{10}}{30} \approx 0.42164$$
$$\cos\beta = \frac{8-1}{3\sqrt{10}} = \frac{7\sqrt{10}}{30} \approx 0.73786$$
$$\cos\gamma = \frac{3-8}{3\sqrt{10}} = \frac{-5\sqrt{10}}{30} \approx -0.52705$$

Because $\cos\gamma$ is negative, γ is between 90 and 180°. If the positive direction had been P_2 to P_1, the signs of the direction cosines would all have been changed.

13-7 DIRECTION NUMBERS

If a point moves from A to B along the line shown in Fig. 13-13, it moves 3 units in the negative x direction, 6 units in the positive y direction, and 2 units in the positive z direction, as indicated by the dashed segments.

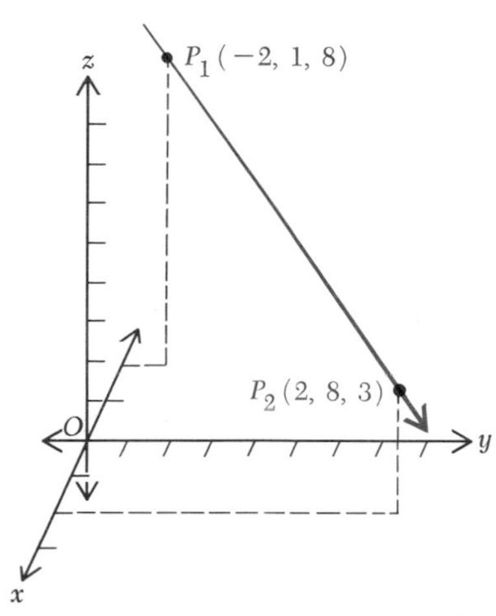

Fig. 13-12 Direction cosines of $\overrightarrow{P_1P_2}$ are $[\frac{4}{30}\sqrt{10}, \frac{7}{30}\sqrt{10}, \frac{-5}{30}\sqrt{10}]$.

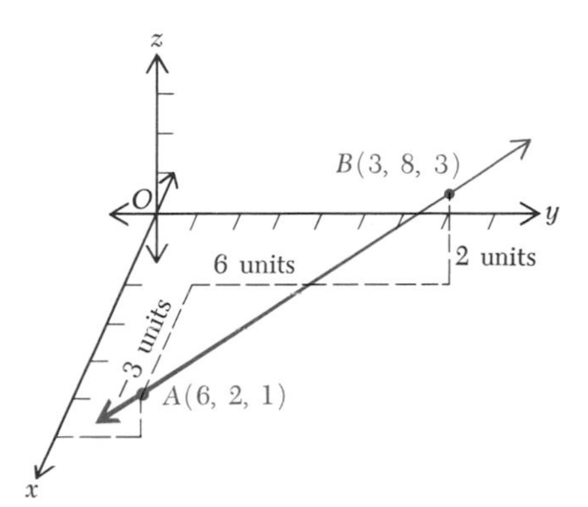

Fig. 13-13 A set of direction numbers for the line directed from A to B is $[3-6, 8-2, 3-1]$.

The numbers -3, 6, and 2, written in the form of an ordered triple

$$[-3, 6, 2]$$

may be used to define the direction of the line directed from A to B. They fix its direction by indicating that for every -3 units moved in the x direction, a point traveling along the line must move 6 units in the y direction and 2 units in the z direction. Any positive multiple of these three numbers gives the same information. Thus if the point moved -6 units in the x direction, it would move 12 units in the y direction and 4 units in the z direction, and the ordered set of numbers $[-6, 12, 4]$ would define the same direction. If the point travels in the opposite direction, that is, from B to A, each of -3, 6, and 2 must have its sign changed; therefore, the ordered set of numbers $[3, -6, -2]$ fixes the direction of the line directed B to A.

Consider now the general case of a line determined by two points $P_1(x_1, y_1, z_1)$ and $P_2(x_2, y_2, z_2)$. The direction of the line in space directed from P_1 to P_2 is determined by the ordered set of numbers

$$[x_2 - x_1, y_2 - y_1, z_2 - z_1] \tag{13-12}$$

Any positive multiple of this ordered triple would serve equally well in fixing the direction of the line P_1 to P_2. Any such set of numbers is called a set of *direction numbers* for the directed line.

If the line is directed from $P_2(x_2, y_2, z_2)$ to $P_1(x_1, y_1, z_1)$ a set of direction numbers is the ordered triple

$$[x_1 - x_2,\ y_1 - y_2,\ z_1 - z_2] \tag{13-13}$$

These are the additive inverses of $[x_2 - x_1,\ y_2 - y_1,\ z_2 - z_1]$ for the lines directed from P_1 to P_2, as was to be expected. Multiplying the numbers in (13-13) by a positive constant results in another set of direction numbers for the line directed P_2 to P_1. Note that multiplying each of the direction numbers in expression (13-12) for the line directed from P_1 to P_2 by a negative constant results in a set of direction numbers for the line directed from P_2 to P_1.

In two-dimensional analytic geometry a line is determined by specifying two points on it, or by specifying one point and the direction (inclination, slope, or direction numbers) of the line. In space, correspondingly, a line may be fixed by specifying two points on it or by specifying one point and a set of direction numbers.

Example 1

Draw the line passing through $A(6, -2, 5)$ with direction numbers $[-1, 2, -1]$.

Solution Starting at A in Fig. 13-14, obtain another point on the line by using any multiple of the direction numbers. In the figure, -3, 6, and -3 units are used, respectively, resulting in a point $B(3, 4, 2)$ which is on the required line.

Any ordered set of three numbers that are not all zero may be regarded as defining the direction of a line in space. They are proportional to the direction cosines in accordance with the following theorem, proof of which is left to the reader.

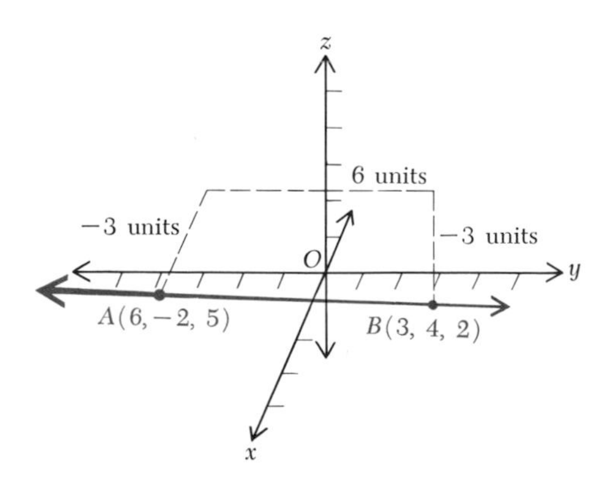

Fig. 13-14 Graphing a line through A(6, −2, 5) with direction numbers [−2, 2, −1].

Theorem 13-1
If a line has as direction numbers the ordered triple [a, b, c], then its direction cosines are

$$\cos\alpha = \frac{a}{\pm\sqrt{a^2+b^2+c^2}}$$
$$\cos\beta = \frac{b}{\pm\sqrt{a^2+b^2+c^2}} \qquad (13\text{-}14)$$
$$\cos\gamma = \frac{c}{\pm\sqrt{a^2+b^2+c^2}}$$

where the denominators are all positive or all negative, depending on which direction along the line is chosen as positive.

Example 2
The line shown in Fig. 13-13 has a set of direction numbers [−3, 6, 2]. Find its direction cosines.

Solution If the positive direction is from A to B, the corresponding directions cosines are, by Eqs. (13-14),

$$\cos\alpha = \frac{-3}{\sqrt{(-3)^2+(6)^2+(2)^2}} = -\frac{3}{7}$$
$$\cos\beta = \tfrac{6}{7}$$
$$\cos\gamma = \tfrac{2}{7}$$

In this problem there was no question of which sign to use, since the direction numbers [−3, 6, 2] were derived from the calculations $3-6 = x_2 - x_1$, $8-2 = y_2 - y_1$, and $3-1 = z_2 - z_1$. If the line were directed from B to A, the direction cosines would be the negatives of those given above.

13-8 THE ANGLE BETWEEN TWO DIRECTED LINES

Two lines drawn at random in space do not, in general, intersect. It is possible, however, to speak of the *angle* formed by such lines.

Definition
The angle formed by directed lines in space is congruent to the angle formed by the positive directions of two lines that are parallel, respec-

tively, to the given lines, have the same positive directions, and intersect at a point A.

Without loss of generality the point A in the definition may be taken as the origin. The problem of finding the angle formed by any two directed lines in space then reduces to that of finding the angle θ formed by the positive directions of two lines l_1 and l_2 through the origin, as shown in Fig. 13-15.

The cosine of this angle may be expressed in terms of the direction cosines of l_1 and l_2 as follows. Choose a point $P_1(x_1, y_1, z_1)$ on l_1 and a point $P_2(x_2, y_2, z_2)$ on l_2; let $|OP_1| = \rho_1$, $|OP_2| = \rho_2$, and $|P_1P_2| = d$. Then, using the law of cosines,

$$d^2 = \rho_1{}^2 + \rho_2{}^2 - 2\rho_1\rho_2 \cos\theta$$

Solving this equation for $\cos\theta$ gives

$$\cos\theta = \frac{\rho_1{}^2 + \rho_2{}^2 - d^2}{2\rho_1\rho_2} \tag{13-15}$$

But $\qquad \rho_1{}^2 = x_1{}^2 + y_1{}^2 + z_1{}^2 \qquad \rho_2{}^2 = x_2{}^2 + y_2{}^2 + z_2{}^2$

and $\qquad d^2 = (x_2 - x_1)^2 + (y_2 - y_1)^2 + (z_2 - z_1)^2$

Substituting these relations into the numerator of Eq. (13-15) and simplifying results in

$$\cos\theta = \frac{x_1x_2 + y_1y_2 + z_1z_2}{\rho_1\rho_2}$$

$$= \frac{x_1}{\rho_1}\frac{x_2}{\rho_2} + \frac{y_1}{\rho_1}\frac{y_2}{\rho_2} + \frac{z_1}{\rho_1}\frac{z_2}{\rho_2} \tag{13-16}$$

Now, if the direction angles of l_1 are α_1, β_1, and γ_1, and those of l_2 are α_2, β_2, and γ_2, it follows that

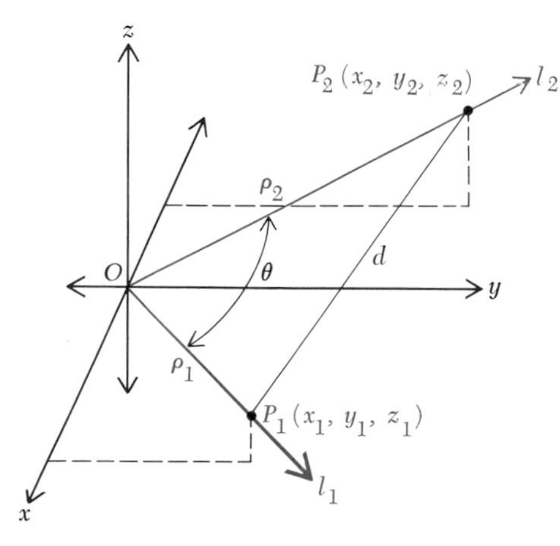

Fig. 13-15 Cosine of the angle θ between l_1 and l_2 is $\cos\alpha_1 \cos\alpha_2 + \cos\beta_1 \cos\beta_2 + \cos\gamma_1 \cos\gamma_2$.

$$\frac{x_1}{\rho_1} = \cos \alpha_1 \qquad \frac{x_2}{\rho_2} = \cos \alpha_2$$
$$\frac{y_1}{\rho_1} = \cos \beta_1 \qquad \text{and} \qquad \frac{y_2}{\rho_2} = \cos \beta_2$$
$$\frac{z_1}{\rho_1} = \cos \gamma_1 \qquad \frac{z_2}{\rho_2} = \cos \gamma_2$$

Making these substitutions in Eq. (13-16) gives

$$\cos \theta = \cos \alpha_1 \cos \alpha_2 + \cos \beta_1 \cos \beta_2 + \cos \gamma_1 \cos \gamma_2 \tag{13-17}$$

Example 1

Find the angle between the line directed from $A(2, 3, 5)$ to $B(6, -2, 2)$ and the line directed from $C(-2, -2, 8)$ to $D(4, 1, 6)$.

Solution For $\overrightarrow{AB}$, by Eq. (13-10),

$$|AB| = \sqrt{(6-2)^2 + (-2-3)^2 + (2-5)^2} = \sqrt{50}$$

and if its direction angles are α_1, β_1, and γ_1, then, by Eqs. (13-11),

$$\cos \alpha_1 = \frac{6-2}{\sqrt{50}} = \frac{4}{\sqrt{50}}$$
$$\cos \beta_1 = \frac{-2-3}{\sqrt{50}} = \frac{-5}{\sqrt{50}}$$
$$\cos \gamma_1 = \frac{2-5}{\sqrt{50}} = \frac{-3}{\sqrt{50}}$$

For $\overrightarrow{CD}$, by Eq. (13-10),

$$|CD| = \sqrt{(4+2)^2 + (1+2)^2 + (6-8)^2} = 7$$

and if its direction angles are α_2, β_2, and γ_2, then, by Eqs. (13-11),

$$\cos \alpha_2 = \frac{4+2}{7} = \frac{6}{7}$$
$$\cos \beta_2 = \frac{1+2}{7} = \frac{3}{7}$$
$$\cos \gamma_2 = \frac{6-8}{7} = -\frac{2}{7}$$

The cosine of the angle θ between these lines is, by Eq. (13-17),

$$\cos \theta = \frac{4}{\sqrt{50}}\frac{6}{7} + \frac{-5}{\sqrt{50}}\frac{3}{7} + \frac{-3}{\sqrt{50}}\frac{-2}{7}$$
$$= \frac{3\sqrt{2}}{14} \approx 0.30304$$

and $\theta \approx 72°22'$

Two lines in space are mutually perpendicular if and only if $\theta = 90°$ (see Fig. 13-15). Since $\cos 90° = 0$, the condition for perpendicularity of l_1 and l_2 is, by Eq. (13-17),

(13-18) $$\cos \alpha_1 \cos \alpha_2 + \cos \beta_1 \cos \beta_2 + \cos \gamma_1 \cos \gamma_2 = 0$$

If, instead of the direction cosines of lines l_1 and l_2, their direction numbers $[a_1, b_1, c_1]$ and $[a_2, b_2, c_2]$, respectively, are specified, we may first compute the direction cosines and then use Eq. (13-17), or we may use the equivalent formula directly:

(13-19) $$\cos \theta = \pm \frac{a_1a_2 + b_1b_2 + c_1c_2}{\sqrt{a_1^2 + b_1^2 + c_1^2}\,\sqrt{a_2^2 + b_2^2 + c_2^2}}$$

The choice of sign depends on the choice of positive directions on l_1 and l_2. In terms of direction numbers, the condition for perpendicularity becomes

(13-20) $$a_1a_2 + b_1b_2 + c_1c_2 = 0$$

Example 2

Is a directed line with direction numbers $[2, 6, -1]$ perpendicular to the directed line whose direction numbers are $[4, -1, 2]$?

Solution From Eq. (13-20),

$$2(4) + 6(-1) + (-1)(2) = 0$$

Hence the lines are perpendicular.

PROBLEMS

1. Graph the line passing through each pair of points and find its direction numbers and direction cosines (the positive direction is as indicated):

   (*a*) $A(2, 1, 1)$ to $B(6, 6, 4)$ (*b*) $P(-2, 3, 4)$ to $Q(0, 1, 3)$
   (*c*) $M(0, -3, 2)$ to $N(0, 1, 5)$ (*d*) $C(-3, -2, -2)$ to $D(6, 0, 3)$
   (*e*) $L(0, 0, -2)$ to $M(4, -4, 0)$ (*f*) $R(7, 4, 1)$ to $S(1, -2, 4)$

2. In each of the following graph the line determined by the given point and the direction numbers and find its direction cosines, assuming the positive direction to be such that $\cos \gamma$ is positive:

   (*a*) $A(2, 5, 1)$; $[2, -2, 1]$ (*b*) $A(-2, 2, 6)$; $[6, 3, -2]$
   (*c*) $A(0, 5, 0)$; $[4, -4, 3]$ (*d*) $A(-4, 0, -2)$; $[2, 0, 1]$

3. Find the direction cosines of each of the coordinate axes.
4. Two of the direction angles of a certain line are $\alpha = 45°$ and $\beta = 60°$. Find its direction cosines if the positive direction is such that γ is an acute angle.
5. A line passes through $A(-4, 6, 1)$ and has direction numbers $[2, -2, 1]$. What are the coordinates of the point at which this line intersects (*a*) the yz plane; (*b*) the xz plane?
6. A line passes through $P(4, 2, 3)$ and has direction numbers $[3, 2, -1]$. What are the coordinates of the point at which it intersects (*a*) the xy plane; (*b*) the xz plane?
7. A line passes through $P(4, 7, 2)$ and makes congruent angles with the coordinate axes. What are the coordinates of the points at which it intersects the coordinate planes?
8. Find the cosine of the angle between the lines whose direction numbers are as follows:

   (*a*) $[2, 1, -2]$; $[3, -6, 2]$ (*b*) $[7, 4, 4]$; $[6, -6, 3]$
   (*c*) $[5, 4, -3]$; $[1, -1, 0]$ (*d*) $[3, -1, -4]$; $[2, -3, 1]$

9. Draw the triangle and show that it is a right triangle; compute the cosine of the angle at A:

   (*a*) $A(2, 7, 1)$, $B(8, 5, 5)$, $C(7, 3, 4)$
   (*b*) $A(-2, 3, 2)$, $B(6, -3, 4)$, $C(1, 3, -1)$

10. A line l_1 is drawn from the origin to $P(4, -2, 4)$; from P a line l_2 is drawn so as to make congruent angles with the coordinate axes. Graph the figure and compute the measure of the angle from l_1 to l_2.
11. What relation must exist between the direction cosines of two lines if the lines are parallel?
12. A force of 42 lb acts along $\overrightarrow{AC}$ as shown in Fig. 13-16. Find its component in the direction of $\overrightarrow{AB}$. *Hint:* The required component is equal to $42 \cos \theta$.

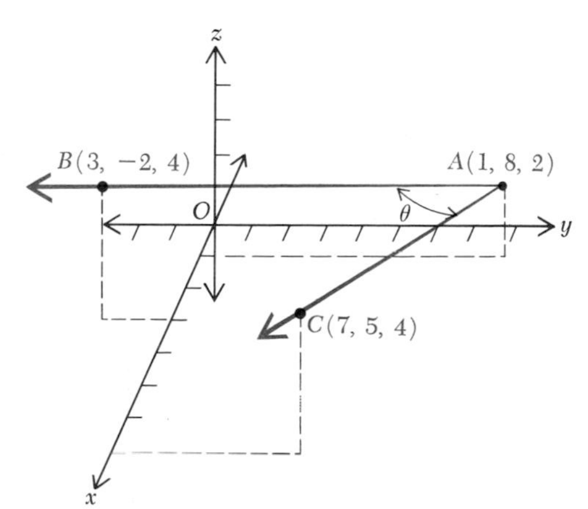

Fig. 13-16

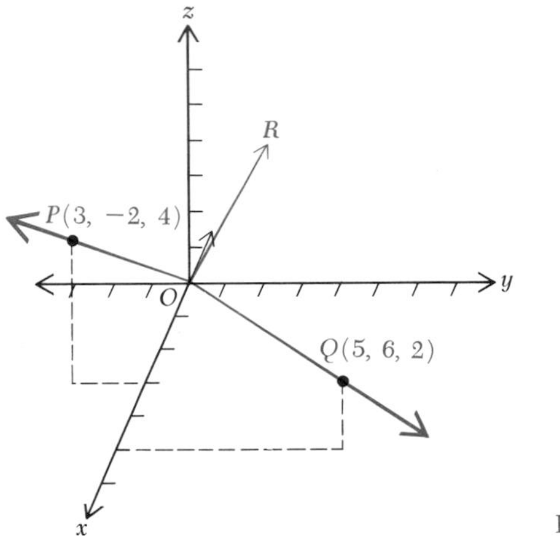

Fig. 13-17

13. A force of 36 lb acts along the line directed from the origin to $P(7, 4, 4)$. Find the component of this force in the direction of a line making congruent angles with the coordinate axes.
14. Find the direction numbers of $\overrightarrow{OR}$ that is perpendicular to the plane of $\overrightarrow{OP}$ and $\overrightarrow{OQ}$ in Fig. 13-17.
15. Graph the triangle and find the direction numbers of the normal to the plane of the triangle; draw the normal from the origin to the plane:

   (*a*) $A(8, 0, 0)$, $B(0, 12, 0)$, $C(0, 0, 8)$
   (*b*) $A(-5, 0, 0)$, $B(3, 8, 2)$, $C(-3, -4, 8)$
   (*c*) $A(2, 7, 1)$, $B(8, 2, 3)$, $C(3, 2, 8)$

SUMMARY

A rectangular coordinate system in three dimensions is established by drawing three mutually perpendicular lines called the *x*, *y*, and *z* axes. These meet at the origin. These axes determine three mutually perpendicular planes called the *xy*, *xz*, and *yz* planes; these also intersect at the origin. Any point *P* is associated with the ordered triple of numbers (x, y, z), called its coordinates, which are defined as follows:

The x coordinate is the directed distance from the yz plane.
The y coordinate is the directed distance from the xz plane.
The z coordinate is the directed distance from the xy plane.

The distance d between two points $P(x_1, y_1, z_1)$ and $Q(x_2, y_2, z_2)$ is analogous to the distance between two points in a plane:

$$d = \sqrt{(x_2 - x_1)^2 + (y_2 - y_1)^2 + (z_2 - z_1)^2}$$

The direction cosines and direction numbers of a line in space are similarly defined in a manner corresponding to that for a line in a plane. For the line directed from point $P(x_1, y_1, z_1)$ to point $Q(x_2, y_2, z_2)$ the direction cosines are the ordered triple $[\cos\alpha, \cos\beta, \cos\gamma]$ with $d = |PQ|$:

$$\cos\alpha = \frac{x_2 - x_1}{d} \qquad \cos\beta = \frac{y_2 - y_1}{d} \qquad \cos\gamma = \frac{z_2 - z_1}{d}$$

The direction numbers of a directed line are proportional to its direction cosines. Hence one ordered triple of direction numbers for line $\overleftrightarrow{PQ}$ is $[x_2 - x_1, y_2 - y_1, z_2 - z_1]$.

For the line directed from Q to P the direction cosines are the additive inverses of those for the line directed from P to Q. A similar statement may be made concerning its direction numbers.

Concepts of vectors in space bear a strong resemblance to corresponding concepts in two dimensions. If, for example, a vector has its initial point at $O(0, 0, 0)$ and terminal point at $P(x, y, z)$, then $\mathbf{OP} = x\mathbf{i} + y\mathbf{j} + z\mathbf{k}$, where $\mathbf{i}$, $\mathbf{j}$, and $\mathbf{k}$ are unit vectors in the x, y, and z directions, respectively.

The angle θ between two directed lines, l_1 with direction angles $[\alpha_1, \beta_1, \gamma_1]$ and l_2 with direction angles $[\alpha_2, \beta_2, \gamma_2]$, is defined as

$$\cos\theta = \cos\alpha_1 \cos\alpha_2 + \cos\beta_1 \cos\beta_2 + \cos\gamma_1 \cos\gamma_2$$

If two lines are perpendicular, $\theta = 90°$ and $\cos\theta = 0$. If two perpendicular lines have triples of direction numbers $[a_1, b_1, c_1]$ and $[a_2, b_2, c_2]$, then $a_1a_2 + b_1b_2 + c_1c_2 = 0$.

REVIEW PROBLEMS

1. For points $A(5, 7, 3)$, $B(-3, 3, 4)$, $C(2, -3, 6)$, and $O(0, 0, 0)$ determine (*a*) $|AB|$; (*b*) $|\mathbf{OC}|$; (*c*) an ordered triple of direction numbers for $\overrightarrow{AC}$; (*d*) the ordered triple of direction cosines for $\overrightarrow{BC}$; (*e*) the direction angles (to the nearest degree) for $\overrightarrow{BA}$.

2. Is it possible to draw ray $\overrightarrow{OP}$ (in space), where O is the origin, for each of the following conditions? Give a reason for your answer.

   (*a*) All direction angles are 90°.
   (*b*) All direction angles are 120°.
   (*c*) All direction angles are congruent.
   (*d*) The ordered triple $[\frac{1}{2}, -\frac{1}{2}, 2]$ is the set of direction cosines.
   (*e*) All direction numbers are negative.

3. (*a*) Prove that the points $P(6, 9, 10)$, $Q(1, -1, -5)$, and $R(6, -11, 0)$ are vertices of a right triangle. (*b*) Find the area of the triangular region.
4. The distance from the point $K(7, 1, -3)$ to the point $J(4, 5, c)$ is 13. Find the value of c.
5. Find the acute angle between the lines $\overleftrightarrow{AB}$ and $\overleftrightarrow{CD}$, for $A(3, 1, -2)$, $B(1, -2, 4)$, $C(-4, 8, 0)$, and $D(5, 2, -2)$.
6. For points $P(7, 6, 3)$, $Q(4, 10, 1)$, $R(-2, 6, 2)$, and $S(1, 2, 4)$ show that $PQRS$ is a rectangle.
7. A point P is 2 units from the yz plane. $\overrightarrow{OP}$ has direction angles $\alpha = 60°$ and $\beta = 45°$. Find the coordinates of P.
8. Describe where point $P(x, y, z)$ is located for each of these conditions:

   (*a*) $x = 0$ (*b*) $y = z = 0$ (*c*) $y = z$, $x = 0$

9. A point $P(x, y, z)$ moves so that its distance from the x axis is always equal to its distance from point $Q(4, 0, 0)$. Find the equation of the surface on which it moves.

FOR EXTENDED STUDY

1. Prove that if $P(x, y, z)$ is any point on $\overleftrightarrow{P_1P_2}$, where P_1 has coordinates (x_1, y_1, z_1) and P_2 has coordinates (x_2, y_2, z_2) and $P_1P/P_1P_2 = k$ for k a real number, then

$$x = x_1 + k(x_2 - x_1)$$
$$y = y_1 + k(y_2 - y_1)$$
$$z = z_1 + k(z_2 - z_1)$$

   These are the equations for the coordinates of the point of division of a line segment in three dimensions that correspond to similar equations developed for two dimensions.
2. Prove that segments with end points which are the midpoints of opposite sides of any quadrilateral (in three dimensions) bisect each other.
3. Prove that it is impossible for a line (in space) to have two direction angles whose measures are each less than 45°. Remember that all direction angles must measure 0° or greater.
4. Show that the direction numbers $[a, b, c]$ of a line l that is perpendicular to each of two nonparallel lines l_1 and l_2 with direction numbers $[a_1, b_1, c_1]$

and $[a_2, b_2, c_2]$, respectively, are

$$a = \begin{vmatrix} b_1 & c_1 \\ b_2 & c_2 \end{vmatrix} \qquad b = \begin{vmatrix} c_1 & a_1 \\ c_2 & a_2 \end{vmatrix} \qquad c = \begin{vmatrix} a_1 & b_1 \\ a_2 & b_2 \end{vmatrix}$$

Hint: The condition that l is perpendicular to l_1 is that $aa_1 + bb_1 + cc_1 = 0$ and the condition that l is perpendicular to l_2 is that $aa_2 + bb_2 + cc_2 = 0$. Solve these equations for a/c and b/c, using determinants.

5. Find the direction numbers for a line through $P(3, 4, 4)$ perpendicular to the plane determined by the points $A(-4, 2, -1)$, $B(2, -3, 0)$, and $C(-3, -3, 5)$.
6. Forces of 44, 35, and 30 lb act away from the origin along $\overrightarrow{OA}$, $\overrightarrow{OB}$, and $\overrightarrow{OC}$, respectively. The coordinates of A, B, and C are $A(9, -2, -6)$, $B(-6, 3, 2)$, and $C(3, 4, 0)$. Find the algebraic sum of the components of these forces along each coordinate axis. Find the magnitude and direction angles of the single force having these components.
7. Is it possible to place an acute triangle ABC with vertex A on the x axis, vertex B on the y axis, and vertex C on the z axis?

PLANES AND LINES 14

14-1 THE EQUATION OF A PLANE

The plane in a three-dimensional coordinate system is analogous to the line in a two-dimensional system. A line separates a plane into two half-planes, whereas a plane separates space into two half-spaces. Moreover, a line is the set of points equidistant from two distinct points on a plane; a plane is the set of points equidistant from two distinct points in space. This last definition may be used to derive the equation of a plane.

The derivation may be started by letting distinct points $P_1(x_1, y_1, z_1)$ and $P_2(x_2, y_2, z_2)$, not on the given plane, be equidistant from any point $P(x, y, z)$ on the plane (Fig. 14-1). Then

$$|P_1P| = |P_2P|$$

Application of the distance formula to this equation results in

$$\sqrt{(x-x_1)^2+(y-y_1)^2+(z-z_1)^2} = \sqrt{(x-x_2)^2+(y-y_2)^2+(z-z_2)^2}$$

If both members are squared and the expression is simplified, the result is

$$(14\text{-}1) \quad 2(x_2-x_1)x + 2(y_2-y_1)y + 2(z_2-z_1)z - [(x_2^2-x_1^2)+(y_2^2-y_1^2)+(z_2^2-z_1^2)] = 0$$

Because undirected distances are being considered (the distances are positive numbers), the argument can be reversed, and any point $P(x, y, z)$ whose coordinates satisfy Eq. (14-1) is equidistant from P_1 and P_2. This equation is of first degree because not all the coefficients of x, y, and z are zero.

Now, make these simplifications of the constants in Eq. (14-1):

$$A = 2(x_2 - x_1)$$
$$B = 2(y_2 - y_1)$$
$$C = 2(z_2 - z_1)$$
$$D = -[(x_2^2 - x_1^2) + (y_2^2 - y_1^2) + (z_2^2 - z_1^2)]$$

In the three-dimensional coordinate system under consideration the equation of a plane becomes

$$Ax + By + Cz + D = 0 \qquad (14\text{-}2)$$

with A, B, and C not all zero. This is called the *general form* of the equation of a plane.

A, B, and C cannot be zero simultaneously because P_1 and P_2 are distinct points, so that $x_2 - x_1$, $y_2 - y_1$, and $z_2 - z_1$ are not all zero. The necessity for this restriction is made clear by noting the inconsistencies that result in letting $A = B = C = 0$. There are two possible cases:

1. If $D \neq 0$, the solution set of the equation $Ax + By + Cz + D = 0$ is the empty set.
2. If $D = 0$, every ordered triple (x, y, z) is a member of the solution set of the equation.

Clearly, neither of these sets is a plane.

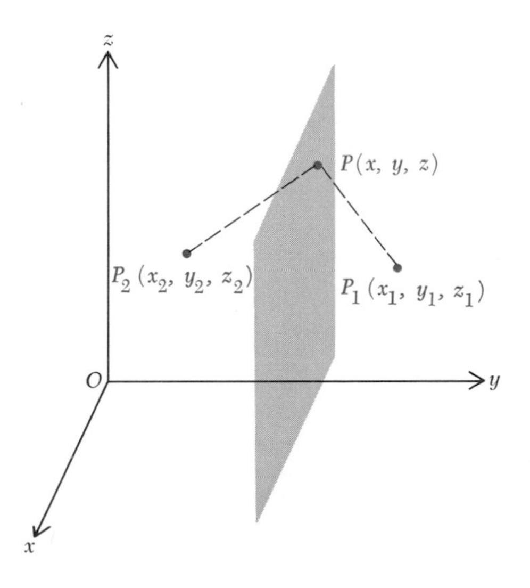

Fig. 14-1 $P_1(x_1, y_1, z_1)$ and $P_2(x_2, y_2, z_2)$ are equidistant from $P(x, y, z)$ on the plane.

14-2 NORMAL EQUATION OF A PLANE

A plane may also be determined by specifying the length and direction of the normal $\overline{ON}$ drawn from the origin to a point N on the plane. In Fig. 14-2 let the directed distance $ON = p$ and let the direction angles of $\overrightarrow{ON}$ be α, β, and γ. To determine the equation of the plane in terms of these data let $P(x, y, z)$ be any point in the plane, let the directed distance OP be ρ, and let the direction angles of $\overrightarrow{OP}$ be α_1, β_1, and γ_1. Then, for θ the angle formed by $\overrightarrow{OP}$ and $\overrightarrow{ON}$, the triangle OPN is a right triangle with hypothenuse $\overline{OP}$ in which

$$OP \cos \theta = ON$$

or

$$\rho \cos \theta = p$$

By making use of the formula for $\cos \theta$ in Sec. 13-8, this can be expressed in the form

$$\rho(\cos \alpha_1 \cos \alpha + \cos \beta_1 \cos \beta + \cos \gamma_1 \cos \gamma) = p$$

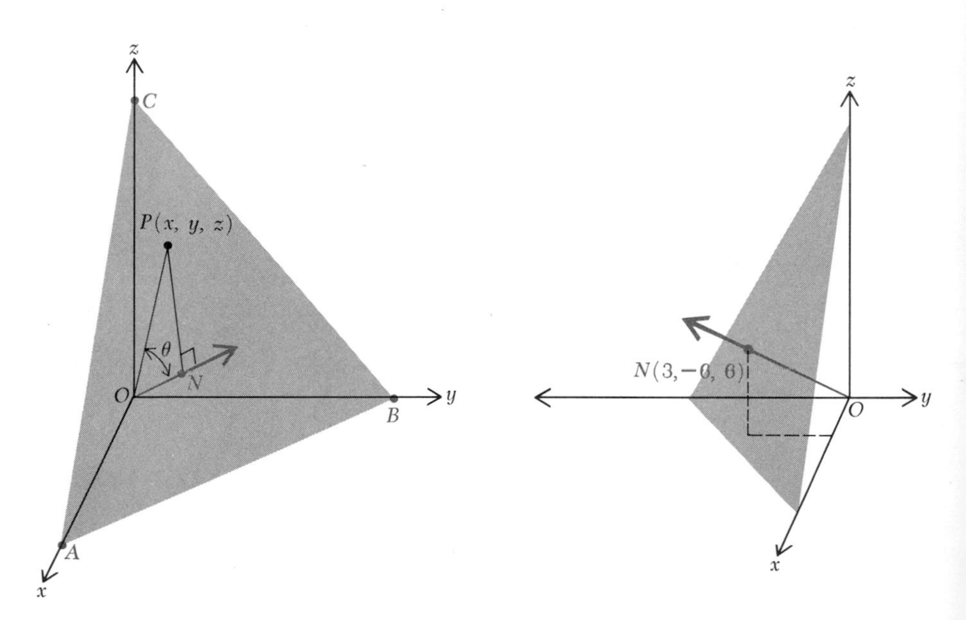

Fig. 14-2 The normal equation of a plane may be determined by specifying the length and direction of the normal $\overrightarrow{ON}$.

Fig. 14-3 Plane through $N(3, -6, 6)$ perpendicular to $\overline{ON}$.

Now, since

$$\rho \cos \alpha_1 = x$$
$$\rho \cos \beta_1 = y$$
$$\rho \cos \gamma_1 = z$$

the equation of the plane can be written in terms of the direction cosines of the normal $\overrightarrow{ON}$ as

$$x \cos \alpha + y \cos \beta + z \cos \gamma = p \tag{14-3}$$

This is called the *normal form* of the equation of a plane. It can be used for writing the equation of a plane when the length and direction of its normal $\overline{ON}$ are known.

Example

Find the equation of the plane through $N(3, -6, 6)$ perpendicular to $\overrightarrow{ON}$.

Solution In this case (see Fig. 14-3)

$$ON = p = \sqrt{3^2 + (-6)^2 + 6^2} = 9$$
$$\cos \alpha = \tfrac{3}{9} = \tfrac{1}{3}$$
$$\cos \beta = -\tfrac{6}{9} = -\tfrac{2}{3}$$
$$\cos \gamma = \tfrac{6}{9} = \tfrac{2}{3}$$

The normal form of the required equation is then, by Eq. (14-3),

$$\tfrac{1}{3}x - \tfrac{2}{3}y + \tfrac{2}{3}z = 9$$

which can be simplified to

$$x - 2y + 2z - 27 = 0$$

Two equations for a plane in space have been derived from two different definitions. It is now possible to relate these equations by proving that in the general equation of a plane [Eq. (14-2)] the normal to the plane has a set of direction numbers

$$[A, B, C]$$

The proof may be outlined as follows. An equation equivalent to Eq. (14-2) is obtained by dividing each term by $\pm\sqrt{A^2 + B^2 + C^2}$ (this is not zero because not all of A, B, and C are zero). The resulting equation is

$$\frac{A}{\pm\sqrt{A^2 + B^2 + C^2}}x + \frac{B}{\pm\sqrt{A^2 + B^2 + C^2}}y + \frac{C}{\pm\sqrt{A^2 + B^2 + C^2}}z + \frac{D}{\pm\sqrt{A^2 + B^2 + C^2}} = 0 \tag{14-4}$$

Compare Eq. (14-4) with the normal form [Eq. (14-3)]; its locus is a *plane* whose normal has the following direction cosines:

$$\cos\alpha = \frac{A}{\pm\sqrt{A^2+B^2+C^2}} \qquad (14\text{-}5)$$

$$\cos\beta = \frac{B}{\pm\sqrt{A^2+B^2+C^2}}$$

$$\cos\gamma = \frac{C}{\pm\sqrt{A^2+B^2+C^2}}$$

It follows that the numbers A, B, and C are direction numbers of the normal. (See the last example, in which the direction cosines were calculated exactly as above. Since $\overline{ON}$ *extends from the origin to point* N, the coordinates of N form a set of direction numbers.)

The equation $Ax + By + Cz + D = 0$ can evidently be expressed in the normal form by dividing through by $\pm\sqrt{A^2+B^2+C^2}$. In doing this the sign of the radical is chosen as follows. The distance from the origin to the plane is

$$p = \frac{-D}{\pm\sqrt{A^2+B^2+C^2}}$$

In order that p may be positive it is agreed to take the sign of the radical opposite that of D if $D \neq 0$. If $D = 0$, the plane passes through the origin, and $p = 0$. In this case we take the sign of C as the sign of the radical; this amounts to choosing the positive direction on the normal so that γ is an acute angle. If $C = D = 0$, then $p = 0$ and $\gamma = 90°$. In this case the sign of B is chosen as the sign of the radical, making β an acute angle.

14-3 DRAWING THE PLANE AND ITS NORMAL

The line in which a given plane intersects a coordinate plane is called the *trace* of the given plane on that coordinate plane. Thus the trace of the plane whose equation is

$$5x + 8y + 10z = 40 \qquad (14\text{-}6)$$

on the xy plane is the line $\overleftrightarrow{AB}$ in Fig. 14-4. Its equation in the xy plane, found by substituting $z = 0$ in Eq. (14-6), is

$$5x + 8y = 40$$

The equations of the traces on the other coordinate planes are found similarly and are shown in the figure. It is often convenient to represent

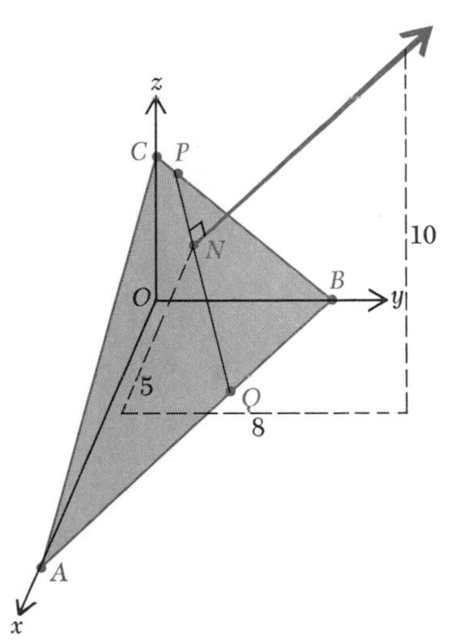

Fig. 14-4 Traces of the plane $5x + 8y + 10z = 40$ on the three coordinate planes are xy: $5x + 8y = 40$; xz: $5z + 10z = 40$; yz: $8y + 10z = 40$.

a plane in space by its traces, as is done in Fig. 14-4.

The *intercept* of a plane on a coordinate axis is the directed distance from the origin to the point of intersection of the plane and the axis. In other words, the x intercept is the x coordinate of the point where the plane meets the x axis. Similar statements can be made for the y and z intercepts. From Fig. 14-4 it can be seen that each intercept is a coordinate of a point of intersection of two traces of the plane. To find the x intercept, for example, set y and z equal to zero in the equation of the plane.

Example

Find the intercepts of the plane whose equation is $5x + 8y + 10z = 40$.

Solution If $x = y = 0$, then $10z = 40$ or $z = 4$. If $x = z = 0$, then $8y = 40$ or $y = 5$. If $y = z = 0$, then $5x = 40$ or $x = 8$. Hence the intercepts are

x intercept: 8
y intercept: 5
z intercept: 4

If a line is drawn from any point P on one trace of a plane to any point Q on another trace, the line $\overleftrightarrow{PQ}$ lies on the plane. Any point N on $\overleftrightarrow{PQ}$ also lies on the plane. A line through point N and normal to the plane may be drawn by considering that its direction numbers are A, B, and C (or any set of numbers proportional to these) in the general equation. In Fig. 14-4 a second point on the normal is found by moving from point N, 5 units in the x direction, 8 units in the y direction, and 10 units in the z, direction.

14-4 PLANE PARALLEL TO ONE OR MORE COORDINATE AXES

The general equation of first degree in x, y, and z is

$$Ax + By + Cz + D = 0 \tag{14-7}$$

in which A, B, and C are not all zero. As indicated above, every such equation represents a plane whose normal has the direction numbers $[A, B, C]$. Suppose now that $C = 0$. In this case $\cos \gamma = 0$ and $\gamma = 90°$; the plane is parallel to the z axis.

Example 1

The equation $3x + 4y = 16$ represents a plane parallel to the z axis, and consequently perpendicular to the xy plane, as shown in Fig. 14-5. On the xy plane its trace is $3x + 4y = 16$, on the xz plane (substitute $y = 0$) its trace is $x = \frac{16}{3}$, and on the yz plane (substitute $x = 0$) its trace is $y = 4$.

Similar considerations apply if A or B is 0. A discussion of the case in which two of the coefficients are zero is given below.

Example 2

Graph the plane whose equation is $x = 2$.

Solution The traces are as follows:

On the xy plane ($z = 0$): $x = 2$
On the xz plane ($y = 0$): $x = 2$
On the yz plane ($x = 0$): None

This plane is parallel to the yz plane and perpendicular to the xz and xy planes (Fig. 14-6). Hence $\alpha = 0°$, $\beta = 90°$, and $\gamma = 90°$.

Note that the equation of a trace of a plane is expressed as the simultaneous solution of the equations of the given plane and a coordinate

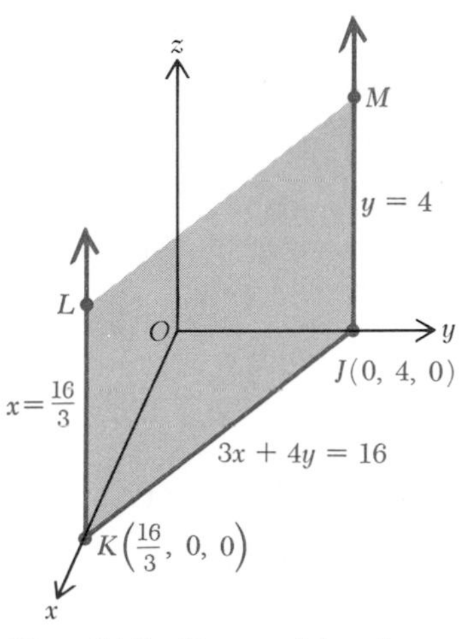

Fig. 14-5 Traces of the plane $3x + 4y = 16$ are $3x + 4y = 16$, $x = \frac{16}{3}$, and $y = 4$.

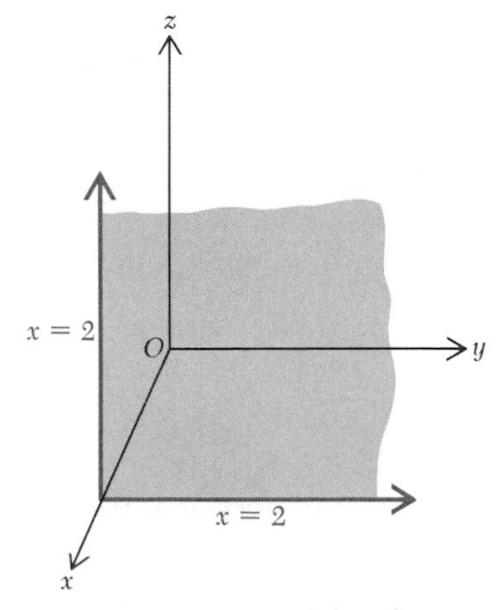

Fig. 14-6 Traces of the plane $x = 2$ on the xy plane and the xz plane are the lines $x = 2$.

plane. In Example 1, for the plane $3x + 4y = 16$ the equations of the traces are simultaneous solutions as follows:

On the xy plane: $$3x + 4y = 16 \qquad z = 0$$

On the xz plane: $$3x + 4y = 16 \qquad y = 0$$

On the yz plane: $$3x + 4y = 16 \qquad x = 0$$

In Fig. 14-5 these are

$$\overleftrightarrow{JK} = \{(x, y, z) \mid (3x + 4y = 16) \cap (z = 0)\}$$
$$\overleftrightarrow{KL} = \{(x, y, z) \mid (3x + 4y = 16) \cap (y = 0)\}$$
$$\overleftrightarrow{JM} = \{(x, y, z) \mid (3x + 4y = 16) \cap (x = 0)\}$$

PROBLEMS

1. Write the equation of the plane:

   (*a*) $\alpha = 60°, \beta = 45°, \gamma = 60°; \quad p = 4$
   (*b*) $\alpha = 135°, \beta = 45°, \gamma = 90°; \quad p = 8$
   (*c*) $\alpha = 90°, \beta = 120°, \gamma = 30°; \quad p = 6$

2. Plot the point P and draw $\overline{OP}$, find the equation of the plane through P perpendicular to $\overrightarrow{OP}$, and represent the plane in the drawing by its traces:

   (*a*) $P(2, 4, 2)$ (*b*) $P(-3, 1, 2)$ (*c*) $P(-4, -3, 0)$

3. Express the equation in the normal form, find the direction cosines of the line normal to the plane and the distance of the plane from the origin, and draw the figure, representing the plane by its traces:

   (*a*) $2x + 3y + 6z + 21 = 0$ (*b*) $x + 2z - 6 = 0$
   (*c*) $8x + 4y - z = 27$ (*d*) $y - 3 = 0$

4. Make a drawing in which the given plane is represented by its traces. Find the equations of the traces. Select a random point on the plane and draw the normal at this point.

   (*a*) $5x + 6y + 10z = 30$ (*b*) $3x - 2y + 2z - 12 = 0$
   (*c*) $3x - 4z + 12 = 0$ (*d*) $x - 2y - 4z + 8 = 0$

5. Find the equation of the plane that passes through the point P and is perpendicular to the line $\overleftrightarrow{AB}$:

   (*a*) $A(-2, 1, 1)$, $B(4, 5, 3)$; $P(2, 1, 2)$
   (*b*) $A(1, 5, 2)$, $B(4, -1, 5)$; $P(2, 2, 2)$

6. A plane passes through the midpoint of the segment having end points $A(-2, 5, 1)$ and $B(6, 1, 5)$ and is perpendicular to the line $\overleftrightarrow{AB}$. Find its equation. Check your results by using a different method to find the equation.
7. (*a*) Find the coordinates of the point P on the segment with end points $A(-2, -1, 2)$ and $B(7, 5, 5)$ that is two-thirds of the way from A to B. (*b*) Find the equation of the plane through P perpendicular to $\overline{AB}$.
8. Find the equation of the locus of points equidistant from the two points $A(-3, -2, -1)$ and $B(7, 0, 5)$.
9. Find the point on the y axis that is equidistant from the two points $A(1, -4, 4)$ and $B(7, 6, 2)$.
10. Find the center and radius of the sphere that passes through the points $A(10, -3, 4)$, $B(11, 7, -3)$, $C(-2, 0, 7)$, and $D(0, -4, 5)$. *Hint:* The center is the point in space that is equidistant from A, B, C, and D; find the equations of the planes that are the perpendicular bisectors of segments $\overline{AB}$, $\overline{BC}$, and $\overline{CD}$ and solve these equations simultaneously.
11. Show that the undirected distance from the plane

$$Ax + By + Cz + D = 0$$

   to the point $P(x_1, y_1, z_1)$ is given by the formula

$$d = \frac{|Ax_1 + By_1 + Cz_1 + D|}{\sqrt{A^2 + B^2 + C^2}}$$

12. Find the distance from the given plane to point P (see Prob. 11):

   (*a*) $2x + 2y + z = 15$; $P(5, 4, 6)$ (*b*) $2x - 3y + 6z = 0$; $P(1, -1, 1)$

13. Sketch the graphs of the following sets of points:

(a) $\{(x, y, z) \mid x + 2y + 3z = 6\}$ (b) $\{(x, y, z) \mid 4x + 5z = 10\}$
(c) $\{(x, y, z) \mid 4x + y = 10\}$ (d) $\{(x, y, z) \mid 6x + 5y - 30 = 0\}$
(e) $\{(x, y, z) \mid 3z - 5 = 0\}$ (f) $\{(x, y, z) \mid 3y - z = 0\}$

14. Find the equation of a plane passing through the point P and parallel to the given plane:

(a) $x - 4y + 5z + 15 = 0$; $P(3, 2, -4)$
(b) $3x - 5y - z - 12 = 0$; $P(-1, 6, 3\frac{1}{2})$

14-5 PLANE DETERMINED BY THREE POINTS

Of the four constants in the equation

$$Ax + By + Cz + D = 0$$

only three are essential, for it is possible to divide through by any one (whose value is not zero) and thus reduce the number of constants to three. Thus if $A \neq 0$, the equation may be written in the form

$$x + by + cz + d = 0 \tag{14-8}$$

where $b = B/A$, $c = C/A$, and $d = D/A$.

Since Eq. (14-8) contains three essential constants, the plane is determined by a set of three independent conditions. One such set of conditions is the requirement that the plane pass through three distinct points not all on the same line. The following example illustrates a procedure for finding the equation of the plane which satisfies this requirement.

Example

Find the equation of the plane passing through the points $(-1, 2, 4)$, $(3, 3, \frac{1}{2})$, and $(4, 1, 1\frac{1}{2})$.

Solution If the plane determined by these points is such that $A \neq 0$ in the equation $Ax + By + Cz + D = 0$, its equation may be written in form (14-8). Substituting the coordinates of the given points, in turn, for x, y, and z results in three simultaneous equations in b, c, and d:

$$\begin{aligned} -1 + 2b + 4c + d &= 0 \\ 3 + 3b + \tfrac{1}{2}c + d &= 0 \\ 4 + b + \tfrac{3}{2}c + d &= 0 \end{aligned} \tag{14-9}$$

Their solution is

$$b - \tfrac{5}{4}$$
$$c = \tfrac{3}{2}$$
$$d = -\tfrac{15}{2}$$

The required equation is then

$$x + \tfrac{5}{4}y + \tfrac{3}{2}z - \tfrac{15}{2} = 0$$

or

$$4x + 5y + 6z - 30 = 0$$

If the plane determined by the three points had been parallel to the x axis (so that $A = 0$), the set of equations (14-9) would have been inconsistent. In this case the equation of the plane may be assumed to be $y + cz + d = 0$. Since A, B, and C are not all zero, it is always possible to divide through by one of them.

14-6 THE ANGLE FORMED BY TWO PLANES

Let two planes intersect along a line as shown in Fig. 14-7. Two supplementary angles of intersection, θ and $\pi - \theta$, are formed. These angles are congruent, respectively, to the two angles formed by the normals drawn to the planes from any point P in space. The problem of finding the angle formed by two planes thus reduces to that of finding the angle formed by the normals to the two planes. If the equations of the planes are

$$A_1x + B_1y + C_1z + D_1 = 0$$

and

$$A_2x + B_2y + C_2z + D_2 = 0$$

then the normals have direction numbers

$$[A_1, B_1, C_1] \qquad \text{and} \qquad [A_2, B_2, C_2]$$

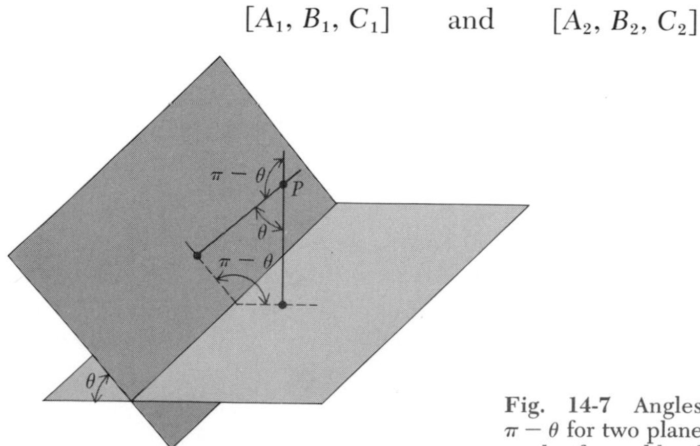

Fig. 14-7 Angles of intersection θ and $\pi - \theta$ for two planes are congruent to the angles formed by the normals.

The angle θ formed by these normals, and consequently that formed by the planes, is given by (see Sec. 13-8)

$$\cos \theta = \pm \frac{A_1A_2 + B_1B_2 + C_1C_2}{\sqrt{A_1^2 + B_1^2 + C_1^2}\sqrt{A_2^2 + B_2^2 + C_2^2}} \tag{14-10}$$

If positive directions are chosen on the normals, the angle between the planes may be defined as that between the positive directions of the normals, and the sign in Eq. (14-10) is chosen accordingly. An alternative agreement would be to choose the sign in each case so as to make $\cos \theta$ positive; in this case Eq. (14-10) always gives the cosine of the acute angle.

The two planes are perpendicular if and only if $\cos \theta = 0$. Hence the condition for perpendicularity of the two planes is

$$A_1A_2 + B_1B_2 + C_1C_2 = 0 \tag{14-11}$$

Two planes are parallel if and only if their normals are parallel, in which case the direction numbers of the normals are proportional. The condition for parallelism of the two planes is then

$$\frac{A_1}{A_2} = \frac{B_1}{B_2} = \frac{C_1}{C_2} \tag{14-12}$$

14-7 VECTOR EQUATION OF A PLANE

A plane may be described by naming one point on it and by identifying a vector normal to the plane (see Fig. 14-8). This information is equiva-

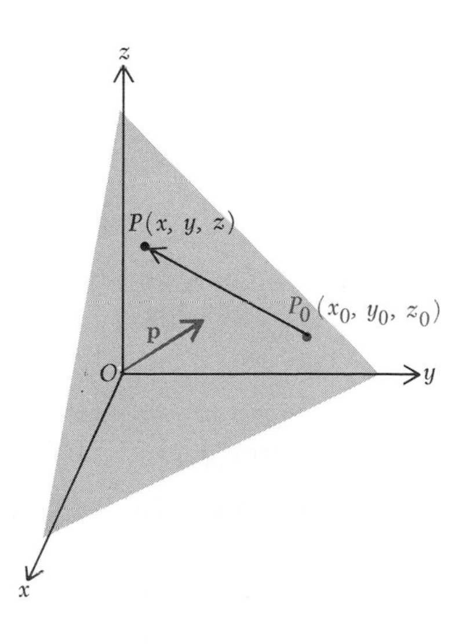

Fig. 14-8 Vector **p** is normal to the plane through $P_0(x_0, y_0, z_0)$.

lent to three independent conditions, since a vector normal to the plane actually specifies a family of planes perpendicular to that vector.

Let the given point on the plane be $P_0(x_0, y_0, z_0)$, and **p** be the given normal from the origin to the plane. Then, if $P(x, y, z)$ is any other point on the plane, **p** is perpendicular to $\mathbf{P_0P}$. That is,

$$\mathbf{P_0P} \cdot \mathbf{p} = 0 \tag{14-13}$$

Equation (14-13) is one form of the vector equation of a plane. Since $\mathbf{P_0P} = \mathbf{OP} - \mathbf{OP_0}$, a second vector form is

$$(\mathbf{OP} - \mathbf{OP_0}) \cdot \mathbf{p} = 0 \tag{14-14}$$

Since

$$\begin{aligned} \mathbf{OP} &= x\mathbf{i} + y\mathbf{j} + z\mathbf{k} \\ \mathbf{OP_0} &= x_0\mathbf{i} + y_0\mathbf{j} + z_0\mathbf{k} \\ \mathbf{p} &= a\mathbf{i} + b\mathbf{j} + c\mathbf{k} \end{aligned}$$

where $[a, b, c]$ is a set of direction numbers for **p**,

$$\begin{aligned} (\mathbf{OP} - \mathbf{OP_0}) \cdot \mathbf{p} &= [(x - x_0)\mathbf{i} + (y - y_0)\mathbf{j} + (z - z_0)\mathbf{k}] \cdot (a\mathbf{i} + b\mathbf{j} + c\mathbf{k}) \\ &= [(x - x_0)\mathbf{i} + (y - y_0)\mathbf{j} + (z - z_0)\mathbf{k}] \cdot a\mathbf{i} \\ &\qquad + [(x - x_0)\mathbf{i} + (y - y_0)\mathbf{j} + (z - z_0)\mathbf{k}] \cdot b\mathbf{j} \\ &\qquad + [(x - x_0)\mathbf{i} + (y - y_0)\mathbf{j} + (z - z_0)\mathbf{k}] \cdot c\mathbf{k} \\ &= a(x - x_0)\mathbf{i} \cdot \mathbf{i} + 0 + 0 + 0 + b(y - y_0)\mathbf{j} \cdot \mathbf{j} \\ &\qquad + 0 + 0 + 0 + c(z - z_0)\mathbf{k} \cdot \mathbf{k} \end{aligned}$$

since $\mathbf{i} \cdot \mathbf{j} = \mathbf{j} \cdot \mathbf{k} = \mathbf{k} \cdot \mathbf{i} = 0$; hence

$$\mathbf{OP} - \mathbf{OP_0} \cdot \mathbf{p} = a(x - x_0) + b(y - y_0) + c(z - z_0) = 0 \tag{14-15}$$

You should recognize this as the equation of a plane through (x_0, y_0, z_0) and with a normal through the origin with direction numbers $[a, b, c]$. This development shows an alternative way of deriving the general equation of a plane through the use of vectors.

Example

Find the equation of a plane through $P_0(3, -2, 4)$ and perpendicular to **ON** if N has coordinates $(2, 2, -1)$.

Solution From Eq. (14-14), $\mathbf{P_0P} \cdot \mathbf{ON} = 0$, where P is a general point in the required plane. Then

$$\begin{aligned} \mathbf{P_0P} \cdot \mathbf{ON} &= [(x - 3)\mathbf{i} + (y + 2)\mathbf{j} + (z - 4)\mathbf{k}] \cdot (2\mathbf{i} + 2\mathbf{j} - \mathbf{k}) \\ &= 2(x - 3) + 2(y + 2) - (z - 4) = 0 \end{aligned}$$

and the desired equation is

$$2x + 2y - z + 2 = 0$$

This is, of course, the same equation that would have been obtained by methods described earlier in the chapter.

PROBLEMS

1. Find the equation of a plane that passes through the given points:

   (*a*) $E(2, 2, 2)$, $F(3, 1, 1)$, $G(6, -4, -6)$
   (*b*) $E(5, 1, -1)$, $F(9, -3, -5)$, $G(-1, 4, 3)$
   (*c*) $E(2, 1, 0)$, $F(3, 0, 2)$, $G(0, 4, 3)$

2. Find the equation of the plane determined by points P, Q, and R; find the intercepts of the plane on the axes and draw the figure:

   (*a*) $P(4, 2, 3)$, $Q(1, 1, 5)$, $R(2, 6, 2)$
   (*b*) $P(-1, 1, \frac{7}{3})$, $Q(\frac{3}{2}, 8, 1)$, $R(-3, 4, 0)$
   (*c*) $P(3, 2, 0)$, $Q(1, 0, -5)$, $R(4, 2, 1)$

3. Show that if the x, y, and z intercepts of a plane are, respectively, a, b, and c with the product abc not zero, the equation of the plane is

$$\frac{x}{a} + \frac{y}{b} + \frac{z}{c} = 1$$

4. A plane passes through the points $(3, 1, 7)$ and $(-3, -2, 3)$ and has x intercept equal to three times its z intercept. What is its equation?
5. A plane has y intercept equal to -4 and passes through the points $(4, 1, 5)$ and $(-1, -2, 4)$. What is its equation?
6. A plane passes through the points $(-3, 5, 4)$ and $(8, -4, -2)$ and is perpendicular to the plane $2x - y - z = 3$. Find its equation.
7. The x and z intercepts of a plane are -4 and 6, respectively. The plane is perpendicular to the plane $x + 2y - 2z = 12$. What is its equation?
8. Determine which of the following pairs of planes are parallel:

   (*a*) $x - 2y + 5z = 4$; $3x - 6y + 10z = 0$
   (*b*) $3x + 4y - 5z = 7$; $-x - \frac{4}{3}y + \frac{5}{3}z = 4$
   (*c*) $x - 3y + 7z + 5 = 0$; $4x - y - z = 1$
   (*d*) $4x + 3y - 7 = 0$; $6x - 8y + 3z - 29 = 0$

9. Determine which pairs of planes in Prob. 8 are perpendicular.
10. Find the acute angle formed by each set of two given planes:

    (*a*) $2x - y + 2z - 10 = 0$; $4x + y + z - 7 = 0$
    (*b*) $5x + 3y - 4z + 14 = 0$; $x - 4y - z + 12 = 0$
    (*c*) $3x + 4y = 16$; $4y - 2z = 5$

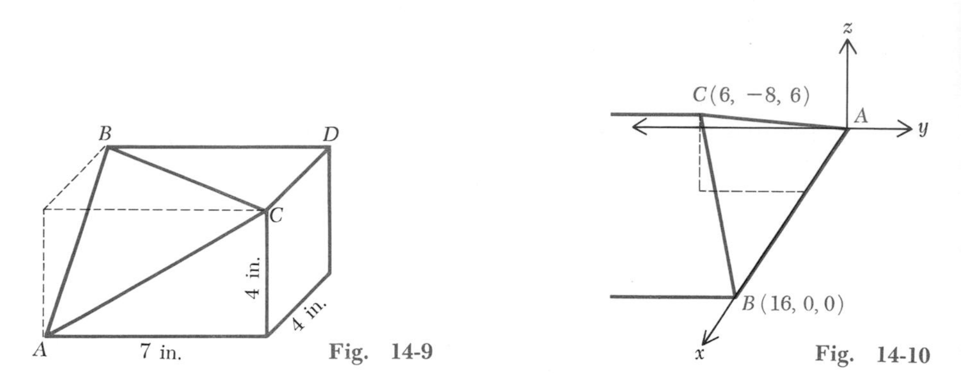

Fig. 14-9

Fig. 14-10

11. Write the vector equation and use it to derive the general equation of each plane:

    (*a*) A plane passing through (7, 1, 2), with [1, −2, −2] as direction numbers of the normal through the origin
    (*b*) A plane passing through (−3, 5, 1), with [−3, 8, 4] as direction numbers of the normal through the origin
    (*c*) A plane passing through (4, −3, −1), with [6, −1, 5] as direction numbers of the normal through the origin

12. One corner of the rectangular block shown in Fig. 14-9 is cut off by a plane passing through points A, B, and C. What angle does this plane make with the top of the block (plane BCD)?
13. The base of a right pyramid 12 in. high is a rectangle 8 in. long and 6 in. wide. Compute the dihedral angle between two of its faces.
14. One end of a roof has the construction shown in Fig. 14-10. Compute the angle formed by the planes that intersect along $\overleftrightarrow{BC}$.
15. A plane passes through the point (−4, −2, 3) and is parallel to the plane $3x + y - 2z = 5$. Find its equation.
16. For what value of k are the planes $x - 6y + 8z - 4 = 0$ and $4x + ky + z - 7 = 0$ perpendicular?
17. A line segment is to be drawn from $P(10, 4, 4)$ to a point Q on the z axis. $\overleftrightarrow{PQ}$ is parallel to the plane $2x + 3y + 4z = 30$. What are the coordinates of Q?

14-8 EQUATIONS OF A LINE IN SPACE: GENERAL FORM

Earlier in this chapter it was shown that the locus of any linear equation in x, y, and z is a plane. Consider now the pair of equations

$$\begin{aligned} A_1x + B_1y + C_1z + D_1 &= 0 \\ A_2x + B_2y + C_2z + D_2 &= 0 \end{aligned} \tag{14-16}$$

Each of these equations represents a plane, and if they are not parallel, these planes intersect along a line $\overleftrightarrow{PQ}$ as shown in Fig. 14-11. The first equation of (14-16) is satisfied by the coordinates of all points on the first plane and by no other points. Similarly, the second equation of (14-16) is satisfied by the coordinates of all points on the second plane and by no other points. It is immediately evident that *the line of intersection of the planes is the locus of all points whose coordinates satisfy both equations.* The *two* equations (14-16) therefore *define* this line.

In order to graph the line, two points may be located on it. Substitute any convenient value for z into Eqs. (14-16) and then solve them simultaneously for the corresponding values of x and y. This gives the coordinates of one point on the line. Obtain a second point in the same way, using a different value for z, and draw the line through the two points.

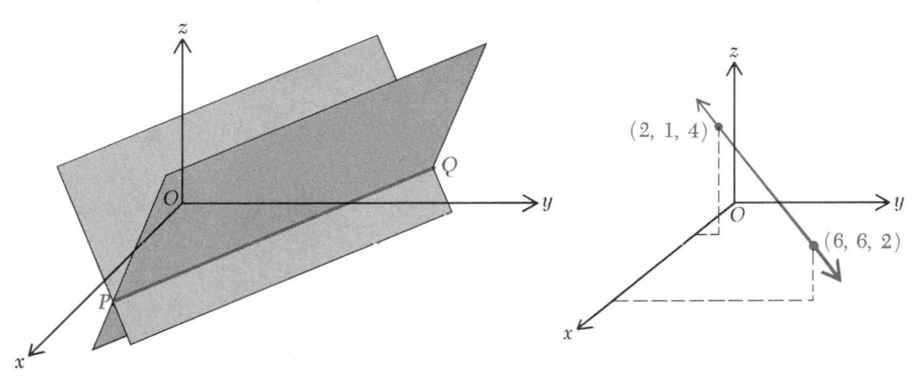

Fig. 14-11 Two planes intersecting along line $\overleftrightarrow{PQ}$.

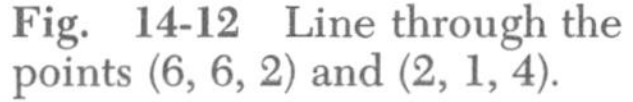

Fig. 14-12 Line through the points (6, 6, 2) and (2, 1, 4).

Example

Draw the line defined by the equations $3x - 2y + z - 8 = 0$ and $2x - 2y - z + 2 = 0$.

Solution Letting $z = 2$, obtain the equations $3x - 2y = 6$ and $x - y = 0$. Their solution is $x = 6$ and $y = 6$, giving the point (6, 6, 2). Let $z = 4$, and obtain the solution $x = 2$ and $y = 1$. Two points on the line are then (6, 6, 2) and (2, 1, 4). The line is shown in Fig. 14-12.

Arbitrary values could have been substituted for x or y instead of z. In particular, letting $y = 0$ locates the point at which the line intersects the xz plane; letting $x = 0$ locates the point at which it intersects the yz plane.

Two simultaneous linear equations constitute the *general form* of the equations of a line. Other forms in which the equations may be given are discussed in the following sections.

14-9 THE PROJECTION FORM

Let a line l be defined by Eqs. (14-16). Each of these equations represents a plane through l (see Fig. 14-11, where l is the line $\overleftrightarrow{PQ}$). Now consider the equation

$$(A_1x + B_1y + C_1z + D_1) + k(A_2x + B_2y + C_2z + D_2) = 0 \tag{14-17}$$

where k is any constant. This equation represents a plane because it is of the first degree. Furthermore, it is satisfied by the coordinates of every point on line l because for every such point both of Eqs. (14-16) are satisfied, and $0 + k0 = 0$. Hence for every value of k, Eq. (14-17) represents a plane through the line l. It is the equation of the family of planes through l, one of which corresponds to each value assigned to k.

Suppose now that k is chosen in Eq. (14-17) so as to eliminate z (that is, let $k = -C_1/C_2$). The resulting equation in x and y is that of the plane through l perpendicular to the xy plane. Similarly, if k is chosen so as to eliminate x or y from Eq. (14-17), the resulting equation in y and z or in x and z is that of the plane through l perpendicular to the yz or xz plane, respectively. These planes are called the *projecting planes* of the line l on the coordinate planes. It is often convenient to define a line by giving the equations of two of its projecting planes instead of two general planes.

Example 1
The equations

$$\begin{aligned} 4x + 7y - 64 &= 0 \\ 2y - 7z + 3 &= 0 \end{aligned} \tag{14-18}$$

define the line $\overleftrightarrow{PQ}$ shown in Fig. 14-13. The first equation is that of the plane through $\overleftrightarrow{PQ}$ perpendicular to the xy plane. The second is that of the plane through $\overleftrightarrow{PQ}$ perpendicular to the yz plane. $\overleftrightarrow{ST}$ and $\overleftrightarrow{MN}$ are the projections of $\overleftrightarrow{PQ}$ on these coordinate planes.

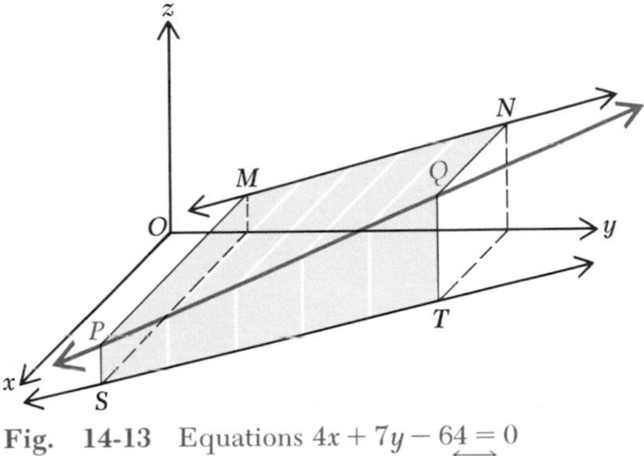

Fig. 14-13 Equations $4x + 7y - 64 = 0$ and $2y - 7z + 3 = 0$ define the line $\overleftrightarrow{PQ}$.

Equations (14-18) defining $\overleftrightarrow{PQ}$ are called the *projection form* of the equations of the line. If the equations of a line are given in the general form, the corresponding projection form can be obtained as illustrated by the following example.

Example 2
Find the projection form of the equations for the line defined by $10x + 3y + 4z = 50$ and $15x + 6y + 4z = 80$ and draw the line.

Solution If z is eliminated by subtracting the first equation from the second, the result is

$$5x + 3y = 30$$

If y is similarly eliminated, the result is

$$5x + 4z = 20$$

The equations

$$5x + 3y = 30$$
$$5x + 4z = 20$$

then constitute the projection form of the line. The traces of these planes are as follows:

Plane $5x + 3y = 30$:

In the xy plane: $5x + 3y = 30$
In the xz plane: $5x = 30$
In the yz plane: $3y = 30$

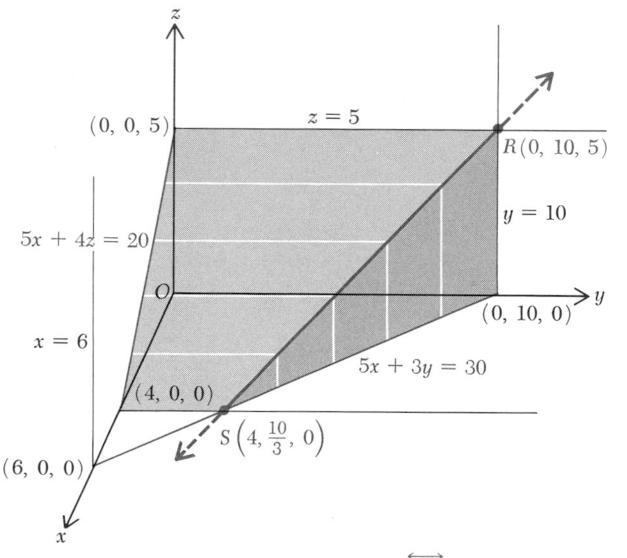

Fig. 14-14 Two projection planes for $\overleftrightarrow{RS}$ are $5x + 3y = 30$ and $5x + 4z = 20$.

Plane $5x + 4z = 20$:

In the xy plane:	$5x = 20$
In the xz plane:	$5x + 4z = 20$
In the yz plane:	$4z = 20$

These traces are graphed in Fig. 14-14. Note that points R and S are points on both planes because they are on the traces of each. Then points R and S are also on the line defined as the intersection of those planes, and $\overleftrightarrow{RS}$ is the required line.

Of course, x could have been eliminated and the resulting equation in y and z used as one of the two equations defining the line.

14-10 THE SYMMETRIC FORM

A line is determined by two points on it, or by one point and the direction cosines or direction numbers of the line. From such data the equations of the line can easily be written in the *symmetric form.* Let $P_1(x_1, y_1, z_1)$ be one point on a line l (Fig. 14-15) whose directions angles are $[\alpha, \beta, \gamma]$, and let $P(x, y, z)$ be any other point on the line. Denote the distance P_1P by d. Then

$$\cos \alpha = \frac{x - x_1}{d}$$

$$\cos \beta = \frac{y - y_1}{d} \tag{14-19}$$

$$\cos \gamma = \frac{z - z_1}{d}$$

If each of these equations is solved for d and the values so obtained are equated, the result is

$$\frac{x - x_1}{\cos \alpha} = \frac{y - y_1}{\cos \beta} = \frac{z - z_1}{\cos \gamma} \tag{14-20}$$

These equations constitute the *symmetric form* of the equations of a line. They can be used for writing the equations of the line when the coordinates of one point on the line and the direction cosines of the line are known (provided, of course, that none of the direction cosines is zero).

Equalities (14-20) still hold if all the denominators are multiplied by a nonzero constant. Thus the direction cosines may be replaced by a set of direction numbers. The equations of the line through the point $P_1(x_1, y_1, z_1)$ with direction numbers $[a, b, c]$ are then

$$\frac{x - x_1}{a} = \frac{y - y_1}{b} = \frac{z - z_1}{c} \tag{14-21}$$

This is also called a symmetric form. It is, of course, equivalent to any two of the equations

$$\frac{x - x_1}{a} = \frac{y - y_1}{b}$$

$$\frac{x - x_1}{a} = \frac{z - z_1}{c} \tag{14-22}$$

$$\frac{y - y_1}{b} = \frac{z - z_1}{c}$$

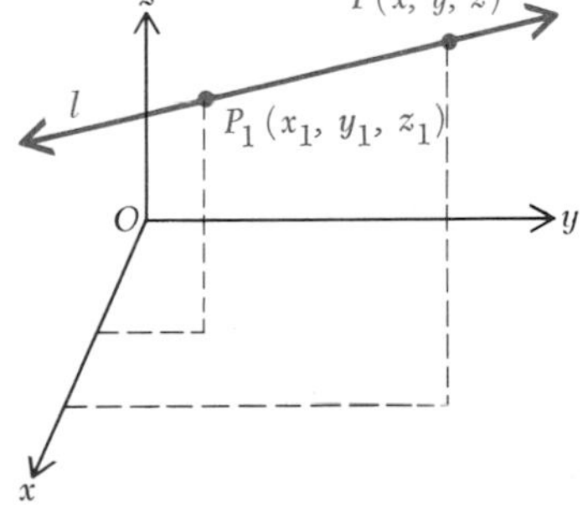

Fig. 14-15 $P_1(x_1, y_1, z)$ and $P(x, y, z)$ are points on line l with direction angles $[\alpha, \beta, \gamma]$.

These equations, when simplified, constitute the *projection form.* The first is the equation of the plane that contains the line and is perpendicular to the xy plane. The second represents the xz projecting plane, and the third represents the yz projecting plane.

To find the equations of a line when two points $P_1(x_1, y_1, z_1)$ and $P_2(x_2, y_2, z_2)$ are given, determine the direction numbers

$$[x_2 - x_1, y_2 - y_1, z_2 - z_1]$$

Then write its equations in symmetric form (14-21).

Example

Find the symmetric and projection equations of the line passing through $A(2, 2, 4)$ and $B(8, 6\frac{1}{2}, 2\frac{1}{2})$.

Solution A set of direction numbers for the line is $[6, 4\frac{1}{2}, -1\frac{1}{2}]$, which may be simplified to $[4, 3, -1]$. The symmetric equations are

$$\frac{x-2}{4} = \frac{y-2}{3} = \frac{z-4}{-1}$$

The equation $(x-2)/4 = (y-2)/3$ simplifies to $3x - 4y + 2 = 0$; the equation $(x-2)/4 = (z-4)/-1$ simplifies to $x + 4z - 18 = 0$. The projection form of the equations is then

$$3x - 4y + 2 = 0$$
$$x + 4z - 18 = 0$$

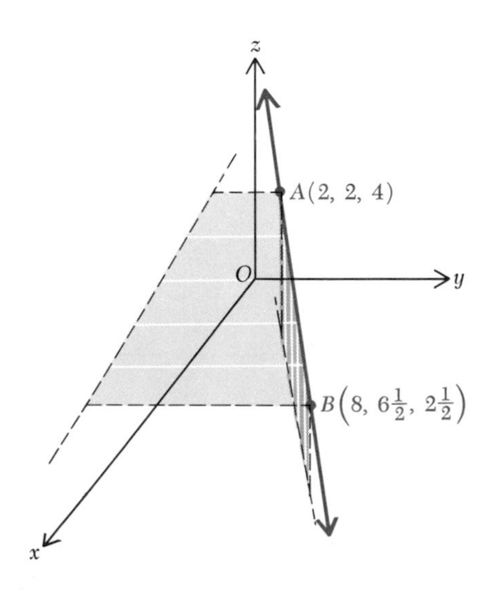

Fig. 14-16 Projection form for $\overleftrightarrow{AB}$ is $3x - 4y + 2 = 0$, $x + 4z - 18 = 0$.

These equations represent the two projecting planes shown in Fig. 14-16.

14-11 THE PARAMETRIC FORM

Let t denote the value of each of the three fractions in the symmetric form (14-21); then

$$\frac{x - x_1}{a} = t$$

(14-23) $$\frac{y - y_1}{b} = t$$

$$\frac{z - z_1}{c} = t$$

Solving these equations for x, y, and z, respectively, results in

(14-24) $$\begin{aligned} x &= x_1 + at \\ y &= y_1 + bt \\ z &= z_1 + ct \end{aligned}$$

These are *parametric equations* of the line with parameter t. Each value of t determines the coordinates of one point on the line.

Example 1
The equations

$$\begin{aligned} x &= 2 + 4t \\ y &= 2 + 3t \\ z &= 4 - t \end{aligned}$$

represent the line shown in Fig. 14-16. The coordinates of a point on the line are obtained by assigning any value to t. Thus, if $t = -1$, it is found that $x = -2$, $y = -1$, and $z = 5$. The point $(-2, -1, 5)$ is on the line.

Example 2
Write a set of parametric equations for the line passing through $P(1, 5, -3)$ with direction numbers $[6, -1, -1]$.

Solution In Eqs. (14-24), $x_1 = 1$, $y_1 = 5$, and $z_1 = -3$. Also, $a = 6$, $b = -1$, and $c = -1$, hence a set of parametric equations for the line is

$$\begin{aligned} x &= 1 + 6t \\ y &= 5 - t \\ z &= -3 - t \end{aligned}$$

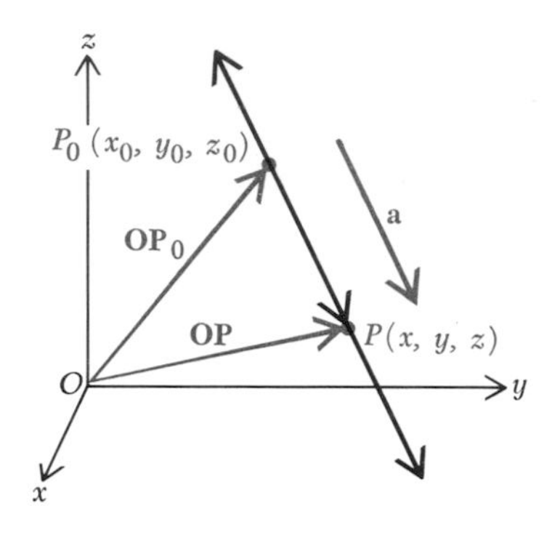

Fig. 14-17 One vector form for the equation of line $\overleftrightarrow{P_0P}$ is $\mathbf{OP} - \mathbf{OP}_0 = m\mathbf{a}$.

14-12 VECTOR EQUATION OF A LINE IN SPACE

To write the equation of line $\overleftrightarrow{P_0P}$ in Fig. 14-17 in vector form, let $\mathbf{a} = A\mathbf{i} + B\mathbf{j} + C\mathbf{k}$ be a vector parallel to $\overleftrightarrow{P_0P}$. Then $\mathbf{P_0P} = m\mathbf{a}$ for m a scalar. Since $\mathbf{P_0P} = \mathbf{OP} - \mathbf{OP}_0$, one vector form of the equation of a line in space is

$$\mathbf{OP} - \mathbf{OP}_0 = m\mathbf{a} \tag{14-25}$$

Since $\mathbf{OP} = x\mathbf{i} + y\mathbf{j} + z\mathbf{k}$ and $\mathbf{OP}_0 = x_0\mathbf{i} + y_0\mathbf{j} + z_0\mathbf{k}$,

$$(x - x_0)\mathbf{i} + (y - y_0)\mathbf{j} + (z - z_0)\mathbf{k} = m(A\mathbf{i} + B\mathbf{j} + C\mathbf{k}) \tag{14-26}$$

The three equations obtained by equating the coefficients in the vector form (14-26) are a set of parametric equations

$$\begin{aligned} x - x_0 &= mA \\ y - y_0 &= mB \\ z - z_0 &= mC \end{aligned} \tag{14-27}$$

Equations (14-27) represent a line through (x_0, y_0, z_0) with direction numbers $[A, B, C]$.

PROBLEMS

1. Locate two points on the line defined by each pair of equations and draw the line:

   (a) $x + 3y + z = 18;\quad 2x + y - 3z = 6$
   (b) $x - y + z - 2 = 0;\quad 8x - 9y + 4z + 10 = 0$
   (c) $x + 2y - 7z = 2;\quad 2x - y + 4z = 6$

2. Find the points at which each defined line intersects the xy and xz planes and draw the graph of the line through these points:

   (a) $\{(x, y, z) \mid (6x - y + 3z - 18 = 0) \cap (2x - 3y - 3z + 10 = 0)\}$
   (b) $\{(x, y, z) \mid (x + y - z - 2 = 0) \cap (x - 3y + z - 10 = 0)\}$
   (c) $\{(x, y, z) \mid (2x + 6y - 6z - 35 = 0) \cap (x - 4y + 8z + 21 = 0)\}$

3. Find the direction cosines of the line defined by the equations $2x - 3y - 2z + 11 = 0$ and $x - 6y + 2z + 10 = 0$ if the positive direction is such that γ is acute.
4. Write the symmetric equations of the line satisfying each set of conditions, and then express the equations in the projection form:

   (a) Passing through $P(2, 1, 6)$, with direction numbers $[4, 3, -2]$
   (b) Passing through $Q(-1, 8, 1)$, with direction numbers $[2, -1, 2]$
   (c) Passing through $R(-2, -5, -2)$ and $S(4, -1, 2)$
   (d) Passing through $T(-2, 0, 0)$ and $U(4, 5, 3)$
   (e) Passing through $V(6, 2, -2)$ parallel to the line through $A(4, -4, 2)$ and $B(1, -2, -1)$

5. Find equations of the line passing through the given point and perpendicular to the given plane:

   (a) $A(8, 10, 8)$; $x + y + 2z = 6$ (b) $P(2, 1, 4)$; $4x - 3y + 2z = 12$

6. Show that the lines

$$\frac{x+5}{3} = \frac{y-7}{4} = \frac{z+2}{2} \quad \text{and} \quad \frac{x-3}{6} = \frac{y+5}{-5} = \frac{z-2}{1}$$

   are perpendicular.
7. A line is drawn through $P(0, 4, 0)$ with direction numbers $[6, 4, 5]$. Write its equations. Another line is to be drawn from point P to a point A on the z axis, and this second line is to be perpendicular to the first. Find the coordinates of A.
8. At what point does the line

$$\frac{2x-9}{6} = \frac{y-3}{1} = \frac{z-5}{4}$$

   intersect the plane $2x + 3y + 3z - 12 = 0$?
9. Draw the line whose parametric equations are $x = 2 - 3t$, $y = 1 + 4t$, and $z = 6 - t$. Find its projection equations.
10. A line is drawn from $A(3, -1, 4)$ to meet at right angles the line whose parametric equations are $x = 4t + 1$, $y = 6 - t$, and $z = 1 + t$. Find the point of intersection.
11. Write an equation of the family of planes (a) containing the origin; (b) parallel to the xy plane; (c) parallel to the x axis; (d) perpendicular to the xz plane.
12. Show that the four points $A(1, 2, 3)$, $B(2, 4, 0)$, $C(-3, -3, 6)$, and $D(0, 0, 4)$ are coplanar.

13. Find the equation of the plane through $P(-4, 6, 1)$ perpendicular to the line whose projection equations are $x - 2z + 2 = 0$ and $2y + z + 5 = 0$.
14. Find the equation of the plane that passes through the point $P(2, -3, 4)$ and contains the line whose projection equations are $2x - y - 3 = 0$ and $x + 3z - 6 = 0$. *Hint:* First write the equation of the family of planes containing the line.
15. Find the equation of the plane that passes through the origin and contains the line

$$\frac{x+1}{3} = \frac{2y+3}{4} = \frac{z+6}{-2}$$

16. Show that the lines

$$\frac{x-1}{3} = \frac{y+2}{5} = \frac{z+6}{13} \quad \text{and} \quad \frac{x-4}{9} = \frac{y-3}{5} = \frac{z-7}{4}$$

determine a plane and find the equation of the plane.
17. Show that the lines

$$\frac{x+2}{2} = \frac{y-1}{-3} = \frac{z+5}{4} \quad \text{and} \quad \frac{x-3}{1} = \frac{2y+5}{-3} = \frac{z+3}{2}$$

are parallel and find the equation of the plane they determine.

SUMMARY

One definition of a plane is a set of points equidistant from two distinct points in three-dimensional space. Some equations for a plane are as follows:

1. The general form $Ax + By + Cz + D = 0$ represents a plane whose normal has direction numbers $[A, B, C]$.
2. The normal form is $x \cos \alpha + y \cos \beta + z \cos \gamma = p$, where p is the directed distance from the origin to the plane and α, β, and γ are the direction angles of this normal.
3. The vector form is $\mathbf{P_0P} \cdot \mathbf{p} = 0$, where P_0 is a certain point in the plane, P is a general point on the plane, and $\mathbf{p}$ is a vector normal to the plane.

The formula for the angle formed by two planes $A_1x + B_1y + C_1z + D_1 = 0$ and $A_2x + B_2y + C_2z + D_2 = 0$ is

$$\cos \theta = \pm \frac{A_1A_2 + B_1B_2 + C_1C_2}{\sqrt{A_1^2 + B_1^2 + C_1^2}\,\sqrt{A_2^2 + B_2^2 + C_2^2}}$$

These two planes are perpendicular if and only if $A_1A_2+B_1B_2+C_1C_2=0$ and are parallel if and only if $A_1/A_2 = B_1/B_2 = C_1/C_2$.

A line in space may be represented in various ways by sets of equations. For the general form

$$A_1x + B_1y + C_1z + D_1 = 0$$
$$A_2x + B_2y + C_2z + D_2 = 0$$

the line may be thought of as determined by the intersection of two planes.

In the projection form, as in the general form, the line is represented by the equations of two planes, but this time the planes chosen are two projecting planes. A projecting plane for a line is a plane through the line and perpendicular to one of the coordinate planes.

The symmetric form, $(x - x_1)/\cos\alpha = (y - y_1)/\cos\beta = (z - z_1)/\cos\gamma$, represents a line with direction angles $[\alpha, \beta, \gamma]$ and passing through the point (x_1, y_1, z_1).

The parametric form $x = x_1 + at$, $y = y_1 + bt$, and $z = z_1 + ct$ represents a line with direction numbers $[a, b, c]$ and passing through the point (x_1, y_1, z_1).

The vector form $\mathbf{OP} - \mathbf{OP}_0 = m\mathbf{a}$ is the equation of a line parallel to $\mathbf{a}$ passing through P_0 and P.

REVIEW PROBLEMS

1. Write the equation of the plane passing through the point $P(1, 2, 1)$ and perpendicular to $\overleftrightarrow{AB}$ for $A(-1, 3, 2)$ and $B(2, 5, -1)$.
2. (*a*) Express the equation $x + 3y - z + 2 = 0$ in normal form. (*b*) Draw the graph, representing the plane by its traces.
3. Find the equation of the locus of points in space equidistant from $A(1, -2, 4)$ and $B(-1, 1, 2)$.
4. Find the equation of the plane passing through points $P(3, 1, 0)$, $Q(4, 1, 5)$, and $R(-1, 1, 2)$.
5. Find the acute angle formed by the two planes in Probs. 1 and 4.
6. Find two points on the line defined by the equations

$$x + y + z - 2 = 0 \qquad \text{and} \qquad 2x - y + z + 1 = 0$$

and draw the line.

7. For a line through $P(-1, 3, 8)$ with direction numbers $[1, -1, 5]$ (*a*) write the equations in symmetric form; (*b*) express the symmetric equations in the projection form; (*c*) write a set of parametric equations for the line.
8. Does the line in Prob. 7 lie on the plane in Prob. 4?
9. Find the equations of a line through $A(6, 0, 4)$ and perpendicular to $x - y + 7z + 2 = 0$.

FOR EXTENDED STUDY

1. Prove that if line l has the parametric representation $x = x_1 + at$, $y = y_1 + bt$, and $z = z_1 + ct$ and if P_1 and P_2 are points on l for which $t = t_1$ and $t = t_2$, respectively, then

$$|P_1P_2| = \sqrt{a^2 + b^2 + c^2}\,|t_2 - t_1|$$

2. Find the equations of two planes that intersect in the line whose parametric equations are as shown. Explain in each case how you know both planes contain the line.

   (*a*) $x = 3 + t$, $y = 2$, $z = 4 - 3t$ (*b*) $x = 1 - t$, $y = 4 + t$, $z = 3$

3. Write the equation of a plane containing the point $P(9, 12, -3)$ and the line $\overleftrightarrow{MN}$ if

$$\overleftrightarrow{MN} = \{(x, y, z) \mid (x + y + z - 14 = 0) \cap (4x + y - 2z - 2 = 0)\}$$

4. Describe in words the location of each of these sets of points:

   (*a*) $P = \{(x, y, z) \mid x \geq 2\}$ (*b*) $Q = \{(x, y, z) \mid |x| \leq 2\}$
   (*c*) $R = \{(x, y, z) \mid (|x| \leq 2) \cap (|y| \leq 2)\}$

5. Determine a so the points $(a, 3, -1)$, $(3, a, 0)$, and $(5, 0, a)$ are collinear.

6. The angle that a line $\overleftrightarrow{AB}$ makes with a plane may be defined as the acute angle ϕ between $\overleftrightarrow{AB}$ and its projection $\overleftrightarrow{A'B'}$ on the plane (Fig. 14-18). If the direction numbers of $\overleftrightarrow{AB}$ are $[a_1, b_1, c_1]$ and those of the normal to the plane are $[a_2, b_2, c_2]$ show that

$$\sin \phi = \left| \frac{a_1a_2 + b_1b_2 + c_1c_2}{\sqrt{a_1^2 + b_1^2 + c_1^2}\,\sqrt{a_2^2 + b_2^2 + c_2^2}} \right|$$

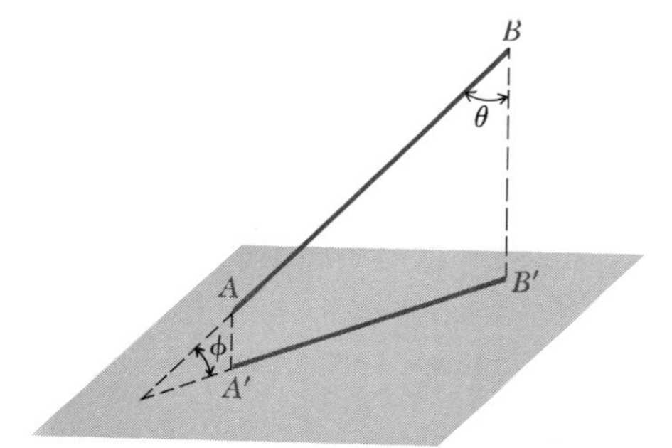

Fig. 14-18

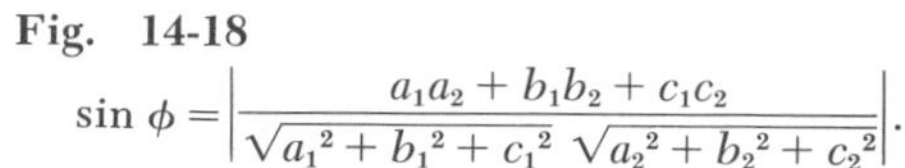
$$\sin \phi = \left| \frac{a_1a_2 + b_1b_2 + c_1c_2}{\sqrt{a_1^2 + b_1^2 + c_1^2}\,\sqrt{a_2^2 + b_2^2 + c_2^2}} \right|.$$

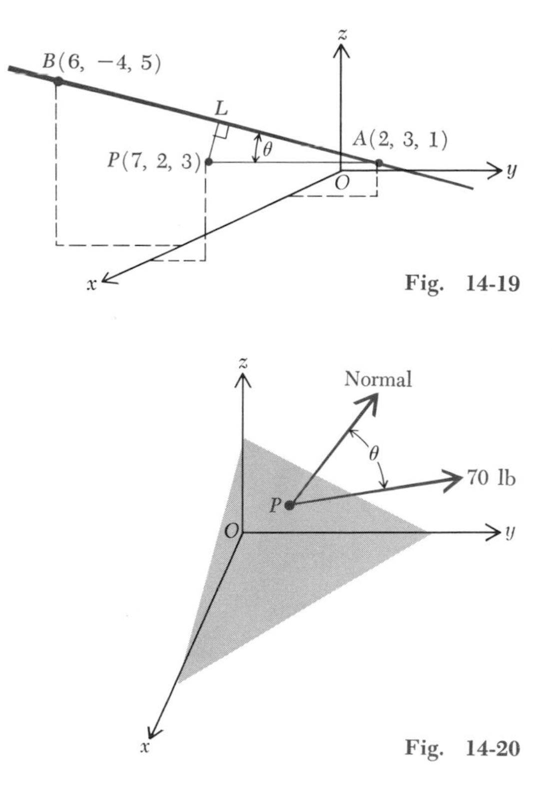

Fig. 14-19

Fig. 14-20

7. Compute the shortest distance from $P(7, 2, 3)$ to the line through $A(2, 3, 1)$ and $B(6, -4, 5)$. *Hint:* In Fig. 14-19 $PL = AP \sin \theta$.
8. At a point P on the plane $2x + 3y + 6z = 18$ a force of 70 lb is applied along the line whose projection equations are $3x - 4y + 2 = 0$ and $3x - 6z + 2 = 0$. Compute the component of the force normal to the plane (see Fig. 14-20).
9. Show that if $P_1(x_1, y_1, z_1)$ and $P_2(x_2, y_2, z_2)$ are two distinct points in the plane with equation $Ax + By + Cz + D = 0$, then every point of $\overleftrightarrow{P_1P_2}$ is in the plane. *Hint:* Express the equation of the line in parametric form.
10. Show that direction numbers of the line determined by the two equations $A_1x + B_1y + C_1z + D_1 = 0$ and $A_2x + B_2y + C_2z + D_2 = 0$ are

$$\left[\begin{vmatrix} B_1 & C_1 \\ B_2 & C_2 \end{vmatrix}, \begin{vmatrix} C_1 & A_1 \\ C_2 & A_2 \end{vmatrix}, \begin{vmatrix} A_1 & B_1 \\ A_2 & B_2 \end{vmatrix}\right]$$

11. A plane is determined by the points $A(-3, 7, -2)$, $B(8, 1, 3)$, and $P(4, 6, 1)$. A second plane is determined by points A, B, and $Q(-3, 7, 2)$. A third plane passes through the point $R(2, 3\frac{1}{2}, -1)$ and is perpendicular to the first two planes. What is its equation?

SURFACES AND CURVES 15

15-1 INTRODUCTION

In the previous chapter the locus of an equation of first degree in x, y, and z was identified as a plane. Moreover, it was found that two such equations taken simultaneously determine a line. The extension to equations of higher degree is made in this chapter. The locus of a single such equation is, in general, a surface. Two such equations, considered simultaneously, will be used to represent a curve in space. For example, the intersection of two spheres may be a circle.

15-2 QUADRIC SURFACES

The locus of an equation of second degree in x, y, and z is called a *quadric surface*. Common quadric surfaces are spheres, cones, and cylinders. The intersection of a plane with any quadric surface is a conic section (except for limiting cases). In the following sections the *standard equations* of several quadric surfaces are discussed.

15-3 THE SPHERE

By use of the distance formula it may be shown that the equation of a sphere with center at $C(h, k, l)$ and radius r is

$$(x-h)^2 + (y-k)^2 + (z-l)^2 = r^2 \tag{15-1}$$

This is the *standard form* of the equation of the sphere. By performing the indicated operations on the left-hand member, we can express the equation in the *general form*

$$x^2 + y^2 + z^2 + Gx + Hy + Iz + K = 0 \tag{15-2}$$

Conversely, any equation of form (15-2) can be expressed in form (15-1) by completing the squares. Such an equation therefore represents a sphere, a point sphere, or no locus, depending upon whether the constant obtained [the right-hand member of Eq. (15-1)] is positive, zero, or negative, respectively. The situation is entirely analogous to that of the circle.

Graphing an equation such as $x^2 + y^2 + z^2 = 9$ (the spherical surface with the center at the origin and radius 3) by determining and plotting isolated points would be a tiresome task. As with the plane, the graphing of quadric surfaces is facilitated by graphing their traces, the intersection of the surfaces with the coordinate planes.

The traces of $x^2 + y^2 + z^2 = 9$ are the intersections of the following sets of equations:

Trace on xy plane: $\quad x^2 + y^2 + z^2 = 9$
$\qquad\qquad z = 0$

Trace on xz plane: $\quad x^2 + y^2 + z^2 = 9$
$\qquad\qquad y = 0$

Trace on yz plane: $\quad x^2 + y^2 + z^2 = 9$
$\qquad\qquad x = 0$

Note that if the second of each set of equations is substituted in the first, the result is a circle in the coordinate plane. In drawing traces only the curves in the first octant are ordinarily shown (Fig. 15-1).

In some cases sketches of *sections* parallel to the coordinate planes aid in visualizing the surface. Equations of sections parallel to the xz plane are determined by letting $y = k$, where $-3 < k < 3$ in this case, in the equation of the surface (Fig. 15-1). All sections parallel to the xz plane are circles because the equations of the curve in those sections are of the form $x^2 + z^2 = m^2$.

Sections parallel to the xy plane and yz plane can be investigated by letting $z = k$ and $x = k$, respectively, in Eq. (15-1), the equation of the sphere. It is clear that these sections are also circles.

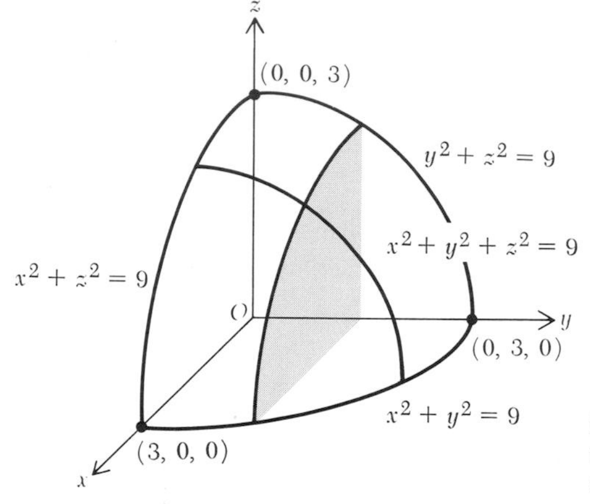

Fig. 15-1 Graph of $x^2 + y^2 + z^2 = 9$ in the first octant, showing traces and a section parallel to the xy plane.

15-4 SURFACES OF REVOLUTION

The surface generated by revolving a plane curve about a line in its plane is called a *surface of revolution.* The line about which the curve is revolved is called the *axis of revolution.* The sphere, for example, is a surface of revolution generated by revolving a circle whose equation is $x^2 + y^2 = r^2$ about any diameter as an axis. The equation of the resulting sphere is $x^2 + y^2 + z^2 = r^2$, as may be shown by procedures similar to those discussed below.

Ordinarily the problem of finding the equation of a surface of revolution is simplified by a suitable choice of coordinate axes. One such choice places the curve in one of the coordinate planes such that the axis of revolution is one of the coordinate axes. For example, consider finding the equation of the surface generated by revolving the parabola with equation

$$3z = 24 - x^2$$

about the z axis. As shown in Fig. 15-2a, any point on the parabola, such as $(x, 0, z)$, generates a circle. In Fig. 15-2b, point A generates a circle of radius r (in this case $r = |QA|$) when revolved. For any point $P(x, y, z)$ on this circle, then,

$$x^2 + y^2 = r^2$$

But the coordinates of point A must satisfy the equation

$$3z = 24 - r^2$$

Hence

$$r^2 = 24 - 3z$$

Substitution of this expression for r^2 in the equation $x^2 + y^2 = r^2$ results in

$$x^2 + y^2 = 24 - 3z \tag{15-3}$$

This is the equation of the surface. The surface is called a *paraboloid of revolution;* the part in the first octant is shown in Fig. 15-2*b*.

It may be observed that the procedure for finding the equation of a surface of revolution for a curve in the xz plane revolved about the z axis amounts to replacing x in the equation of the given curve by $\sqrt{x^2 + y^2}$.

A surface of revolution may often be visualized by studying its sections parallel to the coordinate planes. Consider the paraboloid of revolution whose equation is $x^2 + y^2 = 24 - 3z$.

Sections parallel to the xy plane If z is replaced by 3 in Eq. (15-3), the resulting equation is

$$x^2 + y^2 = 15$$

This is the equation of the intersection of the surface and the plane $z = 3$. In this case the section is the circle with center at Q, one quadrant of which is shown in Fig. 15-2*b*. In general, if $z = k$ is substituted in the equation $x^2 + y^2 = 24 - 3z$, the result is

$$x^2 + y^2 = 24 - 3k$$

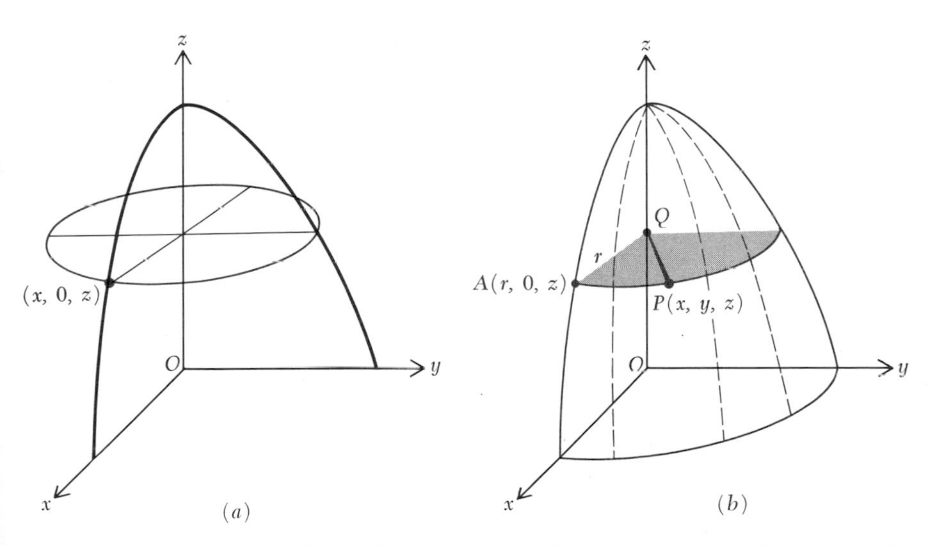

Fig. 15-2 (*a*) A point on the parabola $3z = 24 - x^2$ generates a circle when revolved about the z axis. (*b*) Graph of the paraboloid of revolution $x^2 + y^2 = 24 - 3z$ in the first octant.

These sections are circles of radius $\sqrt{24-3k}$, for $k<8$. As k decreases (as the surface is cut by planes farther and farther below the plane $z=8$), the radius of the circle becomes greater.

Sections parallel to the yz plane If $x=k$ for $-\sqrt{24}<k<\sqrt{24}$ is substituted in Eq. (15-3), it becomes

$$y^2 = 24 - 3z - k^2$$

Thus the intersections of the surface with planes parallel to the yz plane are parabolas. In particular, the section cut by the yz plane is the parabola $y^2 = 24 - 3z$ (Fig. 15-2*a*). If we recall that the shape of the parabola $z = ay^2 + by + c$ is completely determined by the value of a alone, we can immediately conclude that the above sections are all *congruent* parabolas.

Sections parallel to the xz plane can be investigated similarly by putting $y=k$. These are also parabolas.

15-5 THE ELLIPSOID

The surface defined by the equation

$$\frac{x^2}{a^2}+\frac{y^2}{b^2}+\frac{z^2}{c^2}=1 \qquad \text{for } abc \neq 0 \tag{15-4}$$

is called an *ellipsoid* (Fig. 15-3). It has certain recognizable properties:

1. All traces are ellipses. In the xy plane $z=0$, and the equation of the trace is

$$\frac{x^2}{a^2}+\frac{y^2}{b^2}=1$$

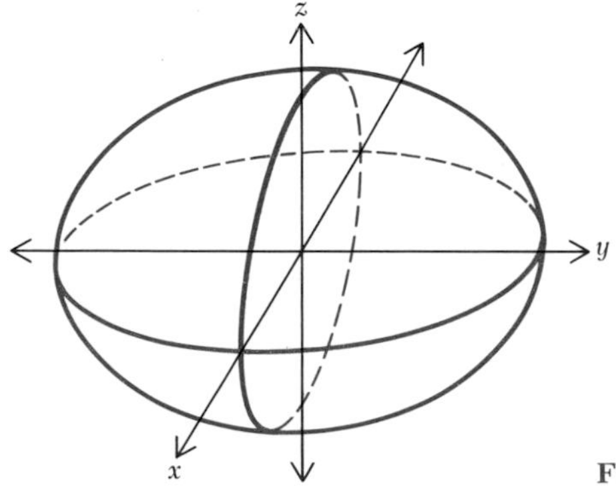

Fig. 15-3 Ellipsoid $x^2/a^2 + y^2/b^2 + z^2/c^2 = 1$.

2. Sections parallel to the coordinate planes are ellipses. A section parallel to the yz plane is the intersection of

$$\frac{x^2}{a^2}+\frac{y^2}{b^2}+\frac{z^2}{c^2}=1$$

and

$$x=k \qquad \text{for } |k| \leqslant a$$

3. If $a=b=c$, the surface is a sphere.

4. In general, an ellipsoid is not a surface of revolution, since the cross sections are ellipses rather than circles. But if any two of the constants a, b, and c in Eq. (15-4) are equal, the surface is an ellipsoid of revolution, sometimes called a *spheroid*.

PROBLEMS

1. Find the equation of the sphere satisfying each set of conditions:

    (*a*) Center at $A(4, -2, 1)$ and radius 6
    (*b*) Center at $B(2, 0, 0)$ and radius 2
    (*c*) A diameter $\overline{AB}$, with $A(-4, 5, 1)$ and $B(4, 1, 7)$
    (*d*) Center at $A(4, 2, 5)$ tangent to the xy plane

2. Show that the sphere given by Eq. (15-2) has center at $(-\frac{1}{2}G, -\frac{1}{2}H, -\frac{1}{2}I)$ and radius $\frac{1}{2}\sqrt{G^2+H^2+I^2-4K}$. What are the conditions under which the equation represents (*a*) a sphere; (*b*) a point sphere; (*c*) no locus?
3. Find the equation of the plane that is tangent to the sphere

$$x^2+y^2+z^2-4x+6y-4z-32=0$$

at $Q(8, -1, 5)$. *Hint:* The tangent plane is perpendicular to the radius.
4. A sphere has its center at $C(0, 4, 0)$ and passes through the point $P(7, 0, 4)$. Write the equation of the tangent plane and normal line at this point (the normal line is a line perpendicular to the tangent plane at the point of tangency).
5. Sketch the graph of each set of points in the first octant and show the traces and sections parallel to the coordinate axes:

    (*a*) $\{(x, y, z) \mid x^2+y^2+z^2=25\}$
    (*b*) $\{(x, y, z) \mid 6x^2+6y^2+6z^2-18=0\}$
    (*c*) $\{(x, y, z) \mid 4x^2+9y^2+z^2-36=0\}$
    (*d*) $\{(x, y, z) \mid x^2+2y^2=4(4-z^2)\}$

In Probs. 6 to 13 the equation of a curve lying in one of the coordinate planes is given. Find the equation of the surface generated when the curve is revolved about the specified coordinate axis and draw the figure.

6. $x^2=4z$; z axis
7. $x^2+z^2=a^2$; x axis
8. $z^2+1=x$; x axis
9. $y=2x$; y axis

10. $\dfrac{x^2}{a^2}+\dfrac{y^2}{b^2}=1$; x axis

11. $\dfrac{x^2}{25}+\dfrac{z^2}{9}=1$; z axis

12. $z=\sin x$; x axis

13. $z=e^{-x^2}$; z axis

In Probs. 14 to 17 determine the nature of the curves cut from the given surface by planes parallel to the coordinate planes. Determine whether or not each given surface could be generated by revolving a curve about one of the coordinate axes.

14. $x^2+2y^2+2z^2=16$
15. $4x^2+4y^2-z^2=0$
16. $2x^2+3y^2+4z^2=12$
17. $x^2+z^2-y=4$

15-6 THE HYPERBOLOID OF ONE SHEET

The surface defined by the equation

$$\frac{x^2}{a^2}+\frac{y^2}{b^2}-\frac{z^2}{c^2}=1 \qquad \text{for } abc\neq 0 \tag{15-5}$$

is called a *hyperboloid of one sheet* (Fig. 15-4). The following properties of this surface may be identified:

1. Its trace in the xy plane is the ellipse

$$\frac{x^2}{a^2}+\frac{y^2}{b^2}=1$$

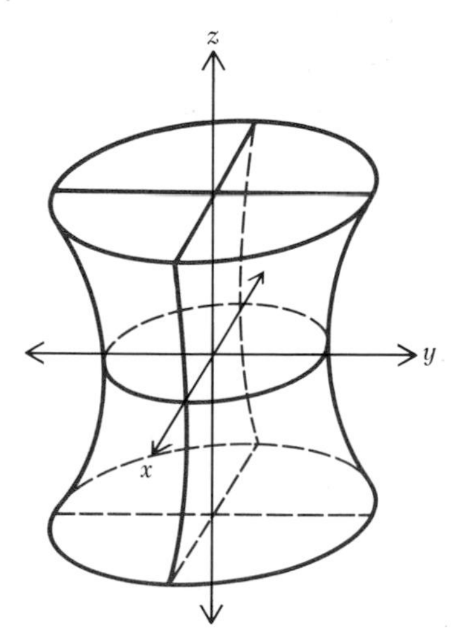

Fig. 15-4 Hyperboloid of one sheet $x^2/a^2+y^2/b^2-z^2/c^2=1$.

Its traces in the xz plane and the yz plane are hyperbolas whose equations are

$$\frac{x^2}{a^2} - \frac{z^2}{c^2} = 1 \qquad \text{and} \qquad \frac{y^2}{b^2} - \frac{z^2}{c^2} = 1$$

respectively.

2. Sections parallel to the xy plane are ellipses, while those parallel to the other coordinate planes are hyperbolas.

3. If $a = b$, the surface is a hyperboloid of revolution and sections parallel to the xy plane are circles.

15-7 THE HYPERBOLOID OF TWO SHEETS

The locus of the equation

$$\frac{x^2}{a^2} - \frac{y^2}{b^2} - \frac{z^2}{c^2} = 1 \qquad \text{for } abc \neq 0 \tag{15-6}$$

is called a *hyperboloid of two sheets* (Fig. 15-5). Sections parallel to the xy and xz planes are hyperbolas. The trace in the yz plane is the *imaginary ellipse* $y^2/b^2 + z^2/c^2 = -1$. The sections cut by the planes $x = a$ and $x = -a$ are *point ellipses,* while those cut by planes $|x| > a$ are *ellipses.* If $b = c$, these ellipses are circles and the surface is a hyperboloid of revolution.

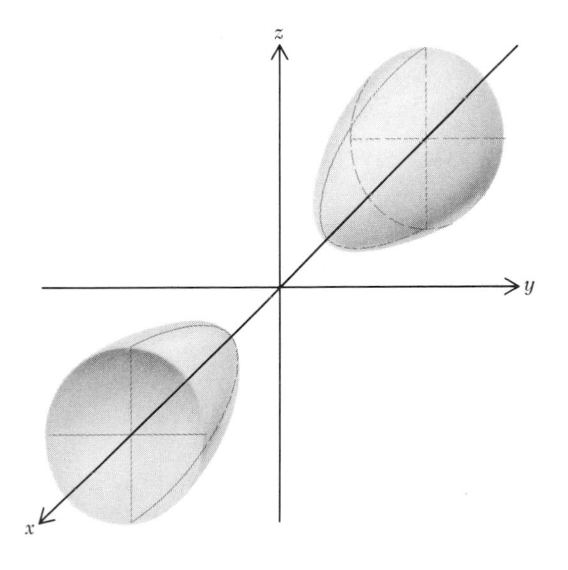

Fig. 15-5 Hyperboloid of two sheets $x^2/a^2 - y^2/b^2 - z^2/c^2 = 1$.

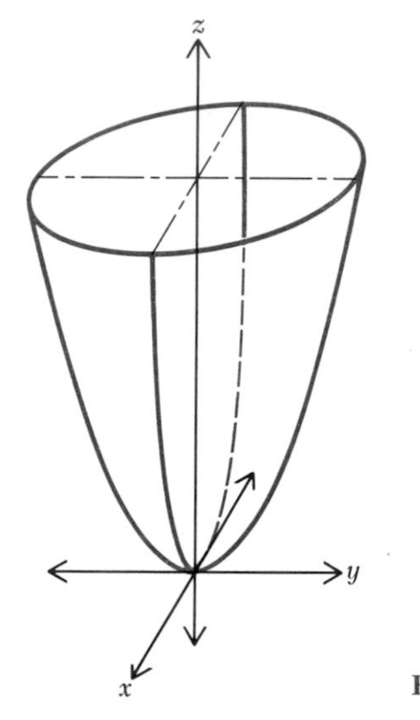

Fig. 15-6 Elliptic paraboloid $x^2/a^2 + y^2/b^2 = z$.

15-8 THE ELLIPTIC PARABOLOID

The surface defined by the equation

$$\frac{x^2}{a^2} + \frac{y^2}{b^2} = z \qquad \text{for } ab \neq 0 \tag{15-7}$$

is an *elliptic paraboloid* (Fig. 15-6). Sections parallel to and above the xy plane are ellipses. Sections parallel to the other coordinate planes are parabolas. If $a = b$, the elliptic sections are circles and the surface is a paraboloid of revolution.

15-9 THE HYPERBOLIC PARABOLOID

The surface whose equation is

$$\frac{x^2}{a^2} - \frac{y^2}{b^2} = z \qquad \text{for } ab \neq 0 \tag{15-8}$$

is called a *hyperbolic paraboloid* (Fig. 15-7). Its trace in the xy plane consists of the two lines $x^2/a^2 - y^2/b^2 = 0$; sections parallel to and above the xy plane are hyperbolas with transverse axes parallel to the x axis; sections parallel to and below the xy plane are hyperbolas with transverse axes parallel to the y axis.

Sections parallel to the other coordinate planes are parabolas. Those

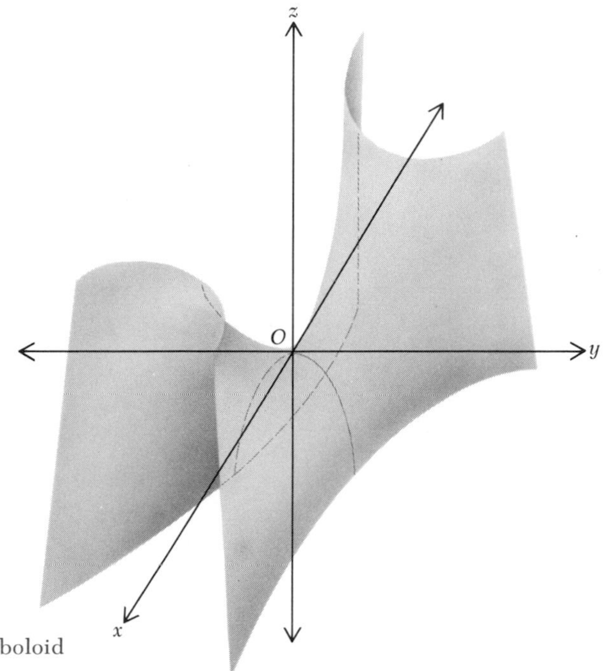

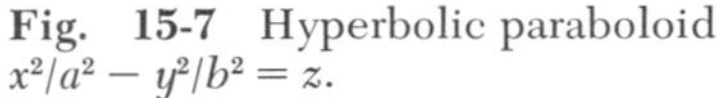
Fig. 15-7 Hyperbolic paraboloid $x^2/a^2 - y^2/b^2 = z$.

parallel to the xz plane open upward, while those parallel to the yz plane open downward.

15-10 THE ELLIPTIC CONE

The locus of the equation

$$\frac{x^2}{a^2} + \frac{y^2}{b^2} - \frac{z^2}{c^2} = 0 \qquad \text{for } abc \neq 0 \tag{15-9}$$

is an *elliptic cone* (Fig. 15-8). Its trace in the xy plane is the *point ellipse* $x^2/a^2 + y^2/b^2 = 0$. Sections parallel to the xy plane are ellipses. Its trace in each of the other two coordinate planes is a pair of lines through the origin; sections parallel to these coordinate planes are hyperbolas.

If $a = b$, the elliptic sections are circles and the surface is a *circular cone*. The circular cone could be generated by revolving the line with equations $z = mx$ and $y = 0$ about the z axis.

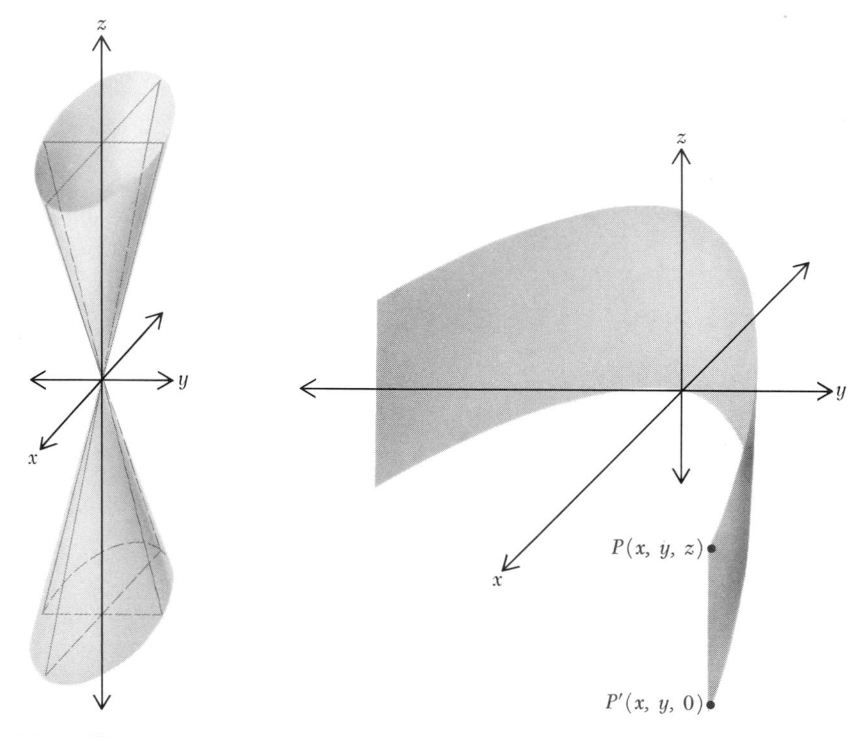

Fig. 15-8 Elliptic cone $x^2/a^2 + y^2/b^2 - z^2/c^2 = 0$.

Fig. 15-9 Parabolic cylinder $y^2 = 4x$.

15-11 CYLINDRICAL SURFACES

The surface generated by a line which moves so that it is always parallel to a fixed line and always intersects a fixed curve is called a *cylindrical surface* or a *cylinder*. Any position of the generating line is called an *element* of the cylinder, while the fixed curve is the *directrix* of the cylinder.

Consider now the equation

$$y^2 = 4x \tag{15-10}$$

In the xy plane the graph of this equation is the parabola (Fig. 15-9). Let a cylindrical surface be generated by a line which moves so that it always intersects this parabola while remaining parallel to the z axis. It is easy to show that Eq. (15-10) is satisfied by the coordinates of every point on this surface. Thus let $P(x, y, z)$ be any point on the surface and let $P'(x, y, 0)$ be the projection of P on the xy plane. Then, if the coordinates of P' satisfy Eq. (15-10), those of P must also satisfy it, because *both points have the same x and y coordinates.* Conversely, every point whose coordinates satisfy the equation $y^2 = 4x$ lies on the cylindrical surface. Equation (15-10) is then the equation of the surface.

The trace of the surface in each of the coordinate planes is the locus of points whose coordinates satisfy two equations:

In the xy plane: $y^2 = 4x$, $z = 0$

In the yz plane: $y^2 = 4x$, $x = 0$

In the xz plane: $y^2 = 4x$, $y = 0$

Thus the trace in the xy plane is the parabola whose equation is $y^2 = 4x$, and the trace in each of the other coordinate planes is the z axis.

All sections parallel to the xy plane are parabolas. The upper parabola in Fig. 15-9 is the section of the surface cut by the plane $z = 4$. All points on this parabola satisfy the two equations $y^2 = 4x$ and $z = 4$. Sections parallel to the yz plane and xz plane are straight lines, obtained by replacing x by $|k|$ and y by k, respectively, in $y^2 = 4x$.

By applying this reasoning to the general case of an equation of the form $\phi(x, y) = 0$, the following theorem may be proved.

Theorem 15-1

If an equation that represents a surface does not contain the variable z, then the surface is a cylinder with elements parallel to the z axis. The locus of the given equation, considered as a set of points in the xy plane, is the directrix of the cylinder.

Similarly, if the variable x (or y) is absent, the locus is a cylindrical surface whose elements are parallel to the x (or y) axis.

Cylinders with circular and elliptical sections are discussed in the following examples.

Example 1

Graph $x^2 + y^2 = 16$.

Solution In the xy plane the locus of the equation $x^2 + y^2 = 16$ is a circle. In space the locus of this equation is a *circular cylinder* (Fig.

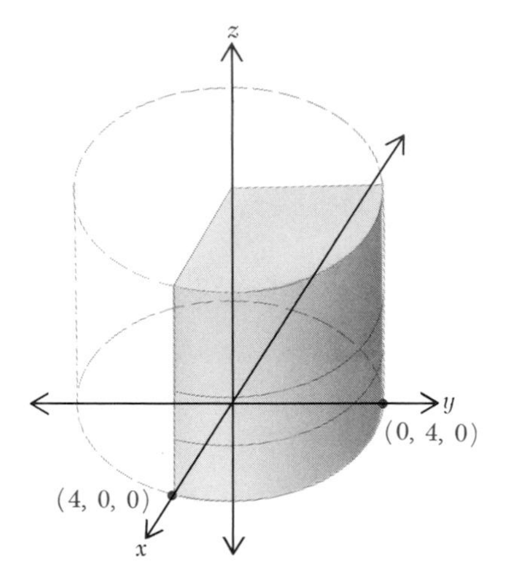

Fig. 15-10 Circular cylinder $x^2 + y^2 = 16$.

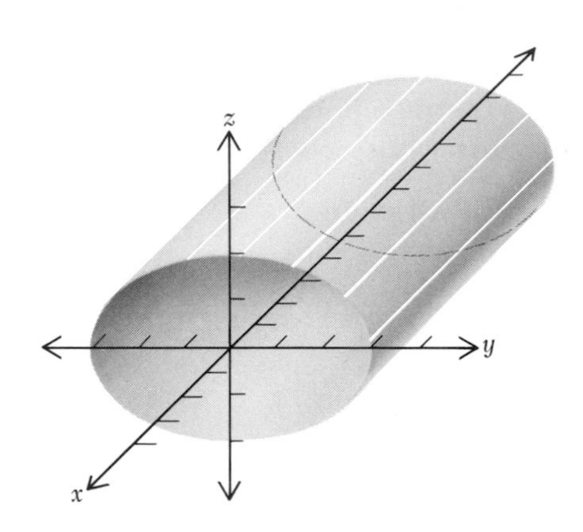

Fig. 15-11 Elliptic cylinder $4y^2 + 9z^2 = 36$.

15-10). Its traces are

In the xy plane, the circle with equation $x^2 + y^2 = 16$
In the xz plane, two lines with equations $x = \pm 4$
In the yz plane, two lines with equations $y = \pm 4$

Sections parallel to the xy plane are circles of radius 4.

Example 2
Graph $4y^2 + 9z^2 = 36$.

Solution In the yz plane the locus of the equation $4y^2 + 9z^2 = 36$ is an ellipse. In space the locus of this equation is the elliptic cylinder having this ellipse as its directrix and having its elements parallel to the x axis (Fig. 15-11).

Not all cylinders are quadric surfaces. The graphs in space of equations such as $y = x^3$, $y = \log x$, and $y = \sin x$ are cylinders, but are not quadric surfaces.

PROBLEMS

1. The point that is symmetrical to $P(x, y, z)$ with respect to the xy plane has the coordinates $(x, y, -z)$. It follows that if the equation of a surface is unal-

tered when z is replaced with $-z$, the surface is symmetrical with respect to the xy plane. State corresponding tests for symmetry with respect to the other coordinate planes. Which of the quadric surfaces discussed in the text are symmetrical with respect to all three coordinate planes?

2. The point that is symmetrical to $P(x, y, z)$ with respect to the x axis has the coordinates $(x, -y, -z)$. It follows that if the equation of a surface is unaltered when y and z are replaced with $-y$ and $-z$, respectively, the surface is symmetrical with respect to the x axis. State corresponding tests for symmetry with respect to the other coordinate axes. Is the hyperbolic paraboloid [Eq. (15-8)] symmetrical with respect to one of the axes? Which one?

3. Sketch the following cylindrical surfaces:

(*a*) $y^2 + z^2 = 25$	(*b*) $x^2 + y^2 = 8x$	(*c*) $x^2 + z = 4$
(*d*) $xy + 4 = 0$	(*e*) $z = e^{-x^2}$	(*f*) $y + \log z = 0$
(*g*) $x^2 + z^2 = 9$	(*h*) $x^2 + y^2 = 6x$	(*i*) $z^2 + x = 9$

4. Identify each of the following sets of points and sketch it:

(*a*) $\{(x, y, z) \mid x^2 - 3y^2 - z^2 - 4 = 0\}$
(*b*) $\{(x, y, z) \mid 9x^2 + 16z^2 - 36y = 0\}$
(*c*) $\{(x, y, z) \mid x^2 + y^2 = z^2\}$
(*d*) $\{(x, y, z) \mid 3x^2 + 4y^2 - 6z = 0\}$
(*e*) $\{(x, y, z) \mid x^2 + 3y^2 = z^2\}$
(*f*) $\{(x, y, z) \mid 4x^2 = 25(y^2 + z^2)\}$
(*g*) $\{(x, y, z) \mid x^2 = 4(y - z^2)\}$
(*h*) $\{(x, y, z) \mid x^2 + 6(y^2 + z^2) = 24\}$

In Probs. 5 to 18 identify and sketch each surface:

5. $z = x^2 + y^2$	6. $y^2 + z^2 = 8 - x$
7. $z + 8 = 2(x^2 + y^2)$	8. $4x^2 + 4y^2 - z^2 = 0$
9. $4x^2 - y = 4 - z^2$	10. $y^2 - 9x^2 = 4z^2$
11. $4x^2 + 4y^2 - z^2 = 16$	12. $x^2 - 2y^2 - z^2 = 4$
13. $y^2 + z^2 = x^2 + 4$	14. $x^2 + 4y^2 + 4z^2 = 8x$
15. $y^2 + z^2 = 2(x^2 - 8)$	16. $x^2 - y^2 = 4z$
17. $2x^2 + 4y^2 + z^2 = 16$	18. $x^2 - 2y^2 = 2z$

19. Discuss the sections of the hyperboloid $x^2/a^2 + y^2/b^2 - z^2/c^2 = 1$ made by the plane $x = k$. Show in particular that the transverse axis of this hyperbola is parallel to the y axis for $|k| < a$ but parallel to the z axis for $|k| > a$. Discuss the case in which $k = a$.

20. The square of the distance to a point from the z axis is two-thirds its distance from the xy plane. Find the equation of the locus and sketch it.

21. Find the equation of the locus of a point whose undirected distance from $P(1, 0, 0)$ is one-half its undirected distance from the plane $x = 2$. Sketch the surface.

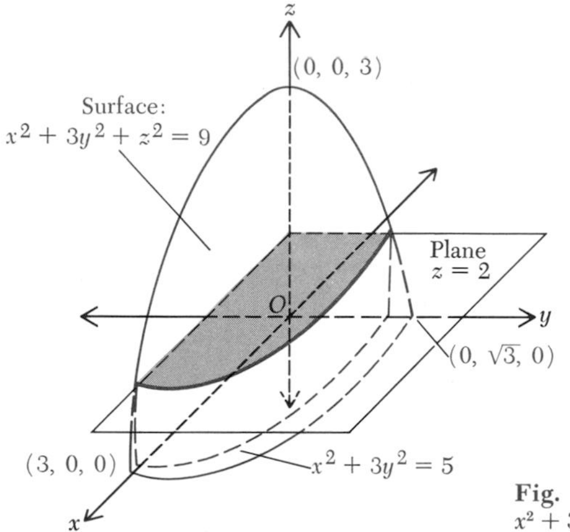

Fig. 15-12 Graph of curve $x^2 + 3y^2 + z^2 = 9$ and $z = 2$ in the first octant.

15-12 CURVES IN SPACE

In the study of traces and sections of surfaces the equations of these curves were expressed as the simultaneous solutions of the equations of two surfaces. Now consider the two equations

$$x^2 + 3y^2 + z^2 = 9 \qquad z = 2 \tag{15-11}$$

The locus of the first equation is an ellipsoid, whereas that of the second is a plane that intersects the ellipsoid in a curve (Fig. 15-12). The points whose coordinates satisfy both of Eqs. (15-11) are those lying on this curve. This pair of equations is therefore one set of equations representing the curve. In this special case the curve of intersection lies entirely on the plane $z = 2$.

It is possible to project the space curve of Eqs. (15-11) onto the xy plane by dropping a perpendicular from each point of the curve to the plane. These perpendiculars or projectors form a cylindrical surface whose equation is obtained by eliminating z from the given equations. In the present case the equation of this *projecting cylinder* is

$$x^2 + 3y^2 = 5$$

A curve in space is often defined by stating the equations of two of its projecting cylinders. This corresponds to the method of defining a line in space by stating the equations of two of its projecting planes.

The equation of the projecting cylinder of Eqs. (15-11) in the xz plane and in the yz plane is $z = 2$ (Fig. 15-12), since this plane is perpendicular to both of these coordinate planes.

A curve in space may be represented by the intersections of many pairs of surfaces. In addition to Eqs. (15-11), the curve in Fig. 15-12 is represented by the following pairs (among others) of equations:

$$x^2 + 3y^2 = 5 \qquad x^2 + 3y^2 = \tfrac{5}{4}z^2 \qquad 2x^2 + 6y^2 + z^2 = 14$$
$$z = 2 \qquad\qquad z = 2 \qquad\qquad x^2 + 3y^2 - z^2 = 1$$

and above the xy plane

It is often necessary to find the volume of a region bounded by a set of surfaces. In attacking such a problem a sketch of these bounding surfaces is almost a necessity.

Example

Sketch the region in the first octant bounded by the cylinders $x^2 + y^2 = r^2$ and $x^2 + z^2 = r^2$.

Solution The two surfaces are circular cylinders, one perpendicular to the xy plane and the other perpendicular to the xz plane. In Fig. 15-13 the curve of intersection of the surfaces is shown by a heavy curve and the bounding surfaces are shaded. The equation of the projecting cylinder of the curve of intersection in the xy plane is $x^2 + y^2 = r^2$, in the xz plane it is $x^2 + z^2 = r^2$, and in the yz plane it is $y = z$. This last equation was obtained by solving $x^2 + y^2 = r^2$ and $x^2 + z^2 = r^2$ simultaneously. The result is the equation $y^2 = z^2$; only $y = z$ is given, since the curve of intersection in only the first octant is being considered.

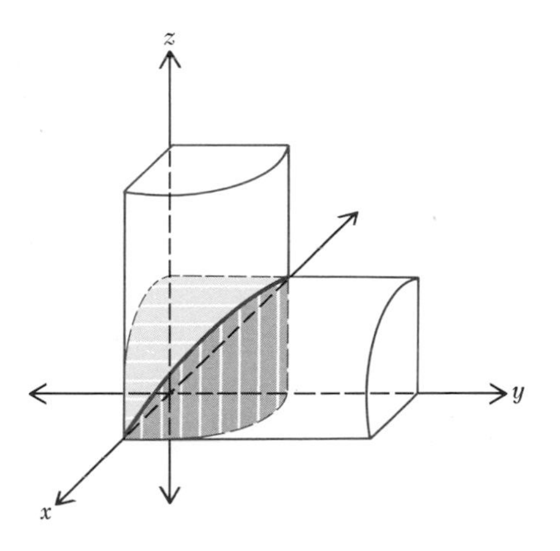

Fig. 15-13 The region in the first octant bounded by the cylinders $x^2 + y^2 = r^2$ and $x^2 + z^2 = r^2$.

15-13 PARAMETRIC EQUATIONS OF A CURVE IN SPACE

The line through (x_1, y_1, z_1) with direction numbers $[a, b, c]$ has the parametric equations

$$\begin{aligned} x &= x_1 + at \\ y &= y_1 + bt \\ z &= z_1 + ct \end{aligned}$$

In each of these equations the right-hand member is a linear function of t.

In general, the three equations

$$\begin{aligned} x &= f_1(t) \\ y &= f_2(t) \\ z &= f_3(t) \end{aligned} \tag{15-12}$$

are parametric equations of a *curve* in space, with parameter t. The result of eliminating t between the first two equations is the xy projecting cylinder of the curve; the other projecting cylinders are obtained similarly.

Example

Graph the curve with parametric equations

$$\begin{aligned} x &= 4 \cos t \\ y &= 4 \sin t \\ z &= \tfrac{2}{3}t \end{aligned} \tag{15-13}$$

Solution When $t = 0$ these equations become $x = 4$, $y = 0$, and $z = 0$. These are the coordinates of the point A in Fig. 15-14. If $t = \frac{1}{2}$, then $x = 0$, $y = 4$, and $z = \frac{1}{3}$. These values correspond to point B in the figure. Continuing in this manner, we may plot the curve as shown. It is called a *helix*, and its parametric equations have the general form

$$\begin{aligned} x &= a \cos t \\ y &= a \sin t \\ z &= bt \end{aligned} \tag{15-14}$$

The thread on a bolt and the handrailing on a circular staircase are examples of curves of this kind.

If t is eliminated between the first two of the given Eqs. (15-14), the result is

$$x^2 + y^2 = a^2$$

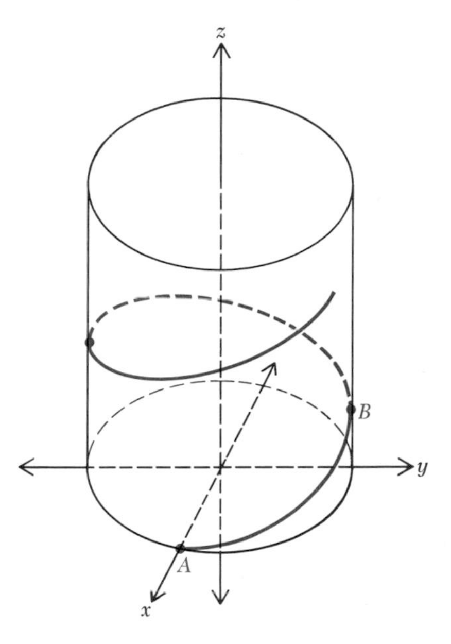

Fig. 15-14 Graph of helix: parametric equations are $x = 4 \cos t$, $y = 4 \sin t$, and $z = \frac{2}{3}t$.

Since the x and y coordinates of every point on the curve satisfy this equation, the curve must lie on this cylinder.

15-14 CYLINDRICAL COORDINATES

In the study of plane analytic geometry it was found that some types of curves have simpler equations in polar than in rectangular coordinates. Similarly, some of the problems of space geometry can be solved more easily by means of coordinate systems other than rectangular. In this section and the following one the *cylindrical* and *spherical* coordinate systems are discussed briefly. Both these systems are of great value in the study of more advanced topics in mathematics.

Let P be a point in space whose rectangular coordinates are (x, y, z) and let A be the foot of the perpendicular from P to the xy plane (Fig. 15-15). Let (r, θ) be the polar coordinates in the xy plane of the point A.* Then the three numbers r, θ, and z comprise the cylindrical coordinates of P.

The relations that enable us to change the equation of a surface from rectangular to cylindrical coordinates are seen from Fig. 15-15 to be

$$
\begin{aligned}
x &= r \cos \theta \\
y &= r \sin \theta \qquad (15\text{-}15) \\
z &= z
\end{aligned}
$$

* (r, θ) is used here instead of (ρ, θ) because the radius vector $\overline{OP}$ will be denoted by ρ.

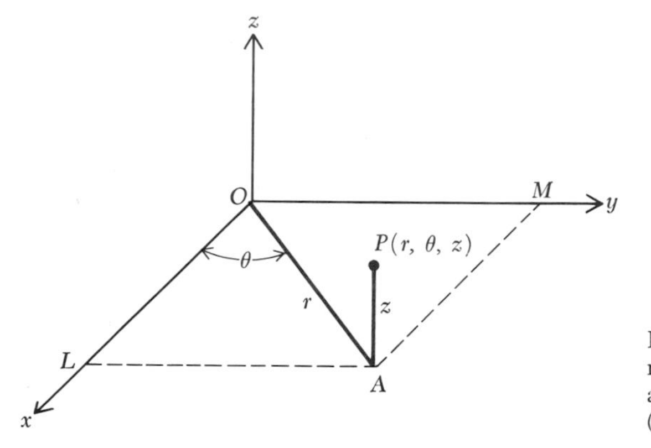

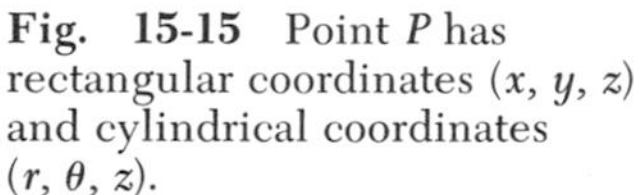
Fig. 15-15 Point P has rectangular coordinates (x, y, z) and cylindrical coordinates (r, θ, z).

Those for the transformation from cylindrical to rectangular coordinates may also be determined from Fig. 15-15.

Example

Write the equation in cylindrical coordinates for the surfaces

$$x^2 + y^2 - z^2 = 0$$
$$x^2 + y^2 - z^2 = 16$$

and

$$x^2 + y^2 = a^2$$

Solution The equation $x^2 + y^2 - z^2 = 0$, which represents a circular cone, becomes $r^2 - z^2 = 0$ or $z = \pm r$ in cylindrical coordinates. Similarly, the equation $x^2 + y^2 + z^2 = 16$ becomes $r^2 + z^2 = 16$, and the right circular cylinder with equation $x^2 + y^2 = a^2$ has the equation $r = a$ in cylindrical coordinates (hence the name).

15-15 SPHERICAL COORDINATES

Let P be a point in space whose rectangular coordinates are (x, y, z), and let A be the foot of the perpendicular from P to the xy plane. Then, as shown in Fig. 15-16, let

$$OP = \rho$$
$$\text{angle } LOA = \theta$$
$$\text{angle } NOP = \phi$$

The three numbers ρ, θ, and ϕ comprise the spherical coordinates of P.

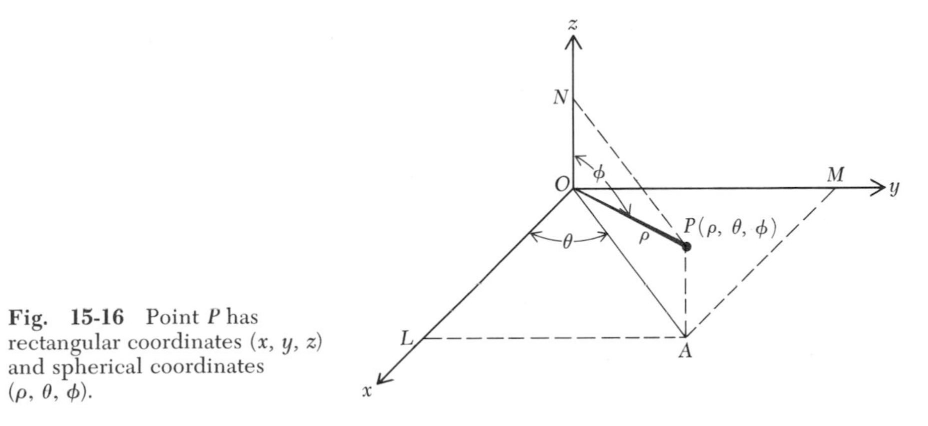

Fig. 15-16 Point P has rectangular coordinates (x, y, z) and spherical coordinates (ρ, θ, ϕ).

This coordinate system is similar to that used in locating points on the earth's surface by means of longitude and latitude. The angle θ corresponds to the longitude and ϕ to the colatitude; the latitude is actually the angle AOP.

In the spherical-coordinate system θ may have any nonnegative value not greater than 2π; ϕ is identical with the direction angle γ and is restricted to nonnegative values not greater than π. It is customary to restrict ρ to positive values, but it is possible to interpret negative values of ρ as was done in dealing with polar coordinates in the plane.

The relations that enable us to transform the equation of a surface from rectangular to spherical coordinates are easily found. Constructing $OA = NP$ (Fig. 15-16) shows that each is equal to $\rho \sin \phi$, while $ON = \rho \cos \phi$; it can be deduced that

$$x = \rho \sin \phi \cos \theta \qquad (15\text{-}16)$$
$$y = \rho \sin \phi \sin \theta$$
$$z = \rho \cos \phi$$

Example

The equation $x^2 + y^2 = 4z$ represents a paraboloid of revolution. The equation of this surface in spherical coordinates is

$$\rho^2 \sin^2 \phi(\cos^2 \theta + \sin^2 \theta) = 4\rho \cos \phi$$
$$\rho \sin^2 \phi = 4 \cos \phi$$
$$\rho = 4 \cos \phi \csc^2 \phi$$

As the name "spherical coordinates" might indicate, the equation of a sphere is expressed very simply in this system. The equation $x^2 + y^2 + z^2 = a^2$ transforms to $\rho = a$.

PROBLEMS

In Probs. 1 to 4 write two other pairs of equations whose intersection is the same curve as that given; sketch each curve.

1. $x^2 + y^2 + z^2 = 16; \quad z = 3$
2. $y^2 + z^2 = x^2; \quad x = 1$
3. $x^2 + z^2 = 25; \quad y = 0$
4. $z^2 = 4(x^2 + y^2); \quad x + z = 4$

In Probs. 5 to 14 sketch the curve represented by the given equations.

5. $x^2 + y^2 + z^2 = 25; \quad x^2 + y^2 = 9$
6. $x^2 + y^2 + z^2 = 36; \quad x^2 + y^2 = 6x$
7. $x^2 + y^2 + z^2 = 25; \quad 9z^2 = 16(x^2 + y^2)$
8. $x^2 + y^2 + z^2 = 16; \quad 2x + z = 4$
9. $x^2 + y^2 = 1 - z; \quad x + y = 1$
10. $2z = x^2 + y^2; \quad x + y + 2z = 6$
11. $x = 5 \cos t$, $y = 5 \sin t$, $z = t$
12. $x = t - 4$, $y = t$, $z = \frac{1}{2}t^2$
13. $x = 4 \cos^2 t$, $y = 4 \sin^2 t$, $z = t^2$
14. $x = 4 \cos t$, $y = 4 \sin t$, $z = 8 \sin t$

In Probs. 15 to 17 sketch the region in the first octant bounded by the coordinate planes and the given surfaces.

15. Inside the cylinder $x^2 + y^2 = 25$ and inside the cylinder $x^2 + z^2 = 25$.
16. Inside the cylinder $x^2 + y^2 = 16$ and under the plane $x + y + z = 10$.
17. Inside the sphere $x^2 + y^2 + z^2 = 9$ and inside the cylinder $x^2 + y^2 = 3x$.
18. Derive a formula for the distance between two points in terms of their spherical coordinates.
19. Describe in spherical coordinates the locus of each of the following equations:

    (*a*) $\rho = k$ (*b*) $\theta = k$ (*c*) $\phi = k$

20. Describe in spherical coordinates the locus of each of the following equations:

    (*a*) $\rho \sin \phi = k$ (*b*) $\rho \cos \phi = k$

Express each of the equations in Probs. 21 to 28 in both cylindrical and spherical coordinates and sketch the locus:

21. $x^2 + y^2 = 6$
22. $x^2 + y^2 = 8xy$
23. $xy = 2$
24. $x^2 + y^2 - \frac{1}{2}z^2 = 0$
25. $x^2 + y^2 + z^2 = 16$
26. $x^2 + y^2 + z^2 = 6z$
27. $x^2 - y^2 - 2z^2 = 4$
28. $4z = x^2 - y^2$

SUMMARY

The locus of an equation of second degree in x, y, and z is a quadric surface. The equations of several quadric surfaces are given in the following table:

Surface	Standard Equation ($abc \neq 0$)
Sphere	$(x-h)^2+(y-k)^2+(z-l)^2=r^2$
Ellipsoid	$\frac{x^2}{a^2}+\frac{y^2}{b^2}+\frac{z^2}{c^2}=1$
Spheroid	$\frac{x^2}{a^2}+\frac{y^2}{a^2}+\frac{z^2}{b^2}=1$
Hyperboloid of one sheet	$\frac{x^2}{a^2}+\frac{y^2}{b^2}-\frac{z^2}{c^2}=1$
Hyperboloid of two sheets	$\frac{x^2}{a^2}-\frac{y^2}{b^2}-\frac{z^2}{c^2}=1$
Elliptic paraboloid	$\frac{x^2}{a^2}+\frac{y^2}{b^2}=z$
Hyperbolic paraboloid	$\frac{x^2}{a^2}-\frac{y^2}{b^2}=z$
Elliptic cone	$\frac{x^2}{a^2}+\frac{y^2}{b^2}-\frac{z^2}{c^2}=0$
Circular cone	$\frac{x^2}{a^2}+\frac{y^2}{a^2}-\frac{z^2}{c^2}=0$

The surface generated by revolving a plane curve about an axis in its plane is a surface of revolution. The equation of a surface of revolution about the z axis can be found by replacing x in the equation of the given curve in the xz plane by $\sqrt{x^2+y^2}$.

The surface generated by a line moving so that it is always parallel to a fixed line and always intercepting a fixed curve is a cylinder. If an equation representing a surface does not contain one of the variables x, y, or z, then the surface is a cylinder with elements parallel to the axis named by the missing variable.

Curves in space can be expressed as the simultaneous solution of equations of two surfaces. Often the two equations chosen are those of projecting cylinders. Parametric equations of a curve in space consist of a set of three simultaneous equations of the form

$$x = f_1(t)$$
$$y = f_2(t)$$
$$z = f_3(t)$$

Two special coordinate systems, cylindrical coordinates and spherical coordinates, are often useful in solving the problems of three-dimensional analytical geometry. The equations that enable us to change the equation of a surface from rectangular to cylindrical or spherical coordinates, respectively, are $x = r \cos \theta$, $y = r \sin \theta$, and $z = z$ and $x = \rho \sin \phi \cos \theta$, $y = \rho \sin \phi \sin \theta$, and $z = \rho \cos \phi$.

REVIEW PROBLEMS

1. Identify the surfaces represented by the following equations:

   (*a*) $\frac{x^2}{9} + \frac{y^2}{4} - z^2 = 0$ (*b*) $\frac{x^2}{9} - \frac{y^2}{4} - z^2 = 1$

   (*c*) $\frac{x^2}{9} + \frac{y^2}{9} + z^2 = 0$ (*d*) $(x - 9)^2 + (y - 4)^2 + z^2 = 1$

   (*e*) $\frac{x^2}{9} + \frac{y^2}{9} - z^2 = 0$ (*f*) $\frac{x^2}{9} + \frac{y^2}{4} = z$

   (*g*) $\frac{x^2}{9} + \frac{y^2}{4} + z^2 = 1$ (*h*) $\frac{x^2}{9} - \frac{y^2}{4} = z$

   (*i*) $\frac{x^2}{9} + \frac{y^2}{4} - z^2 = 1$ (*j*) $x^2 + y^2 + z^2 = 1$

2. For each equation in Prob. 1 sketch the graph of the surface.
3. Find the equation of the sphere with center at $P(3, 1, -2)$ and radius 5.
4. Find the equation of the surface generated when the curve with equation $x^2 - 9z$ in the xz plane is revolved about the z axis.
5. Sketch the cylindrical surface $x^2 + y^2 = 6x$.
6. Identify and sketch each set of points:

   (*a*) $\{(x, y, z) \mid x^2 - y^2 - z^2 - 2 = 0\}$
   (*b*) $\{(x, y, z) \mid 2x^2 + 3y^2 - 4z = 0\}$
   (*c*) $\{(x, y, z) \mid y^2 + 3x^2 = 2z^2\}$

7. Sketch the curve $x^2 + y^2 + z^2 = 16$ and $x^2 + y^2 = 4$.
8. Sketch the curve $x = 2 \cos t$, $y = 2 \sin t$, and $z = t$.
9. Express the equation $x^2 - y^2 = 4z$ in (*a*) cylindrical coordinates; (*b*) spherical coordinates. Sketch the locus.

FOR EXTENDED STUDY

1. Which of the three systems of space coordinates establishes a one-to-one correspondence between points in space and triples of numbers? Justify your answer.
2. Describe and sketch a suitable portion of the surfaces whose equations are

   (*a*) $y = \sin x$ (*b*) $y = e^x$

3. The graph of $x^2/a^2 - y^2/b^2 + z^2/c^2 = 0$, for $abc \neq 0$, is a cone. Sections of the graph parallel to the yz plane are ellipses. Prove that all these ellipses have the same eccentricity.
4. State a test for symmetry of a surface with respect to the origin. Consider this test and the tests in Probs. 1 and 2, pages 388–389. Are they both necessary and sufficient, or merely sufficient? *Hint:* If replacing z with $-z$ does not alter the equation, the surface is symmetrical with respect to the xy plane; hence the test is sufficient. But if replacing z with $-z$ does change the equation, might the surface be symmetrical to the xy plane anyway?
5. A curve in the xy plane satisfies the equations $f(x, y) = 0$ and $z = 0$. Prove that if $y \geqslant 0$, the equation of the surface obtained by revolving the curve about the x axis is $f(x, \sqrt{y^2 + z^2}) = 0$.
6. Write the equations expressing cylindrical coordinates r, θ, and z in terms of spherical coordinates ρ, θ, and ϕ.
7. A sphere of radius 6 intersects a circular cylinder of diameter 6 so that an element of the cylinder contains a diameter of the sphere. Choose the most suitable axes and write the equations of these surfaces in (*a*) rectangular; (*b*) cylindrical; and (*c*) spherical coordinates.

APPENDIXES

A TRIGONOMETRIC IDENTITIES AND FORMULAS

B NUMERICAL TABLES

I Common Logarithms
II Trigonometric Functions
III Powers and Roots
IV Natural Logarithms
V Exponential and Hyperbolic Functions

TRIGONOMETRIC IDENTITIES AND FORMULAS | A

The Fundamental Identities

(*a*) $\sin^2 \theta + \cos^2 \theta = 1$

(*b*) $1 + \tan^2 \theta = \sec^2 \theta$

(*c*) $1 + \cot^2 \theta = \csc^2 \theta$

(*d*) $\tan \theta = \dfrac{\sin \theta}{\cos \theta}$

(*e*) $\csc \theta = \dfrac{1}{\sin \theta}$

(*f*) $\sec \theta = \dfrac{1}{\cos \theta}$

(*g*) $\cot \theta = \dfrac{1}{\tan \theta}$

Functions of $-\theta$

(*a*) $\sin (-\theta) = -\sin \theta$

(*b*) $\cos (-\theta) = \cos \theta$

(*c*) $\tan (-\theta) = -\tan \theta$

Reduction Formulas

(a) $\sin(\frac{1}{2}\pi - \theta) = \cos\theta$ $\quad$ $\sin(\frac{1}{2}\pi + \theta) = \cos\theta$
(b) $\cos(\frac{1}{2}\pi - \theta) = \sin\theta$ $\quad$ $\cos(\frac{1}{2}\pi + \theta) = -\sin\theta$
(c) $\tan(\frac{1}{2}\pi - \theta) = \cot\theta$ $\quad$ $\tan(\frac{1}{2}\pi + \theta) = -\cot\theta$
(d) $\sin(\pi - \theta) = \sin\theta$ $\quad$ $\sin(\pi + \theta) = -\sin\theta$
(e) $\cos(\pi - \theta) = -\cos\theta$ $\quad$ $\cos(\pi + \theta) = -\cos\theta$
(f) $\tan(\pi - \theta) = -\tan\theta$ $\quad$ $\tan(\pi + \theta) = \tan\theta$

Addition Formulas: Double and Half-angle Formulas

(a) $\sin(\alpha + \beta) = \sin\alpha\cos\beta + \cos\alpha\sin\beta$
(b) $\cos(\alpha + \beta) = \cos\alpha\cos\beta - \sin\alpha\sin\beta$
(c) $\tan(\alpha + \beta) = \dfrac{\tan\alpha + \tan\beta}{1 - \tan\alpha\tan\beta}$
(d) $\sin(\alpha - \beta) = \sin\alpha\cos\beta - \cos\alpha\sin\beta$
(e) $\cos(\alpha - \beta) = \cos\alpha\cos\beta + \sin\alpha\sin\beta$
(f) $\tan(\alpha - \beta) = \dfrac{\tan\alpha - \tan\beta}{1 + \tan\alpha\tan\beta}$

For the case in which $\beta = \alpha$, formulas (a), (b), and (c) reduce to

(g) $\sin 2\alpha = 2\sin\alpha\cos\alpha$
(h) $\cos 2\alpha = \cos^2\alpha - \sin^2\alpha$
(i) $\tan 2\alpha = \dfrac{2\tan\alpha}{1 - \tan^2\alpha}$

From formula (h) it can readily be shown that

(j) $\sin\frac{1}{2}\alpha = \pm\sqrt{\dfrac{1 - \cos\alpha}{2}}$
(k) $\cos\frac{1}{2}\alpha = \pm\sqrt{\dfrac{1 + \cos\alpha}{2}}$
(l) $\tan\frac{1}{2}\alpha = \pm\sqrt{\dfrac{1 - \cos\alpha}{1 + \cos\alpha}} = \dfrac{1 - \cos\alpha}{\sin\alpha} = \dfrac{\sin\alpha}{1 + \cos\alpha}$

Sine and Cosine Laws

(a) $\dfrac{a}{\sin A} = \dfrac{b}{\sin B} = \dfrac{c}{\sin C}$
(b) $c^2 = a^2 + b^2 - 2ab\cos C$

NUMERICAL TABLES | B

Table I Common Logarithms

N	0	1	2	3	4	5	6	7	8	9
10	0000	0043	0086	0128	0170	0212	0253	0294	0334	0374
11	0414	0453	0492	0531	0569	0607	0645	0682	0719	0755
12	0792	0828	0864	0899	0934	0969	1004	1038	1072	1106
13	1139	1173	1206	1239	1271	1303	1335	1367	1399	1430
14	1461	1492	1523	1553	1584	1614	1644	1673	1703	1732
15	1761	1790	1818	1847	1875	1903	1931	1959	1987	2014
16	2041	2068	2095	2122	2148	2175	2201	2227	2253	2279
17	2304	2330	2355	2380	2405	2430	2455	2480	2504	2529
18	2553	2577	2601	2625	2648	2672	2695	2718	2742	2765
19	2788	2810	2833	2856	2878	2900	2923	2945	2967	2989
20	3010	3032	3054	3075	3096	3118	3139	3160	3181	3201
21	3222	3243	3263	3284	3304	3324	3345	3365	3385	3404
22	3424	3444	3464	3483	3502	3522	3541	3560	3579	3598
23	3617	3636	3655	3674	3692	3711	3729	3747	3766	3784
24	3802	3820	3838	3856	3874	3892	3909	3927	3945	3962
25	3979	3997	4014	4031	4048	4065	4082	4099	4116	4133
26	4150	4166	4183	4200	4216	4232	4249	4265	4281	4298
27	4314	4330	4346	4362	4378	4393	4409	4425	4440	4456
28	4472	4487	4502	4518	4533	4548	4564	4579	4594	4609
29	4624	4639	4654	4669	4683	4698	4713	4728	4742	4757
30	4771	4786	4800	4814	4829	4843	4857	4871	4886	4900
31	4914	4928	4942	4955	4969	4983	4997	5011	5024	5038
32	5051	5065	5079	5092	5105	5119	5132	5145	5159	5172
33	5185	5198	5211	5224	5237	5250	5263	5276	5289	5302
34	5315	5328	5340	5353	5366	5378	5391	5403	5416	5428
35	5441	5453	5465	5478	5490	5502	5514	5527	5539	5551
36	5563	5575	5587	5599	5611	5623	5635	5647	5658	5670
37	5682	5694	5705	5717	5729	5740	5752	5763	5775	5786
38	5798	5809	5821	5832	5843	5855	5866	5877	5888	5899
39	5911	5922	5933	5944	5955	5966	5977	5988	5999	6010
40	6021	6031	6042	6053	6064	6075	6085	6096	6107	6117
41	6128	6138	6149	6160	6170	6180	6191	6201	6212	6222
42	6232	6243	6253	6263	6274	6284	6294	6304	6314	6325
43	6335	6345	6355	6365	6375	6385	6395	6405	6415	6425
44	6435	6444	6454	6464	6474	6484	6493	6503	6513	6522
45	6532	6542	6551	6561	6571	6580	6590	6599	6609	6618
46	6628	6637	6646	6656	6665	6675	6684	6693	6702	6712
47	6721	6730	6739	6749	6758	6767	6776	6785	6794	6803
48	6812	6821	6830	6839	6848	6857	6866	6875	6884	6893
49	6902	6911	6920	6928	6937	6946	6955	6964	6972	6981
50	6990	6998	7007	7016	7024	7033	7042	7050	7059	7067
51	7076	7084	7093	7101	7110	7118	7126	7135	7143	7152
52	7160	7168	7177	7185	7193	7202	7210	7218	7226	7235
53	7243	7251	7259	7267	7275	7284	7292	7300	7308	7316
54	7324	7332	7340	7348	7356	7364	7372	7380	7388	7396
N	0	1	2	3	4	5	6	7	8	9

Table I Common Logarithms (*Continued*)

N	0	1	2	3	4	5	6	7	8	9
55	7404	7412	7419	7427	7435	7443	7451	7459	7466	7474
56	7482	7490	7497	7505	7513	7520	7528	7536	7543	7551
57	7559	7566	7574	7582	7589	7597	7604	7612	7619	7627
58	7634	7642	7649	7657	7664	7672	7679	7686	7694	7701
59	7709	7716	7723	7731	7738	7745	7752	7760	7767	7774
60	7782	7789	7796	7803	7810	7818	7825	7832	7839	7846
61	7853	7860	7868	7875	7882	7889	7896	7903	7910	7917
62	7924	7931	7938	7945	7952	7959	7966	7973	7980	7987
63	7993	8000	8007	8014	8021	8028	8035	8041	8048	8055
64	8062	8069	8075	8082	8089	8096	8102	8109	8116	8122
65	8129	8136	8142	8149	8156	8162	8169	8176	8182	8189
66	8195	8202	8209	8215	8222	8228	8235	8241	8248	8254
67	8261	8267	8274	8280	8287	8293	8299	8306	8312	8319
68	8325	8331	8338	8344	8351	8357	8363	8370	8376	8382
69	8388	8395	8401	8407	8414	8420	8426	8432	8439	8445
70	8451	8457	8463	8470	8476	8482	8488	8494	8500	8506
71	8513	8519	8525	8531	8537	8543	8549	8555	8561	8567
72	8573	8579	8585	8591	8597	8603	8609	8615	8621	8627
73	8633	8639	8645	8651	8657	8663	8669	8675	8681	8686
74	8692	8698	8704	8710	8716	8722	8727	8733	8739	8745
75	8751	8756	8762	8768	8774	8779	8785	8791	8797	8802
76	8808	8814	8820	8825	8831	8837	8842	8848	8854	8859
77	8865	8871	8876	8882	8887	8893	8899	8904	8910	8915
78	8921	8927	8932	8938	8943	8949	8954	8960	8965	8971
79	8976	8982	8987	8993	8998	9004	9009	9015	9020	9025
80	9031	9036	9042	9047	9053	9058	9063	9069	9074	9079
81	9085	9090	9096	9101	9106	9112	9117	9122	9128	9133
82	9138	9143	9149	9154	9159	9165	9170	9175	9180	9186
83	9191	9196	9201	9206	9212	9217	9222	9227	9232	9238
84	9243	9248	9253	9258	9263	9269	9274	9279	9284	9289
85	9294	9299	9304	9309	9315	9320	9325	9330	9335	9340
86	9345	9350	9355	9360	9365	9370	9375	9380	9385	9390
87	9395	9400	9405	9410	9415	9420	9425	9430	9435	9440
88	9445	9450	9455	9460	9465	9469	9474	9479	9484	9489
89	9494	9499	9504	9509	9513	9518	9523	9528	9533	9538
90	9542	9547	9552	9557	9562	9566	9571	9576	9581	9586
91	9590	9595	9600	9605	9609	9614	9619	9624	9628	9633
92	9638	9643	9647	9652	9657	9661	9666	9671	9675	9680
93	9685	9689	9694	9699	9703	9708	9713	9717	9722	9727
94	9731	9736	9741	9745	9750	9754	9759	9763	9768	9773
95	9777	9782	9786	9791	9795	9800	9805	9809	9814	9818
96	9823	9827	9832	9836	9841	9845	9850	9854	9859	9863
97	9868	9872	9877	9881	9886	9890	9894	9899	9903	9908
98	9912	9917	9921	9926	9930	9934	9939	9943	9948	9952
99	9956	9961	9965	9969	9974	9978	9983	9987	9991	9996
N	0	1	2	3	4	5	6	7	8	9

Table II Trigonometric Functions

Angles	Sines		Cosines		Tangents		Cotangents		Angles
	Nat.	Log.	Nat.	Log.	Nat.	Log.	Nat.	Log.	
0° 00′	.0000	∞	1.0000	**0**.0000	.0000	∞	∞	∞	90° 00′
10	.0029	**7**.4637	1.0000	0000	.0029	**7**.4637	343.77	**2**.5363	50
20	.0058	7648	1.0000	0000	.0058	7648	171.89	2352	40
30	.0087	9408	1.0000	0000	.0087	9409	114.59	0591	30
40	.0116	**8**.0658	.9999	0000	.0116	**8**.0658	85.940	**1**.9342	20
50	.0145	1627	.9999	0000	.0145	1627	68.750	8373	10
1° 00′	.0175	**8**.2419	.9998	**9**.9999	.0175	**8**.2419	57.290	**1**.7581	89° 00′
10	.0204	3088	.9998	9999	.0204	3089	49.104	6911	50
20	.0233	3668	.9997	9999	.0233	3669	42.964	6331	40
30	.0262	4179	.9997	9999	.0262	4181	38.188	5819	30
40	.0291	4637	.9996	9998	.0291	4638	34.368	5362	20
50	.0320	5050	.9995	9998	.0320	5053	31.242	4947	10
2° 00′	.0349	**8**.5428	.9994	**9**.9997	.0349	**8**.5431	28.636	**1**.4569	88° 00′
10	.0378	5776	.9993	9997	.0378	5779	26.432	4221	50
20	.0407	6097	.9992	9996	.0407	6101	24.542	3899	40
30	.0436	6397	.9990	9996	.0437	6401	22.904	3599	30
40	.0465	6677	.9989	9995	.0466	6682	21.470	3318	20
50	.0494	6940	.9988	9995	.0495	6945	20.206	3055	10
3° 00′	.0523	**8**.7188	.9986	**9**.9994	.0524	**8**.7194	19.081	**1**.2806	87° 00′
10	.0552	7423	.9985	9993	.0553	7429	18.075	2571	50
20	.0581	7645	.9983	9993	.0582	7652	17.169	2348	40
30	.0610	7857	.9981	9992	.0612	7865	16.350	2135	30
40	.0640	8059	.9980	9991	.0641	8067	15.605	1933	20
50	.0669	8251	.9978	9990	.0670	8261	14.924	1739	10
4° 00′	.0698	**8**.8436	.9976	**9**.9989	.0699	**8**.8446	14.301	**1**.1554	86° 00′
10	.0727	8613	.9974	9989	.0729	8624	13.727	1376	50
20	.0756	8783	.9971	9988	.0758	8795	13.197	1205	40
30	.0785	8946	.9969	9987	.0787	8960	12.706	1040	30
40	.0814	9104	.9967	9986	.0816	9118	12.251	0882	20
50	.0843	9256	.9964	9985	.0846	9272	11.826	0728	10
5° 00′	.0872	**8**.9403	.9962	**9**.9983	.0875	**8**.9420	11.430	**1**.0580	85° 00′
10	.0901	9545	.9959	9982	.0904	9563	11.059	0437	50
20	.0929	9682	.9957	9981	.0934	9701	10.712	0299	40
30	.0958	9816	.9954	9980	.0963	9836	10.385	0164	30
40	.0987	9945	.9951	9979	.0992	9966	10.078	0034	20
50	.1016	**9**.0070	.9948	9977	.1022	**9**.0093	9.7882	**0**.9907	10
6° 00′	.1045	**9**.0192	.9945	**9**.9976	.1051	**9**.0216	9.5144	**0**.9784	84° 00
10	.1074	0311	.9942	9975	.1080	0336	9.2553	9664	50
20	.1103	0426	.9939	9973	.1110	0453	9.0098	9547	40
30	.1132	0539	.9936	9972	.1139	0567	8.7769	9433	30
40	.1161	0648	.9932	9971	.1169	0678	8.5555	9322	20
50	.1190	0755	.9929	9969	.1198	0786	8.3450	9214	10
7° 00′	.1219	**9**.0859	.9925	**9**.9968	.1228	**9**.0891	8.1443	**0**.9109	83° 00′
10	.1248	0961	.9922	9966	.1257	0995	7.9530	9005	50
20	.1276	1060	.9918	9964	.1287	1096	7.7704	8904	40
30	.1305	1157	.9914	9963	.1317	1194	7.5958	8806	30
40	.1334	1252	.9911	9961	.1346	1291	7.4287	8709	20
50	.1363	1345	.9907	9959	.1376	1385	7.2687	8615	10
8° 00′	.1392	**9**.1436	.9903	**9**.9958	.1405	**9**.1478	7.1154	**0**.8522	82° 00′
10	.1421	1525	.9899	9956	.1435	1569	6.9682	8431	50
20	.1449	1612	.9894	9954	.1465	1658	6.8269	8342	40
30	.1478	1697	.9890	9952	.1495	1745	6.6912	8255	30
40	.1507	1781	.9886	9950	.1524	1831	6.5606	8169	20
50	.1536	1863	.9881	9948	.1554	1915	6.4348	8085	10
9° 00′	.1564	**9**.1943	.9877	**9**.9946	.1584	**9**.1997	6.3138	**0**.8003	81° 00′
	Nat.	Log.	Nat.	Log.	Nat.	Log.	Nat.	Log.	
Angles	Cosines		Sines		Cotangents		Tangents		Angles

Table II Trigonometric Functions (*Continued*)

Angles	Sines		Cosines		Tangents		Cotangents		Angles
	Nat.	Log.	Nat.	Log.	Nat.	Log.	Nat.	Log.	
9° 00′	.1564	9.1943	.9877	9.9946	.1584	9.1997	6.3138	0.8003	81° 00′
10	.1593	2022	.9872	9944	.1614	2078	6.1970	7922	50
20	.1622	2100	.9868	9942	.1644	2158	6.0844	7842	40
30	.1650	2176	.9863	9940	.1673	2236	5.9758	7764	30
40	.1679	2251	.9858	9938	.1703	2313	5.8708	7687	20
50	.1708	2324	.9853	9936	.1733	2389	5.7694	7611	10
10° 00′	.1736	9.2397	.9848	9.9934	.1763	9.2463	5.6713	0.7537	80° 00′
10	.1765	2468	.9843	9931	.1793	2536	5.5764	7464	50
20	.1794	2538	.9838	9929	.1823	2609	5.4845	7391	40
30	.1822	2606	.9833	9927	.1853	2680	5.3955	7320	30
40	.1851	2674	.9827	9924	.1883	2750	5.3093	7250	20
50	.1880	2740	.9822	9922	.1914	2819	5.2257	7181	10
11° 00′	.1908	9.2806	.9816	9.9919	.1944	9.2887	5.1446	0.7113	79° 00′
10	.1937	2870	.9811	9917	.1974	2953	5.0658	7047	50
20	.1965	2934	.9805	9914	.2004	3020	4.9894	6980	40
30	.1994	2997	.9799	9912	.2035	3085	4.9152	6915	30
40	.2022	3058	.9793	9909	.2065	3149	4.8430	6851	20
50	.2051	3119	.9787	9907	.2095	3212	4.7729	6788	10
12° 00′	.2079	9.3179	.9781	9.9904	.2126	9.3275	4.7046	0.6725	78° 00′
10	.2108	3238	.9775	9901	.2156	3336	4.6382	6664	50
20	.2136	3296	.9769	9899	.2186	3397	4.5736	6603	40
30	.2164	3353	.9763	9896	.2217	3458	4.5107	6542	30
40	.2193	3410	.9757	9893	.2247	3517	4.4494	6483	20
50	.2221	3466	.9750	9890	.2278	3576	4.3897	6424	10
13° 00′	.2250	9.3521	.9744	9.9887	.2309	9.3634	4.3315	0.6366	77° 00′
10	.2278	3575	.9737	9884	.2339	3691	4.2747	6309	50
20	.2306	3629	.9730	9881	.2370	3748	4.2193	6252	40
30	.2334	3682	.9724	9878	.2401	3804	4.1653	6196	30
40	.2363	3734	.9717	9875	.2432	3859	4.1126	6141	20
50	.2391	3786	.9710	9872	.2462	3914	4.0611	6086	10
14° 00′	.2419	9.3837	.9703	9.9869	.2493	9.3968	4.0108	0.6032	76° 00′
10	.2447	3887	.9696	9866	.2524	4021	3.9617	5979	50
20	.2476	3937	.9689	9863	.2555	4074	3.9136	5926	40
30	.2504	3986	.9681	9859	.2586	4127	3.8667	5873	30
40	.2532	4035	.9674	9856	.2617	4178	3.8208	5822	20
50	.2560	4083	.9667	9853	.2648	4230	3.7760	5770	10
15° 00′	.2588	9.4130	.9659	9.9849	.2679	9.4281	3.7321	0.5719	75° 00′
10	.2616	4177	.9652	9846	.2711	4331	3.6891	5669	50
20	.2644	4223	.9644	9843	.2742	4381	3.6470	5619	40
30	.2672	4269	.9636	9839	.2773	4430	3.6059	5570	30
40	.2700	4314	.9628	9836	.2805	4479	3.5656	5521	20
50	.2728	4359	.9621	9832	.2836	4527	3.5261	5473	10
16° 00′	.2756	9.4403	.9613	9.9828	.2867	9.4575	3.4874	0.5425	74° 00′
10	.2784	4447	.9605	9825	.2899	4622	3.4495	5378	50
20	.2812	4491	.9596	9821	.2931	4669	3.4124	5331	40
30	.2840	4533	.9588	9817	.2962	4716	3.3759	5284	30
40	.2868	4576	.9580	9814	.2994	4762	3.3402	5238	20
50	.2896	4618	.9572	9810	.3026	4808	3.3052	5192	10
17° 00′	.2924	9.4659	.9563	9.9806	.3057	9.4853	3.2709	0.5147	73° 00′
10	.2952	4700	.9555	9802	.3089	4898	3.2371	5102	50
20	.2979	4741	.9546	9798	.3121	4943	3.2041	5057	40
30	.3007	4781	.9537	9794	.3153	4987	3.1716	5013	30
40	.3035	4821	.9528	9790	.3185	5031	3.1397	4969	20
50	.3062	4861	.9520	9786	.3217	5075	3.1084	4925	10
18° 00′	.3090	9.4900	.9511	9.9782	.3249	9.5118	3.0777	0.4882	72° 00′
	Nat.	Log.	Nat.	Log.	Nat.	Log.	Nat.	Log.	
Angles	Cosines		Sines		Cotangents		Tangents		Angles

Table II Trigonometric Functions (*Continued*)

Angles	Sines		Cosines		Tangents		Cotangents		Angles
	Nat.	Log.	Nat.	Log.	Nat.	Log.	Nat.	Log.	
18° 00′	.3090	9.4900	.9511	9.9782	.3249	9.5118	3.0777	0.4882	72° 00′
10	.3118	4939	.9502	9778	.3281	5161	3.0475	4839	50
20	.3145	4977	.9492	9774	.3314	5203	3.0178	4797	40
30	.3173	5015	.9483	9770	.3346	5245	2.9887	4755	30
40	.3201	5052	.9474	9765	.3378	5287	2.9600	4713	20
50	.3228	5090	.9465	9761	.3411	5329	2.9319	4671	10
19° 00′	.3256	9.5126	.9455	9.9757	.3443	9.5370	2.9042	0.4630	71° 00′
10	.3283	5163	.9446	9752	.3476	5411	2.8770	4589	50
20	.3311	5199	.9436	9748	.3508	5451	2.8502	4549	40
30	.3338	5235	.9426	9743	.3541	5491	2.8239	4509	30
40	.3365	5270	.9417	9739	.3574	5531	2.7980	4469	20
50	.3393	5306	.9407	9734	.3607	5571	2.7725	4429	10
20° 00′	.3420	9.5341	.9397	9.9730	.3640	9.5611	2.7475	0.4389	70° 00′
10	.3448	5375	.9387	9725	.3673	5650	2.7228	4350	50
20	.3475	5409	.9377	9721	.3706	5689	2.6985	4311	40
30	.3502	5443	.9367	9716	.3739	5727	2.6746	4273	30
40	.3529	5477	.9356	9711	.3772	5766	2.6511	4234	20
50	.3557	5510	.9346	9706	.3805	5804	2.6279	4196	10
21° 00′	.3584	9.5543	.9336	9.9702	.3839	9.5842	2.6051	0.4158	69° 00′
10	.3611	5576	.9325	9697	.3872	5879	2.5826	4121	50
20	.3638	5609	.9315	9692	.3906	5917	2.5605	4083	40
30	.3665	5641	.9304	9687	.3939	5954	2.5386	4046	30
40	.3692	5673	.9293	9682	.3973	5991	2.5172	4009	20
50	.3719	5704	.9283	9677	.4006	6028	2.4960	3972	10
22° 00′	.3746	9.5736	.9272	9.9672	.4040	9.6064	2.4751	0.3936	68° 00′
10	.3773	5767	.9261	9667	.4074	6100	2.4545	3900	50
20	.3800	5798	.9250	9661	.4108	6136	2.4342	3864	40
30	.3827	5828	.9239	9656	.4142	6172	2.4142	3828	30
40	.3854	5859	.9228	9651	.4176	6208	2.3945	3792	20
50	.3881	5889	.9216	9646	.4210	6243	2.3750	3757	10
23° 00′	.3907	9.5919	.9205	9.9640	.4245	9.6279	2.3559	0.3721	67° 00′
10	.3934	5948	.9194	9635	.4279	6314	2.3369	3686	50
20	.3961	5978	.9182	9629	.4314	6348	2.3183	3652	40
30	.3987	6007	.9171	9624	.4348	6383	2.2998	3617	30
40	.4014	6036	.9159	9618	.4383	6417	2.2817	3583	20
50	.4041	6065	.9147	9613	.4417	6452	2.2637	3548	10
24° 00′	.4067	9.6093	.9135	9.9607	.4452	9.6486	2.2460	0.3514	66° 00′
10	.4094	6121	.9124	9602	.4487	6520	2.2286	3480	50
20	.4120	6149	.9112	9596	.4522	6553	2.2113	3447	40
30	.4147	6177	.9100	9590	.4557	6587	2.1943	3413	30
40	.4173	6205	.9088	9584	.4592	6620	2.1775	3380	20
50	.4200	6232	.9075	9579	.4628	6654	2.1609	3346	10
25° 00′	.4226	9.6259	.9063	9.9573	.4663	9.6687	2.1445	0.3313	65° 00′
10	.4253	6286	.9051	9567	.4699	6720	2.1283	3280	50
20	.4279	6313	.9038	9561	.4734	6752	2.1123	3248	40
30	.4305	6340	.9026	9555	.4770	6785	2.0965	3215	30
40	.4331	6366	.9013	9549	.4806	6817	2.0809	3183	20
50	.4358	6392	.9001	9543	.4841	6850	2.0655	3150	10
26° 00′	.4384	9.6418	.8988	9.9537	.4877	9.6882	2.0503	0.3118	64° 00′
10	.4410	6444	.8975	9530	.4913	6914	2.0353	3086	50
20	.4436	6470	.8962	9524	.4950	6946	2.0204	3054	40
30	.4462	6495	.8949	9518	.4986	6977	2.0057	3023	30
40	.4488	6521	.8936	9512	.5022	7009	1.9912	2991	20
50	.4514	6546	.8923	9505	.5059	7040	1.9768	2960	10
27° 00′	.4540	9.6570	.8910	9.9499	.5095	9.7072	1.9626	0.2928	63° 00′
	Nat.	Log.	Nat.	Log.	Nat.	Log.	Nat.	Log.	
Angles	Cosines		Sines		Cotangents		Tangents		Angles

Table II Trigonometric Functions (*Continued*)

Angles	Sines		Cosines		Tangents		Cotangents		Angles
	Nat.	Log.	Nat.	Log.	Nat.	Log.	Nat.	Log.	
27° 00′	.4540	9.6570	.8910	9.9499	.5095	9.7072	1.9626	0.2928	63° 00′
10	.4566	6595	.8897	9492	.5132	7103	1.9486	2897	50
20	.4592	6620	.8884	9486	.5169	7134	1.9347	2866	40
30	.4617	6644	.8870	9479	.5206	7165	1.9210	2835	30
40	.4643	6668	.8857	9473	.5243	7196	1.9074	2804	20
50	.4669	6692	.8843	9466	.5280	7226	1.8940	2774	10
28° 00′	.4695	9.6716	.8829	9.9459	.5317	9.7257	1.8807	0.2743	62° 00′
10	.4720	6740	.8816	9453	.5354	7287	1.8676	2713	50
20	.4746	6763	.8802	9446	.5392	7317	1.8546	2683	40
30	.4772	6787	.8788	9439	.5430	7348	1.8418	2652	30
40	.4797	6810	.8774	9432	.5467	7378	1.8291	2622	20
50	.4823	6833	.8760	9425	.5505	7408	1.8165	2592	10
29° 00′	.4848	9.6856	.8746	9.9418	.5543	9.7438	1.8040	0.2562	61° 00′
10	.4874	6878	.8732	9411	.5581	7467	1.7917	2533	50
20	.4899	6901	.8718	9404	.5619	7497	1.7796	2503	40
30	.4924	6923	.8704	9397	.5658	7526	1.7675	2474	30
40	.4950	6946	.8689	9390	.5696	7556	1.7556	2444	20
50	.4975	6968	.8675	9383	.5735	7585	1.7437	2415	10
30° 00′	.5000	9.6990	.8660	9.9375	.5774	9.7614	1.7321	0.2386	60° 00′
10	.5025	7012	.8646	9368	.5812	7644	1.7205	2356	50
20	.5050	7033	.8631	9361	.5851	7673	1.7090	2327	40
30	.5075	7055	.8616	9353	.5890	7701	1.6977	2299	30
40	.5100	7076	.8601	9346	.5930	7730	1.6864	2270	20
50	.5125	7097	.8587	9338	.5969	7759	1.6753	2241	10
31° 00′	.5150	9.7118	.8572	9.9331	.6009	9.7788	1.6643	0.2212	59° 00′
10	.5175	7139	.8557	9323	.6048	7816	1.6534	2184	50
20	.5200	7160	.8542	9315	.6088	7845	1.6426	2155	40
30	.5225	7181	.8526	9308	.6128	7873	1.6319	2127	30
40	.5250	7201	.8511	9300	.6168	7902	1.6212	2098	20
50	.5275	7222	.8496	9292	.6208	7930	1.6107	2070	10
32° 00′	.5299	9.7242	.8480	9.9284	.6249	9.7958	1.6003	0.2042	58° 00′
10	.5324	7262	.8465	9276	.6289	7986	1.5900	2014	50
20	.5348	7282	.8450	9268	.6330	8014	1.5798	1986	40
30	.5373	7302	.8434	9260	.6371	8042	1.5697	1958	30
40	.5398	7322	.8418	9252	.6412	8070	1.5597	1930	20
50	.5422	7342	.8403	9244	.6453	8097	1.5497	1903	10
33° 00′	.5446	9.7361	.8387	9.9236	.6494	9.8125	1.5399	0.1875	57° 00′
10	.5471	7380	.8371	9228	.6536	8153	1.5301	1847	50
20	.5495	7400	.8355	9219	.6577	8180	1.5204	1820	40
30	.5519	7419	.8339	9211	.6619	8208	1.5108	1792	30
40	.5544	7438	.8323	9203	.6661	8235	1.5013	1765	20
50	.5568	7457	.8307	9194	.6703	8263	1.4919	1737	10
34° 00′	.5592	9.7476	.8290	9.9186	.6745	9.8290	1.4826	0.1710	56° 00′
10	.5616	7494	.8274	9177	.6787	8317	1.4733	1683	50
20	.5640	7513	.8258	9169	.6830	8344	1.4641	1656	40
30	.5664	7531	.8241	9160	.6873	8371	1.4550	1629	30
40	.5688	7550	.8225	9151	.6916	8398	1.4460	1602	20
50	.5712	7568	.8208	9142	.6959	8425	1.4370	1575	10
35° 00′	.5736	9.7586	.8192	9.9134	.7002	9.8452	1.4281	0.1548	55° 00′
10	.5760	7604	.8175	9125	.7046	8479	1.4193	1521	50
20	.5783	7622	.8158	9116	.7089	8506	1.4106	1494	40
30	.5807	7640	.8141	9107	.7133	8533	1.4019	1467	30
40	.5831	7657	.8124	9098	.7177	8559	1.3934	1441	20
50	.5854	7675	.8107	9089	.7221	8586	1.3848	1414	10
36° 00′	.5878	9.7692	.8090	9.9080	.7265	9.8613	1.3764	0.1387	54° 00′
	Nat.	Log.	Nat.	Log.	Nat.	Log.	Nat.	Log.	
Angles	Cosines		Sines		Cotangents		Tangents		Angles

Table II Trigonometric Functions (*Continued*)

Angles	Sines Nat.	Sines Log.	Cosines Nat.	Cosines Log.	Tangents Nat.	Tangents Log.	Cotangents Nat.	Cotangents Log.	Angles
36° 00′	.5878	9.7692	.8090	9.9080	.7265	9.8613	1.3764	0.1387	54° 00′
10	.5901	7710	.8073	9070	.7310	8639	1.3680	1361	50
20	.5925	7727	.8056	9061	.7355	8666	1.3597	1334	40
30	.5948	7744	.8039	9052	.7400	8692	1.3514	1308	30
40	.5972	7761	.8021	9042	.7445	8718	1.3432	1282	20
50	.5995	7778	.8004	9033	.7490	8745	1.3351	1255	10
37° 00′	.6018	9.7795	.7986	9.9023	.7536	9.8771	1.3270	0.1229	53° 00′
10	.6041	7811	.7969	9014	.7581	8797	1.3190	1203	50
20	.6065	7828	.7951	9004	.7627	8824	1.3111	1176	40
30	.6088	7844	.7934	8995	.7673	8850	1.3032	1150	30
40	.6111	7861	.7916	8985	.7720	8876	1.2954	1124	20
50	.6134	7877	.7898	8975	.7766	8902	1.2876	1098	10
38° 00′	.6157	9.7893	.7880	9.8965	.7813	9.8928	1.2790	0.1072	52° 00′
10	.6180	7910	.7862	8955	.7860	8954	1.2723	1046	50
20	.6202	7926	.7844	8945	.7907	8980	1.2647	1020	40
30	.6225	7941	.7826	8935	.7954	9006	1.2572	0994	30
40	.6248	7957	.7808	8925	.8002	9032	1.2497	0968	20
50	.6271	7973	.7790	8915	.8050	9058	1.2423	0942	10
39° 00′	.6293	9.7989	.7771	9.8905	.8098	9.9084	1.2349	0.0916	51° 00′
10	.6316	8004	.7753	8895	.8146	9110	1.2276	0890	50
20	.6338	8020	.7735	8884	.8195	9135	1.2203	0865	40
30	.6361	8035	.7716	8874	.8243	9161	1.2131	0839	30
40	.6383	8050	.7698	8864	.8292	9187	1.2059	0813	20
50	.6406	8066	.7679	8853	.8342	9212	1.1988	0788	10
40° 00′	.6428	9.8081	.7660	9.8843	.8391	9.9238	1.1918	0.0762	50° 00′
10	.6450	8096	.7642	8832	.8441	9264	1.1847	0736	50
20	.6472	8111	.7623	8821	.8491	9289	1.1778	0711	40
30	.6494	8125	.7604	8810	.8541	9315	1.1708	0685	30
40	.6517	8140	.7585	8800	.8591	9341	1.1640	0659	20
50	.6539	8155	.7566	8789	.8642	9366	1.1571	0634	10
41° 00′	.6561	9.8169	.7547	9.8778	.8693	9.9392	1.1504	0.0608	49° 00′
10	.6583	8184	.7528	8767	.8744	9417	1.1436	0583	50
20	.6604	8198	.7509	8756	.8796	9443	1.1369	0557	40
30	.6626	8213	.7490	8745	.8847	9468	1.1303	0532	30
40	.6648	8227	.7470	8733	.8899	9494	1.1237	0506	20
50	.6670	8241	.7451	8722	.8952	9519	1.1171	0481	10
42° 00′	.6691	9.8255	.7431	9.8711	.9004	9.9544	1.1106	0.0456	48° 00′
10	.6713	8269	.7412	8699	.9057	9570	1.1041	0430	50
20	.6734	8283	.7392	8688	.9110	9595	1.0977	0405	40
30	.6756	8297	.7373	8676	.9163	9621	1.0913	0379	30
40	.6777	8311	.7353	8665	.9217	9646	1.0850	0354	20
50	.6799	8324	.7333	8653	.9271	9671	1.0786	0329	10
43° 00′	.6820	9.8338	.7314	9.8641	.9325	9.9697	1.0724	0.0303	47° 00′
10	.6841	8351	.7294	8629	.9380	9722	1.0661	0278	50
20	.6862	8365	.7274	8618	.9435	9747	1.0599	0253	40
30	.6884	8378	.7254	8606	.9490	9772	1.0538	0228	30
40	.6905	8391	.7234	8594	.9545	9798	1.0477	0202	20
50	.6926	8405	.7214	8582	.9601	9823	1.0416	0177	10
44° 00′	.6947	9.8418	.7193	9.8569	.9657	9.9848	1.0355	0.0152	46° 00′
10	.6967	8431	.7173	8557	.9713	9874	1.0295	0126	50
20	.6988	8444	.7153	8545	.9770	9899	1.0235	0101	40
30	.7009	8457	.7133	8532	.9827	9924	1.0176	0076	30
40	.7030	8469	.7112	8520	.9884	9949	1.0117	0051	20
50	.7050	8482	.7092	8507	.9942	9975	1.0058	0025	10
45° 00′	.7071	9.8495	.7071	9.8495	1.0000	0.0000	1.0000	0.0000	45° 00′
Angles	Cosines Nat.	Cosines Log.	Sines Nat.	Sines Log.	Cotangents Nat.	Cotangents Log.	Tangents Nat.	Tangents Log.	Angles

Table III Powers and Roots

No.	Sq.	Sq. Root	Cube	Cube Root	No.	Sq.	Sq. Root	Cube	Cube Root
1	1	1.000	1	1.000	**51**	2,601	7.141	132,651	3.708
2	4	1.414	8	1.260	**52**	2,704	7.211	140,608	3.733
3	9	1.732	27	1.442	**53**	2,809	7.280	148,877	3.756
4	16	2.000	64	1.587	**54**	2,916	7.348	157,464	3.780
5	25	2.236	125	1.710	**55**	3,025	7.416	166,375	3.803
6	36	2.449	216	1.817	**56**	3,136	7.483	175,616	3.826
7	49	2.646	343	1.913	**57**	3,249	7.550	185,193	3.849
8	64	2.828	512	2.000	**58**	3,364	7.616	195,112	3.871
9	81	3.000	729	2.080	**59**	3,481	7.681	205,379	3.893
10	100	3.162	1,000	2.154	**60**	3,600	7.746	216,000	3.915
11	121	3.317	1,331	2.224	**61**	3,721	7.810	226,981	3.936
12	144	3.464	1,728	2.289	**62**	3,844	7.874	238,328	3.958
13	169	3.606	2,197	2.351	**63**	3,969	7.937	250,047	3.979
14	196	3.742	2,744	2.410	**64**	4,096	8.000	262,144	4.000
15	225	3.873	3,375	2.466	**65**	4,225	8.062	274,625	4.021
16	256	4.000	4,096	2.520	**66**	4,356	8.124	287,496	4.041
17	289	4.123	4,913	2.571	**67**	4,489	8.185	300,763	4.062
18	324	4.243	5,832	2.621	**68**	4,624	8.246	314,432	4.082
19	361	4.359	6,859	2.668	**69**	4,761	8.307	328,509	4.102
20	400	4.472	8,000	2.714	**70**	4,900	8.367	343,000	4.121
21	441	4.583	9,261	2.759	**71**	5,041	8.426	357,911	4.141
22	484	4.690	10,648	2.802	**72**	5,184	8.485	373,248	4.160
23	529	4.796	12,167	2.844	**73**	5,329	8.544	389,017	4.179
24	576	4.899	13,824	2.884	**74**	5,476	8.602	405,224	4.198
25	625	5.000	15,625	2.924	**75**	5,625	8.660	421,875	4.217
26	676	5.099	17,576	2.962	**76**	5,776	8.718	438,976	4.236
27	729	5.196	19,683	3.000	**77**	5,929	8.775	456,533	4.254
28	784	5.291	21,952	3.037	**78**	6,084	8.832	474,552	4.273
29	841	5.385	24,389	3.072	**79**	6,241	8.888	493,039	4.291
30	900	5.477	27,000	3.107	**80**	6,400	8.944	512,000	4.309
31	961	5.568	29,791	3.141	**81**	6,561	9.000	531,441	4.327
32	1,024	5.657	32,768	3.175	**82**	6,724	9.055	551,368	4.344
33	1,089	5.745	35,937	3.208	**83**	6,889	9.110	571,787	4.362
34	1,156	5.831	39,304	3.240	**84**	7,056	9.165	592,704	4.380
35	1,225	5.916	42,875	3.271	**85**	7,225	9.220	614,125	4.397
36	1,296	6.000	46,656	3.302	**86**	7,396	9.274	636,056	4.414
37	1,369	6.083	50,653	3.332	**87**	7,569	9.327	658,503	4.431
38	1,444	6.164	54,872	3.362	**88**	7,744	9.381	681,472	4.448
39	1,521	6.245	59,319	3.391	**89**	7,921	9.434	704,969	4.465
40	1,600	6.325	64,000	3.420	**90**	8,100	9.487	729,000	4.481
41	1,681	6.403	68,921	3.448	**91**	8,281	9.539	753,571	4.498
42	1,764	6.481	74,088	3.476	**92**	8,464	9.592	778,688	4.514
43	1,849	6.557	79,507	3.503	**93**	8,649	9.644	804,357	4.531
44	1,936	6.633	85,184	3.530	**94**	8,836	9.695	830,584	4.547
45	2,025	6.708	91,125	3.557	**95**	9,025	9.747	857,375	4.563
46	2,116	6.782	97,336	3.583	**96**	9,216	9.798	884,736	4.579
47	2,209	6.856	103,823	3.609	**97**	9,409	9.849	912,673	4.595
48	2,304	6.928	110,592	3.634	**98**	9,604	9.899	941,192	4.610
49	2,401	7.000	117,649	3.659	**99**	9,801	9.950	970,299	4.626
50	2,500	7.071	125,000	3.684	**100**	10,000	10.000	1,000,000	4.642

Table IV Natural Logarithms

N	0	1	2	3	4	5	6	7	8	9
1.0	0.0 000	100	198	296	392	488	583	677	770	862
1.1	953	*044	*133	*222	*310	*398	*484	*570	*655	*740
1.2	0.1 823	906	989	*070	*151	*231	*311	*390	*469	*546
1.3	0.2 624	700	776	852	927	*001	*075	*148	*221	*293
1.4	0.3 365	436	507	577	646	716	784	853	920	988
1.5	0.4 055	121	187	253	318	383	447	511	574	637
1.6	700	762	824	886	947	*008	*068	*128	*188	*247
1.7	0.5 306	365	423	481	539	596	653	710	766	822
1.8	878	933	988	*043	*098	*152	*206	*259	*313	*366
1.9	0.6 419	471	523	575	627	678	729	780	831	881
2.0	931	981	*031	*080	*129	*178	*227	*275	*324	*372
2.1	0.7 419	467	514	561	608	655	701	747	793	839
2.2	885	930	975	*020	*065	*109	*154	*198	*242	*286
2.3	0.8 329	372	416	459	502	544	587	629	671	713
2.4	755	796	838	879	920	961	*002	*042	*083	*123
2.5	0.9 163	203	243	282	322	361	400	439	478	517
2.6	555	594	632	670	708	746	783	821	858	895
2.7	933	969	*006	*043	*080	*116	*152	*188	*225	*260
2.8	1.0 296	332	367	403	438	473	508	543	578	613
2.9	647	682	716	750	784	818	852	886	919	953
3.0	986	*019	*053	*086	*119	*151	*184	*217	*249	*282
3.1	1.1 314	346	378	410	442	474	506	537	569	600
3.2	632	663	694	725	756	787	817	848	878	909
3.3	939	969	*000	*030	*060	*090	*119	*149	*179	*208
3.4	1.2 238	267	296	326	355	384	413	442	470	499
3.5	528	556	585	613	641	669	698	726	754	782
3.6	809	837	865	892	920	947	975	*002	*029	*056
3.7	1.3 083	110	137	164	191	218	244	271	297	324
3.8	350	376	402	429	455	481	507	533	558	584
3.9	610	635	661	686	712	737	762	788	813	838
4.0	863	888	913	938	962	987	*012	*036	*061	*085
4.1	1.4 110	134	159	183	207	231	255	279	303	327
4.2	351	375	398	422	446	469	493	516	540	563
4.3	586	609	633	656	679	702	725	748	770	793
4.4	816	839	861	884	907	929	951	974	996	*019
4.5	1.5 041	063	085	107	129	151	173	195	217	239
4.6	261	282	304	326	347	369	390	412	433	454
4.7	476	497	518	539	560	581	602	623	644	665
4.8	686	707	728	748	769	790	810	831	851	872
4.9	892	913	933	953	974	994	*014	*034	*054	*074
5.0	1.6 094	114	134	154	174	194	214	233	253	273

If given number $n = N \times 10^m$, then $\log_e n = \log_e N + m \log_e 10$. Find $m \log_e 10$ from the following table:

Multiples of $\log_e 10$

$\log_e 10 = 2.3026$	$-\log_e 10 = 7.6974 - 10$
$2 \log_e 10 = 4.6052$	$-2 \log_e 10 = 5.3948 - 10$
$3 \log_e 10 = 6.9078$	$-3 \log_e 10 = 3.0922 - 10$
$4 \log_e 10 = 9.2103$	$-4 \log_e 10 = 0.7897 - 10$
$5 \log_e 10 = 11.5129$	$-5 \log_e 10 = 9.4871 - 20$

Table IV Natural Logarithms (*Continued*)

N	0	1	2	3	4	5	6	7	8	9
5.0	1.6 094	114	134	154	174	194	214	233	253	273
5.1	292	312	332	351	371	390	409	429	448	467
5.2	487	506	525	544	563	582	601	620	639	658
5.3	677	696	715	734	752	771	790	808	827	845
5.4	864	882	901	919	938	956	974	993	*011	*029
5.5	1.7 047	066	084	102	120	138	156	174	192	210
5.6	228	246	263	281	299	317	334	352	370	387
5.7	405	422	440	457	475	492	509	527	544	561
5.8	579	596	613	630	647	664	681	699	716	733
5.9	750	766	783	800	817	834	851	867	884	901
6.0	918	934	951	967	984	*001	*017	*034	*050	*066
6.1	1.8 083	099	116	132	148	165	181	197	213	229
6.2	245	262	278	294	310	326	342	358	374	390
6.3	405	421	437	453	469	485	500	516	532	547
6.4	563	579	594	610	625	641	656	672	687	703
6.5	718	733	749	764	779	795	810	825	840	856
6.6	871	886	901	916	931	946	961	976	991	*006
6.7	1.9 021	036	051	066	081	095	110	125	140	155
6.8	169	184	199	213	228	242	257	272	286	301
6.9	315	330	344	359	373	387	402	416	430	445
7.0	459	473	488	502	516	530	544	559	573	587
7.1	601	615	629	643	657	671	685	699	713	727
7.2	741	755	769	782	796	810	824	838	851	865
7.3	879	892	906	920	933	947	961	974	988	*001
7.4	2.0 015	028	042	055	069	082	096	109	122	136
7.5	149	162	176	189	202	215	229	242	255	268
7.6	281	295	308	321	334	347	360	373	386	399
7.7	412	425	438	451	464	477	490	503	516	528
7.8	541	554	567	580	592	605	618	631	643	656
7.9	669	681	694	707	719	732	744	757	769	782
8.0	794	807	819	832	844	857	869	882	894	906
8.1	919	931	943	956	968	980	992	*005	*017	*029
8.2	2.1 041	054	066	080	090	102	114	126	138	150
8.3	163	175	187	199	211	223	235	247	258	270
8.4	282	294	306	318	330	342	353	365	377	389
8.5	401	412	424	436	448	460	471	483	494	506
8.6	518	529	541	552	564	576	587	599	610	622
8.7	633	645	656	668	679	691	702	713	725	736
8.8	748	759	770	782	793	804	815	827	838	849
8.9	861	872	883	894	905	917	928	939	950	961
9.0	972	983	994	*006	*017	*028	*039	*050	*061	*072
9.1	2.2 083	094	105	116	127	137	148	159	170	181
9.2	192	203	214	225	235	246	257	268	279	289
9.3	300	311	322	332	343	354	364	375	386	396
9.4	407	418	428	439	450	460	471	481	492	502
9.5	513	523	534	544	555	565	576	586	597	607
9.6	618	628	638	649	659	670	680	690	701	711
9.7	721	732	742	752	762	773	783	793	803	814
9.8	824	834	844	854	865	875	885	895	905	915
9.9	925	935	946	956	966	976	986	996	*006	*016
10.	2.3 026	036	046	056	066	076	086	096	106	115

Table V Exponential and Hyperbolic Functions

x	e^x	e^{-x}	**sinh** x	**cosh** x	**tanh** x
.00	1.000	1.000	.000	1.000	.000
.01	1.010	.990	.010	1.000	.010
.02	1.020	.980	.020	1.000	.020
.03	1.030	.970	.030	1.000	.030
.04	1.041	.961	.040	1.001	.040
.05	1.051	.951	.050	1.001	.050
.06	1.062	.942	.060	1.002	.060
.07	1.073	.932	.070	1.002	.070
.08	1.083	.923	.080	1.003	.080
.09	1.094	.914	.090	1.004	.090
.1	1.105	.905	.100	1.005	.100
.2	1.221	.819	.201	1.020	.197
.3	1.350	.741	.305	1.045	.291
.4	1.492	.670	.411	1.081	.380
.5	1.649	.607	.521	1.128	.462
.6	1.822	.549	.637	1.185	.537
.7	2.014	.497	.759	1.255	.604
.8	2.226	.449	.888	1.337	.664
.9	2.460	.407	1.027	1.433	.716
1.0	2.718	.368	1.175	1.543	.762
1.1	3.004	.333	1.336	1.669	.800
1.2	3.320	.301	1.509	1.811	.834
1.3	3.669	.273	1.698	1.971	.862
1.4	4.055	.247	1.904	2.151	.885
1.5	4.482	.223	2.129	2.352	.905
1.6	4.953	.202	2.376	2.577	.922
1.7	5.474	.183	2.646	2.828	.935
1.8	6.050	.165	2.942	3.107	.947
1.9	6.686	.150	3.268	3.418	.956
2.0	7.389	.135	3.627	3.762	.964
2.1	8.166	.122	4.022	4.144	.970
2.2	9.025	.111	4.457	4.568	.976
2.3	9.974	.100	4.937	5.037	.980
2.4	11.023	.091	5.466	5.557	.984
2.5	12.182	.082	6.050	6.132	.987
2.6	13.464	.074	6.695	6.769	.989
2.7	14.880	.067	7.406	7.473	.991
2.8	16.445	.061	8.192	8.253	.993
2.9	18.174	.055	9.060	9.115	.994
3.0	20.086	.050	10.018	10.068	.995
3.1	22.20	.045	11.08	11.12	.996
3.2	24.53	.041	12.25	12.29	.997
3.3	27.11	.037	13.54	13.57	.997
3.4	29.96	.033	14.97	15.00	.998
3.5	33.12	.030	16.54	16.57	.998
3.6	36.60	.027	18.29	18.31	.999
3.7	40.45	.025	20.21	20.24	.999
3.8	44.70	.022	22.34	22.36	.999
3.9	49.40	.020	24.69	24.71	.999
4.0	54.60	.018	27.29	27.31	.999
4.1	60.34	.017	30.16	30.18	.999
4.2	66.69	.015	33.34	33.35	1.000
4.3	73.70	.014	36.84	36.86	1.000
4.4	81.45	.012	40.72	40.73	1.000
4.5	90.02	.011	45.00	45.01	1.000
4.6	99.48	.010	49.74	49.75	1.000
4.7	109.95	.0090	54.97	54.98	1.000
4.8	121.51	.0082	60.75	60.76	1.000
4.9	134.29	.0074	67.14	67.15	1.000
5.0	148.41	.0067	74.20	74.21	1.000
6.0	403.4	.0025	201.7		1.000
7.0	1096.6	.00091	548.3		1.000
8.0	2981.0	.00034	1490.5		1.000
9.0	8103.1	.00012	4051.5		1.000
10.0	22026.5	.000045	11013.2		1.000

ANSWERS

CHAPTER 1

Page 4

1. Yes.
5. (*a*) Example: $a = 3 + \sqrt{5}$, $b = 3 - \sqrt{5}$, $a + b = 6$ (a rational number); (*b*) Irrational.
7. (*a*) -5 and 5; (*b*) -2 and 8.
9. 4 and 6.5.
11. $x \geqslant 3.5$; $x \leqslant 3.5$.
13. (*a*) $x = 3$ or $x = -11$
    (*b*) $x > \frac{77}{2}$ or $x < -\frac{77}{2}$
    (*c*) $x = -\frac{1}{2}$
    (*d*) $x > -\frac{4}{5}$ or $x < -4$, $x \neq 0$
    (*e*) $x > \frac{1}{3}$ or $x < -\frac{1}{3}$, $x \neq 0$
    (*f*) No solution
15. $3x + 1$.

Page 11

1. (*a*) ——— P_1 P_2 P_3 (points below the line)
   (*b*) P_1 P_3 P_2 (points above the line) ———
3. (*a*) 7; (*b*) -6; (*c*) 6; (*d*) -9; (*e*) 2; (*f*) -2.
5. $5 - y$.
7. (*a*) $\sqrt{34}$; (*b*) $\sqrt{97}$; (*c*) $\sqrt{101}$; (*d*) $\sqrt{61}$; (*e*) $\sqrt{2}$; (*f*) $\sqrt{37}$.
13. $|CD| = |CE| = |CF| = 5$.
15. $(4\frac{1}{9}, 0)$.
17. $(5.8, 11.6)$.

Page 16

3. (*a*) $(3, 60°)$ or $(3, 300°)$
   (*b*) $(5, 105°)$ or $(5, 345°)$
   (*c*) $(0, 0)$ or $(5\sqrt{3}, \pi/3)$
5. (*a*) $\pi/2$; (*b*) $\pi/4$.
7. (*a*) $(0, 3)$; (*b*) $(7, 0)$; (*c*) $(-3, 0)$; (*d*) $(3\sqrt{2}/2, -3\sqrt{2}/2)$; (*e*) $(5\sqrt{2}/2, -5\sqrt{2}/2)$; (*f*) $(\sqrt{2}, \sqrt{2})$.
9. (*a*) No; (*b*) yes; (*c*) no; (*d*) yes.

Page 22

1. (*a*) $(1, -4)$; (*b*) $(-3, \frac{7}{2})$; (*c*) $(\sqrt{2}/2, -3\sqrt{3})$.
3. $(\frac{11}{2}, -\frac{3}{4})$; $(3, \frac{3}{2})$; $(\frac{1}{2}, \frac{15}{4})$.
7. $(2, \frac{2}{3})$.
9. (*a*) 23; (*b*) 38; (*c*) 40; (*d*) 9.
11. (*a*) 60; (*b*) 36.5.
15. 0.6.
17. Points on $\overline{P_1P_2}$ whose coordinates are rational numbers.

Review Problems for Chap. 1

1. (*a*) $x = -10$ or $x = 4$
   (*b*) $x = 3\frac{1}{4}$ or $x = 2\frac{3}{4}$
   (*c*) $-\frac{5}{3} \leqslant x \leqslant 3$

5. $(-2, -2)$.

7. (a) $(6, 90^\circ)$; (b) $(-6, -90^\circ)$; (c) $(6, 90^\circ + n360^\circ)$, n an integer $\geqslant 0$; (d) $(6, -270 + n360^\circ)$, n an integer $\leqslant 0$.

Extended Study Problems for Chap. 1

3. (a) $(8, 0)$, $(4, 4\sqrt{3})$, $(-4, 4\sqrt{3})$, $(-8, 0)$, $(-4, -4\sqrt{3})$, $(4, -4\sqrt{3})$.
(b) $(8, 0)$, $(8, \pi/3)$, $(8, 2\pi/3)$, $(8, \pi)$, $(8, 4\pi/3)$, $(8, 5\pi/3)$.

5. All points in the interior of a circle, center at O and radius 5.

9. $(\frac{11}{5}, \frac{27}{5})$.

11. $\sqrt{58}$.

CHAPTER 2

Page 33

1. (a) $-\sqrt{3}$; (b) undefined; (c) $\sqrt{3}$; (d) $1/\sqrt{3}$; (e) 1; (f) $-1/\sqrt{3}$.

3. (a) No; (b) yes; (c) no; (d) yes.

7. (a) $\frac{14}{3}, -\frac{4}{3}, \frac{2}{3}$; (b) $-\frac{22}{3}, -\frac{2}{15}, \frac{5}{3}$.

9. (a) $\frac{1}{9}$; (b) $\frac{1}{8}$; (c) $-\frac{1}{2}$; (d) impossible.

Page 39

1. For $\overrightarrow{P_1P_2}$: (a) $[3, 2]$, $[3/\sqrt{13}, 2/\sqrt{13}]$; (b) $[-10, -7]$, $[-10/\sqrt{149}, -7\sqrt{149}]$; (c) $[5, -6]$, $[5/\sqrt{61}, -6\sqrt{61}]$; (d) $[12, 4]$, $[3/\sqrt{10}, 1/\sqrt{10}]$

For $\overrightarrow{P_2P_1}$: (a) $[-3, -2]$, $[-3/\sqrt{13}, -2/\sqrt{13}]$; (b) $[10, 7]$, $[10/\sqrt{149}, 7/\sqrt{149}]$; (c) $[-5, 6]$, $[-5/\sqrt{61}, 6/\sqrt{61}]$; (d) $[-12, -4]$, $[-3/\sqrt{10}, -1/\sqrt{10}]$

3. (a) $\frac{3}{2}$; (b) $-\frac{3}{4}$; (c) $-3/\sqrt{7}$; (d) bc/ad.

5. (a) $-\sqrt{3}$; (b) $[\frac{1}{2}, -\sqrt{3}/2]$; (c) $[-\frac{1}{2}k, \sqrt{3}/2\, k]$, $k > 0$.

7. (a), (c), (d).

9. (a) $[45^\circ, 45^\circ]$; (b) $[135^\circ, 45^\circ]$; (c) $[135^\circ, 135^\circ]$.

11. (a) $-\sqrt{3}/2$; (b) $\sqrt{15}/4$.

13. (a) $[0, 1]$; (b) $[0, -1]$.

Page 44

1. $-\sqrt{3}/3$ and $-1/\sqrt{3}$; $-\sqrt{3}$ and $-3/\sqrt{3}$; $3/\sqrt{3}$ and $\tan[\text{Arctan}(1/\sqrt{3})]$.

3. (a) $\sqrt{3}$; (b) $-1/\sqrt{3}$.

5. 3.

7. Altitudes: $\frac{5}{2}, -\frac{13}{4}, \frac{1}{5}$; medians: $\frac{14}{11}, -\frac{1}{14}, -\frac{16}{17}$.

11. (a) $\tan\theta = -\frac{33}{10}$; (b) $\tan\theta = \frac{12}{5}$.

13. (a) $\tan A = \frac{3}{5}$; B measures 90°; $\tan C = \frac{5}{3}$.
(b) $\tan P = -\frac{34}{19}$; $\tan Q = \frac{17}{30}$; $\tan R = \frac{17}{28}$.

15. 0.8.

Page 53

1. (a) $\sqrt{17}$, 4; (b) $\sqrt{2}$, 1; (c) $4\sqrt{5}$, 2.

Page 58

3. (a) $\frac{15}{2}$; (b) 0; (c) $(21\sqrt{3})/2$.

Review Problems for Chap. 2

1. (*a*) $\frac{4}{7}$; (*b*) $-\frac{1}{4}$; (*c*) $-\frac{2}{3}$; (*d*) -2 or $\frac{1}{2}$; (*e*) $-\sqrt{3}/3$; (*f*) undefined; (*g*) zero (0).

7. 2.

9. $\tan F = \frac{4}{7}$, $\tan E = \frac{7}{4}$.

Extended Study Problems for Chap. 2

1. α_1 and α_2 are supplementary, β_1 and β_2 are supplementary.

5. $\frac{56}{33}$, $-\frac{12}{5}$.

CHAPTER 3

Page 69

1. $x = y$.

3. $y = 5$.

33. $V = (s/6)^{3/2}$.

35. $L = gT^2/4\pi^2$.

37. $AP = x + 3$.

Page 75

17. $A = \pi r^2$.

19. $V = x(18 - 2x)(12 - 2x)$.

21. $A = \frac{5}{6}(12y - y^2)$.

23. $V = 2\pi x^2(6 - x)$.

Page 81

1. $(x + 3)^2 + (y + 1)^2 = 9$.

3. $(x + 5)^2 + (y - 5)^2 = 50$.

5. $3x - 4y - 7 = 0$.

7. $y^2 = 8x - 16$.

9. $x^2 + 8y = 16$.

11. $x^2 + y^2 + 8x - 48 = 0$.

13. $y = x$, $x \neq 1$.

15. $4x^2 + 3y - 32y + 64 = 0$.

17. $16x^2 + 25y^2 = 400$.

19. $5x^2 + 9y^2 - 40x - 100 = 0$.

21. Valid; valid.

23. Valid; valid.

25. Valid; valid.

Page 86

1. (*a*) $(\frac{3}{2}, 3\sqrt{3}/2)$; (*b*) $(5\sqrt{2}/2, -5\sqrt{2}/2)$; (*c*) $(3\sqrt{2}/2, -3\sqrt{2}/2)$; (*d*) $(3\sqrt{3}, -3)$.

3. (*a*) 7; (*b*) $\sqrt{65}$.

Page 88

17. $-2 < x < 4$, x real.

19. $-6 \leqslant x \leqslant 2$, x an integer.

21. $|x| > 1$, x real; $|x - 1| \leqslant 3$, x real; $|x| \geqslant 1$ and $x < 0$, x real; $|x + 2| \leqslant 4$, x an integer; $|x + 1| > 4$, x real.

Review Problems for Chap. 3

5. (3, 4), (−4, 3).

7. $V = x(4 - 2x)(8 - 2x)$.

9. (*a*) $\sqrt{125}$; (*b*) $P(\frac{5}{2}, -5\sqrt{3}/2)$, $Q(5\sqrt{3}, 5)$.

Extended Study Problems for Chap. 3

1. $\{(x, y) \mid (0 < x < 6) \cap (0 < y < 6), x \text{ and } y \text{ real}\}$.

3. $\{(6, y) \mid y \text{ real}\}$.

5. $\{(x, y) \mid x < 3, x \text{ and } y \text{ real}\}$.

7. $\{(x, a) \mid x \text{ real}\}$.

CHAPTER 4

Page 96

1. (*a*) $2x - 3y + 19 = 0$ (*b*) $x + 2y + 5 = 0$
(*c*) $2x + y = 13$ (*d*) $x - 3y + 23 = 0$

3. (*a*) $3x - 10y - 41 = 0$ (*b*) $3x + 4y + 12 = 0$
(*c*) $9x + 7y = 12$ (*d*) $7x - 6y + 4 = 0$

5. (*a*) $y = \frac{2}{3}x - 4$ (*b*) $y = -\frac{1}{2}x + 6$ (*c*) $y = -\frac{5}{2}x - 8$
(*d*) $4x - 3y - 6 = 0$ (*e*) $7x - 10y + 35 = 0$ (*f*) $4x - 3y = 2$

7. $x - 11y + 64 = 0$, $11x + y = 28$.

9. (*a*) $3x - 2y = 8$ (*b*) $3x + 5y + 22 = 0$

11. $10x + 3y = 0$.

13. (*a*) $2x + 13y = 16$ (*b*) $13x - 4y + 11 = 0$

15. $(0, 1)$.

Page 99

1. (*a*) $m = \frac{3}{4}$, $b = -\frac{9}{2}$, $y = \frac{3}{4}x - \frac{9}{2}$ (*b*) $m = 0.2$, $b = -3.2$, $y = .2x - 3.2$
(*c*) $m = \frac{2}{3}$, $b = \frac{4}{3}$, $y = \frac{2}{3}x + \frac{4}{3}$ (*d*) $m = \frac{3}{5}$, $b = -3$, $y = \frac{3}{5}x - 3$

3. $A = 7$, $B = -1$, $C = -13$. **5.** $a = 3$, $b = -2$.

7. (*a*) parallel to x axis; (*b*) parallel to y axis; (*c*) passes through origin.

9. (*a*) $[2, 4]$; (*b*) $[-4, 12]$; (*c*) $[-4, -7]$; (*d*) $[4, -9]$.

11. (*a*) $[1/\sqrt{5}, 2/\sqrt{5}]$ or $[-1/\sqrt{5}, -2/\sqrt{5}]$ (*b*) $[-1/\sqrt{10}, 3/\sqrt{10}]$ or $[1/\sqrt{10}, -3/\sqrt{10}]$
(*c*) $[-4/\sqrt{65}, -7/\sqrt{65}]$ or $[4/\sqrt{65}, 7/\sqrt{65}]$
(*d*) $[4/\sqrt{97}, -9/\sqrt{97}]$ or $[-4/\sqrt{97}, 9/\sqrt{97}]$

13. (*a*) $x - 2y = 0$ (*b*) $6x - y + 17 = 0$
(*c*) $x - 4y + 23 = 0$ (*d*) $b_1x - a_1y - b_1x_1 + a_1y_1 = 0$

Page 106

1. (*a*) $A = -\frac{2}{3}$, $A = 6$ (*b*) $A = -\frac{24}{5}$, $A = \frac{15}{2}$
(*c*) $B = 15$, $B = -\frac{5}{3}$ (*d*) $B = -3$, $B = \frac{1}{3}$

3. $3x + 2y - 12 = 0$, $x - 3y - 15 = 0$.

5. $\arctan \frac{3}{2}$. **7.** 13.

9. (*a*) 3; (*b*) $3\sqrt{2}$; (*c*) $\frac{33}{13}$; (*d*) $2/\sqrt{34}$.

11. $5x - 12y - \pm 65$. **13.** $3\sqrt{5}$.

17. (*a*) 15; (*b*) $-\frac{15}{2}$. **19.** (*a*) $\frac{9}{2}$; (*b*) $-\frac{3}{4}$.

Page 115

1. l_1 and l_3.

3. Sample answers: (*a*) $x = 3 - 5t$, $y = 7 - 3t$ (*b*) $x = 2t$, $y = -5t$
(*c*) $x = -1 - t$, $y = -3 + 11t$ (*d*) $x = 4$, $y = 6 - 13t$

5. Sample answers: (*a*) $x = t$, $y = 3t - 5$ (*b*) $x = -4 - 12t$, $y = 4 + 14t$
(*c*) $x = -6 + 3t$, $y = -4 + 4t$ (*d*) $x = t$, $y = -c_1/b_1 - a_1t/b_1$
(*e*) $x = 7 - 9t$, $y = 3 + 9t$ (*f*) $x = t$, $y = -4$

9. $\sqrt{(e + ft_1 - a - bt_1)^2 + (g + ht_1 - c - dt_1)^2}$.

11. Point A: (*a*) upward to the left; (*b*) 3, 4; (*d*) 5; (*e*) $4x + 3y - 40 = 0$; (*f*) $x = 4 - 3t$,

$y = 8 + 4t$. Point B: (*a*) upward to the right; (*b*) 12, 5; (*d*) 13; (*e*) $5x - 12y - 36 = 0$; (*f*) $x = 12t$, $y = -3 + 5t$.

13. (*a*) Point is moving up to the right at the rate of 25 units/sec; when $t = 5$ it is at (33, 125). (*b*) It is moving straight up at 6 units per unit of time; when $t = 5$ it is at (2, 30). (*c*) It is moving up to the left at a speed of $\sqrt{10}$ units per unit of time; it is at (−14, 1) at $t = 5$.

15. $x = -2 + 5t$, $y = 3 - 4t$.

17. $x = 6$, $y = \frac{4}{5}t$.

19. $x = 2 + 7t$, $y = -18 + 24t$.

21. $x = -5 + 3t$, $y = 5 - 4t$.

Page 119

5. (*a*) $\mathbf{OP} = 3\mathbf{i} + 5\mathbf{j}$ (*b*) $\mathbf{OP} = 3\mathbf{i} + 8\mathbf{j}$ (*c*) $\mathbf{OP} = 2\mathbf{i} + 9\mathbf{j}$ (*d*) $\mathbf{OP} = 7\mathbf{i} + 10\mathbf{j}$

Page 124

1. (*a*) slope of $\frac{3}{2}$; (*b*) x intercept of 3; (*c*) point (−2, 3); (*d*) x intercept of 3; (*e*) y intercept twice the x intercept; (*f*) x intercept 4 more than y intercept.

3. $y - 4 = m(x - 6)$, $9x - 4y = 38$.

5. $x - y = a$, $x - y = \pm 10$.

7. $y = mx - 4$, $y = 4x - 4$ or $y = -4x - 4$.

11. (*a*) (5, 2); (*b*) (3, −2).

13. The first contains the point (0, 0); the second does not.

Page 127

21. $x - y + 2 \leqslant 0$, $x + y - 16 \leqslant 0$, $x - y + 10 \geqslant 0$, $x + y - 8 \geqslant 0$.

23.

\$2 favors	1	1	1	1	1	2	2	2	3
\$1 favors	1	2	3	4	5	1	2	3	1

Page 133

3. 4 at (0, 0), 16 at (4, 0), 6 at (4, 2), −10 at (2, 4), −16 at (0, 4); 16 is max., −16 is min.

5. 300 cows, 200 horses.

7. $x = 0$, $y = 5$.

9. 120 boxes of Quality and 60 of Grade A.

11. 144 of Quality.

13. Twenty-four 3-min and six 2-min problems.

Review Problems for Chap. 4

1. (*a*) $y = -1$ (*b*) $x/5 + y/{-3} = 1$ (*c*) $y - 2 = -\frac{1}{4}(x - 4)$ (*d*) $y + 5 = \frac{7}{4}(x - 2)$ (*e*) $y + 4 = x - 2$ (*f*) $y - 3 = -\frac{2}{7}(x + 2)$ (*g*) $y - 8 = -(x + 3)$

3. (*a*) $5/\sqrt{13}$ (*b*) $18/\sqrt{13}$

7. $3x - y + 5\sqrt{10} + 4 = 0$, $3x - y + 4 - 5\sqrt{10} = 0$.

9. $y + 6 = 4(x - 2)$, $[(x\mathbf{i} + y\mathbf{j}) - (2\mathbf{i} - 6\mathbf{j})] \cdot (-4\mathbf{i} + \mathbf{j}) = 0$.

Extended Study Problems for Chap. 4

1. (*a*) l_3: $b_1x - a_1y + c_1 = 0$, l_4: $b_2x - a_2y + c_2 = 0$; (*b*) $\theta = \phi$.

5. (*a*) circle with center at (0, 0) and radius 5; (*b*) points outside circle in (*a*); (*c*) points inside circle in (*a*).

7. (*a*) $(-\sqrt{2}/2)x + (\sqrt{2}/2)y = 8$
(*b*) $y = 3$
(*c*) $-x/2 - (\sqrt{3}/2)y = 6$
(*d*) $x + 5 = 0$

9. $x = \frac{14}{3}$, $y = \frac{4}{9}$.

CHAPTER 5

Page 144

1. (*a*) $x' = x - 4$, $y' = y + 2$
(*b*) $x = x' + 7$, $y = y' - 3$

3. $x' - y' = 3$.

5. $y' = \frac{2}{3}x'$.

7. $y' = 2x'^2 + 13x'$.

9. $y' = x'^3$.

11. $y' - x'^4 - 4x'^2$.

13. $y' = 8/(x'^2 + 4)$.

15. $A(4\sqrt{3} - 3, -3\sqrt{3} - 4)$, $B(-2\sqrt{3} - 2, 2 - 2\sqrt{3})$, $C(5, 5\sqrt{3})$.

17. $x' = 4/\sqrt{5}$.

19. $x'^2 - 2x'y' + y'^2 - x' - y' = 0$.

21. $x' = x \cos\theta + y \sin\theta$, $y' = y\cos\theta - x\sin\theta$.

Review Problems for Chap. 5

1. $y' = x'^2$.

3. Any point on $3x + y = 7$.

5. (*a*) $y' = 0$; (*b*) $y' = -4$; (*c*) $y' = 7$; (*d*) $y' = x'$.

7. $\sqrt{13}y' - 7 = 0$.

Extended Study Problems for Chap. 5

1. $x'^2 - 4y'^2 - \frac{3}{4} = 0$, $x' = x + \frac{1}{2}$, $y' = y + \frac{1}{2}$.

5. $2y'^2 + a^2 = 2\sqrt{2}ax'$.

CHAPTER 6

Page 153

1. (*a*) $(x - 4)^2 + (y + 2)^2 = 9$
Inside: $(x - 4)^2 + (y + 2)^2 < 9$
Outside: $(x - 4)^2 + (y + 2)^2 > 9$

(*b*) $(x + 2)^2 + (y + 6)^2 = 4$
$(x + 2)^2 + (y + 6)^2 < 4$
$(x + 2)^2 + (y + 6)^2 > 4$

(*c*) $x^2 + (y + 5)^2 = 25$
Inside: $x^2 + (y + 5)^2 < 25$
Outside: $x^2 + (y + 5)^2 > 25$

(*d*) $(x - 2)^2 + y^2 = 16$
$(x - 2)^2 + y^2 < 16$
$(x - 2)^2 + y^2 > 16$

(*e*) $(x - 4)^2 + (y - 1)^2 = 16$
Inside: $(x - 4)^2 + (y - 1)^2 < 16$
Outside: $(x - 4)^2 + (y - 1)^2 > 16$

(*f*) $(x + 3)^2 + (y - 4)^2 = 25$
$(x + 3)^2 + (y - 4)^2 < 25$
$(x + 3)^2 + (y - 4)^2 > 25$

(*g*) $(x - 3)^2 + (y - 3)^2 = 9$
Inside: $(x - 3)^2 + (y - 3)^2 < 9$
Outside: $(x - 3)^2 + (y - 3)^2 > 9$

(*h*) $(x + 5)^2 + (y - 2)^2 = 49$
$(x + 5)^2 + (y - 2)^2 < 49$
$(x + 5)^2 + (y - 2)^2 > 49$

3. (*b*) $x^2 + y^2 = 2ry$.

7. $x^2 + y^2 - 32x + 192 = 0$.

9. $x^2 + y^2 - 19x - 13y + 110 = 0$.

11. $x^2 + y^2 - cx - cy = 0$.

13. $(x-h)^2+(y-2h)^2=5h^2$, $x^2+y^2-4x-8y=0$.

15. Two points.

17. $(x-5)^2+(y+3)^2=9$.

19. $(x-3)^2+(y+5)^2=25$.

21. $(x+2)^2+(y+1)^2=5$, $(x-10)^2+(y+5)^2=125$.

Page 157

1. $x^2+y^2=9$.

3. $x^2+y^2-4x=0$.

5. $x^2+y^2-5y=0$.

7. $x^2+y^2-6x+4y=0$.

9. $x^2+y^2-4y=12$.

13. $(x-2)^2+y^2=16$.

15. $(x+1)^2+(y-1)^2=9$.

Page 161

1. $F(0, \pm 4)$, $x^2/9+y^2/25<1$, $e=\frac{4}{5}$.

3. $F(0, \pm\sqrt{15})$, $x^2+y^2/16<1$, $e=\sqrt{15}/4$.

5. $F(0, \pm\sqrt{3})$, $4x^2+3y^2<36$, $e=\frac{1}{2}$.

7. $F(\pm 2\sqrt{5}, 0)$, $2x^2/3+3y^2/2<24$, $e=\frac{1}{3}\sqrt{5}$.

9. $x^2/108+y^2/144=1$.

11. $x^2/25+y^2/16=1$.

13. $16x^2/441+16y^2/245=1$.

15. $\sqrt{3}/2$, $3\sqrt{11}/10$.

Page 165

1. (*a*) $C(2, 1)$; (*b*) 6 and 2; (*c*) $2\sqrt{2}/3$.

3. (*a*) $C(-2, 2)$; (*b*) 8 and 6; (*c*) $\sqrt{7}/4$.

5. (*a*) $C(-5, -3)$; (*b*) 20 and 16; (*c*) $\frac{3}{5}$.

7. $(x-6)^2/16+(y+4)^2/12=1$.

9. $(x-4)^2/16+(y+4)^2/32=1$.

13. $(x-3)^2/36+y^2/27=1$.

15. $5x^2+9y^2-108x+324=0$.

17. $x^2/256+y^2/64=1$.

19. $(x-3)^2/4+(y+1)^2/9=1$.

21. $x^2/16+(y-4)^2/4=1$.

23. $(x+\frac{3}{2})^2+(y-2)^2=\frac{25}{4}$.

25. $(x-2)^2/3+(y-3)^2/2=1$.

27. $(3, 2.4)$; $(-4, -1.8)$.

31. Circle.

33. Circle.

35. Circle.

Page 171

1. $x=20\cos\theta$, $y=20\sin\theta$.

3. $x=20\cos 50\pi t$, $y=20\sin 50\pi t$.

5. 3 rps counterclockwise, radius of 8, beginning at (8, 0).

7. 2 rps counterclockwise; radius of 4, beginning at (0, 4).

9. $x=10\cos(3\pi/2+50\pi t)$, $y=10\sin(3\pi/2+50\pi t)$.

11. $x=10\sin(-\pi/4+2\pi\lambda t)$, $y=10\cos(-\pi/4+2\pi\lambda t)$; λ is the number of revolutions per second.

Review Problems for Chap. 6

1. (*a*) Circle, $C(3, -2)$, $r=4$

(*b*) Circle, $C(0, 0)$, $r=4$

(*c*) Ellipse, $C(0, 0)$, $a=\frac{1}{5}$, $b=\frac{1}{6}$

(*d*) Ellipse, $C(-2, 3)$, $a=4$, $b=2$

(*e*) Circle, $C(0, \frac{7}{2})$, $r=\frac{7}{2}$

(*f*) Ellipse $C(0, 0)$, $a=\sqrt{3}$, $b=\frac{3}{2}$

(*g*) No graph

(*h*) No graph

3. 10.

7. (a) $9x'^2 + 16y'^2 = 144$ (b) $x'^2 + y'^2 = \frac{49}{4}$

9. (a) $x^2/36 + y^2/35 = 1$ (b) $(x+2)^2/1 + (y - \frac{1}{2})^2/\frac{3}{4} = 1$
 (c) $x^2/9 + (y + \frac{1}{2})^2/\frac{45}{4} = 1$

11. Approximately 13.9 ft.

Extended Study Problems for Chap. 6

1. $y = 2x - 2$. 5. $(-1, \sqrt{3}), (-1, -\sqrt{3})$. 7. $(-1, 3), (2, -2)$.

CHAPTER 7

Page 183

1. (a) $F(\frac{3}{2}, 0), x = -\frac{3}{2}$ (b) $F(-1, 0), x = 1$
 (c) $F(0, -2), y = 2$ (d) $F(0, \frac{1}{16}), y = -\frac{1}{16}$
 (e) $F(\frac{5}{12}, 0), x = -\frac{5}{12}$ (f) $F(0, -0.1), y = 0.1$
 (g) $F(\frac{1}{16}, 0), x = -\frac{1}{16}$ (h) $F(-\frac{3}{32}, 0), x = \frac{3}{32}$
 (i) $F(0, \frac{9}{16}), y = -\frac{9}{16}$

3. (a) $(y-2)^2 = \frac{25}{4}(x-4)$ (b) $(x-5)^2 = \frac{16}{5}(y+3)$
 (c) $x^2 = \frac{32}{3}(y+6)$

7. (a) $y^2 + 6y - 2x - 4 = 0$ (b) $y^2 - 6y - 2x + 8 = 0$
 (c) $3x = y^2 - 12$

9. $10\frac{2}{3}$ in. 15. (c).

17. (a) $(y-4)^2 = 3(x-2)$ (b) $(y-2)^2 = 6(x + \frac{7}{3})$
 (c) $(x+1)^2 = 4(y+3)$ (d) $(x-4)^2 = 4(y+2)$
 (e) $(x-6)^2 = -1(y-36)$ (f) $(y-2)^2 = 4(x+1)$
 (g) $(y - \frac{3}{2})^2 = 3(x-2)$ (h) $(x - \frac{5}{2})^2 = \frac{3}{2}(y + \frac{3}{2})$

19. 120 by 300 ft.

Page 192

1. (a) $V(\pm 10, 0), F(\pm 5\sqrt{5}, 0)$ (b) $V(0, \pm 6), F(0, \pm 2\sqrt{21})$
 (c) $V(\pm 3, 0), F(\pm 5, 0)$ (d) $V(\pm 3, 0), F(\pm \sqrt{13}, 0)$
 (e) $V(\pm 6, 0), F(\pm 3\sqrt{5}, 0)$ (f) $V(\pm 6, 0), F(\pm 2\sqrt{15}, 0)$

5. (a) $x^2/36 - y^2/28 = 1$ (b) $y^2/16 - x^2/9 = 1$
 (c) $y^2/16 - x^2/20 = 1$ (d) $x^2/64 - y^2/64 = 1$

7. $(x-5)^2/9 - y^2/16 = 1$. 9. Yes; $x^2/36 - y^2/64 = 1$.

11. Yes; $16x^2 - 96y - 9x^2 = 0$.

13. New origin $(-2, 1)$; translated equation $x'y' = -3$.

15. $0'(-\frac{3}{2}, \frac{3}{2}), x'y' = -\frac{19}{4}$. 17. $0'(-\frac{1}{2}, -\frac{3}{2}), x'y' = \frac{7}{4}$.

Page 201

1. (a) Circle; (b) ellipse; (c) hyperbola; (d) parabola; (e) ellipse; (f) hyperbola.

3. (a) Ellipse, 63°26′; (b) hyperbola, 18°26′; (c) hyperbola, 36°52′; (d) ellipse, 45°; (e) hyperbola, 14°2′; (f) ellipse, 36°52′.

7. (*a*) $B^2 - 4AC = 25$, two intersecting lines; (*b*) $B^2 - 4AC = 0$, parabola; (*c*) $B^2 - 4AC = 9$, two intersecting lines; (*d*) $B^2 - 4AC = -3$, ellipse; (*e*) $B^2 - 4AC = 0$, two intersecting lines.

9. $(x-4)^2/4 + y^2/3 = 1$. 11. $(y+2)^2/16 - x^2/48 = 1$. 13. $(\frac{1}{4}, \pm 1/\sqrt{2})$.

Page 205

1. $x + y = 3$.
3. $2x - 3y = 12$.
5. $x + 4 = 0$.
7. $x - 3y - 6 = 0$.
9. $y^2 + x^2 + 3y = 0$.
11. $x^2 + y^2 - 8x - 6y = 0$.
13. $x^2 + y^2 - 8x + 4y = 11$.
15. $4x^2 + 3y^2 + 16y = 64$.
17. $7x^2 + 16y^2 + 72x = 144$.
19. $9x^2 + 5y^2 + 24y = 36$.
21. $36x^2 + 20y^2 - 72y = 81$.
23. $9y^2 = 64 - 48x$.
25. $9x^2 = 64 - 48y$.
27. $5x^2 - 4y^2 + 36x + 36 = 0$.
29. $9x^2 - 27y^2 + 192y - 256 = 0$.
31. $y^2 = 6x$.
33. $9x^2 + 16y^2 = 144$.

Page 207

1. $x^2 - 4x + y = 0$.
3. $x^2 + y = 1, -1 \leq x \leq 1, 0 \leq y \leq 1$.
5. $3x^2 - 4y - 12x + 4 = 0, 0 \leq x \leq 4, -2 \leq y \leq 1$.
7. $80\sqrt{2}$ ft/sec.

Review Problems for Chap. 7

1. (*a*) $V(0, 0), F(\frac{5}{2}, 0); x = -\frac{5}{2}$ (*b*) $V(3, -4), F(3, -6); y = -2$
(*c*) $V(1, -2), F(\frac{17}{12}, -2); x = \frac{7}{12}$
3. $7x^2 + 6y - 25x + 12 = 0$.
5. $\frac{1}{2}$ sec.
7. (*a*) $x^2/81 - y^2/9 = 1$ (*b*) $(y-4)^2/4 - (x-4)^2/32 = 1$
9. Hyperbola; $2x'^2 - y'^2 = 2$, $26°34'$.

Extended Study Problems for Chap. 7

1. 288 ft, 72 ft.

CHAPTER 8

Page 217

1. (*a*) 4; (*b*) not a polynomial; (*c*) not a polynomial; (*d*) 3.
11. $A = x(900 - 3x)/2$, 150 by 225 ft.
13. 50 ft.
15. $V = \frac{5}{3}x^2(9 - x)$; about 6 in.

Page 228

19. $V = 12y^2(y - 12)$.

Review Problems for Chap. 8

	Intercepts	Asymptotes
5.	(0, 0)	$x=3,\ x=-3,\ y=0$
7.	(0, 0)	$x=3,\ x=-2$
9.	(0, 0), (2, 0)	$x=2$

Extended Study Problems for Chap. 8

3. $\left(\dfrac{[b^2-a^2]x_1-2aby_1-2ac}{a^2+b^2},\ \dfrac{-2abx_1+[a^2-b^2]y_1-2bc}{a^2+b^2}\right)$.

CHAPTER 9

Page 240

1. Circle. **3.** Ellipse. **5.** Hyperbola.

Page 244

5. $x=t(4-t^2),\ y=t^2(4-t^2)$.

7. $x=2(3-t^2)/(1+t^2),\ y=2t(3-t^2)/(1+t^2)$.

17. (9) $x^4+y^3=4x^2y$
(10) $y^2=(x-1)(x-5)^2$
(11) $y=\frac{1}{8}([x-y]/2)^3-(x-y)/2$
(12) $(4-x)^{\frac{1}{2}}(2.8-1.6x)$
(13) $4x^4-8x^3y+xy^4=0$
(14) $x^4+2x^2y^2-16x^2y+y^4=0$
(15) $x^3+4x^2-4y^2-y^3=0$
(16) $x^2-x^2y-2x+2xy-y+2=0$

Page 247

1. $(x-1)^2+(y-2)^2=1$.

3. $x^2+(y-2)^2=4$.

5. $(x+2)^2/9+(y-2)^2/25=1$.

7. $x^2/9+(y-1)^2=1$.

Page 257

23. $(\frac{12}{5},\ \text{Arctan}\ \frac{3}{4})$; (0, 0).

25. $(2\sqrt{2},\ \frac{1}{4}\pi)$; $(2\sqrt{2},\ \frac{3}{4}\pi)$.

27. $(\frac{1}{2}\sqrt{3},\ \frac{1}{3}\pi)$; (0, 0).

29. $(\frac{3}{2},\ \frac{1}{3}\pi)$; $(\frac{3}{2},\ \frac{5}{3}\pi)$; (0, 0).

Review Problems for Chap. 9

5. $y=2/[(x-2)(x+2)]$. **7.** $y=1-2x^2$. **11.** $x=4m,\ y=4m^2+2$.

Extended Study Problems for Chap. 9

7. $\rho\cos^4\theta=\sin^2\theta(3\cos\theta-2\sin\theta)$.

CHAPTER 10

Page 267

1. $\sin x$, odd; $\cos 3x$, even; 2^x, neither; x^3, odd; x^2-3x, neither.

3. (*a*) 0, 1.308, 2.473, 3.464, 3.888, 3.637, 3.464; (*b*) 4 when $t=\frac{3}{4}\pi$.

5. (*a*) $2(\sqrt{3}-1)$; (*b*) $2(-\sqrt{3}-1)$.

	Amp	Per
7.	5	4
9.	2	$\frac{4}{3}\pi$
11.	2.6	8
13.	3.4	2
15.	10	$\frac{8}{5}$

17. 120π. **19.** 12 sec.

Page 275

25. $(0, 0), (\frac{1}{3}\pi, \frac{1}{2}\sqrt{3}), (\pi, 0), (\frac{5}{3}\pi, -\frac{1}{2}\sqrt{3})$. **27.** $(0, 0), (\frac{2}{3}\pi, \sqrt{3}), (\frac{10}{3}\pi, -\sqrt{3}), (4\pi, 0)$.

Page 281

1. $\frac{3}{5}$. **3.** $\frac{1}{2}\sqrt{3}$. **5.** $\frac{5}{4}$. **7.** 0.
9. $2\sqrt{5}$. **11.** $\frac{24}{7}$. **13.** $\frac{1}{9}$. **15.** $2/(\sqrt{2}+1)$.
17. $\frac{4}{5}$. **19.** $-\frac{3}{5}$. **23.** $\frac{1}{6}$.

Extended Study Problems for Chap. 10

1. 9.5; 4.5; $x = \frac{3}{4}\pi$; $x = \frac{1}{4}\pi$. **5.** $\theta = \arctan(5/x) - \arctan(3/x)$.

CHAPTER 11

Page 291

25. $a = \ln 2$, $c = e/8$. **27.** $e^{4.27x}$. **29.** ln 10.
31. $547. **33.** (*a*) 54.9 mg; (*b*) 2,300 years.

Review Problems for Chap. 11

9. $e^{2.4x \ln 3}$.
11. (*a*) $104; (*b*) $111; (*c*) $117; (*d*) approx. 3.5 years.

Extended Study Problems for Chap. 11

1. (*a*) About 69; (*b*) $e^{0.01t} = 2$. **9.** $x = \frac{1}{2}\ln(1+y)/(1-y)$.

CHAPTER 12

Page 305

1. $y = 0.431x + 5.49$. **3.** $y = 0.584x - 1.88$.
5. (*a*) Approximately 0.108; (*b*) approximately 0.117; (*c*) line in Prob. 1.
9. (*a*) $y - 14.104 = \frac{2.326}{45.145}(x - 29.58)$; (*b*) 14.90.
11. (*a*) $C = .0312N + 138.60$; (*b*) $450.60.

Page 308

1. $y = 0.22x^2 + 0.25x - 5.53$.
3. (a) $y = 0.18x^2 + 0.48x - 5.66$.
5. $y = -0.105x^2 + .450x - 1.83$.

Page 314

1. $y = 0.8746x^{1.9979}$.
3. $y = 2.445x^{0.5859}$.
5. $y = 2.002(10)^{0.2171x} = 2.002e^{0.5200x}$.
7. $y = (40.67)10^{-0.1952x} = 40.67e^{-0.4495x}$.
9. $t = 0.2532s^{0.5178}$.
11. $R = 0.07495V^{2.011}$.
13. $\alpha = t^{0.6667}$.

Page 319

1. $y = 1.303x - 5.16$.
3. $y = 0.59x + 2.325$.
5. $P = 0.1696t + 23.72$.
7. $y = 2.483x^{1.6513}$.
9. $y = 2.605(10)^{0.1806x}$.

Review Problems for Chap. 12

1. $y = 1.37x + 1.86$.
3. $y = 3.792x^{0.3538}$.
5. (a) $P = 16{,}000Y + 85{,}000$; (c) \$165,000.

CHAPTER 13

Page 333

1. (a) $\cos\alpha = \cos\beta = \frac{1}{6}\sqrt{6}$; $\cos\gamma = \frac{1}{3}\sqrt{6}$.
   (b) $\cos\alpha = -\frac{12}{13}$; $\cos\beta = -\frac{3}{13}$; $\cos\gamma = \frac{4}{13}$.
   (c) $\cos\alpha = \frac{2}{13}\sqrt{26}$; $\cos\beta = -\frac{3}{26}\sqrt{26}$; $\cos\gamma = \frac{1}{26}\sqrt{26}$.
   (d) $\cos\alpha = \cos\beta = 1/\sqrt{3}$; $\cos\gamma = -1/\sqrt{3}$.
   (e) $\cos\alpha = -\frac{3}{7}$; $\cos\beta = \frac{2}{7}$; $\cos\gamma = -\frac{6}{7}$.
   (f) $\cos\alpha = \frac{2}{3}$; $\cos\beta = -\frac{1}{3}$; $\cos\gamma = -\frac{2}{3}$.
3. $\frac{1}{3}\sqrt{3}$.
5. (a) $(-x, -y, -z)$; (b) $(x, y, -z)$; (c) $(x, -y, -z)$.
7. (a) $\sqrt{41}$, $\sqrt{50}$, $\sqrt{141}$; (b) $\sqrt{43}$, $\sqrt{62}$, 5.
11. $x^2 + y^2 + z^2 = 25$.
13. $y^2 + z^2 - 8x + 16 = 0$.
15. $9x^2 + 25y^2 + 25z^2 = 225$.
17. (a) $F_x = F_y = 66\frac{2}{3}$ lb; $F_z = 33\frac{1}{3}$ lb.
   (b) $F_x = \frac{840}{13}$ lb; $F_y = -\frac{280}{13}$ lb; $F_z = -\frac{210}{13}$ lb.
   (c) $F_x = 36$ lb; $F_y = 0$; $F_z = -48$ lb.
   (d) $F_x = -120\sqrt{70}/7$ lb; $F_y = 60\sqrt{70}/7$ lb; $F_z = 100\sqrt{70}/7$ lb.
19. $F_x = 20$ lb; $F_y = 36$ lb; $F_z = 8$ lb.

Page 342

1. (a) [4, 5, 3], $\cos\alpha = 4/\sqrt{50}$, $\cos\beta = 5/\sqrt{50}$, $\cos\gamma = 3/\sqrt{50}$.
   (b) [2, −2, −1], $\cos\alpha = \frac{2}{3}$, $\cos\beta = -\frac{2}{3}$, $\cos\gamma = -\frac{1}{3}$.
   (c) [0, 4, 3], $\cos\alpha = 0$, $\cos\beta = \frac{4}{5}$, $\cos\gamma = \frac{3}{5}$.
   (d) [9, 2, 5], $\cos\alpha = \frac{9}{110}\sqrt{110}$, $\cos\beta = \frac{1}{55}\sqrt{110}$, $\cos\gamma = \frac{1}{22}\sqrt{110}$.

(*e*) $[2, -2, 1]$, $\cos\alpha = \frac{2}{3}$, $\cos\beta = -\frac{2}{3}$, $\cos\gamma = \frac{1}{3}$.
(*f*) $[-6, -6, 3]$, $\cos\alpha = -\frac{2}{3}$, $\cos\beta = -\frac{2}{3}$, $\cos\gamma = \frac{1}{3}$.

3. $[\pm 1, 0, 0]$; $[0, \pm 1, 0]$; $[0, 0, \pm 1]$.

5. $(0, 2, 3)$, $(2, 0, 4)$.

7. $(2, 5, 0)$, $(-3, 0, -5)$, $(0, 3, -2)$.

9. (*a*) $\frac{5}{14}\sqrt{7}$; (*b*) $3\sqrt{13}/26$.

13. $20\sqrt{3}$ lb.

15. (*a*) $[3, 2, 3]$; (*b*) $[6, -5, -4]$; (*c*) $[5, 8, 5]$.

Review Problems for Chap. 13

1. (*a*) 9; (*b*) 7; (*c*) $[3, 10, -3]$; (*d*) $[5/\sqrt{65}, -6/\sqrt{65}, 2/\sqrt{65}]$; (*e*) $\alpha \approx 27°$, $\beta \approx 65°$, $\gamma \approx 85°$.

3. $25\sqrt{21}$.

5. Arccos $\frac{12}{77}$.

7. $(2, 2\sqrt{2}, \pm 2)$.

9. $x = 4$.

Extended Study Problems

5. $[5, 7, 5]$.

7. Yes.

CHAPTER 14

Page 355

1. (*a*) $x + \sqrt{2}y + z = 8$
(*b*) $-x + y = 8\sqrt{2}$
(*c*) $y - \sqrt{3}z + 12 = 0$

3. (*a*) $-\frac{2}{7}x - \frac{3}{7}y - \frac{6}{7}z = 3$
(*b*) $x/\sqrt{5} + 2z/\sqrt{5} - 6/\sqrt{5} = 0$
(*c*) $\frac{8}{9}x + \frac{4}{9}y - \frac{1}{9}z = 3$
(*d*) $y = 3$

5. (*a*) $3x + 2y + z = 10$
(*b*) $x - 2y + z = 0$

7. (*a*) $(4, 3, 4)$
(*b*) $3x + 2y + z = 22$

9. $(0, \frac{14}{5}, 0)$.

Page 361

1. (*a*) $x + 2y - z = 4$
(*b*) $x - 2y + 3z = 0$
(*c*) $9x + 7y - z = 25$

5. $10x - 21y + 13z - 84 = 0$.

7. $6x - 7y - 4z + 24 = 0$.

9. (*c*) and (*d*).

11. (*a*) $x - 2y - 2z = 1$
(*b*) $-3x + 8y + 4z = 53$
(*c*) $6x - y + 5z = 22$

13. Arccos $(1/\sqrt{170})$; $85°40'$ (approx.).

15. $3x + y - 2z + 20 = 0$.

17. $(0, 0, 12)$.

Page 370

3. $[\frac{6}{7}, \frac{2}{7}, \frac{3}{7}]$.

5. (*a*) $x - 8 = y - 10 = (z - 8)/2$
(*b*) $(x - 2)/4 = (y - 1)/-3 = (z - 4)/2$

7. $(0, 0, \frac{16}{5})$.

9. $4x + 3y - 11 = 0$, $x - 3z + 16 = 0$.

11. (*a*) $Ax + By + Cz = 0$
(*b*) $z = D$
(*c*) $By + Cz = D$
(*d*) $Ax + Bz = C$

13. $4x - y + 2z + 20 = 0$.

15. $6x - 8y + z = 0$.

Review Problems for Chap. 14

1. $3x + 2y - 3z = 4$.
3. $4x - 6y + 4z = 15$.
5. Arccos $(2/\sqrt{22})$.
7. (*a*) $(x + 1)/1 = (y - 3)/-1 = (z - 8)/5$ (*b*) $y + x = 2$, $z - 5x = 13$
   (*c*) $x = -1 + t$, $y = 3 - t$, $z = 8 + 5t$
9. $(x - 6)/1 = y/-1 = (z - 4)/7$.

Extended Study Problems for Chap. 14

3. $3x + 4y + 5z - 60 = 0$.
5. $a = 2$.
7. 3.86 (approx.).
11. $11x - 6y + 5z + 4 = 0$.

CHAPTER 15

Page 381

1. (*a*) $(x - 4)^2 + (y + 2)^2 + (z - 1)^2 = 36$ (*b*) $(x - 2)^2 + y^2 + z^2 = 4$
   (*c*) $x^2 + (y - 3)^2 + (z - 4)^2 = 29$ (*d*) $(x - 4)^2 + (y - 2)^2 + (z - 5)^2 = 25$
3. $6x + 2y + 3z = 61$.
7. $x^2 + y^2 + z^2 = a^2$.
9. $4x^2 + 4z^2 - y^2 = 0$.
11. $9x^2 + 9y^2 + 25z^2 = 225$.
13. $\sqrt{x^2 + y^2} + \ln z = 0$.

Page 388

5. Elliptic paraboloid.
7. Circular paraboloid.
9. Elliptic paraboloid.
11. Hyperboloid of one sheet.
13. Hyperboloid of one sheet.
15. Hyperboloid of two sheets.
17. Ellipsoid.
21. $\sqrt{(x - 1)^2 + y^2 + z^2} = \frac{1}{2}|x - 2|$.

Page 396

19. Sphere; cone; cone.
21. $r^2 = 6$; $\rho^2 \sin^2 \phi = 6$.
23. $r^2 \sin 2\theta = 4$; $\rho^2 \sin^2 \phi \sin 2\theta = 4$.
25. $r^2 + z^2 = 16$; $\rho = 4$.
27. $r^2 \cos 2\theta - 2z^2 = 4$; $\rho^2(\sin^2 \phi \cos 2\theta - 2 \cos^2 \phi) = 4$.

Review Problems for Chap. 15

1. (*a*) Elliptic cone; (*b*) hyperboloid of two sheets; (*c*) one point; (*d*) sphere; (*e*) circular cone; (*f*) elliptic paraboloid; (*g*) ellipsoid; (*h*) hyperbolic paraboloid; (*i*) hyperboloid of one sheet; (*j*) sphere.
3. $(x - 3)^2 + (y - 1)^2 + (z + 2)^2 = 25$.
9. (*a*) $r^2 \cos 2\theta = 4z$ (*b*) $\rho^2 \sin^2 \phi \cos 2\theta = 4\rho \cos \theta$

Extended Study Problems for Chap. 15

1. Rectangular coordinates.

INDEX

INDEX